TREATISE ON CONTROLLED DRUG DELIVERY

TREATISE ON CONTROLLED DRUG DELIVERY

Fundamentals • Optimization • Applications

edited by
Agis Kydonieus

ConvaTec
A Bristol-Myers Squibb Company
Princeton, New Jersey

Marcel Dekker, Inc.　　　　　New York • Basel • Hong Kong

Library of Congress Cataloging-in-Publication Data

Treatise on controlled drug delivery: fundamentals, optimization,
 applications / edited by Agis Kydonieus.
      p.   cm.
   Includes bibliographical references and index.
   ISBN 0-8247-8519-3
   1. Drugs--Controlled release. 2. Controlled release preparations.
I. Kydonieus, Agis F.
   [DNLM: 1. Delayed-Action Preparations. 2. Dosage Forms. 3. Drug
Carriers.   QV 785 T784]
RS201.C64T74   1991
615'.6--dc20
DNLM/DLC
For Library of Congress   91-24646
                                                                  CIP

This book is printed on acid-free paper.

Marcel Dekker, Inc.
270 Madison Avenue, New York, New York 10016

Current printing (last digit):
10   9   8   7   6   5   4   3   2   1

PRINTED IN THE UNITED STATES OF AMERICA

Preface

Controlled-release delivery of chemicals is practiced in nature on a routine basis. Examples include the oxygenation of blood by the diffusion of oxygen through the alveolar walls, the flow of nutrients and waste through cellular membranes, and the control of body temperature by the permeation and evaporation of water from the epidermis. The development of controlled-release devices is an attempt to simulate nature's processes so as to deliver a drug to a target organ at a specified rate for a specified period of time, and to accomplish the optimum therapeutic effect while keeping undesirable side effects to a minimum. With the development of *targeted* and *self-regulating* therapeutic systems, we are now attempting to mimic nature. Although we still have a long way to go, solid advances have been made in both the scientific and commercial development of controlled-release systems. Transdermal nitroglycerin devices control half a billion dollars of the angina pectoris market worldwide, and once-per-day oral tablets based on osmotic pressure and erodible-system design are being introduced into the marketplace at an ever-increasing rate.

In this book we have attempted to present a concise, readable, and in-depth presentation of the fundamentals, applications, and optimization of controlled-release systems. Each chapter starts with the basics and progresses into the most difficult thoughts and mathematical concepts. Interspersed within the chapters are examples and solved problems to aid in the understanding of the material. Specific attention was paid to the methodologies for modifying drug release and to the application of release kinetics to the design and optimization of controlled-release formulations. Separate chapters are presented discussing the mechanisms, characteristics, and mathematics of erodible devices, as well as the development and performance testing of implantable and oral osmotic pumps. Particularly strong are the chapters on controlled-release fundamentals, including those on the kinetics of solute release and on diffusion controlled matrix systems, written by scientists instrumental in molding these areas.

The second part of the book examines in depth the applications of controlled release. The most important routes of drug administration—oral, parenteral, transdermal, and nasal—are discussed in separate chapters. The pharmacokinetics, pharmacodynamics, and biological and biopharmaceutical parameters pertinent to each route of administration are presented for both peptide and nonpeptide drugs. How polymers, adhesives, control membranes, and other formulation parameters are used to design optimal systems are discussed, including polymers for osmotic, erodible, and diffusion-controlled systems. Chapters on veterinary and pesticide applications are also included to aid in the cross-fertilization of ideas between the delivery of drugs and other biologically active agents.

The volume provides an introductory but detailed treatise on the principles and applications of controlled release and, with nearly 1000 references, it can be used as a valuable source of the most recent literature for the expert in the field. However, the book is written in a readable, basic, didactic format, with solved examples and end-of-chapter problems, allowing for this volume to be used as an educational tool or text for the practicing scientist and the student alike.

Several friends and associates should be given credit for their helpful suggestions and criticisms. Special thanks should go to Dr. Bret Berner for reviewing several parts of the manuscript and providing invaluable suggestions. I am indebted to my associates Dr. John Wille and Ms. Stefanie Decker for reviewing and contributing to the chapter on transdermal delivery. I am also indebted to the chapter authors for their guidance and encouragement, as well as for their cooperation in adhering to strict manuscript specifications. A lot of thanks are also due to Joan VanDerveer for her efforts in typing and assisting in the editorial and administrative endeavors.

Agis Kydonieus

Contents

3. Diffusion-Controlled Matrix Systems   **155**
Ping I. Lee

4. Erodible Systems   **199**
Eyal Ron and Robert Langer

5. Osmotic Drug Delivery   **225**
Robert L. Jerzewski and Yie W. Chien

6. Oral Controlled-Release Delivery   **255**
Pardeep K. Gupta and Joseph R. Robinson

Contributors

Bret Berner, Ph.D.   Director, Basic Pharmaceutics Research, CIBA-GEIGY, Ardsley, New York

John R. Cardinal, Ph.D.   Senior Director, Pharmaceutical Development, Merck Sharp & Dohme Research Laboratories, Rahway, New Jersey

Shyi-Feu Chang, Ph.D.   Research Scientist, Department of Pharmaceutics/Drug Delivery, Amgen, Inc., Thousand Oaks, California

Yie W. Chien, Ph.D.   Parke-Davis Professor of Pharmaceutics, Controlled Drug-Delivery Research Center, Rutgers—The State University of New Jersey, Piscataway, New Jersey

Steven M. Dinh, Sc.D.   Manager, Basic Pharmaceutical Research, CIBA-GEIGY, Ardsley, New York

Thomas J. Franz, M.D.   Associate Professor, Department of Dermatology, University of Arkansas for Medical Sciences, Little Rock, Arkansas

Pardeep K. Gupta, Ph.D.   Assistant Professor, Pharmaceutics Department, Philadelphia College of Pharmacy and Science, Philadelphia, Pennsylvania

George Janes, Ph.D.   President, Marine Test Stations, Akron, Ohio

Robert L. Jerzewski, M.S.   Research Investigator, Pharmaceutics R&D, Bristol-Myers Squibb Pharmaceutical Research Institute, New Brunswick, New Jersey

Robert C. Koestler, Ph.D.   Senior Research Scientist, Agrichemicals Division, Atochem North America, Bryan, Texas

Wei-Youh Kuu, Ph.D.   Research Scientist, Pharmaceutical R&D, Baxter Healthcare Corporation, Round Lake, Illinois

Agis Kydonieus, Ph.D.   Vice President, Research and Development, ConvaTec, A Bristol-Myers Squibb Company, Princeton, New Jersey

Robert Langer, Sc.D.   Kenneth J. Germeshauser Professor of Chemical and Biochemical Engineering, Massachusetts Institute of Technology, Cambridge, Massachusetts

Ping I. Lee, Ph.D.   Professor, Faculty of Pharmacy, University of Toronto, Toronto, Ontario, Canada

J. Allen Miller, Ph.D.   Research Agricultural Engineer, Knipling-Bushland U.S. Livestock Insects Research Laboratory, U.S. Department of Agriculture, Kerrville, Texas

Joseph R. Robinson, Ph.D.   Professor, School of Pharmacy, University of Wisconsin, Madison, Wisconsin

Eyal Ron, Ph.D.   Pharmaceutical Research and Development, Genetics Institute, Cambridge, Massachusetts

Theodore J. Roseman, Ph.D.   Director, Pharmaceutical Research and Development, I.V. Systems Division, Pharmaceutical R&D, Baxter Healthcare Corporation, Round Lake, Illinois

Kishore R. Shah, Ph.D.   Associate Director, Corporate R&D, ConvaTec, A Bristol-Myers Squibb Company, Princeton, New Jersey

S. Esmail Tabibi, Ph.D.   Vice President, Research and Development, Micro Vesicular Systems, Inc., Nashua, New Hampshire

Thomas R. Tice, Ph.D.   Head, Control Release Division, Southern Research Institute, Birmingham, Alabama

Kakuji Tojo, Ph.D.   Professor, Department of Biochemical Engineering, Kyushu Institute of Technology, Iizuka, Japan

Leonore C. Witchey-Lakshmanan, Ph.D.   Pharmaceutical Development, Merck Sharp & Dohme Research Laboratories, Rahway, New Jersey

Ray W. Wood, Ph.D.   Manager, Drug Delivery, Pharmaceutical R&D, Baxter Healthcare Corporation, Round Lake, Illinois

TREATISE ON CONTROLLED DRUG DELIVERY

1

Fundamental Concepts in Controlled Release

Bret Berner and Steven Dinh   *CIBA-GEIGY, Ardsley, New York*

BASIC CONCEPTS

Controlled-release drug delivery combines well-characterized, reproducible dosage-form design with clinical pharmacology, in particular, steady-state pharmacology. The steady-state biology defines the required input or the desired drug delivery profile. A systems approach to experimental design allows the necessary quantitation of the biological response.

In the simplest model (Fig. 1), the fate of the drug may be characterized by a single compartment, which is described by the plasma concentration of drug with time. The concentration of drug bound to the receptor site versus time reflects the pharmacodynamics or biological response, whether it be related to drug efficacy or adverse reactions. With constant drug inputs into the plasma compartment, steady-state conditions are achieved, and the pharmacokinetic/pharmacodynamic problem simplifies to an equilibrium distribution between the plasma and the receptors. At steady state, the plasma concentrations will help predict the biological response, and the required drug delivery profile may be related to that input rate resulting in those steady-state plasma levels associated with the ED_{50} for efficacy while minimizing side effects.

Controlled-release drug delivery design involves the application of physical and polymer chemistry to dosage-form design to produce a well-characterized and reproducible dosage form that controls drug entry into the body within the specifications of the required drug delivery profile. This design typically includes additional characterization of the drug's permeation through the appropriate biological membrane and any first-pass metabolic effects prior to entry of the drug into systemic circulation. Conventional tablets fail this definition in their lack of characterization and control of absorption. The controlled-release dosage form should be tailored so that variations in component characteristics lead

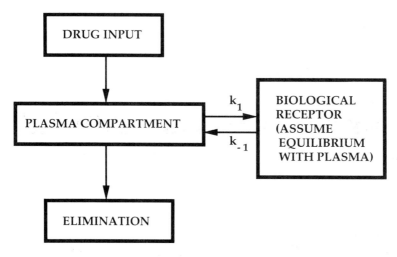

Figure 1 A simple pharmacokinetic model

to predictable alterations in release profiles. This release profile should also include appropriate control over absorption of the drug into the body.

In a similar fashion, *sustained release* indicates merely a dosage form that has its dissolution profile extended in time rather than designed and characterized to achieve a desired drug delivery profile. The classical example of a sustained-release dosage form is a wax matrix of phenylpropanolamine, which extends the period of drug release in the gastrointestinal tract beyond the short half-life (4.9 h) of the drug in the plasma [1]. An osmotic pump for phenylpropanolamine that produces (1) an initial loading burst of drug, (2) followed by a constant delivery rate of drug, and (3) a rapid exponential decay in the release rate timed for a mean gastrointestinal transit time is an example of a *controlled-release* dosage form for the same drug [2]. Plasma levels achieved by sustained-release dosage forms for phenylpropanolamine generally peak well above therapeutic levels, while controlled-release dosage forms, when properly designed, do not. For this controlled-release dosage form for phenylpropanolamine, the drug input rate was deconvoluted from the plasma levels, and the rate calculated from these levels agreed well with the theoretical and experimental in-vitro drug-delivery profile obtained in dissolution testing. The desired profile allows phenylpropanolamine to function as an appetite suppressant while minimizing the classical stimulant side effects of the drug. A suitable profile also extends the duration of efficacy and allows for a more convenient dosage form. In contrast, the sustained-release dosage form achieves the convenience of an extended duration of action [3], but its delivery is less reproducible, and it fails to avoid undesired side effects.

Controlled-release drug delivery currently involves control of either the time course or location of drug delivery. While control of the time course of drug delivery is the more classical approach, *site-specific* or *targeted delivery*, which involves drug delivery to a specific organ or class of cells or physiological compartment, e.g., to cancer cells or to hepatocytes or across the blood-brain barrier, is possible. Currently, this approach is mostly in the research stage.

Delivery rates from temporal controlled-release systems may be characterized in terms of their kinetics and physical processes. Of particular interest are zero-order systems, those systems for which the release of drug is constant with time, t, or proportional to

t^0. Typical zero-order systems involve diffusion of the drug through a membrane, where the steady-state flux determines the constant delivery rate. In practice, these membrane systems are either initially loaded with drug in the membrane to provide an initial burst followed by a period of constant release, or they have a time lag prior to steady-state diffusion across the membrane.

Perhaps the most common controlled-release delivery devices exhibit delivery rates that initially are proportional to $t^{-1/2}$. This square-root-of-time dependence reflects mass transport by sorption/desorption in monolithic systems [4,5].

Zero-order systems can, in principle, allow for selection of precise efficacious plasma levels after titration for interindividual variation. At steady state, plots of drug levels in plasma should reflect the biological response, and studies over a range of plasma levels should resemble in-vitro enzyme kinetics. For example, angiotensin converting enzyme (ACE) converts angiotensin I (AI) to angiotensin II (AII). Through multiple mechanisms, this translates into an increase in mean arterial pressure (MAP). One of the most direct means of controlling hypertension is to use ACE inhibitors [6]. By giving constant simultaneous infusions of the enzyme substrate (AI) and the ACE inhibitor (a competitive inhibitor of ACE, benazeprilat) and, at the same time, measuring the mean arterial pressure, one obtains a curve of the in-vivo response versus delivery rate (Fig. 2), which is an in-vivo competitive inhibition curve [7]. By curve-fitting these data to a competitive inhibition model, one may calculate an in-vivo IC_{50}. The desired drug delivery rate is approximately twice the rate that corresponds to this IC_{50}. A zero-order delivery system could then be

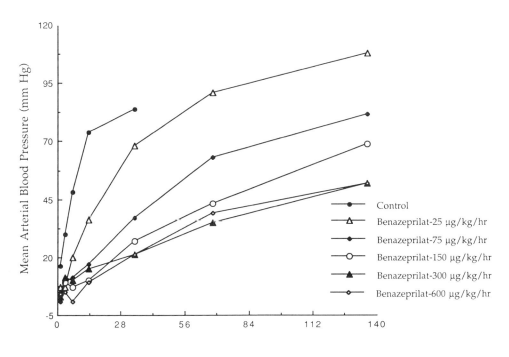

Angiotensin I Infusion Rate (ng/min)

Figure 2 Increase in mean arterial blood pressure in the SHR in response to the administration of angiotensin I, during the steady-state infusion of benazeprilat.

designed to deliver at this rate, and constant plasma levels of drug would result. In a conventional dosage form for an instantaneous input of a drug with a moderate half-life, the initial plasma level overshoots the IC_{50} to compensate for the subsequent decline of the plasma levels due to drug elimination with a falling input rate of drug. Consequently, the initial level of a drug with a short half-life might be 10 times the IC_{50}.

With zero-order delivery, it may be important to select plasma levels that avoid adverse drug reactions. Side effects and efficacy may be associated with different sets of biological receptors, and each set may have very different IC_{50}s or local drug distributions; i.e., side effects and efficacy may be associated with steady-state plasma levels that are quite different. If the therapeutic window is sufficiently large, i.e., the plasma level for the side effect is sufficiently higher than that for the desired efficacy, then a zero-order dosage form may permit drug efficacy without that side effect. The classical pharmacological case is use of scopolamine to prevent motion sickness. The desired drug delivery rate is 5 μg/h, while delivery rates for the side effects of dry mouth and drowsiness are 15 and 35 μg/h, respectively [8]. Based on these results, the transdermal system for scopolamine was designed as a zero-order dosage form that delivers 5 μg/h. More generally, the drug clearance [9,10] from plasma can be calculated from the AUC, the area under the curve to infinite time, for a curve of plasma concentration versus time for a fully bioavailable dosage form, preferably an intravenous infusion, as follows:

$$\text{clearance} = \frac{\text{dose}}{\text{AUC}} \tag{1}$$

The desired steady-state drug input rate may then be obtained from

$$\text{drug input rate} = (\text{target plasma level}) \times (\text{clearance}) \tag{2}$$

A second potential advantage of zero-order dosage forms is improved efficiency of delivery of the drug. Steady-state delivery of the drug may be more efficient when the distribution of the drug into the receptor compartment is much slower than elimination from the receptor compartment; in Fig. 1 $k_1 \ll k_{-1}$. The extent of the added efficiency may be surprisingly large. For example, the dopamine autoreceptor agonist, CGS 15855A, requires 0.4 mg/kg by i.p. injection in rats to lower striatal dopamine metabolites [11]. However, in rats given 40 μg/day by continuous infusion from an osmotic minipump, the effect of lowering striatal dopamine metabolites is saturated [12]. This example emphasizes a concern of zero-order drug delivery. Within 2 weeks of continuous infusion of the drug in rats, the effect of even a 400-μg/day delivery rate of CGS 15855A on dopamine metabolites disappears; i.e., there is complete tolerance.

This issue of tolerance has led to much controversy relative to treatment of angina with nitrates, so that nitrate-free intervals are proposed [13]. More complex delivery profiles may be used to balance tolerance and intervals of efficacious protection. Modulatable controlled-release profiles may be important for drugs with large circadian rhythms. Iontophoresis, phonophoresis, and light-activated systems are being considered for complex time-dependent drug delivery profiles. While the classical chronotherapy problem is enhanced delivery of insulin before a meal, many of these technologies are awaiting application to a therapy. Light-activated systems may have applications in localized or targeted cancer treatments and are being applied to the use of psoralens in skin cancers [14].

ROUTES OF DRUG DELIVERY

The justification for a controlled-release dosage form over a conventional tablet is either to optimize therapy or to circumvent problems in drug absorption or metabolism. Without this therapeutic rationale, a controlled-release dosage form would be unlikely to succeed as a product. The variety of routes available for drug delivery corresponds to the list of biological membranes in the human body: buccal, nasal, and vaginal mucosa, the gastrointestinal tract including the colon and rectum, the eye, and skin. We should further add to this list parenteral delivery, including implants and targeted delivery. Let us briefly review some of the reasons and approaches for each of these routes of administration. While prodrugs can be applied in each of these routes, use of these compounds has been extensively reviewed and will not be discussed in this chapter [15,16].

The GI Tract

Oral drug delivery is the most popular route and, as such, its advantages and limitations define the rationale for the other routes. The drug is absorbed through the various membranes along the gastrointestinal tract as the dosage form migrates through the GI tract. The pH of the stomach is approximately 2. The emptying time from the stomach is highly variable, but its mean is 4 h. Liquids may show a different pattern of emptying and empty in as fast as 15 min [17,18]. The pHs in the duodenum and the ileocecal junction are, respectively, 5.8 ± 0.8 and from 6.5 to 8.5 [19], but the microscopic pH near the membranes may be of the order of 6.1 near the jejunal membranes, as opposed to 7.2 in the bulk fluid [20,21]. Transit through the small intestine is more regular and is of the order of 4–6 h [34]. Entrance into the colon may be accompanied by a 0.5- to 1.0-unit increase in pH, and transit is highly variable giving a total GI transit time varying from less than a day to several days [22]. Drugs may be susceptible to degradation by different means in different regions of the gut—in particular, by hydrolysis in the stomach, by enzymatic digestion such as proteolysis in the small intestinal fluid, by metabolism in the gut wall, and by microorganisms in the colon. It should be noted that many individuals have diseases of the GI tract for which many of the above generalizations are not valid [22].

Other concerns for many drugs are extensive metabolism in the liver which is directly perfused by the portal blood supply leaving the intestine prior to entering the systemic circulation. This first-pass metabolism can lead to highly variable or poor absorption of drug into the systemic circulation. Finally, after absorption, metabolized or intact drug may be excreted into the bile and recirculated into the intestine.

Given these complications, prediction of drug absorption in the GI tract is difficult, but simple models correlating (1) oil solubility or melting point with absorption through a lipid membrane [23,24], or (2) water solubility as an additional variable to account for an "unstirred layer" in series with the intestinal membrane [25], or (3) more complex hydrodynamic models of perfusion [26–28] facilitate the estimation of drug absorption.

The thrust of controlled-release efforts in oral delivery has been limited mostly to four areas: (1) dosage forms with well-defined controlled-release delivery profiles, (2) extension of the transit time of dosage forms either through retention or drug delivery continuing into the colon, (3) enhancers for improved absorption of peptides, and (4) targeted delivery to the colon. Regardless of the approach, extensive in-vitro permeation studies, perfusions or infusions in animals, intubation studies in humans, and studies on transit of the dosage form are needed to characterize an oral dosage form for clinical development.

Oral Controlled-Release Dosage Forms

Varied oral controlled-release dosage forms have been developed, from coated pellets to microcapsules [29], to ion-exchange resins [30]. Hydrogels and osmotic pumps are two of the more important classes of oral controlled-release dosage forms. Hydrogels allow for extremely reproducible loading and delivery with an unusually wide range of specially tailored delivery rates, and also allow relatively easy removal of monomers, cross-linking agents, or other species with unacceptable toxicological profiles. Most hydrogels deliver as monolithic systems; i.e., drug release is proportional to the square root of time once the system is swollen with water. The search for zero-order monolithic oral delivery systems has become a "holy grail" of controlled-release and has included extensive work on novel geometries [31], Case II diffusion [32], and dissolution-controlled systems [33].

Osmotic pumps have seemingly offered the capability of tailoring delivery profiles—in particular, constant infusions—to the desired rate. To date, the tailoring has been limited by practical osmotic pressures, drug solubilities, and the nature of the membrane. The principle of drug release from osmotic pumps is simply one of water diffusing into the capsule down a water activity gradient across a membrane [2]. This water then exerts a constant pressure to force the drug out through a small hole in the capsule. While osmotic pumps for phenylpropanolamine hydrochloride and ascorbic acid are currently marketed, systems for metoprolol, albuterol, nifedipine, and prazosin may soon be available. Early adverse drug reactions related to ulceration after use of an indomethacin osmotic pump led to its withdrawal from the market and the irrational association of the osmotic pump with gastrointestinal side effects. The well-established side effects of ulceration with systemic administration of indomethacin combined with prolonged depression of prostaglandin synthesis for cytoprotection are more appropriate hypotheses.

Temporal control of the dosage form is currently a subject of academic research [34–36]. Several hydrogels display sensitivity to external stimuli. For example, certain hydrogels with mixtures of hydrophobic and hydrophilic groups exhibit temperature behavior analogous to surfactant solutions, that is, lower critical solution temperatures. At these temperatures, the gels exhibit large changes in swelling, which may be utilized for drug loading or release. Certain ionic hydrogels exhibit pH-dependent phase changes, while other gels seem to exhibit continuous swelling changes with pH. Recently, temperature-sensitive gels have been combined with a small fraction of ionic monomers to increase the magnitude of the swelling effect from phase changes with pH [34]. Some of these pH-sensitive gels have been suggested as replacements for enteric coating to prevent emptying in the stomach or to target the gel to the colon where a small pH change may occur. Applications other than oral delivery, such as iontophoresis or an insulin-releasing implant, have been suggested for these polymers [37].

Extension of GI Transit

Retention of dosage forms in the GI tract could be useful to decrease the frequency of dosing to once daily or less or in cases where there is an established absorption window in the GI tract, i.e., a unique region of the GI tract where drug absorption occurs. While carrier mechanisms are being investigated, the presence of absorption windows has often not been verified. Irritation to the colon or bacterial degradation in the colon may be a more frequent rationale for an apparent absorption window.

Bioadhesion to soft mucosa or mucus lining the GI tract has received considerable attention recently [38,39]. These applications are not just for prolonging retention, but also for localizing delivery. In addition to applications to oral dosage forms, bioadhesion to buccal, vaginal, and ocular mucosa has been investigated. Hydrogel adhesion to mucosa, a tissue which is highly swollen with water, may depend on entanglements and hydrogen bonding; for these applications polyacrylic acid and carboxymethylcellulose have received considerable attention [40–42]. Adhesion to mucus is complicated by sloughing of the mucus layer and by changes in the structure of mucus upon isolation. Negative charges from sialic acid with a pK_a of 2.6, in particular, and hydrogen bonding can play a role in adhesion to mucus. Polycarbophil has been shown to double the bioavailability of chlorothiazide in rats due to prolonged gastric retention, but the results were less pronounced in humans and dogs [39,43]. The relationships among the necessary forces, particle size, and release rate remain to be established. Bioadhesion combined with localization in the intestine has been considered for targeting peptides to Peyer's patches, but the absorption characteristics, and distribution between the systemic and lymphatic circulation are not well established. While bioadhesive dosage forms exhibit some promise, extensive technical and toxicological barriers must be overcome.

A second approach to prolonged retention has been to prevent or delay gastric emptying with dosage forms of various size and density or which swell to become too large to pass through the pylorus. However, successful modification of actual emptying has been limited, with these dosage forms exiting within the variability of the emptying time of solids [44,45].

Finally, the dosage form may continue delivering drug to the colon where degradation or poor absorption is not an issue, and once-daily, controlled-release dosage forms may be obtained for at least a substantial proportion of the population.

Oral Protein Delivery

The combined macromolecular and hydrophilic exterior nature of proteins minimizes passive transport across lipophilic barriers, and the size and continuity of aqueous porous pathways in the GI tract are far from established. In addition, proteins and peptides are susceptible to proteolysis. While amino acids and di- or tripeptides are actively transported in the small intestine [46], often with further degradation, the extent of transport of larger oligopeptides is less clear, except that it is substantially diminished. These carriers may play a role in absorption of L-methyldopa or perhaps angiotensin-converting enzyme inhibitors in the diacid form [47,48]. In-vitro methodology to characterize such carrier systems, such as intestinal cell lines [49] and in-situ perfusion models, may be used to evaluate the extent of oligopeptide transport.

Other approaches to peptide delivery have included use of protease inhibitors to minimize degradation in the small intestine and the classical ion-pair absorption [50], which has never been verified for any anionic or cationic drug. The use of micelles or lipid vehicle systems [51] is being tested to take advantage of the chylomicron flux into the lymphatic system or as a means of overcoming hepatic first-pass metabolism. However, the absorption of drugs from the lymphatic to the systemic circulation is not well characterized.

Enhancers of oral absorption may act by increasing passive transport across the small intestine, colon, or rectum [52–58]. Such alterations in gross barrier properties may require interactions of a magnitude that may cause cellular disruption, inflammation, or, at a more minor level, biochemical changes. Enhancers might be used in localized bursts, where the damage is contained, and repair mechanisms can handle alterations within the realm of normal physiological wear. At steady state, enhancers might be used at sufficiently low concentrations to avoid damage. The steady-state approach suffers from the further problem that the enhancer may be absorbed much more rapidly than the drug, or the enhancer may be diluted and provide only a local effect. Two classes of compounds have been studied as enhancers: (1) surfactants such as bile salts, nonionics, long-chain acyl carnitines, and nonionic glycerides; and (2) chelating agents, such as EDTA, salicylates, and phosphate derivatives. Careful toxicological evaluation of these enhancers will be required. While 60 mg/rat of sodium 5-methoxysalicylate increased the oral bioavailabilities of a hexapeptide somatostatin analog, insulin, and horseradish peroxidase, this was at the expense of mucosal damage [57], which got progressively worse. Histology is used to test for gross damage, and more sensitive biochemical tests have not been included. While there are intriguing data in the literature, no systematic characterization of the extent, mechanism, or minimization of damage has been reported.

Colonic Targeting

Targeting drugs to the colon has been suggested both to treat diseases of the colon and to deliver proteins orally from a region where there is little protease activity. The assumption of minimal proteolysis in the colon has not been carefully evaluated experimentally. To target insulin or vasopressin to the colon, an approach was borrowed from studies targeting prodrugs to the colon. Drugs are incorporated into azopolymers, which remain intact upon passage through the upper GI tract. The drugs are released in the colon when the azopolymers are degraded by colonic bacteria [59,60]. More recently, colonic bacterial glycosidases were used to release insulin selectively in the colon. A second approach to colonic targeting is to release the drug following passage through the small pH change from the ileum to the colon [19]. While this pH change may be readily demonstrated, it is not yet clear that the magnitude of this change is sufficiently large to overcome inter- and intraindividual variation. Although colonic targeting is an intriguing method for peptide delivery, current absorption data are based on pharmacodynamic measures rather than plasma levels relative to i.v. infusion or direct transport measurements. Consequently, the extent and variability of colonic absorption of peptides are difficult to assess from existing data.

While the use of suppositories and rectal delivery are remarkably unpopular in the United States, the relatively high permeability of the rectal mucosa and the ease of application have led to the investigation of this route for peptide delivery and delivery of drugs with oral-absorption problems. Many of the enhancer studies with the GI tract were performed on the rectal mucosa, and the aforementioned critique is still valid. Rectal delivery can only in part circumvent the hepatic first-pass metabolism. A hydrogel morphine suppository to relieve pain for 24 h was tested and found to be too variable and absorption too slow for use in acute pain [61].

Transdermal Delivery

Transdermal drug delivery has received perhaps the most attention of these alternatives. Commercial systems are available for scopolamine to treat motion sickness [8], nitroglycerin for angina [62], estradiol for postmenopausal syndrome and perhaps eventually to prevent osteoporosis [63], and clonidine as an antihypertensive [64]. New Drug Applications (NDAs) have been filed for transdermal administration of nicotine for smoking cessation [65–67], fentanyl for analgesia and in anesthesia for relief of moderate to severe surgical pain for an extended duration [68–70], and testosterone for male hormonal insufficiencies. A transdermal system for a D2 or autoreceptor agonist for Parkinson's disease is well into clinical trials [71]. The regimen for these transdermal systems varies from 1 day to 1 week, although skin tolerability is a concern for the longer-duration patches. Recently, trials relating efficacy of transdermal nitroglycerin to time have suggested that chronic daily application leads to tolerance, which may be avoided by wearing the transdermal patch for 10–14 h and then removing it at night.

To pass transdermally, the drug must diffuse across the epidermis to the dermis, where it is carried into the systemic circulation. This direct route into the systemic circulation avoids hepatic first-pass metabolism. The outermost 10–20 μm of the epidermis, the stratum corneum [72], is composed of sheets of keratinized squamous epithelium with the cells joined by tight junctions. There is virtually no metabolic activity in the stratum corneum. The stratum corneum comprises at least 90–95% of the diffusional resistance of the skin. The viable epidermis is the metabolizing layer, approximately 100 μm thick, from which the stratum corneum originates. It has been suggested that this layer is the dominant barrier for passage of extremely lipophilic compounds, although this has never been conclusively demonstrated. To measure skin permeation in vitro, diffusion of a drug across human stratum corneum should be determined from a saturated solution, i.e., unit activity, in an infinite donor reservoir; in-vitro skin permeation that is measured in this manner may be used to predict in-vivo drug delivery. This simple arrangement allows the physical and electrochemistry developed for classical membranes to be utilized without change for skin permeation.

A main function of the integument is to isolate the interior body tissue from its environment, and this is achieved by a combination of the remarkable barrier properties of stratum corneum and by its exceptional immune system. The barrier properties limit transdermal delivery to readily permeable, highly potent drugs. Skin permeation classically may be related to solubility in oil [73], that is, ideal solution theory [74]. The flux of a drug across skin may decrease exponentially with its molecular volume [23,75]. While until recently it appeared that the permeation of polar compounds through stratum corneum was negligible, recent work has indicated that water solubility may also play a role in skin permeation [76,77].

Much work has concentrated on finding enhancers to improve skin permeation [78]. Surfactants may alter protein structure or fluidize lipids as reflected by the discontinuous increase in skin permeation with surfactant concentration or by the FTIR spectra with and without these surfactants [79,80]. Absorption of polar compounds may be enhanced by surfactants with head groups of greater hydrophilicity, or nonpolar and polar compounds when surfactants are mixed with other solvents, such as diols or ethanol [78,81]. In general, there is an excellent correlation between the ability to increase skin permeation with a surfactant and skin irritation. While hairless mouse skin and other animal skins may be qualitative models for aqueous skin permeation in humans, with enhancers the correlation is not nearly as good [82]. Ethanol can increase permeation of nonpolar species by increasing their solubility in the stratum corneum without altering their diffusion constants [83–85]. Ethanol is the only solvent used in a commercial transdermal system where the relationship between the flux of the drug, estradiol, and that of the solvent has been characterized [73,86].

For many drugs the limiting factor in transdermal delivery is not poor permeation, but skin tolerability. Primary irritation, consisting of erythema (redness), edema (swelling), or pain, the classical inflammatory trio, may result from a variety of drugs. While solvents or surfactants may also cause irritation, the skin irritation of drugs may be instigated by local pH changes in the epidermis caused by a permeating drug. For drugs that permeate skin at comparable rates, the pK_a of the drug may help predict relative skin irritation [87,88].

A second, more serious skin-tolerability concern is contact sensitization resulting from repeated exposure to an antigen, which is the drug or system component; the immune system becomes alerted to a hapten, an antigen-protein complex, and quite severe erythema, edema, generalized rash, etc., may ensue. The classical example of a sensitizing agent is poison ivy [89,90]. Transdermal clonidine has become the most-studied transdermal sensitizer to date [91]. While guinea pig or mouse-ear-swelling models may generally be used to predict positive human sensitization responses, they do not indicate that clonidine is a sensitizer in humans with an incidence close to 20%. To date, only one instance has been reported where an individual who was sensitized to clonidine by the transdermal route was also sensitized to clonidine given orally. A recent patent application attempts to address this issue by incorporating corticosteroids into the patch, but given the extensive systemic effects of corticosteroids, an appropriate risk-benefit evaluation should be conducted [92].

The design of transdermal systems falls into two basic types: membrane-controlled and monolithic [93]. When the skin is rate-controlling, both types of systems exhibit zero-order delivery. With individuals with abrasions or exceptionally high skin permeability, the membrane systems may offer a safety advantage. Elegant systems that allow some temporal control of delivery are in development. Some systems allow activation of membrane permeation by diffusion of water from the skin, the transepidermal water loss or TEWL [94], or by mechanical activation by a breakable heat seal [95]. The common elements of current transdermal systems are a drug reservoir, a backing which is impermeable to drug over the system lifetime, an adhesive and release liner to attach the system to skin with minimal tolerability concerns, and appropriate pouching to achieve the desired shelf life.

Iontophoresis, or the use of electrical current to drive drugs across the skin, although noted in the last century, has received much attention recently as a means of increasing skin permeation and for temporal variation in drug delivery [96,97]. Careful attention to the use of nonpolarizable electrodes, well-characterized redox reactions, the use of constant current, etc., allows a fairly simple picture to emerge [98]. With the exception of some long-time relaxations, which may be modeled in a fashion similar to those observed in collagen and bone [99], the equivalent circuit for stratum corneum is a solution resistance in series with parallel membrane capacitor and resistor. The membrane impedance reflects diffusion in bulk solution within a restricted volume fraction and weak ionic selectivity. Current ranges need to be restricted to limit skin irritation, and at higher levels, pain [100,101]. In systems with polarizable electrodes, irritation may be related to pH changes. Substantial increases in drug permeation may be observed at small currents. Extensive use of iontophoresis exists in "sports medicine" applications, but its therapeutic application has not yet begun.

Given the limitations and concerns over transdermal delivery, it is its convenience, accessibility, and simplicity that have led to its early development compared to other routes.

Buccal Delivery

Buccal delivery of drugs, at first glance, seems to offer a combination of the advantages of transdermal and oral delivery. A buccal device or a sublingual tablet offers the easy application and removal of transdermal delivery without the excellent barrier properties of the stratum corneum and with less immune activity than the epidermis. Furthermore, hepatic first-pass metabolism may be avoided. The buccal route is restricted by its limited surface area, concerns of taste and comfort in a highly innervated area, the difficulties of adhesion to a mucosal surface for extended periods without the danger of swallowing or choking on a device, and potential bacterial growth or blockage of salivary glands associated with prolonged occlusion. Sublingual or buccal dosage forms have been developed for nitroglycerin [102], ergotamine [103], nicotine [104], buprenorphine [105], methyl testosterone [102], and nifedipine [123]. Even small peptides, such as oxytocin, have been shown to cross the buccal mucosa [106]. While the gingiva and periodontal pockets have somewhat different anatomy and physiology, it is convenient to mention dental applications of controlled release, including release of fluorides or release of antibiotics from fibers placed in periodontal pockets [107].

While human oral mucosa, like skin, is a stratified epithelium, the hard palate tends to have a cornified epithelial layer, but the buccal mucosa is a nonkeratinized tissue with an average thickness of 580 μm [108]. Keratohyalin granules have been identified in human buccal tissue, but in contrast to many rodent buccal models, there is no stratum corneum. Intercellular connections in buccal tissue are characterized by desmosomes and gap junctions, and the tissue is a somewhat leaky epithelium. Although it is believed that the buccal mucosa has a weak immune response compared to skin, buccal tissue has a reduced number of Langerhans cells, which appear to function in the immune response to antigens in oral mucosa in the same fashion as in the viable epidermis [109].

Most of the in-vivo data on buccal permeation of drugs is derived from the experimental paradigm developed by Beckett and Triggs [110]. With this method, a known concentration and volume of drug are kept in the subject's mouth for a fixed amount of time, spit out, and the drug absorption is calculated as the difference in the total amount of drug. In-situ perfusion cells [111] have also been developed, but both techniques suffer from the same drawbacks: (1) absorption is not distinguished from adsorption or metabolism; and (2) the drug's affinity for water may dominate the absorption measurement. This second aspect of the methodology has led to the conclusion that un-ionized species dominate buccal permeation and that transport follows the partition coefficient [112].

The polar shunt pathway has been largely neglected in the understanding of buccal permeation. Sodium diclofenac in its ionized form has been shown to permeate buccal mucosa extremely rapidly in dogs and in humans [113,114]. Dipolar ions, such as angiotensin-converting enzyme inhibitors [115] and TRH [111,116], also permeate quite extensively. The measurement of these permeants has relied on both in-vivo studies from saturated solutions where plasma AUC was compared with intravenous administration or in-vitro methods, such as Ussing chambers. Permeation through buccal mucosa is very rapid compared to skin, but perhaps an order of magnitude or more slower than through intestinal or nasal tissue [113,117].

Peptide and protein permeation through buccal mucosa has been reported based on pharmacodynamic measurements. The bioavailability of oxytocin given transbuccally seemed to be on the order of a few percent. The usual aminopeptidases are present in buccal tissue, and degradation may be a concern [118,119]. Furthermore, the extent of peptide permeation is unclear, and some histological papers suggest barrier layers for high-molecular-weight markers such as horseradish peroxidase [120]. A buccal tissue culture model is in development, but it is from the hamster cheek pouch, which contains a cornified layer [121]. Careful study of the transport and biochemistry of buccal mucosa will continue with some newer, more appropriate models.

Nasal Delivery

In comparison to the aforementioned routes, nasal delivery currently has a greater capability of delivering peptides and poorly orally absorbed drugs with reasonable bioavailabilities and without encountering hepatic first-pass metabolism. Absorption of TSH, ACTH, buserelin, growth hormone, and LHRH have been measured by the nasal route using pharmacodynamic responses relative to i.v. administration [122]. The nasal delivery of nafarelin acetate, an LHRH analog, has been further quantitated by plasma levels versus i.v. administration [123]. Through the use of adjuvants, such as sodium deoxycholate, insulin has been absorbed intranasally [124]. Such adjuvants may both limit the aggregation of insulin and act as enhancers on the tissue. With these enhancers, extensive toxicological tests must be performed. To date, such toxicology has been limited to histological studies of necropsy tissue, but more sensitive biochemical studies of nasal mucosa are needed.

Nasal anatomy is designed to provide sizable resistance to air flow and to control the humidity and temperature of inhaled air. The nasal septum divides the nasal passage into two, and each passage is convoluted by the turbinates. Large particles have difficulty passing the nasal valve, a constriction. The mucosal lining contains both the goblet cells, which secrete the mucous that flows slowly to the nasopharynx and is swallowed, and the cilia, which aid this transport, plus submucosal glandular structures. The nasal mucosa has been studied in vitro in Ussing chambers and is a leaky mucosa [125]. Transport appears to be by parallel lipoidal and shunt pathways with oil and water solubilities and molecular weight providing an approximate indication of structure-permeation relationships [126]. Large proteins appear unable to cross the mucosa to any significant degree. Small peptides, such as oxytocin and somatostatin analogs, can show bioavailabilities comparable to those with i.v. injection [117].

The in-situ rat perfusion model, in which the inlet and outlet concentrations are measured, is a popular model for nasal drug delivery [127]. While this method has been used to predict nasal delivery of certain compounds, other animal models, such as sheep or primate, may be preferred, even though they are less convenient [128].

Drugs may be administered intranasally as aerosols, powders, or small particulates. Sterility, stability, toxicology, and effects on delivery of adjuvants should be evaluated. Design of a nebulizer device is also a key to reproducible dosing, and there is an extensive literature on the design of these devices. Plasticizers and other excipients in plastic or gaskets may interact with the formulation, so that their effects must be evaluated [129]. Although nasal delivery presents a route for achieving relatively high bioavailabilities of small peptides, there is an opportunity for innovation in reproducible controlled-release nasal formulations.

Ocular Delivery

Ocular delivery involves localized treatment of the eye, unlike the previous routes, which are meant for systemic drug delivery. Ocular prodrugs are the simplest example of site-specific delivery, but as new chemical entities, they require a drug discovery program. The so-called soft drugs [130], compounds that are active locally when applied topically to the eye but that degrade rapidly to an inactive form in the systemic circulation, are examples of new chemical entities designed for localized delivery.

Drug applied to the eye is continually drained by the lachrymal fluid, which has a mean volume of 7 μl per eye and a pH of 7.4. Inside the cornea is the aqueous humor, a 0.01–0.02% protein solution with a volume of some 300 μl and a turnover rate of approximately 180 μl/h. The permeation barrier, the cornea, is classically viewed as a hydrophilic layer, the stroma, sandwiched between two hydrophobic layers, the epithelium and the endothelium [131,132]. A partition coefficient near unity is necessary for transport through the cornea. Indeed, the lack of transport of drugs with partition coefficients differing from unity may result from the long time lag for permeation and not from low permeability [133]. Parallel shunt pathways through the multilaminate may also operate. Melting point, as a predictor of the ideal solubility and molecular weight, can provide an order-of-magnitude estimate of the flux across the cornea [134].

Most in-vivo studies of ocular absorption involve application of a finite dose of drug in a non-steady-state measurement. In a recently developed paradigm [135,136], after a brief application, the eye is washed. Under the near sink conditions that hold on both sides of the cornea, the amount absorbed equals the product of the surface area, the time of application, and the steady-state flux; i.e., steady-state transport properties are easily inferred from the non-steady-state measurement. In other in-vivo models, the eye is perfused [137]. Diffusion measurements are also routinely performed in vitro in diffusion cells, but the electrical properties of this tissue are only recently being defined [138].

Irritation, including erythema, edema, and severe pain, must be considered with any drug, excipient, or enhancer to be applied to the eye. Like the skin, the eye is a well-defined site for the use of an enhancer for drug transport, provided both drug and enhancer are nonirritating.

Most drug products are drops or ointments, and retention in the eye is a concern for controlled release. To prevent drainage, ointments with a large viscosity may be formulated with such hydrophilic polymeric thickeners as certain cellulose or starch derivatives. A hydrophobic formulation may also be useful. Ocular controlled-release devices have been designed, but patient comfort and convenience must be carefully considered. A small membrane-controlled reservoir insert, Ocusert, which delivers pilocarpine to treat glaucoma, has met with limited success. Hydrogel contact lenses may be used as drug-delivery devices, with release proportional to the square root of time. These devices are reproducibly loaded with drug in the laboratory or clinic, but difficulties in home use might be an issue unless disposable lenses become common. Soft contact lenses have also been used for controlling release of pilocarpine.

Intrauterine and Intravaginal Delivery

Intrauterine and intravaginal delivery are the two routes used primarily for fertility control [139]. While numerous intrauterine devices (IUDs) have previously been on the market and may have functioned to provide a sustained release of copper and to irritate the uterine endometrium, one of the few remaining commercial devices in the United States is a membrane-controlled device, Progestasert; it releases progesterone at the rate of 65 μg/day for 1 year. With this low-release rate, endometrial proliferation is suppressed [140].

Intravaginal delivery may have applications beyond birth control. The two existing vaginal implants are (1) silicone elastomer vaginal rings releasing the contraceptive gestagen, medroxyprogesterone acetate, from a dispersed monolithic matrix; and (2) estrogen-progestin vaginal implants using a similar design, but with polyethylene glycol dispersed throughout the silicone matrix [141,142]. The linear release of these devices arises from the lipophilicity of the steroids and the unstirred water layer in series with the rather permeable vaginal membrane.

The human vagina is lined along its 10- to 15-cm length with a mucosal membrane and is lubricated by vaginal secretions. While the pH is generally between 4 and 5, it is cyclically dependent through the influence of estrogen on glycogen deposition. Among body tissues, 3% glycogen content is second only to that in the liver [142]. Estrogen also causes cornification and thickening of the vaginal mucosa, and consequently, transport may be cyclic. A variety of microorganisms and potential pathogens may proliferate in the vagina, and this may be important for understanding drug metabolism and the constraints of the device design. The permeability of the vaginal membrane has been suggested to be closer to nasal mucosa than to buccal mucosa [143]. Given its convenience for self-implantation, the high permeability, and its ease of controlling the delivery, peptide delivery by the vaginal route is being investigated.

Parenteral Delivery

For life-threatening diseases or those in which the quality of life is drastically impaired, parenteral controlled-release administration either by injectables or by implants may be appropriate. It is appropriate to distinguish between depot formulations, which typify sustained release, and true controlled-release administration. There are numerous examples of depot formulations where drug delivery is controlled by solubility either through insoluble salts, oil emulsions, or other approaches [144]. Another route to parenteral controlled release is through the use of infusion pumps of varying degrees of sophistication. Direct infusion into physiological compartments, for example, intrathecal infusion of baclofen or neuropeptides [145], may permit therapies in debilitating diseases, but diffusion of the drug within the target organ is still being investigated [146]. Miniaturization of mechanical devices and improved biocompatibility of materials will make these devices more acceptable.

For most pharmaceutical applications, subcutaneous implants must answer both long-term toxicological concerns and ease of removal to terminate drug therapy. There have been two basic approaches to the use of polymers, conventional and biodegradable. While biodegradation avoids removal of the system after its use, metabolism of the polymer, its monomer, its dimer, and low-molecular-weight polymers must be extensively evaluated, and biodegradation should occur within a time frame that is slow compared to drug release to allow for system removal.

Acute and chronic inflammation and foreign-body reactions to implants are another major concern [147]. Extensive work has been done relating surface free energy, steric hindrance, surface mobility, and other approaches to minimize protein adsorption [148]. Investigations of the blood compatibility of materials have been adapted to accommodate subcutaneous implants, and similar tests of pathology and durability may be developed.

Given the aforementioned issues, development of implants as commercial products is quite slow. Several varieties of implants with different geometries, i.e., beads, fibers, etc., are being investigated for cancer chemotherapy; e.g., a polyanhydride-based device containing carnustine is used to treat brain tumors and a HEMA copolymer hydrogel uses 5-fluorouracil [149]. Several implantable systems have been described containing narcotic antagonists for use in heroin addicts [150]. Examples of biodegradable implants are poly(l-lactic acid) fibers to release levonorgestrel as a contraceptive [149] and poly(orthoesters) to release 5-fluorouracil or LHRH analogs [151].

Some of the more glamorous research in this area involves the development of self-regulating implants, especially for administration of insulin. Two of the approaches used include immobilization of the enzyme, glucose oxidase [152], to form a bioresponsive membrane and a glycosylated insulin con A complex in which glucose binding to con A [153] displaces insulin. The great potential of implants has yet to be achieved. A subcutaneous implant that releases levonorgestrel through silicone rubber capsules is one of the closest to reaching the marketplace [149].

The element of targeting or site-specific delivery is often linked with new controlled-release delivery systems. Such targeting is currently only in the research stage. In perhaps the simplest form, some sort of carrier such as liposomes, colloidal particles, albumin, nanoparticles, or water-soluble polymer conjugates is employed [154,155]. Particle filtration can be used to localize larger particles in the lungs [156], but it is not clear that this localization aids the distribution of the drug to the appropriate target cells [157]. Many particles injected into the circulation are rapidly taken up by the reticuloendothelial system (RES), specifically, the Kupffer cells of the liver [158]. The uptake or kinetics of uptake of such particles may be altered by the use of surface coatings such as poly(oxyethylene) or poloxamers [159]. Water-soluble macromolecular conjugates may include antibodies to target the conjugate, and then the drug may be released by lysosomal enzymes after being internalized into the cell [155,159]. Numerous strategies for targeting drugs including antibodies, rheoviruses, and immunotoxins, have been suggested [154], but this field is largely in its infancy and needs specific problems identified for it to be developed. The difficulties in the analytical methodology have made quantitation and kinetic/dynamic studies in this field slow in coming. One of the more advanced applications is the use of liposomes to deliver immunomodulators for cancer and antiviral therapy, but even this approach is just beginning in the clinic [160].

POLYMER PROPERTIES INFLUENCING DRUG PERMEATION

Controlled drug release is based on diffusion through polymers, erosion of polymers, and special polymer characteristics such as osmotic and ion-exchange properties. In a unidirectional reservoir system, for instance, understanding diffusion through polymers can guide material selection for the system components to regulate the fluxes of active ingredients. The next two sections review some of the important properties of polymers that influence drug permeation, and how these properties can be characterized.

Mass Transfer

Let us begin by examining the migration of a species, such as an active ingredient, across a polymeric film by passive diffusion. The activity gradient of the species across the film provides the driving force. For convenience, however, activity is replaced by concentration following the relationship

$$\text{activity} = (\text{activity coefficient}) \times (\text{concentration}) \qquad (3)$$

In the simple case of constant mass transfer properties, the flux of a species through a polymeric film [4] is given by

$$\text{flux} = \left(\frac{\text{area}}{\text{length}}\right) \times (\text{permeability}) \times (\text{concentration difference}) \qquad (4)$$

Area is the surface through which a species is diffusing, and length is the film thickness. Permeability is given by the product of partition coefficient and diffusivity:

$$\text{permeability} = (\text{partition coefficient}) \times (\text{diffusivity}) \qquad (5)$$

The concentration difference in Eq. (4), which is measurable, refers to concentrations of a diffusing species located on both external surfaces of the film. However, the diffusional driving force, when expressed in terms of concentration, comes from the concentration gradient on both interior surfaces of the film. The partition coefficient relates the concentration in the interior to that at the exterior surface of the film, by the ratio of their respective activity coefficients. The implicit assumption made is that equilibrium is established at the interface. Partition coefficients from a saturated solution simplify the interpretation. From a saturated solution, the partition coefficient is a ratio of the solubilities in the solvent and the film. Physically, the partition coefficient describes the relative affinities of the diffusing species for the solvent and the polymeric film. For instance, nonpolar species tend to partition into nonpolar materials. The amount partitioned can be measured from sorption/desorption experiments. Diffusivity is the component of permeability that accounts for the geometric constraints encountered by the diffusing species in weaving across the polymeric film. Consequently, diffusivity increases as the free volume of the polymer increases relative to the dimensions of the diffusing species. In turn, the internal structure of a polymer depends on molecular weight, chain regularity, chain flexibility, and the strengths of the intermolecular and intramolecular forces. We will investigate these factors in more detail, and examine how they affect the polymer bulk properties.

Molecular Weight and Molecular-Weight Distribution

Polymers are macromolecules that are made up of many repeat units. The chemical composition of a repeat unit or monomer is the basic building block, which can be characterized by elemental analysis and various spectroscopic methods. The number of repeat units gives the molecular weight of a single polymer chain. However, reactants usually do not polymerize at the same time and at the same rate during synthesis. Consequently, a distribution of chain lengths is formed. This means that molecular weight is an average quantity made up of contributions from all polymer chains, but there is no unique way to assign this value. Therefore, a number of averaging methods [161] have been introduced, each attempting to emphasize a certain population of chain length.

The first moment-average or number-average molecular weight is sensitive to the low-molecular-weight species, and is defined by

$$M_n = \frac{\Sigma N_i M_i}{\Sigma N_i} \tag{6}$$

where N_i is the number of polymer chains (i), each having a molecular weight (M_i). Colligative properties are related to M_n through their dependencies on the number of species. This also suggests that experimental methods for determining colligative properties, such as osmometry, ebulliometry (boiling-point elevation), and cryoscopy (melting-point depression), can be used to obtain M_n.

The second moment-average or weight-average molecular weight is more sensitive to the higher-molecular-weight species, and is defined by

$$M_w = \frac{\Sigma N_i M_i^2}{\Sigma N_i M_i} \tag{7}$$

Light scattering is commonly used to measure the weight-average molecular weight. Many properties, such as toughness or mechanical properties associated with large deformations, can be correlated to M_w.

Since the number-average and weight-average molecular weights describe opposite ends of the molecular-weight spectrum, their ratio provides an estimate of the breadth of the molecular-weight distribution. The polydispersity index is the ratio of M_w/M_n. The polydispersity index equals 1 for a monodisperse polymer, and is greater than 1 otherwise.

The third moment-average or z-average molecular weight adds even more weight toward the high-molecular-weight species, and is defined by

$$M_z = \frac{\Sigma N_i M_i^3}{\Sigma N_i M_i^2} \tag{8}$$

Experimentally, electron microscopy and centrifugation have been used to measure the z-average molecular weight. Physical properties, such as elasticity, have been shown to depend on M_z. In addition to M_z, average molecular weights with higher weighting factors can be defined accordingly. For completeness, there is also a viscosity-average molecular weight (M_v), which is obtained from viscosity measurements. In a polydisperse polymer, the relative magnitude of these average molecular weights follows

$$M_z > M_w > M_v > M_n \tag{9}$$

Molecular weight reflects the size and size distribution of a polymer. While we might expect qualitatively that diffusivity decreases with increasing molecular weight, the critical factor that governs diffusivity is determined by the overall microstructure of the polymer in the presence of diffusing and other foreign species. Let us first consider the macroscopic behavior of a polymer and then investigate how the microstructure of a polymer is constructed.

Material Characteristic Time

From a phenomenological perspective, the physical behavior of a material can be predicted from the characteristic time of the material (λ), and the process time scale (θ) to which the material is subjected. The material characteristic time is an intrinsic property of a material. However, it is an average quantity from a spectrum of characteristic times that reflects the distribution of polymer chain lengths within a polymer. These material characteristic times can be determined experimentally from rheological techniques. The process time scale is the duration of a process, such as diffusion or a mechanical deformation. Reiner [162] coined this dimensionless ratio of time scales, the Deborah number (De $= \lambda/\theta$). When the Deborah number is very small (De $<< 1$), the response is termed "viscous," according to the nomenclature in fluid and solid mechanics. In the corresponding case of diffusion, the polymer molecular relaxation is much faster than the duration of the diffusion. In this situation, diffusion can be modeled by Fick's law, with a diffusivity that can depend on temperature, concentration, and molecular weight. At the other extreme of a very large Deborah number De $>> 1$), the response is classified as "elastic." For this case, polymer diffusion takes place in a glassy polymer, and Fick's law has been used quite successfully to model diffusion in this regime. Consequently, diffusion can be characterized by Fick's law in two situations, where the Deborah number is either very small or very large. In both cases, the amount of penetrant absorbed in or desorbed from a polymer film is proportional to the square root of time. This type of diffusion is designated as case I diffusion. When the Deborah number approaches unity (De ≈ 1), the relaxation and the diffusion time scales are of the same order of magnitude. In this case, diffusion in polymers exhibits a number of anomalous behaviors. For instance, the amount of penetrant absorbed or desorbed in a polymer film no longer follows a square-root-of-time dependency. Instead, it approaches a linear relationship with time. The extreme case of non-Fickian diffusion is known as case II diffusion. While we have seen the importance of determining the material characteristic time, and the utility of the Deborah number of classify Fickian and non-Fickian diffusion, let us investigate how the polymer microstructure may give rise to these macroscopic behaviors.

Polymer Microstructures

The conditions and routes of synthesis can lead to three general classes of polymer topologies. These are the linear, branched, and network polymers [163], which can be characterized experimentally. For instance, branching can be detected by IR, NMR, and radiolysis. The cross-link density in network polymers can be determined from swelling data, sol-gel analysis, and viscometry. Linear polymers, as sketched in Fig. 3a, contain two free ends. Branched polymers contain more than two free ends, and thus can form numerous topologies as shown in Fig. 3b. Network polymers are formed by cross-linking polymer segments, as shown in Fig. 3c. Some of these networks have interesting structures and unique properties, such as the high thermal stability of ladder polymers. Hydrogels are of particular interest for controlled-release applications, because their intriguing networks enable this material to pick up enormous amounts of water while remaining water-insoluble and retaining their mechanical strength [164].

The interactions within a polymer chain and between polymer chains depend on the chemical composition and ordering of the functional groups. The chemical makeup of a polymer gives rise to intramolecular forces that determine the local stiffness and flexibility

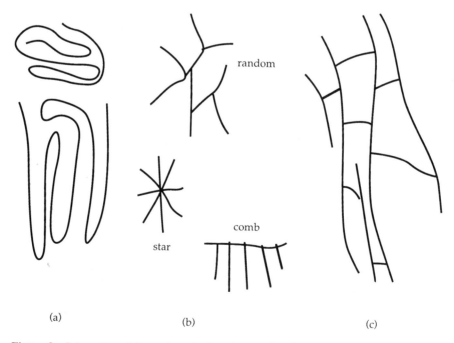

Figure 3 Schematics of linear, branched, and network polymers: (a) linear polymers, (b) branched polymers; and (c) network polymers.

of polymer segments. For instance, double bonds and ring structures along a polymer backbone stiffen the polymer chain, while insertion of oxygen along the chain increases flexibility. In general, diffusivity is higher in polymers containing flexible chains. Molecular interactions also occur between polymer chains. In this case, the polarity of functional groups gives rise to repulsive and attractive forces, which control the positioning of the groups and the separation of the chains.

The ordering of the functional groups can also come about from stereoisomerism, such as cis-trans isomers and tacticity, produced during synthesis. Tacticity is the arrangement of functional groups along a carbon chain. The *isotactic* and *syndiotactic* structures exhibit regularities along a carbon backbone, while random placement of the groups gives rise to an *atactic* structure. The effect of tacticity on macroscopic properties increases as the distance between polymer chains decreases to form crystal lattices. As an example, the densities of isotactic, syndiotactic, and atactic polymethyl methacrylate are, respectively, 1.22, 1.19, and 1.188 g/cm³. The isotactic form has the highest degree of regularity, increases the packing density, and therefore the bulk density. Qualitatively, diffusivity decreases as density increases because of the reduced amount of free volume.

The net effect of these molecular interactions gives rise to different polymer morphologies [165], which are generally classified as amorphous, semicrystalline, or crystalline. The popular analogy for an amorphous polymer is a bucket of worms, because the polymer segments are in constant motion. This phase is usually marked by disorder, and consequently does not give clear X-ray diffractions. Since intermolecular-chain interaction is minimal, the polymer can be deformed and shaped into films or other useful geometries. Diffusivity is normally high in an amorphous polymer because of its relatively high free volume. Semicrystalline polymers contain a mixture of amorphous and crystalline phases. Consequently, these polymers can still be deformed, particularly by raising temperature to enhance molecular motion. Crystalline polymers are characterized by a high degree of order among the polymer chains. As a result, they produce colorful patterns in cross-polarized light. However, they are not readily deformable by thermal energy, and may decompose instead of melting. Consequently, a solvent is needed to partially dissolve these polymers, so that they can be shaped into useful objects. An additional processing step may then be required to remove the solvent by either evaporation or extraction. Diffusivity is low in the crystalline phase because of the reduced free volume.

As expected, polymers can be in different morphological phases at equilibrium, depending on pressure and temperature. The effect of pressure is negligible in most cases. However, temperature can significantly alter the mobility of the polymer segments by thermal energy. Let us investigate the effects of temperature

Transition Temperatures

The melting temperature (T_m) and the glass transition temperature (T_g) are two important transition temperatures of polymers. Above the melting temperature, a polymer is in a molten state, and is completely amorphous. The melting temperature, a first-order transition, is characterized by a sharp change in specific volume. However, not all polymers exhibit a melting temperature. Highly crystalline polymers usually do not melt, because their collective intermolecular forces are comparable or exceed their covalent bond strengths. Very often, degradation can occur before melting, because covalent and secondary bonds are severed.

As temperature decreases below the melting temperature, the polymer chain mobility decreases. If the polymer chains can be ordered even on a local scale, then crystallization can take place. The rate of crystallization then increases with decreasing temperature, but eventually reaches a maximum. With further drop in temperature, the rate of crystallization decreases because of decreasing polymer chain mobility. Ultimately, a certain temperature is reached where virtually no segmental motion can occur. This temperature is the glass transition temperature. The polymer is in a glassy state below the glass transition temperature.

The glass transition temperature is a second-order transition because the associated change in specific volume is not discontinuous as it is at the melting temperature. Furthermore, the glass transition temperature depends on the history and time scale of the experiment, and changes slightly according to the rate of heating or cooling of a sample. Since the glass transition temperature depicts segmental mobility, factors that hinder motion would increase T_g. For instance, stiff chains, ring structures, bulky side groups, and polar groups in a polymer would all contribute to a higher glass transition temperature. This brings up an interesting problem on how to modify the glass transition temperature, particularly relative to room temperature or body temperature, where most of these polymers are to be used.

The glass transition temperature can be modified by changing the strength of the secondary forces within the polymer. This can be accomplished by introducing an additive, such as a plasticizer, to the polymer. For instance, addition of dioctyl phthalate lowers the glass transition of polyvinyl chloride and softens this polymer, which by itself is brittle at room temperature. However, the relationship between the glass transition temperature of the mixture and the amount of plasticizer needs to be determined experimentally. Another approach to modifying the glass transition temperature is to form a copolymer. The glass transition temperature of the copolymer can be estimated from the glass transition temperatures of each homopolymer by the empirical relationship

$$\frac{1}{(T_g)_{copolymer}} = \frac{\omega_1}{(T_g)_1} + \frac{\omega_2}{(T_g)_2} \tag{10}$$

where the temperatures are in kelvins, and ω_1 and ω_2 are the weight fractions of the homopolymers.

The ability to modify the glass transition temperature provides an interesting approach to develop systems with some temporal control of delivery [94]. As described earlier, the delayed onset of drug permeation from a system can be triggered by the back-diffusion of water from the skin. In this case, a drug normally could not permeate across a glassy polymeric membrane. However, when the system is applied on skin, water from the skin plasticizes the membrane and activates drug permeation.

Solubility Parameter

The addition of a second component, such as a drug or a solvent, to a polymer can change the strength of the polymer intermolecular forces, and therefore the physical properties of the polymer. Diffusivity would then depend on the availability of free volume in this multicomponent system. The strength of the intermolecular forces of a polymer is measured by its cohesive energy density (CED). The solubility parameter of a polymer (δ) is an alternative description of the intermolecular forces; it is related to the cohesive energy density as follows:

$$\delta = (CED)^{1/2} \tag{11}$$

The solubility parameter can be determined by various methods, such as intrinsic viscosity and swelling measurements. The solubility parameters for many polymers are tabulated in handbooks [166]. The compatibility between a polymer and another polymer, or a polymer and a solvent, can be estimated from their respective solubility parameters. For instance, a polymer would swell or dissolve in a solvent that has a similar solubility parameter, and would precipitate from a solvent with a significantly different solubility parameter. A poor solvent is also known as a theta solvent. Similarly, the theta temperature is the temperature at which a solvent becomes a theta solvent for a polymer. The extent of interaction between a polymer and a solvent is characterized by the Flory-Huggins interaction parameter (χ). This interaction parameter is proportional to the difference in solubility parameters, and is given by the following equation:

$$\chi = \alpha + \frac{V_1}{RT}(\delta_1 - \delta_2)^2 \tag{12}$$

where α is a lattice constant for the polymer, the subscripts 1 and 2 refer to the polymer and the solvent, respectively, V_1 is the polymer molar volume, R is the gas constant and T is temperature. Two materials are likely to be compatible if the value of χ is small.

Diffusivity

The concept of free volume has been widely used to model diffusivity [167] in amorphous polymers above and below the glass transition temperature. The transport mechanism is modeled by a migration of the penetrate through holes or free volume within the polymer. Therefore, the rate of transport is based on the probability of creating a hole of sufficient size to accommodate the penetrant, and the probability for this penetrant to have sufficient energy to enter this hole. At temperatures significantly higher than the glass transition temperature, both probabilities must be taken into account to estimate diffusivity. However, near the glass transition temperature, the amount of free volume is small, so the probability of encountering holes of sufficient size dominates mass transfer. Below the glass transition temperature the amount of free volume is small, and the redistribution of holes within the polymer is negligible because segmental motion is virtually nonexistent. The size and the distribution of the holes at any temperature below the glass transition temperature are the same as those that were "frozen in" at the glass transition temperature. Consequently, the rate of cooling controls the diffusivity directly. Numerous mathematical models have been proposed to estimate diffusivity based on this diffusion mechanism. All of these models are semiempirical because of the difficulties in predicting free volume from first principles.

POLYMER CHARACTERIZATION

This section provides an overview of experimental methods used to characterize the properties of polymers.

Molecular Weight and Molecular-Weight Distribution

Most methods developed to measure molecular weight involve dissolving the polymer in a solvent. Dilute solutions are often used to avoid additional complications from viscoelastic effects. The experimental methods are then divided into those that measure absolute molecular weights and those that provide relative molecular weights. Both approaches are presented below for the measurement of the various average molecular weights defined earlier.

End-group analysis is used to obtain the number-average molecular weight up to approximately 20,000. The key to this method depends on the availability of analytical techniques to detect and count functional groups on a polymer chain. For instance, hydroxyl, carboxyl, and amino end groups can be titrated.

Osmometry can also be used to determine the number-average molecular weight. The key to this method depends on the availability of a semipermeable membrane to measure the osmotic pressure in a polymer solution. The osmotic pressure is then related to M_n by

$$\pi = \frac{RT}{M_n} C + BC^2 \tag{13}$$

where C is concentration and B is a virial coefficient. The number-average molecular weight is then obtained from the intercept of a plot of π/CRT versus C.

Ebulliometry (or boiling elevation) and cryoscopy (or freezing depression) are based on the Clausius-Clapeyron equation, to relate changes in boiling and freezing points to the number-average molecular weight. This relation is given by

$$M_n = \frac{RT^2V}{\Delta H} \left(\frac{C}{\Delta T} \right)_{C \to 0} \tag{14}$$

where V and ΔH are the specific volume and heat of transition, respectively. Experimentally the boiling and freezing points of a polymer solution are compared sequentially to that of a solvent by differential thermometers or thermistors arranged in a Wheatstone bridge circuit. Both methods are limited to M_n up to approximately 30,000 because of foaming and crystallization problems with high-molecular-weight polymers.

Light scattering provides a means of measuring the weight-average molecular weight by applying Lord Rayleigh's electromagnetic theory, which shows that the intensity of scattering is proportional to the square of the particle mass. Experimentally, the intensity of scattered light from a incident source, such as a laser, is measured at different angles from polymer solutions of different concentrations. In dilute solutions, the Debye equation relates intensity, concentration, and M_w as follows:

$$\frac{HC}{\tau} = \frac{1}{M_w} + \frac{2\beta C}{M_w} \tag{15}$$

where H is determined experimentally from refractive index (n) and specific refractive increment (dn/dC) measurements. H is constant for a polymer in a solvent at a given temperature, and is calculated by the following equation:

$$H = \frac{32\pi^3n^2}{N_0\lambda^4} \left(\frac{dn}{dC} \right)^2 \tag{16}$$

where N_0 is Avogadro's number, and λ is the wavelength of the incident source. The turbidity (τ) is calculated from the intensity of scattering at $0°$ and $90°$, and β is a constant. The weight-average molecular weight is then obtained from the intercept of a plot of HC/τ versus C.

Ultracentrifugation can be used to measure the weight-average and z-average molecular weights. This method takes advantage of an equilibrium that is established between centrifugal and diffusional forces in a dilute polymer solution. For a given angular velocity (Ω), concentration, refractive index, and specific refractive increment are measured at two locations (r_1 and r_2) within the ultracentrifugation cell. The weight-average molecular weight is related to concentration measurements by

$$M_w = \frac{2\,RT\,\ln(C_2/C_1)}{(1 - \bar{V}\rho)\Omega^2\,(r_2^2 - r_1^2)} \tag{17}$$

where $\bar{V}$ is the polymer partial specific volume and ρ is the solution density. The z-average molecular weight is related to refractive index and specific refractive increment by

$$M_z = \frac{RT}{(1 - \bar{V}\rho)\Omega^2}\,\frac{[(1/r)(dn/dr)]_2 - [(1/r)(dn/dr)]_1}{(n_2 - n_1)} \tag{18}$$

The solution viscosity reflects the space that a polymer occupies in a solvent, and therefore provides a measure of molecular weight. The viscosity of a dilute polymer solution (η) can be conveniently determined in capillary viscometers of the Ostwald-Fenske or Ubbelohde type by recording the time required for a volume of solution to flow through a capillary tube. By repeating this measurement in various solution concentrations, the intrinsic viscosity ($[\eta]$) can be extrapolated at infinite dilution from the following relationship:

$$[\eta] = \left(\frac{\eta - \eta_s}{\eta_s C}\right)_{C=0} \tag{19}$$

where η_s is the solvent viscosity. The intrinsic viscosity is then related to viscosity by the Mark-Houwink equation, which is an empirical relationship with two constants (K') and (a):

$$[\eta] = K'M^a \tag{20}$$

The exponent (a) varies from a value of 0.5 in a poor solvent to a maximum of 1. The values of both constants, which are functions of temperature, solvent, and polymer, are tabulated in polymer handbooks [168]. Based on the intrinsic viscosity measurement, Schaefgen and Flory [169] defined a viscosity-average molecular weight by

$$M_v = \left(\frac{\sum_{i=1}^{\infty} N_i M_i^{1+a}}{\sum_{i=1}^{\infty} N_i M_i}\right)^{1/a} \tag{21}$$

From this definition, $M_v < M_w$; except for the case of $a = 1$, $M_v = M_w$.

The above methods for measuring absolute average molecular weights are time-consuming. For a rapid evaluation of molecular weight, gel-permeation chromatography (GPC), or more precisely liquid exclusion chromatography, provides a relative measure of molecular weight [170]. In GPC, a polymer solution is injected into a column of porous beads. Initially, a concentration gradient is established between the liquid phase and the porous beads, to cause fractions of smaller polymer molecules to diffuse into the beads. Since a solvent is continuously passed through the column, the larger polymer molecules are eluted first. With the larger molecules depleted, the concentration gradient between the beads and the solvent phase causes the smaller polymer molecules to diffuse back out of the beads, and eventually eluted from the packed column. Consequently, the molecular weight and the molecular-weight distribution can be obtained by comparing the elution time of the unknown sample to a set of standards. Although standards of narrow-molecular-weight-distribution polymers at different molecular weights are often not available these measurements are still useful for relative comparisons against an acceptable sample, such as one used for quality control.

The zero-shear-rate viscosity of a polymer melt can also be used as a relative measure of molecular weight. From the molecular theories of polymer melts [171], such as entanglement [172] and reptational models [173], the zero-shear-rate viscosity is proportional to molecular weight raised to an exponent that ranges between 3 and 3.5. Consequently, a relative measure of molecular weight can again be obtained from the ratio of viscosities of a new sample and a known sample.

Rheological Measurement of Material Characteristic Time

In addition to obtaining polymer flow properties and molecular weight, rheological measurements provide a means to determine the material characteristic time of a polymer. Various types of rheometers, such as cone-and-plate and parallel-disk rotational viscometers, can be used to measure the shear viscosity as a function of shear rate at a given temperature [174]. Figure 4 shows a typical response of the dependency of shear viscosity (η) on shear rate ($\dot\gamma$), plotted on a log-log scale. At low shear rate, the zero-shear-rate viscosity (η_0) is independent of shear rate, and depends only on temperature:

$$\eta = \eta_0 \tag{22}$$

The shear viscosity then decreases with increasing shear rate. The linearly decreasing shear viscosity on a log-log plot can be conveniently represented by a power-law model:

$$\eta = m\,\dot\gamma^{(n-1)} \tag{23}$$

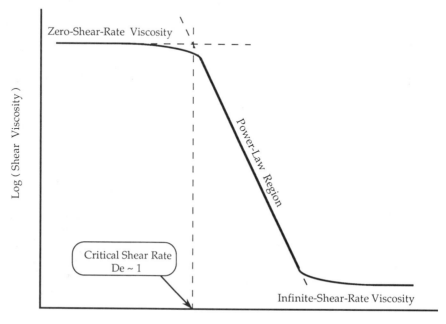

Figure 4 Log-log plot of shear viscosity as a function of shear rate.

where the power-law coefficient (m) is a function of temperature and (n) is the power-law index. For some polymers, the shear viscosity may become independent of shear rate at very high shear rate. This viscosity is the infinite-shear-rate viscosity (η_∞).

The use of the Deborah number simplifies the interpretation of the material response. In shear flow, the process time scale is approximately equal to the reciprocal of shear rate ($1/\dot{\gamma}$). The Deborah number is then given by

$$De = \lambda \dot{\gamma} \tag{24}$$

At low shear rate (De $\ll$ 1), a viscous response is observed, as indicated by a constant shear viscosity. In this region, hydrodynamic forces are extremely weak, and do not interfere with the molecular motion of the polymer, driven primarily by Brownian forces. Consequently, molecular weight radius of gyration, and other molecular parameters can be obtained by combining the low-shear-rate data and molecular models. As shear rate increases, the time scale of the shear flow approaches the polymer relaxation time. Here, the shear viscosity begins to depend on shear rate, and the onset of non-Newtonian behavior sets in. Since De $\approx$ 1, the longest material characteristic time (λ_L) can be estimated at the critical shear rate ($\dot{\gamma}_c$) where the zero-shear-rate viscosity intersects the shear viscosity from the power-law model:

$$\lambda_L = \frac{1}{\dot{\gamma}_c} \tag{25}$$

Hence the material characteristic time can be estimated from a simple shear flow measurement. More elaborate experiments and analyses would be required to map out the spectrum of material characteristic times [175].

Transition Temperatures

Differential thermal analysis (DTA) and differential scanning calorimetry (DSC) are two commonly used techniques to determine transition temperatures [176]. In differential thermal analysis, the temperature difference between a sample and a reference material is recorded when both specimens are heated or cooled at a controlled rate. Figure 5 shows a schematic DTA curve, in which the temperature difference is plotted against the temperature of the environment where the sample and the reference material are tested. In differential scanning calorimetry, the energy required to maintain a zero temperature difference between a sample and a reference material is recorded when temperature is scanned at a controlled rate. The output from a DSC experiment is similar to a DTA experiment, except that the ordinate of a DSC curve shows the amount of energy required to maintain a zero temperature difference. This energy can also be represented in the form of a heat capacity, as shown in Fig. 5. The area under a peak in a DSC curve is related directly to a change in enthalpy, whereas the area under a peak in a DTA curve is a complex function of sample geometry, heat capacity, and heat losses.

In general, narrow peaks are indicative of first-order transitions, such as the melt temperature, in both DTA and DSC curves. Second-order transitions, such as the glass transition temperature, occur at inflection points in both curves; and chemical reactions are depicted by broad peaks. The specific heat capacity (c_p) can also be calculated from

$$c_p = \frac{\Delta Q}{m \, \Delta T} \tag{26}$$

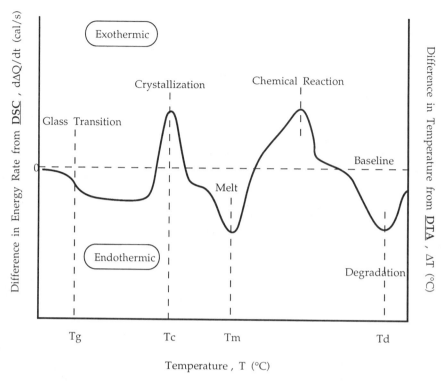

Figure 5 Schematic of a DTA or DSC curve.

where ΔQ is the energy required to change the temperature of a sample of mass (m) by ΔT. Additional thermodynamic properties can be determined from the heat capacity data. Enthalpy is given by

$$\Delta H = \int_0^T c_p \, dT \tag{27}$$

entropy is given by

$$\Delta S = \int_0^T \frac{c_p}{T} \, dT \tag{28}$$

and Gibbs free energy is given by

$$\Delta G = \Delta H - T \, \Delta S \tag{29}$$

Lastly, an estimate of the degree of crystallinity can be obtained from

$$\delta_c = \frac{(c_p)_a - c_p}{(c_p)_a - (c_p)_c} \tag{30}$$

where $(c_p)_a$ and $(c_p)_c$ are, respectively, the specific heats of the amorphous and crystalline standards.

Permeation in Polymers

As discussed earlier, permeability is a lumped parameter, consisting of the product of partition coefficient and diffusivity, that is used to characterize the transport of a species

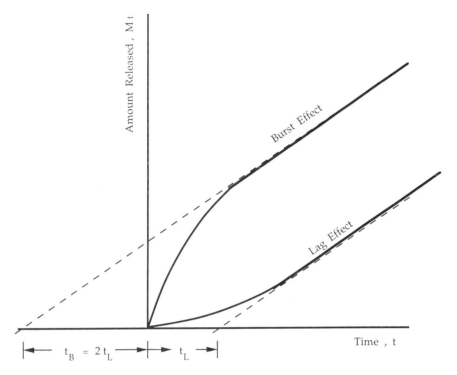

Figure 6 Amount permeated across a membrane as a function of time.

across a polymer. The partition coefficient and the diffusivity can be measured from a simple permeation experiment. A polymeric film is clamped between a donor cell that contains a saturated solution and a receiver cell in which the concentration of the permeating species is measured as a function of time. Both cells are well stirred to minimize the boundary-layer mass transfer resistances. Figure 6 shows a typical plot of the amount released (M_t) versus time [177]. The amount released is initially nonlinear and then approaches steady state. The initial nonlinearity exhibits a burst or a lag effect, depending on whether the membrane is initially saturated or devoid of the diffusing species, respectively. Diffusivity ($\mathscr{D}$) can be obtained from the time to reach steady state in both cases, since

$$\mathscr{D} = -\frac{L_m^2}{3t_B} = \frac{L_m^2}{6t_L} \tag{31}$$

where the burst time (t_B) and the lag time (t_L) are intercepts on the time axis extrapolated from the amount released at steady state, and L_m is the membrane thickness. Permeability ($\mathscr{P}$) is evaluated from the slope at steady state, which is constant, and is given by

$$\mathscr{P} = \frac{L_m}{A\,\Delta C}\left(\frac{dM_t}{dt}\right)_{\text{steady state}} \tag{32}$$

where A and ΔC are, respectively, the membrane surface area and the bulk concentration difference between the donor and receiver cells. The partition coefficient ($\mathscr{K}$) is then calculated from

$$\mathscr{K} = \frac{\mathscr{P}}{\mathscr{D}} \tag{33}$$

In addition to permeation experiments, partition coefficient and diffusivity can be determined by sorption/desorption measurements, radioactively tagged penetrants, and a variety of spectroscopic methods.

CONCLUSION

Within the voluminous literature and list of technologies regarding controlled drug delivery, interesting applications can be found for almost any conceivable biological membrane or therapeutic area, with most academic and industrial pharmaceutical institutions contributing. However, few applications have yet found their way to the market. This observation may be surprising even in light of the long development time in the pharmaceutical industry and the effort required to change established R&D processes. Perhaps the greatest drawback to the success of controlled drug delivery is the isolation of the technology from the therapeutics. Coordination of technology and therapeutics is needed in a rational drug development process that is initially governed by the steady-state pharmacology and later by the clinical pharmacology.

PROBLEMS

1. A very potent drug has a half-life of 3 min and an extensive hepatic first-pass metabolism. Discuss the relative merits of an oral, a sublingual immediate-release formulation, a sustained-release sublingual formulation in which some of the drug is swallowed, and a transdermal patch.
2. For a drug to be delivered by a zero-order device, the desired effective plasma level is 10 ng/ml and the clearance is 1 liter/min. Calculate the required steady-state delivery rate of drug.
3. Construct a phase diagram for copolymers of poly(acrylic acid) and poly(methacrylate) with T_g's of 379 and 278 K, respectively.
4. The fluxes from saturated solutions of drug in either water or propylene glycol were measured across a membrane. Assume that neither the drug nor the solution affects drug permeation across the membrane, i.e., ideal membrane approximation. If the solubilities of the drug in water and propylene glycol are 10 and 1 mg/ml, respectively, what is the ratio of their fluxes across the membrane?
5. Given a minimum dosing interval for an implant of 6 months and a maximum drug loading due to size constraints of 0.5 g, calculate the required drug potency for use in an implant. Comment on the toxicological concerns of using an implant.
6. Discuss the relative merits of oral, buccal, nasal, and transdermal delivery of the small peptide, oxytocin.
7. Discuss the effects on membrane permeation of (a) glass transition, (b) cross-linking density, (c) plasticizer content, (d) molecular-weight distribution of the polymer, (e) molecular volume of the drug, (f) solubility parameters of the drug and membrane, and (g) surface topology of the membrane.
8. Two batches of the same copolymer are processed under different temperature conditions to yield membranes with different permeation properties. Discuss some of the methods that might be used to characterize this phenomenon.

ANSWERS

1. The large first-pass metabolism would cause an oral dosage form to produce erratic plasma levels, and the half-life would necessitate a controlled-release

dosage form. If the therapeutic window were extremely large, then an oral controlled-release dosage might be viable. The sublingual immediate-release formulation could be used only for very short-term effects, and a sublingual sustained-release formulation would circumvent both concerns. A transdermal formulation might be possible for a very potent drug, would also circumvent the metabolism and half-life issues, and would probably have better patient acceptance than the buccal route.

2. From Eq. (2),

$$\text{drug input rate} = (10 \text{ ng/ml})(1000 \text{ ml/min})(60 \text{ min/h}) = 0.6 \text{ mg/h}$$

3. From Eq. (10), the glass transition temperatures for any copolymer composition will be linear with the weight fraction, and this line will connect the two T_g's for the two homopolymers.

4. Saturated solutions are at unit activity and are in equilibrium with the solid drug regardless of the choice of solvent. Since the driving force for diffusion is the activity gradient, for an ideal membrane, the flux from any saturated solution will be the same. This is, then, a method of testing for solvent effects on membranes.

5. The required drug potency for an implant may be estimated as $(0.5 \text{ g})/180 \text{ days}$) or approximately 2.7 mg/day. That is, extreme potency is required, and this estimate may be somewhat large. An implant needs to be capable of being removed rapidly in emergencies, and inflammation and foreign body reactions are of particular concern.

6. Oxytocin requires rapid delivery and would be broken down by peptidases in the gut. An oral dosage form is therefore unlikely. Peptide permeation through skin is unlikely in sufficient doses, although iontophoresis might be a possibility. However, this may be too elaborate a technology for this task. A sublingual formulation may be possible, but the bioavailability is somewhat poor and variable. Nasal delivery would provide the most bioavailable and rapidly absorbed formulation of oxytocin.

7. Membrane permeation (a) decreases with increasing T_g, (b) decreases with increasing cross-linking density, (c) increases with plasticizer content, (d) typically decreases with polymers or (e) drugs of increasing molecular weight or volume, (f) is largest when the two solubility parameters are identical, and (g) can increase or become variable due to irregular surface topography.

8. To analyze this problem, SEM and TEM will allow one to visualize differences in the membrane, and either DSC or thermomechanical analysis will help in understanding the thermal properties. If crystallinity is an issue, cross-polarizers may help visualize changes and X-ray diffraction may prove useful.

REFERENCES

1. W. D. Mason and E. W. Amick, *J. Pharm. Sci.*, *70*: 707 (1981).
2. R. W. Duncan and L. W. Seymour, *Controlled-Release Technologies*, Elsevier Advanced Technology, Oxford, p. 58 (1989).
3. E. R. Barnhart, *Physician's Desk Reference for Nonprescription Drugs*, 9th ed., Medical Economics, Oradell, N.J., pp. 536, 713 (1988).
4. J. Crank, *The Mathematics of Diffusion*, 2nd ed., Clarendon Press, Oxford, p. 44 (1975).

5. H. S. Carslaw and J. C. Jaeger, *Conduction of Heat in Solids*, 2nd ed., Clarendon Press, Oxford, p. 99 (1959).

6. S. H. Kubo and R. J. Kody, *Clin. Pharmacokin.*, *10*:377 (1985).

7. B. Berner and S. LeRoy Steady-state PK/PD and converting enzyme (ACE) inhibitor, *Proc. Int. Symp. Control. Rel. Bioact. Mater.*, *15*:324 (1988).

8. J. E. Shaw, M. P. Cramer, and R. Gale, *Transdermal Delivery of Drugs*, Vol. I (A. F. Kydonieus and B. Berner, Eds.), CRC Press, Boca Raton, Fla., p. 106 (1987).

9. M. Gibaldi, *Biopharmaceutics and Clinical Pharmacokinetics*, 3rd ed., Lea & Febiger, Philadelphia, pp. 131–155 (1984).

10. J. G. Wagner, *Fundamentals of Clinical Pharmacokinetics*, Drug Intelligence Publications, Hamilton, Ill., p. 342 (1975).

11. C. A. Altar, W. C. Boyar, and P. L. Wood, *Eur. J. Pharmacol.*, *134*:303–311 (1987).

12. C. A. Altar, B. Berner, P. Beall, S. F. Carlsen and W. C. Boyar, *Mol. Pharmacol.*, *33*:690 (1988).

13. J. O. Parker, *Hospital Practice*, 63 (1988).

14. N. L. Krinick, B. Rihova, K. Ulbrich, and J. Kopecek, Targetable photoactivatable polymeric drugs, *Proc. Int. Symp. Control. Rel. Bioact. Mater.*, *16*:138 (1989).

15. H. Bundgaard, *Design of Prodrugs*, Elsevier, Amsterdam, (1985).

16. T. Higuchi and V. Stella, *Prodrugs as Novel Delivery Systems*, American Chemical Society, Washington, D.C. (1975).

17. K. A. Kelly, *Am. J. Physiol.*, *239*:671 (1980).

18. S. S. Davis, F. Norring-Christensen, R. Khosla, and L. C. Feely, *J. Pharm. Pharmacol.*, *40*:205 (1988).

19. P. Mojaverian, K. Chan, A. Desai, and V. John, *Pharmaceut. Res.*, *6*:719 (1989).

20. M. Lucas, *Gut*, *24*:734 (1983).

21. M. L. Lucas and J. A. Blair, *Proc. R. Soc. London Ser. B*, *200*:27 (1978).

22. H. W. Davenport, *Physiology of the Digestive Tract*, 5th ed., Year Book Medical Publishers, Chicago (1982).

23. E. R. Cooper and G. Kasting, *J. Contr. Rel.*, *6*:23 (1987).

24. S. H. Yalkowsky, *Techniques of Solubilization of Drugs* (S. H. Yalkowsky, Ed.), Marcel Dekker, New York, p. 1 (1981).

25. J. B. Dressman, G. L. Amidon, and D. Fleisher, *J. Pharm. Sci.*, *74*:588 (1985).

26. G. L. Amidon, J. Kou, R. L. Elliott, and E. W. Lightfoot, *J. Pharm. Sci.*, *69*:1369 (1980).

27. D. A. Johnson and G. L. Amidon, *J. Theor. Biol.*, *131*:93 (1988).

28. R. L. Elliot, G. L. Amidon, and E. W. Lightfoot, *J. Theor. Biol.*, *87*:757 (1980).

29. R. M. Gilley, J. H. Eldridge, J. L. Optiz, L. K. Hanna, J. K. Staas, and T. R. Tice, Development of secretory and systemic immunity following oral administration of microencapsulated antigens, *Proc. Int. Symp. Control. Rel. Bioact. Mater.*, *15*:123 (1988).

30. L. S. Lilienfield and E. J. Zapolski, *Curr. Therap. Res.*, *33*:692 (1983).

31. R. Langer, *Chem. Eng. Commun.*, *6*:1 (1980).

32. S. K. Singh and L. T. Fan, *Biotech. Prog.*, *2*:145 (1986).

33. N. A. Peppas, *J. Biomed. Mater. Res.*, *17*:1079 (1983).

34. L. Dong, A. S. Hoffman, and P. Sadurni, pH Sensitive hydrogels based on thermally reversible gels for enteric drug delivery, *Proc. Int. Symp. Control. Rel. Bioact. Mater.*, *16*:95 (1989).

35. I. C. Kwon, Y. H. Bae, T. Okano, B. Berner, and S. W. Kim, *Makromol. Chem.* Macromol. Symp., *33*:265 (1990).

36. S. J. Trank and E. L. Cussler, *Chem. Eng. Sci.*, *42*:381 (1987).

37. R. A. Siegel and B. A. Firestone, Progress toward an implantable, self-regulating, mechanochemical insulin pump, *Proc. Int. Symp. Control. Rel. Bioact. Mater.*, *15*:164 (1988).

38. K. Park, H. S. Ch'ng, and J. R. robinson, *Recent Advances in Drug Delivery Systems* (J. M. Anderson and S. W. Kim, Eds.), Plenum Press, New York, p. 163 (1984).

39. M. A. Longer, H. S. Ch'ng, and J. R. Robinson, *J. Pharm. Sci.*, *74*:406 (1985).

40. H. Park and J. R. Robinson, *Pharm. Res.*, *4*:457 (1987).

41. J. D. Smart, I. W. Kellaway, and H. E. C. Worthington, *J. Pharm. Pharmacol.*, *36*:295 (1984).

42. A. Silberberg and F. A. Meyer, *Adv. Exp. Med. Biol.* *144*:53 (1982).

43. J. R. Robinson, U.S. Patent 4,615,697 (October 7, 1986).

44. H. Bechgaard and K. Ladefoged, *J. Pharm. Pharmacol.*, *30*:690 (1978).

45. S. S. Davis, J. G. Hardy, and J. W. Fara, *Gut*, *27*:886 (1986).

46. D. M. Matthews and D. Burston, *Clin. Sci.*, *67*:541 (1984).

47. M. Hu, P. Subramanian, H. I. Mosberg, and G. L. Amidon, *Pharmaceut. Res.*, *6*:66 (1989).

48. D. I. Friedman and G. L. Amidon, *Pharmaceut. Res.*, *6*:1043 (1989).

49. M. Rousset, *Biochim.*, *68*:1035 (1986).

50. J. H. Jonkman and C. A. Hunt, *Pharm. Weekblad.*, *Sci. Ed.*, *5*:41 (1983).

51. N. Muranishi, *Int. J. Pharmaceut.*, *4*:279 (1980).

52. J. A. Fix, K. Engle, P. A. Porter, P. S. Leppert, S. J. Selk, C. R. Gardner, and J. Alexander, *Am. J. Physiol.*, *251*:6332 (1986).

53. K. Higaki, N. Takechi, M. Hashida, and H. Sezaki, *Chem. Pharm. Bull.*, *36*:1214 (1988).

54. M. Murakami, K. Takada, T. Fujii, and S. Muranishi, *Biochim. Biophys. Acta*, *939*:238 (1988).

55. T. Nishihata, H. Takahagi, and T. Higuchi, *J. Pharm. Pharmacol.*, *35*:124 (1983).

56. H. Okada, I. Yamazaki, Y. Ogawa, S. Hirai, T. Yashiki, and H. Mima, *J. Pharm. Sci.*, *71*:1367 (1982).

57. G. E. Peters, L. E. F. Hutchinson, R. Hyde, C. McMartin, and S. B. Metcalfe, *J. Pharm. Sci.*, *76*:857 (1987).

58. E. J. Van Hoogdalem, H. J. M. Van Kan, A. G. deBoer, and D. D. Breimer, *J. Control. Rel.*, *7*:53 (1988).

59. M. Saffran, G. S. Kumar, C. Savariar, J. C. Burnham, F. Williams, and D. C. Neckers, *Science*, *233*:1081 (1986).

60. J. A. Atchinson, W. E. Grizzle, and D. J. Pillon, *J. Pharmacol. Exp. Ther.*, *248*:567 (1989).

61. M. E. McNeill and N. B. Graham, *Controlled Release Technology*, (P. I. Lee and W. R. Good, Eds.), ACS Symposium, Washington, D.C., *348*:158 (1987).

62. W. R. Good, *Drug Dev. Ind. Pharm.*, *9*:647 (1983).

63. W. R. Good, M. S. Powers, P. Campbell, and L. Schenkel, *J. Control Rel.*, *2*:89 (1985).

64. T. R. MacGregor, M. M. Kandace, J. J. Keiras, R. G. A. van Wayjen, A. van den Ende, and R. G. L. van Tol, *Clin. Pharmacol. Ther.*, *38*:279 (1985).

65. P. Muller, P. R. Imhof, D. Mauli, and D. Milovanonic, *Meth. Find. Exp. Clin. Pharmacol.*, *11*:197 (1989).

66. J. P. Dubois, A. Sioufi, P. Muller, D. Mauli, and P. R. Imhof, *Meth. Find. Exp. Clin. Pharmacol.*, *11*:187 (1989).

67. J. E. Rose, J. E. Herskovic, Y. Trilling, and M. E. Jarvik, *Clin. Pharmacol. Ther.*, *38*:450 (1985).

68. J. R. Varvel, S. L. Shafer, S. S. Hwang, P. A. Coen, and D. R. Stanski, *Anesthesiol*, *70*:928 (1989).

69. G. K. Gourlay, S. R. Kowalski, J. L. Plummer, M. J. Cousins, and P. J. Armstrong, *Anesth. Analg.*, *67*:329 (1988).

70. P. M. Plezia, T. H. Kramer, J. Linford, and S. R. Hameroft, *Pharmacotherapeutics*, *9*:2 (1989).

71. F. J. Grendas-Perez, P. G. Jenner, M. Nomoto, S. Stahl, N. P. Quinn, J. D. Parkes, P. Critchley, and C. D. Marsden, *Lancet*, 906 (1986).

72. R. J. Scheuplein and I. H. Blank, *Physiol. Rev.*, *51*:702 (1971).

73. A. S. Michaels, S. K. Chandrasekaran, and J. E. Shaw, *AIChE J.*, *21*:985 (1975).

74. E. R. Cooper, *J. Control. Rel.*, *1*:153 (1984).

75. B. Berner and E. R. Cooper, *Transdermal Delivery of Drugs*, Vol. II (A. F. Kydonieus and B. Berner, Eds.), CRC Press, Boca Raton, Fla., p. 41 (1987).
76. D. C. Patel, Eur. Patent EP0321870 (1989).
77. G. C. Mazzenga and B. Berner, *J. Control. Rel.* (in press).
78. E. R. Cooper and B. Berner, *Transdermal Delivery of Drugs*, Vol. II (A. F. Kydonieus and B. Berner, Eds.), CRC Press, Boca Raton, Fla., p. 57 (1987).
79. R. P. Oertel, *Biopolymers 16*:232 (1977).
80. L. J. DeNoble, K. Knutson, and T. Kurihara-Bergstrom, *Pharmaceut. Res.(Suppl.)*, *4*:S59 (1987).
81. D. C. Patel and Y. Chan, U.S. Patent 4,863,970 (1989).
82. B. W. Barry Penetration enhancers in skin permeation, *Proc. Int. Symp. Control. Rel. Bioact. Mater.*, *13*:136 (1986).
83. B. Berner, J. H. Otte, G. C. Mazzenga, R. J. Steffens, and C. D. Ebert, *J. Pharm. Sci.*, *78*:314 (1989).
84. B. Berner, G. C. Mazzenga, J. H. Otte, R. J. Steffens, R. H. Juang, and C. D. Ebert, *J. Pharm. Sci.*, *78*:402 (1989).
85. B. Berner, R. H. Juang, and G. C. Mazzenga, *J. Pharm. Sci.*, *78*:472 (1989).
86. P. S. Campbell and S. K. Chandrasekaran, U.S. Patent, 4,379,454 (1983).
87. B. Berner, D. R. Wilson, R. H. Guy, G. C. Mazzenga, F. H. Clarke, and H. I. Maibach, *Pharmaceut. Res.*, *5*:660 (1988).
88. B. Berner, D. R. Wilson, R. H. Guy, G. C. Mazzenga, F. H. Clarke, and H. I. Maibach, *J. Toxicol-Cut Ocular Toxicol.*, *8*:1 (1989).
89. G. Dupuis and C. Benezra, *Allergic Contact Dermatitis to Simple Chemicals*, Marcel Dekker, New York (1982).
90. T. Maurer, *Contact and Photocontact Allergens*, Marcel Dekker, New York (1983).
91. M. R. Holdiness, *Contact Dermatit.*, *20*:3 (1989).
92. A. Amkraut, J. E. Shaw, Eur. Patent EP0282156 (1988).
93. Y. W. Chien, *The Transdermal Delivery of Drugs*, Vol. I (A. F. Kydonieus and B. Berner, Eds.), CRC Press, Boca Raton, Fla., p. 81 (1987).
94. E. S. Lee and S. I. Yum, U.S. Patent 4,781,924 (1988).
95. C. D. Ebert, W. Heiber, R. Andriola, and P. Williams, *J. Control. Rel.*, *6*:107 (1987).
96. A. K. Banga and Y. W. Chien, *J. Control. Rel.*, *7*:1 (1988).
97. R. R. Burnette and B. Ungpipattanakul, *J. Pharm. Sci.*, *76*:765 (1987).
98. J. DeNuzzio and B. Berner, *J. Control Rel.*, *11*:105 (1990).
99. S. R. Eisenberg and A. J. Grodzinsky, *J. Membr. Sci.*, *19*:173 (1984).
100. H. Molitor, *The Merck Report*, *52*:22 (1943).
101. J. G. Webster, *Medical Instrumentation*, Houghton Mifflin, Boston, p. 210 (1978).
102. *Physicians Desk Reference*, 43rd ed., Medical Economics Co., Oradell, N.J., pp. 757, 1573, (1989).
103. V. L. Perrin, *Clin. Pharmacol.*, *10*:334 (1985).
104. M. A. H. Russell, C. Feyerabend, and P. V. Cole, *Br. Med. J.*, *1*:1043 (1976).
105. R. E. S. Bullingham, H. J. McQuay, E. J. B. Porter, M. C. Allen, and R. A. Moore, *Br. J. Clin. Pharm.*, *13*:665 (1982).
106. K. Muller and M. Osler, *Acta. Obstet. Gynecol. Scand.*, *46*:59 (1967).
107. R. Duncan and L. Seymour, *Controlled-Release Technologies*, Elsevier Advanced Technology, Oxford, p. 72 (1989).
108. S. Y. Chen and C. A. Squier, *The Structure and Function of Oral Mucosa* (J. Meyer, C. A. Squier, and S. J. Gerson, Eds.), Pergamon Press, Oxford, p. 7 (1984).
109. J. P. Waterhouse, *The Structure and Function of Oral Mucosa* (J. Meyer, C. A. Squier, and S. J. Gerson, Eds.), Pergamon Press, Oxford, p. 109 (1984).
110. A. H. Beckett and E. F. Triggs, *J. Pharm. Pharmacol. Suppl.*, *19*:31 (1967).

111. M. M. Veillard, M. A. Longer, T. W. Martens, and J. R. Robinson, *J. Control. Rel.*, 6:123 (1987).
112. N. F. H. Ho and C. L. Barsuhn, *Buccal Delivery of Drugs, Proc. Int. Symp. Control. Rel. Bioact. Mater.*, 16:24 (1989).
113. C. D. Ebert, V. A. John, P. T. Beall, and K. A. Rosenzweig, *Controlled-Release Technology* (P. I. Lee and W. R. Good, Eds.), *ACS Symp. Ser.*, 348:310 (1987).
114. J. Cassidy, B. Berner, K. Chan, V. John, S. Toon, B. Holt, and M. Rowland. Buccal delivery of diclofenac sodium in man using a prototype hydrogel delivery device, *Proc. Int. Symp. Control. Rel. Bioact. Mater.*, 16:91 (1989).
115. E. Quadros, J. P. Cassidy, K. A. Rosenzweig, and B. Berner, Buccal permeation of a novel angiotensin converting enzyme (ACE) inhibitor, *Proc. Int. Symp. Control. Rel. Bioact. Mater.*, 15:325 (1988).
116. M. A. Longer, M. E. Dowty, and J. R. Robinson, Characterization of buccal epithelia relevant to peptide drug delivery, *Proc. Int. Symp. Control. Rel.*, 15:68 (1988).
117. S. J. Sjostedt, *Acta. Obstet. Gynecol. Scand.*, 48 (Suppl. 7):1 (1969).
118. R. E. Stratford, Jr. and V. H. L. Lee, *Int. J. Pharmaceut.*, 30:73 (1986).
119. J. Cassidy and E. Quadros, *Pharmaceut. Res.*, 5 (Suppl.):S100 (1988).
120. C. A. Squier, *J. Ultrastruct. Res.*, 43:160 (1973).
121. M. R. Tavakoli-Saberi and K. L. Audus, *Pharmaceut. Res.*, 6:160 (1989).
122. V. Sandow and W. Petri, *Transnasal Systemic Medication* (Y. W. Chien, Ed.), Elsevier, Amsterdam, p. 183 (1985).
123. B. H. Vickery, S. Anik, M. Chaplin, and M. Henzl, *Transnasal Systemic Medication* (Y. W. Chien, Ed.), Elsevier, Amsterdam, p. 201 (1985).
124. J. S. Flier, A. C. Moses, M. C. Carey, G. S. Gordon, and R. S. Silver, *Transnasal Systemic Medication*, Elsevier, Amsterdam, p. 217 (1985).
125. S. J. Hersey and R. T. Jackson, *J. Pharm. Sci.*, 76:876 (1987).
126. C. McMartin and G. Peters, *Delivery Systems for Peptide Drugs* (S. S. Davis, L. Illum, and E. Tomlinson, Eds.), Plenum Press, New York, p. 255 (1986).
127. J. Seki, H. Mukai, and M. Sugiyama, *J. Pharmacobio-Dyn.*, 8:337 (1985).
128. J. P. Longnecker, A. C. Moses, J. S. Flier, R. D. Silver, M. C. Carey, and E. J. Dubovi, *J. Pharm. Sci.*, 76:351 (1987).
129. W. Petri, R. Schmiedel, and J. Sandow, *Transnasal Systemic Medications* (Y. W. Chien, Ed.), Elsevier, Amsterdam, p. 161 (1985).
130. S. Wilk, H. Mizuguchi, and M. Orlovski, *J. Pharmacol. Exp. Ther.*, 206:227 (1978).
131. Y. W. Chien, *Novel Drug Delivery Systems*, Marcel Dekker, New York p. 13 (1982).
132. G. C. Y. Chio and K. Watanabe, *Methods of Drug Delivery* (G. M. Ihler, Ed.), Pergamon Press, New York, p. 203 (1986).
133. B. Berner and E. R. Cooper, *J. Membr. Sci.*, 14:139 (1983).
134. G. M. Grass, E. R. Cooper, and J. R. Robinson, *J. Pharm. Sci.*, 77:24 (1988).
135. J. C. Keister, *J. Control. Rel.*, 3:67 (1986).
136. R. A. Siegel and R. D. Schoenwald, *J. Control. Rel.*, 5:193 (1987).
137. R. A. Ralph, M. G. Doane, and C. H. Dohlman, *Arch. Opthal.*, 93:1039 (1975).
138. Y. Rojanasakul and J. R. Robinson, *Int. J. Pharmaceut.*, 55:237 (1989).
139. Y. W. Chien, *Novel Drug Delivery*, Marcel Dekker, New York, p. 51 (1982).
140. D. Dallenbach-Hellweg, and J. Sievers, *Virchows Arch. [Pathol. Anat. Histol.]*, 368:289 (1975).
141. R. Duncan and L. W. Seymour, *Controlled Release Technologies*, Elsevier Advanced Technology, Oxford, p. 11 (1989).
142. H. M. Goodman, *Medical Physiology*, Vol. II, 13th ed (V. B. Mountcastle, Ed.), C. V. Mosby, St. Louis, p. 1747 (1974).
143. D. C. Corbo, J. C. Liu, and Y. W. Chien, *Pharmaceut. Res.*, 6:848 (1989).

144. Y. W. Chien, *Novel Drug Delivery Systems*, Marcel Dekker, New York, p. 219 (1982).
145. S. Kroin, R. D. Penn, R. C. Beissinger, and R. C. Arzbaecher, *Exp. Brain Res.*, *54*:191 (1984).
146. J. D. Fenstermacher and H. Davson, *Am. J. Physiol.*, *242*:F171 (1982).
147. J. M. Anderson, *Trans. Soc. Artif. Intern. Organs*, *34*:101 (1988).
148. J. D. Andrade, S. Nagaoka, S. Cooper, T. Okano, and S. W. Kim, *Trans. Soc. Artif. Inter. Organs*, *33*:75 (1987).
149. R. Duncan and L. W. Seymour, *Controlled Release Technologies*, Elsevier Advanced Technology, Oxford, p. 143 (1989).
150. S. E. Harrigan and D. A. Downs, *NIDA Res. Monogr. Ser.*, *28*:77 (1980).
151. J. Heller, *J. Control. Rel.*, *2*:167 (1985).
152. T. A. Hurbett, B. D. Ratner, J. Kost, and M. Singh, *Recent Advances in Drug Delivery Systems* (J. M. Anderson and S. W. Kim, Eds.), Plenum Press, New York, p. 209 (1984).
153. S. W. Kim, S. Y. Jeong, S. Sato, J. C. McRea, and J. Feijen, *Recent Advances in Drug Delivery Systems* (J. M. Anderson and S. W. Kim Eds.), Plenum Press, New York, p. 123 (1984).
154. E. Tomlinson and S. Davis, Eds., *Site-Specific Delivery*, John Wiley, New York (1986).
155. R. Duncan and L. W. Seymour, *Controlled-Release Technologies*, Elsevier Advanced Technology, Oxford, p. 21 (1989).
156. I. J. Fidler, A. Raz, W. E. Fogler, K. Krish, P. Bugelski, and G. Poste, *Cancer Res.*, *40*:4460 (1980).
157. J. N. Weinstein, C. D. V. Black, J. Barbet, R. R. Eger, R. J. Parker, O. D. Hulton, J. L. Mulshine, A. M. Keenan, S. M. Larson, J. A. Carrasquillo, S. M. Sieber, and D. G. Covell, *Site-Specific Drug Delivery* (E. Tomlinson and S. S. Davis, Eds.), John Wiley, New York, p. 81 (1986).
158. L. Illum, N. W. Thomas, and S. S. Davis, *J. Pharm. Sci.*, *75*:16 (1986).
159. J. Kopecek and R. Duncan, *J. Control. Rel.*, *6*:315 (1987).
160. I. J. Fidler, *Site-Specific Drug Delivery* (E. Tomlinson and S. S. Davis, Eds.) John Wiley, New York, p. 111 (1986).
161. P. J. Flory, *Principles of Polymer Chemistry*, Cornell University Press, Ithaca, N.Y., p. 266 (1986).
162. M. Reiner, *Physics Today*, 62, (January 1964)
163. F. Rodriguez, *Principles of Polymer Systems*, McGraw-Hill, New York, p. 15 (1970).
164. W. E. Roorda, H. E. Bodde, A. G. De Boer, and H. E. Junginger, *Pharmaceut. Weekblad Sci. Ed.*, *8*:165 (1986)
165. R. B. Seymour and C. E. Carraher, Jr., *Polymer Chemistry*, Marcel Dekker, New York, p. 19 (1988).
166. H. Burrell, Solubility Parameter Values *Polymer Handbook* (J. Brandrup, E. H. Immergut, and W. McDowell, Eds.), John Wiley, New York, p. IV-337 (1975).
167. J. S. Vrentas and J. L. Duda, *Encyclopedia of Polymer Science and Engineering*, Vol. 5, John Wiley, New York, p. 36 (1986)
168. M. Kurata, Y. Tsunashima, M. Iwama, and K. Kamada, Viscosity-molecular weight relationships and unperturbed dimensions of linear chain molecules, *Polymer Handbook* (J. Brandrup, E. H. Immergut, and W. McDowell, Eds.), John Wiley, New York, p. IV-1 (1975).
169. J. R. Schaefgen and P. J. Flory, *J. Am. Chem. Soc.*, *70*:2709 (1948).
170. I. Tomka and G. Vancso, Determination of molecular mass distribution of polymers by a combination of dynamic light scattering, gel permeation chromatography and other methods, *Applied Polymer Analysis and Characterization* (J. Mitchell, Jr., Ed.), Hanser, Munich, p. 237 (1987).
171. R. B. Bird, C. F. Curtiss, R. C. Armstrong, and O. Hassager, *Dynamics of Polymeric Liquids, Vol. 2: Kinetic Theory*, John Wiley, New York (1987).

172. J. D. Ferry, *Viscoelastic Properties of Polymers*, John Wiley, New York (1970).

173. M. Doi and S. F. Edwards, *The Theory of Polymer Dynamics*, Oxford Science Publications, Oxford, p. 188 (1988).

174. R. B. Bird, R. C. Armstrong, and O. Hassager, *Dynamics of Polymeric Liquids, Vol. 1: Fluid Mechanics*, John Wiley, New York, p. 509 (1987).

175. N. W. Tschoegl, *The Phenomenological Theory of Linear Viscoelastic Behavior*, Springer-Verlag, Berlin, p. 69 (1989).

176. J. F. Rabek, *Experimental Methods in Polymer Chemistry*, John Wiley, Chichester, p. 549 (1980).

177. J. H. Richards, The role of polymer permeability in the control of drug release, Polymer Permeability (J. Comyn, Ed.), Elsevier Applied Science Publishers, London, p. 217 (1986).

2

Factors Influencing the Kinetics of Solute Release

Wei-Youh Kuu, Ray W. Wood, and Theodore J. Roseman
Baxter Healthcare Corporation, Round Lake, Illinois

INTRODUCTION

During the past 40 years a substantial number of drug delivery systems have been introduced to the marketplace, and considerable effort has been expended in the development of new delivery concepts. There are a number of reasons for this continued interest in developing new systems, as has been emphasized in Chapter 1 of this book. First, recognition of the possibility of repatenting successful drugs by applying the concepts of controlled drug delivery, coupled with the increasing expense of bringing new drug entities to the marketplace, has provided justification for continued research on and development of these products. Second, the idea of improving the therapeutic efficacy and safety of new drug entities by more precise spatial and temporal placement within the body as compared to conventional delivery systems is attractive. This second goal will result in less total drug being required, thereby decreasing side effects and increasing the efficiency of treatment by reducing frequency of dosing.

There are other reasons for the development of these products beyond the obvious therapeutic advantages of controlled drug release. For example, patient compliance is a chronic problem for all self-administered drugs and is a sufficiently compelling reason in itself to warrant an alternative delivery system. Indeed, in recent years this has been recognized as a very important component of successful drug therapy. Economic justification is sufficient for designing controlled-release products as well. This is particularly true in a critical care, hospital setting in which drugs are administered intravenously. Systems that eliminate the need for manual reconstitution or admixing will result in labor reduction and time savings to the clinical health-care professional, and ultimately result in improved therapy. In addition, drug delivery systems that overcome drug stability problems inherent in conventional dosage forms warrant attention as well. For example, in the case of intravenous drug administration, some drugs cannot be practically formulated in a ready-to-use sterile aqueous solution because of their significant instability in an aqueous environment. In these cases, an in-line drug delivery system in which the drug is formulated in the solid state to improve stability is attractive. The drug would be delivered intravenously in the solution phase by controlling its dissolution kinetics as a large-volume parenteral solution flows through the system via an administration set.

For many years, controlled drug release or delivery was attainable only by intravenous infusion. In this mode of administration, the rate at which a drug of known concentration is delivered to the bloodstream can be controlled or changed according to the therapeutic requirements of the patient. In this form of drug delivery, control is achieved by the use of a flow control pump or, in the case of gravity-flow infusion, by the plumbing/valving of the administration set. More recently, this ideal of controlled drug delivery has been approached through the design of polymer systems in which the drug is dissolved or dispersed, coupled with an increased understanding of the factors influencing the kinetics of drug release. This fundamental concept (drug dissolved or dispersed in a polymer) represents a commonality with regard to the approach taken in designing drug delivery systems for a breadth of administration modes (transdermal patches, bioadhesive platforms, microencapsulated products, monolithic implants, etc.). Typically, the objective in a controlled drug delivery system is to have drug release from the dosage form be the rate-limiting step for drug availability. In this fashion, drug availability is controlled by the kinetics of drug release rather than absorption. Therefore, central to the development of successful controlled delivery systems is the synthesis of the principles of molecular transport in polymeric materials, as well as a firm understanding of the variables that influence the transport processes. This understanding will allow more efficient drug delivery system design once the drug delivery rate requirements are known based on pharmaco-kinetic/pharmacodynamic considerations.

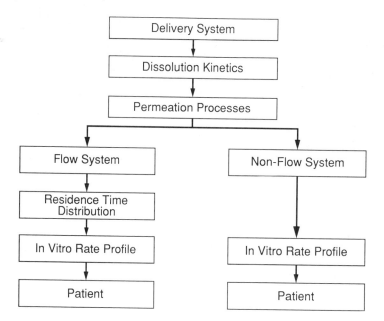

Figure 1 Schematic diagram of solute release from a drug delivery system.

Figure 1 is a generalized flow diagram illustrating the processes involved in solute release from most drug delivery systems. The nonflow system is representative of the situation for most drug delivery systems in which the drug is eluted into a relatively stagnant milieu. The flow system applies, for example, in the case of in-line drug delivery systems in which a large-volume parenteral solution continuously flows through a cartridge containing the formulated drug. Drug is released into this flowing stream and is delivered intravenously in solution phase via the administration set. This chapter, using Fig. 1 as the guide, critically reviews the factors which influence the kinetics of solute release using a quantitative, physical-model approach. It describes the diffusion and partition coefficients of solutes, how these coefficients are measured/computed, and the role they play in defining solute release kinetics. The effect of viscosity and concentration on diffusion is described. Membrane and matrix parameters important to drug release, as well as polymeric drug solubility considerations, are discussed. The chapter also describes the particularly important effect in flow-through intravenous drug delivery systems in which the drug is delivered via an administration set. Finally, approaches for designing and/or optimizing drug delivery systems based on pharmacokinetic/pharmacodynamic requirements are also considered.

PRINCIPAL SYMBOLS, CONSTANTS, AND ABBREVIATIONS

Note: The units presented in the following list are suggested by the authors. In general, any consistent set of units may be used in all equations in this chapter, except the equations derived empirically, such as Eq. (5). These empirical equations are valid only when specific units are used.

A: surface area, cm^2

$\mathring{A}$: angstrom

a: general symbol for the radius of a cylinder, sphere, or the half-thickness of a slab; constant equal to k_m/K, used in Eq. (130)

a_i: radius of a cylindrical or spherical cavity

b: dimensionless constant defined by Eq. (131)

C: time-dependent solute concentration, either in the bulk or in a polymer, mg/ml

C_B': molar density of the solvent, g-mol/cm^3

C_b: solute concentration in the bulk; solute concentration in the receptor solution, used in Eq. (116)

C_{b0}: initial solute concentration in the bulk; initial solute concentration in the receptor solution, used in Eq. (117)

C_{b1}: solute concentration in the solution phase adjacent to face 1 of the polymer, mg/ml

C_{b2}: solute concentration in the solution phase adjacent to face 2 of the polymer, mg/ml

C_b^*: solute concentration in the polymer phase adjacent to the interface between the bulk and the polymer, used in the section on membrane and matrix parameters with the absence of boundary-layer diffusion, mg/ml

CED: cohesive energy

C_M: solute solubility in the polymer, mg/ml

$\bar{C}_M$: mole fraction solubility in polymer, defined by Eq. (158)

C_0: solute concentration in the polymer phase adjacent to the interface between the bulk and the polymer, used in the section on recipient environment and in Fig. 16 with the presence of boundary-layer diffusion, mg/ml

C_s: solute solubility in the liquid phase, mg/ml

C_T: total solute loading dose in the polymer, equal to the amount of solute per unit volume of the entire matrix, mg/ml

D: diffusion coefficient of solute molecules in the liquid phase or solvent, cm^2/s

D(C): diffusion coefficient of solute as a function of concentration, cm^2/s

$\overset{\circ}{D}$: dispersion coefficient, cm^2/s

D_{am}: diffusion coefficient in the amorphous polymer, cm^2/s

D_{calc}: calculated diffusion coefficient, cm^2/s

D_D: drug diffusion coefficient in the donor-side boundary layer, cm^2/s

D_e: effective diffusion coefficient of solute molecules in the polymer phase, cm^2/s

D_{exp}: diffusion coefficient determined by experiment, cm^2/s

D_m: diffusion coefficient in the polymeric membrane or matrix equivalent to De, cm^2/s

D_{M1}: diffusion coefficient of a diffusant of unit molecular weight, used in Eq. (19)

D_0: diffusion coefficient at infinite dilution, cm^2/s

D_R: drug diffusion coefficient in the receptor-side boundary layer, cm^2/s

d: characteristic length, cm

E(t): residence time distribution function of a flow system, 1/min

$E(\tau)$: dimensionless residence time distribution function of a flow system

E_V: molar energy, used in Eq. (157)

$E_{all}(t)$: overall residence time distribution function, 1/min

ΔE: penetration energy

erf(x): error function with argument x

erfc(x): complementary error function = $1 - erf(x)$

F: flow rate, ml/min

f: proportionality constant used in Eq. (112)

$f(\xi_p)$: a function dependent on the mesh size ξ_p

Fa: faraday, 96,500 C/g-equiv

FF: a function defined by Eq. (213)

$FF(R_{L1})$: functional value of FF at $R = R_{L1}$

$FF(R_{L2})$: functional value of FF at $R = R_{L2}$

$FF(R_{R1})$: functional value of FF at $R = R_{R1}$

$FF(R_{R2})$: functional value of FF at $R = R_{R2}$

$FF1_{1/2}$: functional value of FF at the halfway point, determined in the first step of the half-interval iterations

G(C): dimensionless variable defined by Eq. (175)

g_1: physical parameter, used in Eq. (12)

g_2: physical parameter, used in Eq. (12)

g_3: constant defined by Eq. (15)

ΔH_f: molar heat of fusion

ΔH_f^*: energy required for dissociation of the solute molecules from their crystal lattice

ΔH_f^{**}: energy required for solvation of the solute molecules into polymer structure

h: boundary-layer thickness, cm

h_D: boundary-layer thickness on the donor side, cm

h_0: thickness of the boundary layer having constant viscosity, cm

h_R: boundary-layer thickness on the receptor side, cm

ierfc(x): integrated complementary error function = $1/\pi^{1/2}exp(-x^2) - x\,erfc(x)$

J: diffusion flux, rate per unit area, $mg/(cm^2$-s$)$

$J_0(x)$: Bessel function of the first kind of order 0

$J_1(x)$: Bessel function of the first kind of order 1

J_{max}: permeation flux from a saturated solution, mg/cm^2-s

K: partition coefficient (polymer/bulk), dimensionless

K_D: partition coefficient (polymer/bulk) on the donor side, dimensionless

K_n: partition coefficient for the compound with a chain length of n carbons, used in Eq. (148)

K_0: partition coefficient for the reference homolog, used in Eq. (148)

K_{ps}: partition coefficient of the solute from the pore to the solvent, used in Eq. (7)

K_{rs}: restriction coefficient dependent on the ratio of molecular radius to pore radius, used in Eq. (7)

K_R: partition coefficient (polymer/bulk) on the receptor side, dimensionless

k_D: mass transfer coefficient in the donor cell, cm^2/s

k_m: mass transfer coefficient, cm^2/s

k_R: mass transfer coefficient in the receptor cell, cm^2/s

k_0: infusion rate, mg/min

L: length of a cylinder or tube, cm

l_s: characteristic length of the solute molecule, used in Eq. (13)

ℓ: thickness of membrane, cm

M: molecular weight, dalton

$\bar{M}_c$: number average molecular weights between crosslinks, used in Eq. (12)

$\bar{M}_c^*$: value of $\bar{M}_c$ below which no diffusion of the solute molecules can take place, used in Eq. (12)

M_{it}: cumulative amount of solute released at time t, following the intrinsic kinetics, mg

M_{it}': fictitious amount of drug release at time t, determined from the plot of M_t versus $t^{1/2}$ of the apparent release profile, as illustrated in Fig. 23

M_{i1}': value of M_{it}' at time t_1, determined from the plot of M_t versus $t^{1/2}$ of the apparent release profile, as illustrated in Fig. 23

M_{i2}': value of M_{it}' at time t_2, determined from the plot of M_t versus $t^{1/2}$ of the apparent release profile, as illustrated in Fig. 23

$\bar{M}_n$: number-average molecular weight before cross-linking, used in Eq. (12)

M_0: initial mass used in Eq. (24), mg

M_t: cumulative amount of solute released at time t, mg

$M_t(i)$: cumulative amount of solute released of datum point i

$\hat{M}_t$: estimated (computed) value of M_t at time t

M_∞: cumulative amount of solute released at infinite length of time, mg

m: molality, g-mole per 1000 g of solvent

N_0: Avogadro's number, $6.023 \times 10^{23}/g$-mol

P: permeability coefficient

Q_{it}: intrinsic release rate

Q_m: swelling ratio

Q_0: initial rate; transport rate of the reference diffusant, mg/s

Q_t: diffusion rate, dissolution rate, release rate, or permeation rate at time t, mg/s

$Q_t(i)$: release rate of datum point i

$\hat{Q}_t$: estimated (computed) release rate at time t

Q_∞: diffusion rate, dissolution rate, or permeation rate at steady state, mg/s

R: gas-law constant, equal to 8.314×10^7 g-cm^2/s^2-g mol-K

$R(t)$: radius or distance of receding boundary, cm

Re: Reynolds number, equal to (dU/ν), Eq. (114), dimensionless

R_i: inner radius of a cylinder or sphere

R_{L1}: value of the radius, $R(t)$, on the left-hand side, determined in the first step of the half-interval iterations

R_{L2}: value of the radius, $R(t)$, on the left-hand side, determined in the second step of the half-interval iterations

R_{Lk}: value of the radius, $R(t)$, on the left-hand side, determined in the k-th step of the half-interval iterations

R_{R1}: value of the radius, $R(t)$, on the right-hand side, determined in the first step of the half-interval iterations

R_{R2}: value of the radius, $R(t)$, on the right-hand side, determined in the second step of the half-interval iterations

R_{Rk}: value of the radius, $R(t)$, on the right-hand side, determined in the k-th step of the half-interval iterations

R_o: outer radius of a cylinder or sphere, cm

r_A: radius of diffusant molecules, cm

r: radius of a sphere or a cylinder, cm

r_s: characteristic radius of a solute molecule

Sc: Schmidt number, equal to ν/D, Eq. (115), dimensionless

ΔS_f: entropy of fusion, cal/mol-K

Sh: Sherwood number, equal to $k_m d/D$, Eq. (113), dimensionless

S_0: initial slope of the dimensionless concentration profile defined by Eq. (182)

T: absolute temperature, K

T_f: forming temperature of polymer

T_M: melting temperature, K

$t(i)$: time of datum point i

$\bar{t}$: mean residence time, min

t_{lag}: lag time

$t_{lag}{}'$: apparent lag time

U: mean velocity of flow; effective velocity, cm/s

u_0: maximum velocity of flow, cm/s

V: volume, ml

V_A: molar volume of solute molecule A, cm^3/g-mol

$V_A{}'$: molar volume of solute molecule A at its normal boiling point, cm^3/g-mol

V_B: partial molar volume of a solvent

V_c: volume of central compartment

v_a: volume fraction of equilibrium amorphous polymer

v_c: volume fraction of equilibrium crystalline polymer

v_p: volume fraction of polymer

v_s: volume fraction of solvent or swelling-agent

$W(i)$: weighting factor of datum point i

w: polymer concentration in weight-percent, used in Eq. (10)

Z_+, Z_-: valences of cation and anion, respectively

α: shear rate; pharmacokinetic constant; $V_1/(KV_2)$ defined by Eq. (33)

β: pharmacokinetic constant

γ: the ratio of the solute molecule radius to the pore radius; correction factor used in Eq. (125); activity coefficient

$\gamma_{\pm}$: mean ionic activity coefficient

δ_A: solubility parameter of the solute

δ_p: solubility parameter of the polymer

ϵ: porosity, dimensionless

ζ: a variable defined by Eq. (48)

η: viscosity, cP or g/cm-s

η_0: viscosity of dissolution medium, cP or g/cm-s

θ: half-angle of the pie-shaped cylindrical device

θ_b: dimensionless variable defined by Eq. (47)

ϑ: immobilization factor used in Eq. (17)

Λ: cell constant defined by Eq. (21)

λ: effective volume ratio defined by Eq. (41) through (43), dimensionless; pharmacokinetic parameters

λ_+, λ_-: limiting ionic conductance (cm^2/Ω/g-equiv)

ν: kinematic viscosity, cm^2/s

ν_0: kinematic viscosity at infinite dilution, cm^2/s

ξ: dimensionless variable, equal to M_{it}'/M_t, used in Eq. (144)

ξ_p: mesh size, defined by Eq. (13)

π: constant, characteristic of the membrane and elution solvent, defined by Eq. (148)

ρ: density, g/ml

ρ_m: density of the drug, used in Eq. (162), g/cm^3

σ: solute-polymer interaction constant

τ: dimensionless time; tortuosity

Φ: dimensionless variable defined by Eq. (140), equal to $(dM_t/dt^{1/2})/(dM_{it}/dt^{1/2})$

$\Phi(v)$: free-volume contribution, defined by Eq. (14)

ϕ: association factor used in Eq. (5), dimensionless; dimensionless variable, M_t/M_{it},
   Eq. (139)

Ψ: detour ratio used in Eq. (17)

ψ: variable defined by Eq. (139)

ω: rotational speed, rad/s

DIFFUSION COEFFICIENTS

Diffusion is the process by which particles are transported from one part of a system to another as a result of random movement, in the absence of mixing (by mechanical means or by convection). Diffusion can result from the gradients of pressure, temperature, external force fields, or concentration. Only the concentration gradient as the driving force for diffusion is considered in this chapter. The proportionality constant between the diffusion flux and diffusion potential is termed the diffusion coefficient, or diffusivity.

The diffusion coefficient is one of the most important factors that govern the release kinetics of a solute from a controlled-release (CR) device. Its importance can be exemplified by the kinetics of solute release from a reservoir device or a planar monolithic diffusion-controlled device. The release rate of the former is proportional to the diffusion coefficient, whereas that of the latter is proportional to the square root of the diffusion coefficient. Only the diffusion coefficients of relatively low solute concentrations in liquid and polymeric phases are presented in this section.

Diffusion Coefficients in Liquids

Compared to liquids, molecules of gases are far apart and the intermolecular forces can often be neglected or considered only during collisions. For this reason, the kinetic theory has been successfully utilized to derive the diffusion coefficients of gases in binary and multicomponent mixtures [1]. The prescribed kinetics theory assumes that the diffusion coefficient is directly proportional to the mean molecular velocity and the mean free path.

The molecules presented in the liquid phase exhibit much stronger intermolecular forces than in the gas phase due to the denser arrangement of the molecules. There are two theories available for crude prediction of approximate diffusion coefficients in the liquid phase [2]. In the Eyring theory, the solute molecules are depicted as forming a quasi-crystalline lattice and the analysis is performed more or less as it was for diffusion in a solid. In the hydrodynamic theory, the diffusion coefficient is first related to the force which acts in a sphere moving in a continuum. The force can be evaluated in terms of Stokes' law, and the resulting expression is called the Stokes-Einstein equation. The primary interest in this section is the hydrodynamic approach. Although the hydrodynamic theory is quite idealized and derived only for specific conditions, the *form* of the resulting Stokes-Einstein equation has provided a foundation for development of several useful semiempirical prediction methods. In the following sections, the Stokes-Einstein equation and its modified forms are presented.

There are various semiempirical equations for prediction of the diffusion coefficients [1,3]. In this section, only the Wilke-Chang correlation equation will be discussed. The theory of electrolyte diffusion in the liquid phase is relatively simpler than that of undissociated molecules.

Diffusion Coefficients of Undissociated Molecules

Stokes-Einstein equation. The application of the hydrodynamic theory and the diffusion coefficient in the liquid phase was initiated by Einstein, who applied Stokes' law to describe the drag on large, spherical solute molecules moving through a continuum of small molecules [3]. Thus, for solute molecules that are spherical and large compared to the solvent molecules, the solvent performs as a continuum to the diffusing solute molecules. Under these conditions, the diffusion coefficient, D, of the solute molecules may be estimated by the Stokes-Einstein equations [4–7]

$$D = \frac{RT}{6\pi\eta N_0 r_A} \tag{1}$$

where R is the gas-law constant, equal to 8.314×10^7 g-cm^2/s^2-g mol-K; η is the viscosity of the solvent in g/cm-s; N_0 is Avogadro's number, equal to 6.023×10^{23}/g-mol; T is the absolute temperature in K; and r_A is the radius of the spherical solute molecule.

According to Flynn et al. [5], r_A in Eq. (1) can be estimated by the following equation:

$$r_A = \left(\frac{3V_A}{4\pi N_0}\right)^{1/3} \tag{2}$$

where V_A is the molal volume of the solute. It is important to realize that the radius expressed by Eq. (2) is not that of the bare molecule but of the hydrodynamic molecule, which consists of the diffusant molecule plus any solvent or solute that is bound to the surface of the diffusant [5,8]. The molar volume is frequently difficult to measure by experiment, but it can be estimated with reasonable accuracy, since the molal volume of a compound is an additive property of its constituent atoms and function groups. With this advantage, it is possible to estimate V_A from the chemical structure of the diffusant. Some researchers [4] preferred to calculate the van der Waals' volume of the solute, while others [9,10–14] chose the partial or apparent molal volume; still others [15,16] determined V_A from space-filling molecular models. In view of the approximate nature of atomic and group values and the possibility of solvent incorporation into the hydrodynamic particle, it is difficult to justify a theoretical preference for any of these approaches. The partial molal volume is frequently easiest to calculate and has been used extensively in the estimation of protein hydrodynamic properties [8]. It has also been used successfully in the calculation of micelle surface charge density [17,18] and in the estimation of solubility parameters [19]. The partial molal volumes of some common atoms and groups are listed in Table 1 [5,20].

Sutherland-Einstein equation.   Equation (1) fails to give reliable prediction for smaller solute molecules. To solve this problem, an improvement of the Stokes-Einstein equation was introduced by Sutherland and is termed the Sutherland-Einstein equation [7], as given by

$$D = \frac{RT}{6\pi N_0 r_A \eta} \left(\frac{\beta r_A + 3\eta}{\beta r_A + 2\eta} \right) \tag{3}$$

where β is the coefficient of sliding friction between the solute molecule and the solvent molecule. For large solute molecules, the solvent molecules tend to be dragged along

Table 1   Partial Molal Volumes of Some Common Atoms and Groups [5]

Atom	Partial molal volume (cm³/g-mol)	Group	Partial molal volume (cm³/g-mol)
C	9.9	CH_3	19.3
H	3.1	CH_2	16.2
H^+	−4.5	NH_2	7.7
N	1.5	$N(CH_3)_3{}^+$	66.3
N^+	8.4	COOH	19.0
O(=O or —O—)	5.5	COO^-	11.5
O(—OH)	2.3	C_2H_5	35.3
O(diol)	0.4	C_3H_7	51.7
S	15.5	C_4H_9	67.9
P	17.0	C_6H_{13}	100.3
P^+	28.5	C_8H_{15}	132.7
Li^+	−5.2	$C_{10}H_{21}$	165.1
Na^+	−5.7	$C_{12}H_{25}$	197.5
K^+	4.5	$C_{14}H_{29}$	229.9
Cl^-	22.3	OCH_2CH_2	37.9
Br^-	29.2	One ring	−8.1
I^-	40.8	Two fused rings	−26.4

191.0

with the solute molecules, and a "no-slip" condition between the solute molecule and the solvent molecule exists. For this case, β becomes infinitely large and Eq. (2) reduces to the Stokes-Einstein equation, Eq. (1). If, on the other hand, the solute and solvent molecules are of similar size, the tendency to slip is large and β becomes equal to zero, and Eq. (3) reduces to

$$D = \frac{RT}{4\pi\eta N_0 r_A} \tag{4}$$

The radius r_A is given by Eq. (2), and it can be estimated by the partial molal volume as described earlier. Equations (1) and (4) represent the two limiting cases for Eq. (3), and therefore can only predict the "range" of the diffusion coefficient as a function of the molal volume. This will be discussed on p. 47 of the text.

Wilke-Chang correlation equation.   The lack of widespread quantitative success with the theoretical approaches, as described by Eq. (1), has led to the development of several semiempirical relationships based on Eq. (1). A review of these approaches is given by Reid et al. [1], and a collection of 10 expressions has been presented by Skelland [3].

One of the best-known, most comprehensive, and most convenient empirical equations to use is the Wilke-Chang correlation [1], which is a modification of the Stokes-Einstein equation. The method is a widely used correlation for estimating the diffusion coefficients of small molecules (to be discussed later) in low-molecular-weight solvents, as given by

$$D = 7.4 \times 10^{-8} \frac{(\phi M_B)^{1/2} T}{\eta_B V_A'^{0.6}} \tag{5}$$

where

D = mutual diffusion coefficient of the solute at very low concentrations in the solvent, in cm^2/s

M_B = molecular weight of solvent B

T = absolute temperature

η_B = viscosity of solvent B, cP

V_A' = molal volume of solute A at its normal boiling temperature, cm^3/g-mol

ϕ = association factor of solvent B, dimensionless

According to Wilke and Chang [1], the value of V_A' is best determined at the normal boiling point of the solute T_B. If no experimental data are available, the Le Bas additive volumes given in Table 2 [1,21] may be used.

The parameter ϕ is taken as 2.6 for water, 1.9 for methanol, 1.5 for ethanol, and 1.0 for unassociated solvents. The average error in the use of Eq. (5) for an aqueous system is about 10–15%; considerably greater errors are common in the case of organic solvents.

The size of molecules and various predictive equations.   Listed in Table 3 are the predicted diffusion coefficients for various diffusants at 25°C using the Stokes-Einstein equations and the Wilke-Chang correlation, along with the experimentally determined values. The approximate relationship between the molecular weight and V_A or V_A' can be seen from Table 3. According to Eqs. (1) and (4), the plot of log(D) versus log(V_A)

Table 2 Additive-Volume Increments for the Calculation of Molar
Volume Using Le Bas's Approach [1,21]

	Increment $(cm^3/g\text{-mol})$
Carbon	14.8
Hydrogen	3.7
Oxygen (except as noted below)	7.4
In methyl esters and ethers	9.1
In ethyl esters and ethers	9.9
In higher esters and ethers	11.0
In acids	12.0
Joined to S, P, N	8.3
Nitrogen	
Double-bonded	15.6
In primary amine	10.5
In secondary amine	12.0
Bromine	27.0
Chlorine	24.6
Fluorine	8.7
Iodine	37.0
Sulfur	25.6
Ring, three-membered	-6.0
Four-membered	-8.5
Five-membered	-11.5
Six-membered	-15.0
Naphthalene	-30.0
Anthracene	-47.5

for the predicted diffusion coefficients gives a straight line. This is shown in Fig. 2, along with the experimental values. As seen, the Stokes-Einstein equation, Eq. (1), is applicable only to molal volumes larger than 35 cm^3/g-mol, while Eq. (4) is appropriate for molal volumes smaller than 120 cm^3/g-mol. Also presented in Fig. 3 are the log-log plots of D (predicted and experimental) versus V_A'. They show that the Wilke-Chang equation is accurate for the molal volume ranges from 45 to 600 cm^3/g-mol. For very small molecules (V_A' smaller than 45 cm^3/g-mol), the Wilke-Chang equation gives higher predictions than the experimental values, while for large molecules (V_A' larger than 600 cm^3/g-mol), it gives lower predictions than the experimental values. Based on the data presented in Table 3, it is interesting to note that although the Wilke-Chang correlation was originally derived for liquid diffusants, it predicts surprisingly well for solid compounds.

Diffusion Coefficients of Electrolytes

The diffusion of electrolytes is considerably different from that of undissociated molecules. When dissolved, molecules of electrolytes dissociate into cations and anions. Despite differences between the sizes of cations and anions, both the positive- and negative-charged ions diffuse at the same rate through the dissolution boundary layer. In this way, electroneutrality is maintained.

The theory of diffusion of electrolytes at low concentrations has been developed. For

Table 3　Predicted and Measured Diffusion Coefficients (cm²/s) of Various Diffusants in Water at 25°C

Diffusant	Molecular weight (daltons)	Molal volume V_A by Table 1 (cm³/g-mol)	D_{calc} Eq. (1) ($\times 10^6$) (cm²/s)	D_{calc} Eq. (4) ($\times 10^6$) (cm²/s)	Molal volume V_A' by Table 2 (cm³/g-mol)	D_{calc} Eq. (5) ($\times 10^6$) (cm²/s)	D_{exp} (cm²/s)	Ref.
Methane	16	22.4	17.6	11.7	29.6	22.1	18.8	141
Ethane	30	38.4	14.7	9.8	51.8	15.8	15.2	141
Propane	44	54.8	13.0	8.7	74.0	12.8	12.1	141
Butane	58	71.0	12.0	8.0	96.2	10.9	9.6	141
Methanol	32	24.7	17.0	11.3	37.0	19.4	15.8	142
	32	24.7	17.0	11.3	37.0	19.4	17.0	143
	32	24.7	17.0	11.3	37.0	19.4	13.7	144
Ethanol	46	40.7	14.4	9.6	59.2	14.6	12.4	142
	46	40.7	14.4	9.6	59.2	14.6	12.6	143
n-Propanol	60	57.1	12.9	8.6	81.4	12.1	10.2	142
	60	57.1	12.9	8.6	81.4	12.1	11.5	143
n-Butanol	74	73.3	11.8	7.9	103.6	10.4	9.5	142
	74	73.3	11.8	7.9	103.6	10.4	11.0	143
n-Pentanol	88	89.5	11.1	7.4	125.8	9.3	8.8	144
Isopropanol	60	57.0	12.9	8.6	81.4	12.1	10.2	142
	60	57.0	12.9	8.6	81.4	12.1	10.7	143
Isobutanol	74	73.2	11.8	7.9	103.6	10.4	9.3	142
sec-Butanol	74	73.2	11.8	7.9	103.6	10.4	9.2	142
tert-Butanol	74	73.2	11.8	7.9	103.6	10.4	8.8	142
	74	73.2	11.8	7.9	103.6	10.4	9.8	143
Formamide	45	26.3	16.6	11.1	49.8	16.2	17.2	143
Acetamide	59	42.3	14.2	9.5	72.0	13.0	13.2	143
Propionamide	73	58.7	12.7	8.5	94.2	11.0	12.0	143
Butyramide	87	74.9	11.7	7.8	116.4	9.7	10.7	143
Isobutyramide	87	74.8	11.7	7.8	116.4	9.7	10.2	143
Formic acid	46	22.1	17.6	11.8	46.2	16.9	14.6	145
Acetic acid	60	38.3	14.7	9.8	68.4	13.4	12.0	145

Propanoic acid	74	54.3	13.1	8.7	90.6	11.3	10.1	145
Butyric acid	88	70.7	12.0	8.0	112.8	9.9	9.2	145
Pentanoic acid	102	86.9	11.2	7.5	135.0	8.9	8.2	145
Hexanoic acid	116	103.1	10.6	7.0	157.2	8.1	7.8	145
Isobutyric acid	88	70.6	12.0	8.0	112.8	9.9	9.5	145
Isopentanoic acid	102	86.8	11.2	7.5	135.0	8.9	8.2	145
Chloroacetic acid	95	57.5	12.8	8.5	89.3	11.4	10.0	145
Hydroxyacetic acid	76	40.6	14.4	9.6	75.8	12.6	9.8	145
Oxalic acid	90	38.0	14.7	9.8	85.0	11.7	8.6	145
Succinic acid	118	80.3	11.5	7.6	129.4	9.1	7.9	145
Adipic acid	146	112.7	10.2	6.8	173.8	7.6	7.4	145
α,ω-Octanedioic acid	174	135.2	9.6	6.4	218.2	6.7	7.1	145
Pyridine	79	58.4	12.8	8.5	93.1	11.1	11.4	146
4-Methylpyridine	93	74.6	11.8	7.8	115.0	9.8	10.8	146
2-Ethylpyridine	107	90.6	11.0	7.3	137.5	8.8	9.8	146
4-Ethylpyridine	107	90.6	11.0	7.3	137.5	8.8	10.0	146
2-Propylpyridine	121	107.0	10.4	7.0	159.7	8.1	8.8	146
4-Propylpyridine	121	107.0	10.4	7.0	159.7	8.1	10.0	146
4-tert-Butylpyridine	135	123.1	10.0	6.6	181.9	7.4	9.2	146
4-n-Amylpyridine	149	139.4	9.5	6.4	204.1	7.0	9.5	146
Glycine	75	42.9	14.1	9.4	82.6	12.0	10.6	142
α-Alanine	89	59.1	12.7	8.5	104.8	10.4	9.1	142
β-Alanine	89	59.1	12.7	8.5	104.8	10.4	9.3	142
α-Aminobytyric acid	103	75.0	11.7	7.8	127.0	9.2	8.3	142
α-Aminopentancic acid (norvaline)	117	91.4	11.0	7.3	149.2	8.4	7.7	142
α-Aminohexanoic acid (norleucine)	131	107.6	10.4	6.9	171.4	7.7	7.2	142
α-Aminoisobutyric acid	103	75.2	11.7	7.8	127.0	9.2	8.1	142
α-Aminoisopentanoic acid (valine)	117	91.4	11.0	7.3	149.2	8.4	7.7	142

Table 3, cont'd.

Diffusant	Molecular weight (daltons)	Molal volume V_A by Table 1 (cm^3/g-mol)	D_{calc} Eq. (1) ($\times 10^6$) (cm^2/s)	D_{calc} Eq. (4) ($\times 10^6$) (cm^2/s)	Molal volume $V_A{}'$ by Table 2 (cm^3/g-mol)	D_{calc} Eq. (5) ($\times 10^6$) (cm^2/s)	D_{exp} (cm^2/s)	Ref.
α-Aminoisohexaic acid (leucine)	131	107.6	10.4	6.9	171.4	7.7	7.3	142
Threonine	119	77.4	11.6	7.7	134.4	8.9	8.0	142
Asparagine	132	79.0	11.5	7.7	141.2	8.7	8.3	142
Proline	115	76.8	11.6	7.8	132.3	9.0	8.8	142
Hydroxyproline	131	79.1	11.5	7.7	139.7	8.7	8.3	142
Phenylalanine	165	125.9	9.9	6.6	197.1	7.1	7.0	142
Tryptophan	204	127.3	9.8	6.6	227.2	6.5	6.6	142
Glucose	180	105.5	10.5	7.0	166.2	7.9	6.8	147
Cellobiose	342	202.5	8.4	5.6	314.6	5.4	5.2	147
Triose	504	299.5	7.4	4.9	476.2	4.2	4.2	147
Tetrose	666	396.5	6.7	4.5	631.2	3.5	3.8	147
Pentose	828	493.5	6.3	4.2	786.2	3.1	3.2	147
Hexose	990	590.5	5.9	3.9	941.2	2.8	2.9	147

Source: Part of this table was compiled by Flynn et al. [5].

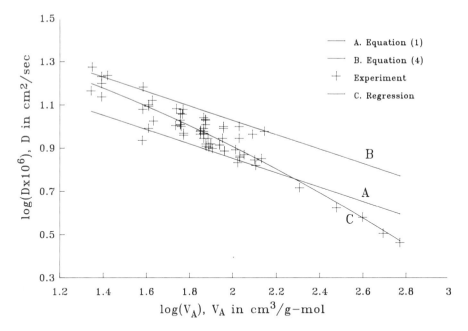

Figure 2   Plot of log(D) versus log(V_A) using Eqs. (1) and (4) along with the experimental data.

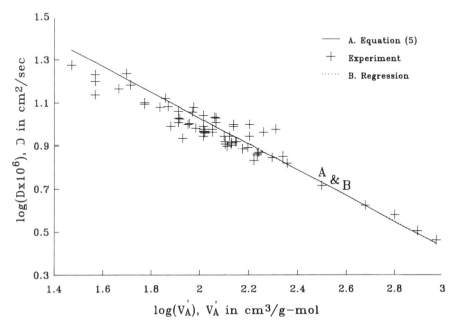

Figure 3   Plot of log(D) versus log(V_A') using Eq. (5) along with the experimental data.

dilute solutions of a single electrolyte, the diffusion coefficient is given by the Nernst-Haskell equation from Ref. 1,

$$D_0 = \frac{RT}{Fa^2} \frac{1/Z_+ + 1/Z_-}{1/\lambda_+ + 1/\lambda_-}$$ (6)

where

D_0 = diffusion coefficient at infinite dilution, cm^2/s

T = temperature, K

R = gas-law constant, equal to 8.314×10^7 $g\text{-}cm^2/s^2\text{-}g$ mol-K

λ_+, λ_- = limiting (zero-concentration) ionic conductances, $(cm^2/\Omega/g\text{-}equiv)$

Z_+, Z_- = valences of cation and anion, respectively

Fa = faraday, 96,500 C/g-equiv

Equations (5) and (6) imply that the diffusion coefficient of the compound's electrolyte should be greater than that of the undissociated molecule due to the fact that the dissociated cation, which diffuses faster, will "drag" the slower anion. The following problem is employed to demonstrate this phenomenon.

Problem 1.   Predict the diffusion coefficients of the undissociated ampicillin and salicylic acid and their sodium salts in water at 25°C using the Wilke-Chang correlation and the Nernst-Haskell equation. The value of λ_+ for sodium at 25°C is 50.1 $(cm^2/\Omega/g\text{-}equiv)$ [1], while the values of λ_- for ampicillin and salicylate are equal to 21.16 and 33.71 $(cm^2/\Omega/g\text{-}equiv)$, respectively [22].

Solution.   The constants and molecular formula that are necessary to solve this problem are listed below:

$\phi = 2.6$
$M_B = 18.0$
$T = 298$ K
$\eta_B = 0.8930$ cP
Molecular formula of ampicillin: $C_{16}H_{19}N_3O_4S$
Molecular formula of salicylic acid: $C_7H_6O_3$

Diffusion coefficient of undissociated ampicillin molecule: The additive volume increments of ampicillin using Le Bas's approach are listed below:

C: 16×14.8	$= 236.8$
H: 19×3.7	$= 70.3$
O (joined to acid): 12.0×2	$= 24.0$
O (other): 7.4×2	$= 14.8$
N (in primary amine): 10.5×1	$= 10.5$
N (in secondary amine): 12.0×1	$= 12.0$
Sulfur: 25.6	25.6
Total	$V_A = 394.0$ cm/g-mol

Thus, the diffusion coefficient D calculated by Eq. (5) gives 4.68×10^{-6} cm²/s, which is in close agreement with the experimental value of 4.58×10^{-6} [23].

Diffusion coefficient of undissociated ampicillin molecule: The additive volume increments of ampicillin using Le Bas's approach are listed below:

C: 7×14.8	$= 103.6$
H: 6×3.7	$= 22.2$
O (joined to acid): 12.0×2	$= 24.0$
O (other): 7.4×1	$= \underline{7.4}$
Total	$V_A = 157.2$ cm³/g-mol

The diffusion coefficient D calculated by Eq. (5) gives 8.12×10^{-6} cm²/s.

Diffusion coefficient D_0 for sodium ampicillin and sodium salicylate: Using the given values of λ_+ for sodium and the values of λ_- for ampicillin and salicylate, the value of D_0 for sodium ampicillin evaluated by Eq. (6) is 7.92×10^{-6} cm²/s, and 10.7×10^{-6} cm²/s for sodium salicylate, which is in excellent agreement with the value of 10.0×10^{-6} cm²/s obtained by Desai et al. [24].

Diffusion Coefficients in Polymers

For the purpose of investigating solute diffusion in polymers, the polymer structure can be classified in terms of the pore size, namely, macroporous, microporous, and nonporous [25]. Recalling the equations for predicting the diffusion coefficients in liquids, such as Eq. (1), (4), or (5), these equations indicate that the relationship among the diffusion coefficient, the viscosity of the diffusion medium, and the size of the solute molecules is relatively simple. The diffusion coefficients of solutes in polymers, however, are highly dependent either on the structure of the pores (for the cases of macro- and microporous polymers) or on the structure of the polymeric network (for the case of nonporous polymers). As a result, the hydrodynamic theory is no longer directly applicable to the polymers, since it is not appropriate to describe the polymer as a "continuum" at the microscopic level [26]. Thus, various relationships developed for prediction are based on semiempirical or empirical approaches.

The mechanisms of diffusion of solute molecules through macroporous and microporous polymers are relatively simpler than those of nonporous polymeric networks, since the solute diffuses only through the solvent-filled pores. It will be seen later that for macroporous polymers it is only necessary to correct the diffusion coefficient estimated in the liquid phase to account for the porosity, the tortuosity, and the partition coefficient. For microporous polymers, additional steric hindrance and frictional resistance of the pores also needs to be included. These parameters usually have to be determined by experiments.

For nonporous polymeric networks, the situation becomes more complex, since both the solute molecules and the structure of the polymeric network will affect diffusion. Thus, prediction of drug diffusion coefficients in these polymeric networks is much more difficult than that in the macro- and microporous polymers. The polymer structure-related factors include the degree of crystallinity, the size of crystallites, the degree of swelling, the molecular weight between crosslinks, and the state of the polymer (whether glassy or rubbery) [27]. The semiempirical relationships for predicting diffusion coefficient were developed based on the free-volume theory. The free-volume theory states that if a molecule acquires sufficient thermal energy and moves in the proper direction during the time interval that the neighboring cell is vacant, the molecule can "jump" from its own cell to a new equilibrium position in the adjacent cell. If another molecule can also jump into the vacancy left by this molecule, the displacement gives rise to the diffusive motion [28].

Semiempirical Relationships

Macroporous and microporous polymeric networks. Macroporous and microporous polymeric networks contain two phases, the polymer phase and the porous phase. For diffusion through pores, the diffusion coefficients refer to solute diffusion through the solvent-filled pores. For macroporous polymers, the pore sizes usually range from 0.1 to 1.0 μm, which are much larger than the pore sizes of diffusant molecules. Evidently, the restriction of the pores on diffusion is not significant. For microporous polymers, the pore sizes range from 50 to 200 Å (angstroms), which are slightly larger than that of the diffusant molecules. Under these conditions, the diffusional restriction by pore walls may be significant.

For convenience, the structure of the pores in the membrane is incorporated into the diffusion coefficient in terms of the porosity, ϵ, (void volume fraction) and the tortuosity, τ, of the polymer. In general, the effective diffusion coefficient D_e can be expressed by

$$D_e = D \frac{\epsilon K_{ps} K_{rs}}{\tau} \tag{7}$$

where

       D = the diffusion coefficient of the solute through the solvent-filled pores

      K_{ps} = the equilibrium partition coefficient (i.e., the ratio of the concentration inside the pore to the concentration outside the pore at equilibrium), if there is any [29]

      K_{rs} = the fractional reduction in diffusion coefficient within the pore which results when the solute and pore sizes are of comparable magnitude [29]

In general, the constants K_{ps} and K_{rs} are more important in the membrane controlled-release system than in the matrix controlled-release system. For the case of a monolithic matrix release system, especially when the drug loading dose is much greater than the drug solubility in the polymer, the values of K_{ps} and K_{rs} are essentially equal to unity, since the porosity of the matrix is much greater than the solute molecules.

In the special case of the microporous membrane system, where the size of the diffusing species is of the same order of magnitude as the diameter of the pore, the following simplified expressions have been developed to describe the transport process.

Microporous and gel-type polymeric networks: The theories by Faxén [30] and Renkin [31] have often been used to describe the pore-solute diffusion process. Both theories were developed based on a hydrodyamic analysis of solute diffusion through porous systems in the absence of pressure gradients. Two factors need to be considered for diffusion within the pores. The first factor is established under the condition that a molecule must pass through the pores' opening without striking the edge of the pore. The second factor corrects the friction between a molecule moving within a pore and the pores' walls. The total restriction to diffusion due to the combined effects of steric hindrance at the entrance to pores and the frictional resistance within the pores is given by the Renkin equation [31],

$$\frac{D_e}{D} = (1 - \gamma)^2(1 - 2.104\gamma + 2.09\gamma^3 - 0.95\gamma^5) \qquad (8)$$

where γ is the ratio of the solute molecule radius to the pore radius. The first portion of Eq. (8) accounts for the statistical likelihood of a molecule entering a pore, whereas the remainder of the equation represents the solvent drag on a solute molecule traveling through a narrow pore. This equation is valid up to about $\gamma = 0.5$. Equation (8) has been applied to solute diffusion through biological membranes [32] and through swollen hydrophilic membranes [33]. For $\gamma < 0.2$, the following equation may be used [5]:

$$\frac{D_e}{D} = (1 - \gamma)^4 \qquad (9)$$

It can be seen from Eqs. (8) and (9) that large molecules (large values of γ) retard to a much greater extent than small ones in their passage through narrow pores.

Davis [34] proposed the following empirical equation for diffusion through polyacrylamide and polyvinylpyrrolidone gels:

$$\frac{D_e}{D} = \exp[-(0.05 + 10^{-6}M)w] \qquad (10)$$

where M is the solute molecular weight and w is the polymer concentration in weight-percent.

Satterfield et al. [29] and Colton et al. [36] proposed the following empirical correlation to describe the diffusion of nonadsorbed solutes through fine pores:

$$\frac{D_e}{D} = \frac{1}{\tau} \exp(-2\gamma) \qquad (11)$$

Nonporous polymers. Unlike the macroporous or microporous polymers, nonporous polymers contain only a homogeneous polymeric phase (or network). The space between the macromolecular chains (mesh), rather than pores, is the only area available for diffusion of solutes [37–39]. These polymers include most types of hydrogels such as water-swollen networks of poly(vinyl alcohol), poly(hydroxyethyl methacrylate), and related polymers, as well as networks of polystyrene and other hydrophobic polymers swollen in appropriate organic solvents [37]. The "macromolecular meshes" in these polymeric networks range from 20 to 100 Å [37,38]. Any specific interaction of the diffusant molecules with the polymeric macromolecules would affect the diffusion coefficients. Therefore, the effect of the polymer's structure on the solute diffusion coefficient is more complex than that of the porous polymers. For this reason, the diffusion coefficients of solutes in the polymers cannot be expressed in terms of ϵ, τ, or γ, as indicated in Eqs. (7)–(11). The "macromolecular meshes" in the cross-linked structure of the polymer create a "screening effect" [25,37,38] on solute diffusion through polymers. For uncross-linked polymers, this "screening" is provided by the meshes formed by entangled chains, whereas in semicrystalline polymers the crystallites act as physical crosslinks.

Cross-linked or uncross-linked rubbery polymers: The solute diffusion coefficient is dependent on the equilibrium polymer volume fraction in the polymer, the cross-linking density and the size of the solute. Recent theoretical analyses by Yasuda and Lamaze [40], Peppas and Meadows [25], and Peppas and Reinhart [37] give the general behavior of the effective solute diffusion coefficient D_e for highly swollen polymers:

$$\frac{D_e}{D} = g_1 \frac{\bar{M}_c - \bar{M}_c^*}{\bar{M}_n - \bar{M}_c} \exp\left(-\frac{g_2 r_s^2}{(Q_m - 1)}\right) \tag{12}$$

where $\bar{M}_c$ and $\bar{M}_n$ are the number-average molecular weights between crosslinks (for the network membrane) and before cross-linking (for original polymer chains), respectively. $\bar{M}_c^*$ is the value of $\bar{M}_c$ below which no diffusion of the solute can take place, r_s is the characteristic radius of the solute, and Q_m is the swelling ratio. The detailed derivations of Eq. (12) are given by Peppas and Reinhart [37]. The final two constants, g_1 and g_2, are physical parameters of the polymer and the swelling agent. Equation (12) indicates that a plot of $\ln(D_e/D)$ versus $1/(Q_m - 1)$ at a constant temperature gives a straight line. The intercept $\ln[g_1(\bar{M}_c - \bar{M}_c^*)/(\bar{M}_n - \bar{M}_c)]$ indicates the dependence of the diffusion coefficient on the mesh size of the network (or the structure of the polymer), whereas the slope $-g_2 r_s^2$ gives the dependence of the diffusion coefficient on the size of the diffusant molecules.

For moderately swollen polymers, Peppas and Monihan [41] modified Eq. (12) to predict the effective diffusion coefficient in terms of measurable quantities:

$$\frac{D_e}{D} = f(\xi_p) \exp[g_3(\bar{M}_n - \bar{M}_c) - \pi r_s^2 l_s \Phi(v)] \tag{13}$$

where $f(\xi_p)$ is a function dependent on the mesh size, ξ_p, and l_s is the characteristic length of the solute molecule. For highly swollen membranes, $f(\xi_p)$ is linear and equal to $(\bar{M}_c - \bar{M}_c^*)/(\bar{M}_n - \bar{M}_c)$, as indicated by Eq. (12). The free-volume contribution, $\Phi(v)$, is defined by the equation

$$\Phi(v) = \frac{v_s - v_p}{(Q_m - 1)v_s^2 + v_s v_p} \tag{14}$$

where v_s refers to the average free volume of the solvent in the polymer, v_p is the free volume of the polymer, and g_3 is a constant defined by

$$g_3 = \frac{-2\,\Delta E}{\bar{M}_n R T_f} \tag{15}$$

where ΔE is the penetration energy, R is the gas-law constant, and T_f is the forming temperature of the polymer. If the dependence of the diffusion coefficient on the mesh size is given, Eq. (13) may be used to predict the influence of the mesh size, the degree of swelling, and the solute size on the screening effect of a polymer. Moynihan et al. [42] verifed Eq. (13) experimentally and determined the function $f(\xi_p)$ using phenylalanine as the diffusant, poly(2-hydroxyethyl methacrylate) (PHEMA) as the membrane material, and water as the swelling agent. The value of $\bar{M}_c$ calculated varied between 1700 and 3425 daltons, corresponding to a mesh size of 24 to 35 Å. The obtained diffusion coefficient of phenylalanine at 37°C varies from 0.17×10^{-6} to 0.97×10^{-6} cm^2/s.

Moderately cross-linked polymers: For moderately cross-linked polymers, Peppas and Meadows [25] modified Eq. (12) to give the following equation:

$$\frac{D_e}{D} = g_1 \xi_p \exp\left(-\frac{g_2 r_s^2 \xi_p}{\bar{M}_n^2 (Q_m - 1)}\right) \tag{16}$$

where ξ_p is the mesh size, which is equivalent to the term $(\bar{M}_c - \bar{M}_c^*)/(\bar{M}_n - \bar{M}_c)$ in Eq. (12). Other parameters are the same as those used in Eq. (12). The mesh size is used to describe the average distance of the space between four tetrafunctional crosslinks or four entanglements in the polymer.

Semicrystalline rubbery polymers: In semicrystalline rubbery polymers, diffusant molecules are further slowed by crystallites in the homogeneous structure of the amorphous polymer. Since the crystals are practically impermeable to the diffusant molecules, the diffusant molecules must travel over longer paths through the amorphous component to reach any new position. Peppas and Meadows [25] introduced the following equation to give the true diffusion coefficient in the amorphous polymer.

$$D_{am} = \frac{\Psi D_e}{\vartheta} \tag{17}$$

where D_{am} is the diffusion coefficient in the amorphous rubbery polymer, D_e is the effective diffusion coefficient observed in the sorption or permeation experiment, Ψ is the ''detour ratio,'' and ϑ is an ''immobilization factor.'' The detour ratio accounts for reduction in solute mobility due to the tortuosity of diffusion paths between crystallites (similar to the tortuosity τ), whereas the immobilization factor describes the physical crosslinking due to the crystallites, which is also known as the blocking factor [43].

A second relationship between D_e and D_{am} was introduced by Harland and Peppas [44,45], and is given by

$$D_e = \frac{(v_a + v_s)D_{am}}{\tau} = \frac{(1 - v_c)D_{am}}{\tau} \tag{18}$$

where τ is the tortuosity, v_a is the volume fraction of equilibrium amorphous polymer, v_c is the volume fraction of equilibrium crystalline polymer, and v_s is the volume fraction of the equilibrium swelling-agent.

Equations (17) and (18) are the general expressions to relate the effective diffusion coefficient in the polymer to the diffusion coefficient in the swollen, amorphous region. They can be combined with Eq. (12) or (13), depending on the degree of swelling, to give the effective diffusion coefficient of a semicrystalline network [44]. The method for combining these relationships is to substitute D_e in Eq. (12) or (13) for D_{am} in Eq. (17) or (18).

Empirical Relationship Between Diffusion Coefficient and Molecular Weight

It should be noted that in order to give a reliable prediction of the diffusion coefficient of a particular solute in a polymer using Eqs. (7)–(18), the key parameters associated with the structure of the polymer, such as τ, ϵ, γ, g_1, g_2, ξ_p, etc., must be known. For porous polymers, the methods for determining porosity and tortuosity have been proposed by Desai et al. [24]. For nonporous polymers, exact expressions of the mesh size, ξ_p, are available only for special cases of polymer networks, which are usually highly swollen with chains exhibiting Gaussian distribution behavior [37]. Experimental determination of the mesh size can be performed by neutron scattering and laser light-scattering techniques [45,46]. If a direct prediction of the diffusion coefficients using the above semiempirical approaches is too difficult due to the complexity of the experiments, an empirical relationship between the diffusion coefficient and the molecular weight of the diffusant molecules may be used. For instance, the following equation has been proposed by several investigators [5,47–49]:

$$D_e = D_{M1}(M)^n \tag{19}$$

where D_{M1} is the diffusion coefficient of a diffusant of unit molecular weight and n is a constant. Equation (19) indicates a linear relationship between log(D) and log(M). The value of n in Eq. (19) varies from a low of approximately 0.5 for water to almost 5 for polystyrene.

Methods for Measuring Diffusion Coefficients

Diffusion Coefficient in Liquids

Rotating disk.   The rotating disk is frequently used for intrinsic dissolution-rate determinations for drugs [22,50–55]. The experimental apparatus is depicted in Fig. 4. Equations for the steady-state mass transfer with a constant diffusion coefficient at the surface of a rotating disk have been solved by Levich [56] for both laminar and turbulent diffusion regimes. The diffusion coefficient of the drug can be evaluated by simple equations from knowledge of the solubility of the drug, viscosity of the dissolution medium, and the rotational speed of the disk. The theoretical derivations have been validated in a large number of drug dissolution studies with constant diffusion coefficients.

For the cases of a constant drug diffusion coefficient in the entire boundary layer, an analytical solution can be obtained by direct integration of the diffusion equation, giving the well-known Levich equation [56],

$$Q_t = 0.62AD^{2/3}v_0^{-1/6}\omega^{1/2}C_s \tag{20}$$

where Q_t is the rate of dissolution, A is the surface area of the drug on the disk, D is the diffusion coefficient of the drug, v_0 is the kinematic viscosity of the dissolution medium, C_s is the drug solubility in the dissolution medium, and ω is the rotational speed of the disk.

Equation (20) indicates that the plot of the dissolution rate against $\omega^{1/2}$ gives a straight line passing through the origin with a slope equal to $0.62AD^{2/3}v_0^{-1/6}C_s$, from which the diffusion coefficient D can be deduced provided that the solubility of the drug and the values of A and v_0 are known. The advantage of this method is the relative simplicity of the experiment.

Two-cell diffusion system.   The two-cell diffusion system for measuring the diffusion coefficient of solute in the liquid phase has been used by several researchers [24,57]. The system consists of a porous sintered glass disk, used as a bridge, mounted between two flasks with side arms. One of the flasks is closed with a ground-glass stopper and the other is closed with a special adaptor. The stirring of the solution can be achieved using impellers or magnetic stirring bars. The entire apparatus is water-jacketed to maintain constant temperature. In this method, the solute transfer rate through a sintered glass disk from one cell to the other is measured as a function of time.

The system is calibrated by a diffusant with a known diffusion coefficient. Provided that the boundary-layer diffusional resistance is negligible compared to that of the bridge, the cell constant, Λ, of this system is determined, using the reference diffusant solutions, by the following relationship:

$$\Lambda = \frac{Q_0}{D_0(C_2 - C_1)} \tag{21}$$

where C_2 and C_1 (with $C_1 \approx 0$) are the diffusant concentrations in cells 2 and 1, respectively, D_0 is the diffusion coefficient of the reference diffusant, and Q_0 is the measured transport rate of the reference diffusant.

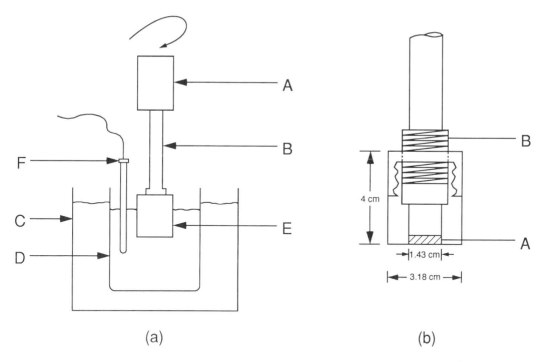

(a)   (b)

Figure 4   (a) Rotating-disk apparatus for dissolution rate studies. Key: A, constant-speed motor; B, rotating shaft; C, water bath; D, reservoir; E, rotating disk; F, conductivity probe. (b) Cross section of rotating disk. Key: A, drug disk; B, thread.

As soon as the transport rates of the drug of interest through the same system are measured, the diffusion coefficient of the drug is then calculated by the equation

$$D = \frac{Q_t}{A \, \Delta C} \tag{22}$$

where Q_t is the measured transport rate of the drug and ΔC is the concentration difference between the two cells. It is important to emphasize that in order to use this method to accurately determine the diffusion coefficient, it is necessary to minimize the boundary diffusional resistance on both sides of the sintered glass using vigorous stirring to ensure that the thickness of the boundary layer is much smaller than that of the bridge.

Taylor diffusion.   This method was suggested by Taylor [58,59] based on observations of the dispersion of soluble matter in a solvent flowing slowly through a small-bore tube. The distribution of concentration is found to be centered on a point which moves with the mean velocity of flow, U, and is symmetrical in spite of the asymmetry of flow. Taylor [58] showed that the distribution is determined by a longitudinal dispersion coefficient, $\overset{\circ}{D}$, which is related to the molecular diffusion coefficient, D, by the relation

$$\overset{\circ}{D} = \frac{r^2 U^2}{48 \, D} \tag{23}$$

Two useful experimental conditions are as follows.

1.   Material of mass, M_0, is concentrated at $x = 0$ at the onset of the experiment. The solution for this is

$$C = 0.5M_0 r^{-2}\pi^{-3/2}(\mathring{D}t)^{-1/2} \exp\left(-\frac{x_1^2}{4\mathring{D}t}\right) \tag{24}$$

and

$$x_1 = x - 0.5u_0 t \tag{25}$$

where u_0 is the maximum velocity on the axis and r is the radius of the tube.

2. Material of constant concentration, C_0, is allowed to enter the tube at a uniform rate at $x = 0$, starting at $t = 0$. Initially, the tube is filled with solvent only ($C = 0$). The solution for this case becomes

$$\frac{C}{C_0} = 0.5 + 0.5 \, \text{erf}\left(\frac{x_1}{2(\mathring{D}t)^{1/2}}\right) \qquad x_1 < 0 \tag{26}$$

$$\frac{C}{C_0} = 0.5 - 0.5 \, \text{erf}\left(\frac{x_1}{2(\mathring{D}t)^{1/2}}\right) \qquad x_1 > 0 \tag{27}$$

In either case, $\mathring{D}$, and hence D, may be deduced by comparing the appropriate mathematical solution with an observed concentration profile. Taylor [59] suggested the necessary conditions for using Eqs. (24)–(27) to be

$$\frac{4L}{r} \geq \frac{Ur}{D} \geq 6.9 \tag{28}$$

where L is the length of tube over which appreciable changes in concentration occur.

Diffusion Coefficients in Polymers

Permeation method using diffusion cells. The permeation method is similar to the two-cell diffusion system described earlier except that the sintered glass is replaced by a membrane, as shown in Fig. 5. This method has been used by many researchers (15,60–63] using one of the following means for stirring: (1) impeller stirring, (2) flow-through

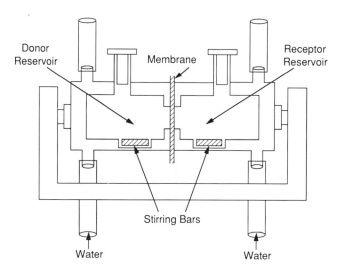

Figure 5 Two-cell system for diffusion coefficient measurement and membrane permeation experiment.

stirring, and (3) magnetic stirring. The advantage of impeller stirring is that it can provide vigorous stirring to minimize the boundary-layer diffusional resistance, and has been used by Colton [61], Smith et al. [62], Reinhart et al. [27], and Bellucci and Drioli [87]. In this method, the impeller is driven by a motor, and the volume of the cell is sufficiently large to accommodate the impeller and the solution. The system consists of two half-cells, each equipped with external jackets for temperature control using a water bath. The degree of mixing is controlled by the geometry of the cell and the rotational speed of the impeller. This method becomes inconvenient when it is necessary to perform several experiments at the same time, or when the amount of fluid in the cell is small.

The flow-through stirring method [15,63] can also provide vigorous stirring to minimize the boundary-layer diffusional resistance. In this method, it is important to control the circulation flow rate so that the boundary-layer diffusional resistance can be eliminated. It is necessary to use two pumps to recirculate the fluids in the two cells. The other advantage of this method is that the amount of fluid in the circulation system can be varied over a wide range, and the disadvantage is that it is not convenient to perform several identical experiments at the same time.

The drawback of the magnetic stirring system is its relatively weaker stirring efficiency. The degree of mixing is limited by the dimensions of the stirring bars as well as the rotational speed and the geometry of the cell, but the method offers a great convenience for performing several identical experiments simultaneously. Also, it is necessary to use only a small amount of fluid in the donor and the receptor cells.

The effect of stirring on the boundary-layer diffusional resistance for the two-cell permeation system will be discussed later. One important issue that should be addressed is that in order to use the permeation method to accurately determine the solute diffusion coefficient in the polymeric membranes, it is preferable to eliminate the boundary-layer diffusional resistance to avoid an additional source of error. From the above discussion, it can be seen that the impeller stirring and flow-through stirring systems are preferred for high-permeability polymeric membranes, whereas magnetic stirring is appropriate only for low-permeability polymers.

To perform the experiment on the two-cell permeation system, a saturated solution containing a large excess of undissolved solute is introduced into the donor compartment, while the pure solvent is introduced into the receptor compartment. The stirring speed or flow rate is varied to determine the speed or flow rate at which the boundary layer diffusional resistance is negligible. Solute then diffuses through the membrane from the donor to the receptor. The amount of solute that has accumulated in the receptor compartment at time t is assayed. The receptor compartment should be emptied and refilled with pure solvent during the course of the experiment or otherwise maintained to assure a perfect sink condition. The permeation data yield a lag time, t_{lag}, which is related to the diffusion coefficient D_m by the equation [28]

$$t_{lag} = \frac{\ell^2}{6D_m} \tag{29}$$

where D_m is the solute diffusion coefficient in the membrane, which is equivalent to De, and ℓ is the thickness of the membrane. The steady-state permeation rate, $(dM_t/dt)_\infty$, can be expressed by [28]

$$\left(\frac{dM_t}{dt}\right)_\infty = \frac{D_m K C_s}{\ell} \tag{30}$$

where K is the partition coefficient (membrane/bulk). Equation (30) shows that after the diffusion coefficient is evaluated by Eq. (29), the solubility of drug in the membrane can be determined using the same set of data. Therefore, the diffusion coefficient and the solubility can be obtained simultaneously.

Wood and Mulski [64] have estimated the diffusion coefficient of water vapor in various polymer systems through use of Eq. (30) after experimentally determining the permeation rate and the solubility of water in each polymer. The disadvantages of this method are that it requires elimination of the boundary-layer diffusion resistance, and mechanically strong materials are needed for the membranes. For diffusants with small diffusion coefficients, the permeation time-lag experiment becomes impractically long.

Direct-release method.   The direct-release method, as proposed by Roseman and Higuchi [65], determines the diffusion coefficient directly from the release experiment, using a curve-fitting technique to fit the release equation to experimental data. For instance, the equation describing the radius of the receding boundary from a cylindrical matrix, assuming matrix-controlled kinetics and negligible boundary-layer diffusional resistance, can be derived as (to be discussed in the next section)

$$\frac{R^2(t)}{2} \ln \frac{R(t)}{R_0} + \frac{R_0{}^2 - R^2(t)}{4} = \frac{C_M D_m t}{C_T} \tag{31}$$

where R_0 is the radius of the cylinder, $R(t)$ is the radius of the receding boundary, C_M is the solubility of the drug in the polymer, C_T is the total loading dose of the drug in the polymer, and D_m is the diffusion coefficient of the drug in the homogeneous monolithic matrix. Provided that the radius of the receding boundary $R(t)$ can be measured as a function of time, the value of D can be deduced from Eq. (31) by a nonlinear least-squares algorithm.

The advantage of this method is that the diffusion coefficient can be determined simultaneously with the release experiment. The disadvantage is that the variability of the resulting diffusion coefficient for several replicates could be high, since the accuracy in measuring the receding boundary radius is, in general, low. The time that is necessary to perform the experiment is also long when the release rate is slow.

Sorption and desorption method from a stirred solution of finite volume.   If a polymeric membrane or matrix is suspended in a finite volume of solution, the concentration of solute in the solution falls as solute enters the polymer. If the solution is well stirred, the concentration in the solution depends only on time, and is determined essentially by the condition that the total amount of solute in the solution and in the polymer remains constant as diffusion proceeds. The experiment is relatively simple, since it is only necessary to measure the solute concentration in the bulk and deduce the diffusion coefficient mathematically. The solutions to the diffusion equation are discussed below.

Computation of Diffusion Coefficients Using Exact Solution by Crank

For a planar, cylindrical, or spherical device initially free of solute and immersed in a limited volume of solute solution, Crank and Park [28] developed the following solutions for the diffusion equation. This form expresses the total amount of solute uptake, M_t, by the polymer at time t as a fraction of M_∞, which is the corresponding quantity after infinite time:

Planar sheet.   The fraction of sorption of the solute to the planar sheet is given by Crank [66] as

$$\frac{M_t}{M_\infty} = 1 - \sum_{n=1}^{\infty} \frac{2\alpha(1 + \alpha)}{1 + \alpha + \alpha^2 q_n^2} \exp\left(-\frac{D_m q_n^2 t}{a^2}\right) \tag{32}$$

where the values of q_n are the nonzero positive roots of

$$\tan q_n = -\alpha q_n \tag{33}$$

and $\alpha = V_1/(KV_2)$, where a is the half-thickness of the sheet, K is the partition coefficient, and V_1 and V_2 are the volumes of the solution and polymer, respectively. Roots of Eq. (33) for various roots of α are given by Crank [66] and by Carslaw and Jaeger [67]. M_∞ of the sheet is given by

$$M_\infty = \frac{2a\alpha C_{b0}}{1 + \alpha} \tag{34}$$

where C_{b0} is the initial concentration of the solute in the bulk.

Cylinder. The cylinder is initially free of solute. The fraction of sorption of the solute to the cylinder can be expressed by the relation [66]

$$\frac{M_t}{M_\infty} = 1 - \sum_{n=1}^{\infty} \frac{4\alpha(1 + \alpha)}{4 + 4\alpha + \alpha^2 q_n^2} \exp\left(-\frac{D_m q_n^2 t}{a^2}\right) \tag{35}$$

where the values of q_n are the positive, nonzero roots of

$$\alpha q_n J_0(q_n) + 2J_1(q_n) = 0 \tag{36}$$

where J_0 and J_1 are the Bessel functions of the first kind of order 0 and 1, respectively [68]. Roots of Eq. (36) for various values of α are given by Crank [66] and by Carslaw and Jaeger [67]. The parameter α is expressed in terms of the final fractional uptake of the solute by the cylinder according to the expression

$$M_\infty = \frac{\pi a^2 \alpha C_{b0}}{1 + \alpha} \tag{37}$$

Sphere. The sphere is initially free of solute. The fractional uptake of the sphere is expressed by [66]

$$\frac{M_t}{M_\infty} = 1 - \sum_{n=1}^{\infty} \frac{6\alpha(\alpha + 1)\exp(-D_m q_n^2 t/a^2)}{9 + 9\alpha + q_n^2 \alpha^2} \tag{38}$$

where the values of q_n are the nonzero roots of

$$\tan q_n = \frac{3q_n}{3 + \alpha q_n^2} \tag{39}$$

The parameter α is expressed in terms of the final fractional uptake of the solute by the cylinder as given by

$$M_\infty = \frac{4\pi a^3 \alpha C_{b0}}{3(1 + \alpha)} \tag{40}$$

The solutions for small times and other conditions for the three geometries may be simplified, as summarized by Crank [66]. Equations (38), (39), and (40) indicate that the values of D_m can be deduced by a trial-and-error method from the measured fractional uptake data. It is noted that in order to perform the experiment more conveniently, it is important to design the value of α carefully. For instance, if the value is too large, the change of solute concentration in the bulk may be very difficult to determine.

The disadvantage of this method is that solutions to the diffusion equation are expressed in terms of infinite series, which may converge slowly during the computations. Also, the solubility of solute in the polymer cannot be determined simultaneously.

Computation of Diffusion Coefficients Using the Refined Integral Method

The refined integral method was proposed by Lee [69,70] to obtain the approximate analytical solutions for sorption from a constant, finite volume in planar, cylindrical, and spherical geometries. The solutions express the dimensionless time τ explicitly in terms of measurable external concentration of the solute. This method is generally applicable to sorption and desorption problems in planar, cylindrical, and spherical geometries. The method has been demonstrated successfully by the above investigator for the absorption of butane, NaCl, and dexamethasone by silica gel and resin.

For a solid sorbant with a uniform solute concentration KC_{b0}' which is in equilibrium with the corresponding solute concentration in the bulk C_{b0}' at $t < 0$, the effective volume ratio, λ, is defined as

Planar: $\quad \lambda = \dfrac{C_{b\infty} - C_{b0}'}{C_{b0} - C_{b\infty}} = \dfrac{V}{2KAa}$ (41)

Cylindrical: $\quad \lambda = \dfrac{V}{K\pi a^2 L}$ (42)

Spherical: $\quad \lambda = \dfrac{3V}{4K\pi a^3}$ (43)

where V is the total liquid volume, A is the area of each side of the planar sheet, a is the half-thickness of the planar sheet or the radius of the sphere or the cylinder, K is the partition coefficient, and L is the length of the cylinder. The solute concentrations in the bulk at $t < 0$, $t = 0$, and $t = \infty$ are expressed by C_{b0}', C_{b0}, and $C_{b\infty}$, respectively. For the case when the polymer is initially free of drug, the concentration C_{b0}' is equal to 0.

The solutions to the diffusion equation describing the concentration distribution, C, in the solid after exposing it to a constant, finite volume of well-stirred solution of concentration C_{b0} at time $t = 0$ are

Planar: $\quad \tau = \dfrac{3\lambda^2}{4}\left[\ln\theta_b + \dfrac{1}{2}\left(\dfrac{1}{\theta_b}\right)^2 - \dfrac{1}{2}\right] = \dfrac{Dt}{a^2}$ (44)

Cylindrical: $\quad \tau = \dfrac{3\lambda^3}{40}\left(\dfrac{1}{\theta_b} - 1\right)^3 + \dfrac{3(3\lambda + 5)\lambda^2}{160}\left(\dfrac{1}{\theta_b} - 1\right)^2$
$\qquad\qquad + \dfrac{3\lambda^3(5 - 3\lambda)}{80}\left(\dfrac{1}{\theta_b} - 1\right) - \dfrac{3\lambda^2(5 - 3\lambda)}{80}\ln\left(\dfrac{1}{\theta_b}\right)$ (45)

Spherical: $\quad \tau = -\dfrac{\zeta}{3} - \left(\dfrac{\lambda + 2}{3}\right)\ln\left(\dfrac{4\lambda + 4 - (\zeta - 2)^2}{4\lambda}\right)$
$\qquad\qquad + \dfrac{2}{3}(1 + \lambda)^{0.5}\ln\left(\dfrac{[2(1 + \lambda)^{0.5} + (\zeta - 2)][(1 + \lambda)^{0.5} + 1]}{[2(1 + \lambda)^{0.5} - (\zeta - 2)][(1 + \lambda)^{0.5} - 1]}\right)$ (46)

with

$$\theta_b = \frac{C_b - C_{b0}'}{C_{b0} - C_{b0}'} \qquad (47)$$

$$\zeta = 2 - 2\left[(\lambda + 1) - \left(\frac{\lambda(1 + \lambda)}{(1 + \lambda) - (M_t/M_\infty)}\right)\right]^{0.5} \qquad (48)$$

The dimensionless time, τ, is defined by

$$\tau = \frac{D_{ii}t}{a^2} \qquad (49)$$

In all the above geometries, the fraction of release, M_t/M_∞, can be expressed in terms of the concentrations as

$$\frac{M_t}{M_\infty} = (1 + \lambda)\frac{C_{b0} - C_b}{C_{b0} - C_{b0}'} \qquad (50)$$

which has been compared to the exact solution given by Carman and Haul [71] (see also Crank [66]). The results indicate that for small values of λ the approximate solutions almost completely coincide with the exact solutions (within 2% error). As the value of λ increases, deviations start to appear at large values of τ.

The advantages of this method are that both the diffusion coefficients and partition coefficients can be obtained from one experiment with a simple apparatus and a wide range of accessible experimental conditions. The geometry and mechanical strength of the samples are not restricted in the experiment. Moreover, the method can be applied to absorption or desorption data.

The disadvantages include the mathematical complexity of final working equations, especially for nonplanar geometries. The boundary-layer diffusion resistance needs to be eliminated by a certain degree of stirring. In order to obtain values of τ from C_b-versus-t data, it is necessary to determine λ from $C_{b\infty}$. The following exercise will elucidate how to apply the refined integral method to determine the diffusion coefficient from the measured concentration-versus-time profile.

The diffusion coefficients of some steroids in various polymers were determined experimentally and are given in Table 4.

Problem 2. Absorption experiments of dexamethasone in polyhydroxyethyl methacrylate [poly(HEMA)] disks are described below. Each absorption experiment was carried out in a rectangular silica ultraviolet cell with 1-cm optical path length. The cell was filled with 3 ml of dexamethasone solution of concentration 5.1×10^{-5} M prepared in a phosphate buffer. It was then placed into a spectrophotometer and stirred with a built-in magnetic stirrer. The temperature of the UV cell was controlled at 25°C. A poly(HEMA) disk, 0.12 cm thick, hanging on a thin stainless steel wire, was introduced into the upper half of the UV cell. The cell was then capped to prevent evaporation of solvent. The change of bulk concentration as the drug was absorbed by the poly(HEMA) disk was recorded continuously at 243 nm for dexamethasone and 265 nm for phenylbutane. The experimental conditions and the drug concentrations in the bulk, C_{b0}, as a function of time were measured as shown in the first two columns of Table 5. With this information, determine the diffusion coefficient of dexamethasone.

Solution. The values of τ at the corresponding values of C_b in Table 5 were computed by Eq. (44) and are listed in column 3. The fractions released, M_t/M_∞, as a function of

Table 4 Diffusion Coefficients of Selected Solutes in Polymers

Solute	Polymer	Diffusion coefficient (cm^2/s)	Ref.
Acetophenone[a]	Polyethylene	3.55×10^{-8}	148
p-Aminoacetophenon	Silicone rubber	2.44×10^{-6}	149
Androstenedione	Silicone rubber	14.8×10^{-7}	150
Androstenedione	Poly(etherurethane)	9.0×10^{-9}	151
Androstenedione	Poly(ethylene vinyl acetate)	5.5×10^{-9}	151
Adrenosterone	Poly(etherurethane)	8.9×10^{-9}	151
Adrenosterone	Poly(ethylene vinyl acetate)	5.7×10^{-9}	151
Benzaldehyde[a]	Polyethylene	3.39×10^{-8}	148
Benzoic acid[a]	Polyethylene	5.29×10^{-10}	148
Chlormadione acetate	Silicone rubber	3.03×10^{-7}	152
Corticosterone	Poly(etherurethane)	6.3×10^{-9}	
Corticosterone	Poly(ethylene vinyl acetate)	5.7×10^{-9}	151
Cortisone	Poly(etherurethane)	5.8×10^{-9}	151
Cortisone	Poly(ethylene vinyl acetate)	2.3×10^{-9}	151
11-Dehydrocorticosterone	Poly(etherurethane)	5.2×10^{-9}	151
11-Dehydrocorticosterone	Poly(ethylene vinyl acetate)	3.6×10^{-9}	151
Delmadinone acetate	Silicone rubber	0.38×10^{-7}	153
Deoxycorticosterone acetate	Silicone rubber	4.94×10^{-7}	154
11-Deoxycorticosterone	Poly(etherurethane)	5.4×10^{-9}	151
11-Deoxycorticosterone	Poly(ethylene vinyl acetate)	4.9×10^{-9}	151
11-Deoxy-17-hydroxycorticosterone	Poly(etherurethane)	5.6×10^{-9}	151
11-Deoxy-17-hydroxycorticosterone	Poly(ethylene vinyl acetate)	4.7×10^{-9}	151
Estriol[b]	Polyurethane ether	2×10^{-9}	155
Estrone	Silicone rubber	2.4×10^{-7}	150
Ethyl-p-aminobenzoate	Silicone rubber	2.67×10^{-6}	149
		1.78×10^{-6}	156
Ethylnodiol diacetate	Silicone rubber	3.79×10^{-7}	157
Fluphenazine	Polymethylmethacrylate	1.74×10^{-17}	158
Fluphenazine	Polyvinyl acetate	1.05×10^{-12}	158
Fluphenazine enanthate	Polymethylmethacrylate	1.12×10^{-17}	158
Fluphenazine enanthate	Polyvinylacetate	1.82×10^{-12}	158
11-Hydroxyandrostenedione	Poly(etherurethane)	8.1×10^{-9}	151
11-Hydroxyandrostenedione	Poly(ethylene vinyl acetate)	3.8×10^{-9}	151
Hydrocortisone	Silicone rubber	4.5×10^{-7}	94
Hydrocortisone	Polycarpolactone	1.58×10^{-10}	159
Hydrocortisone	Ethylene-vinyl acetate	1.18×10^{-11}	159

Table 4, cont'd.

Solute	Polymer	Diffusion coefficient (cm^2/s)	Ref.
Hydrocortisone	Polyvinyl acetate terpolymer	4.31×10^{-12}	159
Hydrocortisone	Poly(etherurethane)	4.8×10^{-9}	151
Hydrocortisone	Poly(ethylene vinyl acetate)	2.8×10^{-9}	151
Hydroquinone	Polymethylmethacrylate	5.75×10^{-15}	158
17α-hydroxyprogesterone	Silicone rubber	5.65×10^{-7}	95
Medroxyprogesterone acetate	Silicone rubber	4.17×10^{-7}	95
4-Methylacetophenone[a]	Polyethylene	1.79×10^{-8}	
4-Methylbenzaldehyde[a]	Polyethylene	1.37×10^{-8}	
6α-Methyl-11β-hydroxyprogesterone	Silicone rubber	2.84×10^{-7}	95
4-Nitro-aniline	Polymethylmethacrylate	3.02×10^{-15}	158
4-Nitro-aniline	Polyvinyl acetate	3.02×10^{-11}	158
Norprogesterone	Silicone rubber	18.5×10^{-7}	159
Procaine	Polymethylmethacrylate	1.35×10^{-15}	158
4-Methylacetophenone[a]	Polyethylene	1.79×10^{-8}	
Procaine	Polyvinyl acetate	1.45×10^{-11}	158
Progesterone	Silicone rubber	5.78×10^{-7}	95
		6.4×10^{-7}	150
Promethazine	Polymethylmethacrylate	1.41×10^{-17}	158
Promethazine	Polyvinyl acetate	1.45×10^{-12}	158
Pyrimethamine	Silicone rubber	1.10×10^{-10}	159
Salicylic acid	Polymethylmethacrylate	9.55×10^{-15}	158
Sallicylic acid	Polyvinyl acetate	4.37×10^{-11}	158
Trinitrophenol	Polymethylmethacrylate	3.55×10^{-16}	158
Trinitrophenol	Polyvinyl acetate	7.59×10^{-12}	158
Steroid-type compound	Silicone rubber (polydimethylsiloxane)	6×10^{-7}	101
	Low-density polyethylene	1×10^{-9}	101
	Poly(co-ethylene-vinyl acetate) 9%	4×10^{-9}	101
	Poly(co-ethylene-vinyl acetate) 18%	3×10^{-9}	101
	Poly(co-ethylene-vinyl acetate) 40%	5×10^{-9}	101
	Poly(co-polytetra methylene ether glycol-diphenylmethane di-isocyanate	9×10^{-10}	101

Note: Temperature is 37°C unless stated otherwise.
[a] At 40°C.
[b] At 30°C.
Source: Part of this table was compiled by Roseman and Cardarelli [92].

68 Kuu et al.

Table 5 Sorption of Dexamethasone into a Poly(HEMA) Disk

pH = 7.0
Temperature = 37°C
Partition coefficient = 51.4
λ = 0.621
Disk diameter = 1.143 cm
Disk thickness = 0.12 cm
C_{b0} = 5.1 × 10^{-5}; M = 0.020015 mg/ml
V = 3.93 ml
C_{b0}' = 0.0
$C_{b\infty}$ = 0.007668 mg/ml

t (h)	C_b (mg/ml)	τ	M/M_∞
0.0	0.02002	0.000	0.000
0.5	0.01427	0.042	0.189
1.0	0.0138	0.052	0.284
1.5	0.01267	0.084	0.347
2.0	0.01188	0.115	0.395
3.0	0.01073	0.178	0.463
4.0	0.00981	0.251	0.521
5.0	0.00921	0.314	0.563
6.0	0.00858	0.398	0.600
7.0	0.00819	0.461	0.637
8.0	0.00775	0.545	0.658
9.0	0.00757	0.586	0.684
10.0	0.00723	0.670	0.705
11.0	0.007	0.733	0.726
12.0	0.00674	0.817	0.747
13.0	0.00659	0.869	0.758
14.0	0.00642	0.932	0.774
15.0	0.00627	0.995	0.784
16.0	0.0061	1.068	0.789
17.0	0.00597	1.131	0.805
18.0	0.00583	1.204	0.816
19.0	0.00571	1.267	0.826
20.0	0.00561	1.330	0.837
21.0	0.00552	1.382	0.847
22.0	0.00547	1.414	0.853

time were computed by Eq. (50) and are listed in the last column of Table 5. As shown in Fig. 6, the plot of τ against time is linear in the range of M_t/M_∞ from 0 to 0.8. The values of τ in this range were then performed by a linear least-squares routine and the slope was determined to be 0.0678 1/h ($D_m t/a^2$), from which the value of D_m was deduced as 0.678 × 10^{-8} cm^2/s.

Other Methods for Determining the Diffusion Coefficients

In addition to the foregoing methods for determining the solute diffusion coefficient in liquids and polymers, other methods are also available for some specific applications.

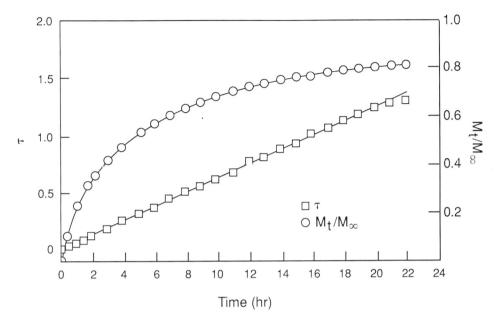

Figure 6 Plot of τ versus time for the sorption experiment of dexamethasone. Key: $\square$, τ; $\bigcirc$, M_t/M_∞.

Berner and Kivelson [72] presented a paramagnetically enhanced relaxation nuclear magnetic resonance (PER-NMR) technique for measuring the diffusion coefficients of paramagnetic solutes in liquids. The paramagnetic solute gives rise to enhanced relaxation of the nuclear spins of the solvent and the enhancement is found to be linear versus solute concentration. The method is particularly useful when the diffusion coefficient varies with position across the membrane and bilayer.

Muramatsu and Minton [73] elucidated the boundary-layer spreading method for rapid determination of diffusion coefficients. In this method, a layer of solvent is deposited onto a solution of an optically absorbing solute. The spreading of the boundary thus formed is monitored as a function of time using an automated absorbance scanning device. The method is particularly useful for biological macromolecules.

Convath et al. [74] developed a radiotracer method to determine the diffusion coefficient of a solute in polymeric gels. The method is based on the use of a multichannel radioactivity counter to obtain the in-situ concentration profile of the diffusing molecule in the gel. The method was applied successfully to the diffusion of testosterone in a silicic gel by these researchers.

Recently, Thomas et al. [75] designed a fiber optic probe and used it to investigate the mass transport of proteins through a polyacrylamide matrix. The basic component of the probe is an optical fiber, employing the evanescent as an excitation source for molecules near its surface. The optical fiber was coated with a very thin film of polyacrylamide, and the rate of diffusion through this matrix for proteins tagged with fluorescein isothiocyanate (FITC) was determined. In addition to the diffusion coefficient, the effect of cross-linking of the matrix on the mass transfer of FITC-albumin was also determined by this method.

MEMBRANE AND MATRIX PARAMETERS

A controlled-release device in general consists of two components, the drug and the matrix (or membrane). The physical properties of the drug, such as the solubility, partition coefficient, and diffusion coefficient, would, at least in part, affect the release rate. These properties, however, are usually fixed once a drug candidate is decided. On the other hand, the membrane and matrix parameters of the device, including the surface area, geometry, and thickness, play important roles in the overall release kinetics. These parameters can, in general, be manipulated to regulate the release rate. In order to determine a quantitative assessment of the effect of these parameters, the release kinetics of each geometry will first be derived based on prescribed conditions and then discussed.

Membranes are used primarily in reservoir devices, in which a concentrated drug solution is placed on one side of the membrane, whereas the other side of the membrane contacts with a recipient fluid. The release kinetics of the reservoir system can be closely simulated by the two-cell permeation apparatus, as depicted by Fig. 5. Thus, it is clear that the release kinetics of a reservoir device is closely related to the transport kinetics of a membrane in the two-cell permeation apparatus. For a given membrane, with known material and permeability, the only critical parameters that will affect drug release are the thickness of the membrane and the conditions in the recipient environment. For this reason, the release kinetics of matrix systems is, in general, more complex than that of the membrane systems. The investigations of the release kinetics in this section will focus on the matrix systems, whereas that of the membrane systems will be discussed in detail later.

Amount of Drug Loading Greater Than Solubility

In monolithic devices, the drug to be released is dispersed uniformly throughout the polymeric matrices. The coordinate systems to describe the release mechanisms for planar, cylindrical, and spherical devices are depicted in Fig. 7. For steady-state or pseudo-steady-state diffusion, the conservation of mass in the diffusion region for various geometries can be expressed by Fick's first law with appropriate boundary conditions.

In order to derive the release kinetics, the following assumptions must be established:

The drug-loading dose, C_T, in the matrix is much greater than the solubility in the matrix C_M.

The matrix-to-bulk partition coefficient, C_b^*/C_b, of the drug is equal to a constant K, where C_b^* is the concentration of the drug in the matrix of membrane adjacent to the bulk.

Drug permeation through the polymeric matrix or membrane is governed solely by Fickian diffusion.

The diffusion coefficient of the drug in the matrix/membrane as well as in the solvent is constant.

There is no interaction between the drug molecules and the matrix/membrane, e.g., chemical reaction or ion exchange.

The concentration of drug in the bulk is constant, equal to C_b in the entire period of release.

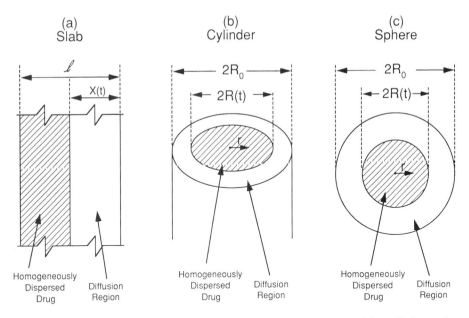

Figure 7 Coordinate systems to study drug release kinetics from the slab, cylinder, and sphere geometries of the monolithic devices: (a) slab; (b) cylinder; (c) sphere.

Pseudo-Steady-State Approximation

Planar geometry. The release of drug from a homogeneous monolithic system was first proposed by Higuchi [76], who used a pseudo-steady-state approximation approach to simplify a complex moving-boundary problem, with the detailed derivations presented below. The diffusion equation can be expressed by

$$Q_t = -AD_m \frac{dC}{dx} \tag{51}$$

where Q_t is the rate of diffusion; A is the cross-sectional area available for diffusion, perpendicular to the direction of diffusion; D_m is the diffusion coefficient of drug molecules in the matrix, which is equivalent to D_e used in Eqs. (7)–(19); C is the concentration of the drug in the polymer; and x is the distance measured from the solvent-matrix interface. The following boundary conditions are necessary to solve Eq. (51):

$$C = C_b K \qquad \text{at } x = 0 \tag{52}$$

and

$$C = C_M \qquad \text{at } x = X(t) \tag{53}$$

where C_b is the drug concentration in the bulk, which is constant over the period of release; C_M is the drug solubility in the matrix; and K is the matrix-to-bulk partition coefficient, equal to C_b^*/C_b.

Equation (53) indicates that the receding boundary length X(t) is a function of time. In order to integrate Eq. (51), it is necessary to know X(t). Higuchi [76] assumed that the diffusion rate is constant across the entire diffusion region; e.g., the concentration profile in the diffusion region is linear at any time for the planar geometry. In other words, although diffusion rate is a function of time, it is essentially independent of the distance x at any given time. This approach is the well-known pseudo-steady-state model.

Integration of Eq. (51), using Eqs. (52) and (53), gives

$$Q_t = - \frac{AD_m(C_M - C_bK)}{X(t)} \tag{54}$$

In Eq. (54), X(t) is not yet known. In order to obtain the solution for Q_t, the following mass balance equation in the diffusion region is applied:

$$\frac{dM_t}{dt} = \frac{d}{dt}\{[C_T - \frac{1}{2}(C_M + C_bK)]AX(t)\} \tag{55}$$

where M_t is the amount of drug released at time t and C_T is the drug-loading dose in the matrix, in mg/ml. Note that since dM_t/dt is equal to Q_t, this allows one to equate the right-hand sides of Eqs. (54) and (55), followed by simplification, to give

$$X(t)\, d[X(t)] = \frac{D_m(C_M - C_bK)}{C_T - \frac{1}{2}(C_M + C_0K)} \tag{56}$$

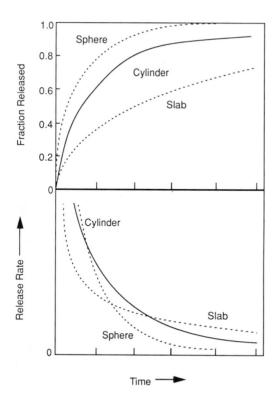

Figure 8 Theoretical release curves from monolithic device (dissolved solute) of different geometries. The solid line represents the complete profile. (From Ref. 49.)

The length of the receding layer can then be obtained by integrating Eq. (56), as

$$X(t) = \left(\frac{2D_m(C_M - C_bK)t}{C_T - \frac{1}{2}(C_M + C_bK)}\right)^{1/2} \tag{57}$$

Substituting Eq. (57) for X(t) in Eq. (54), followed by rearrangement, gives the release rate at time t:

$$\frac{dM_t}{dt} = Q_t = A\left(\frac{D_m(C_M - C_bK)(2C_T - C_M - C_bK)}{4t}\right)^{1/2} \tag{58}$$

The cumulative amount of release is then obtained by integrating Eq. (58):

$$M_t = A[D_m(C_M - C_bK)(2C_T - C_M - C_bK)t]^{1/2} \tag{59}$$

As seen from Eq. (58), the release rate of a device in the slab is inversely proportional to the square root of time. Equations (55), (56), and (57) indicate that both the cumulative amount of drug release and the receding boundary length are proportional to the square root of time, whereas the release rate is inversely proportional to the square root of time. These equations are plotted in Fig. 8. For the case of the slab, the effect of loading dose C_T on the release rate is plotted in Fig. 9, using Eq. (58). Depicted in Fig. 10 is the release profile of chloramphenicol from a slab device [77], which is an excellent example with which to verify the Higuchi model. As seen, the plot of cumulative mass against $t^{1/2}$ gives a straight line until the drug is depleted from the device.

Cylindrical geometry. Using a similar approach, Fick's first law, when applied to the cylindrical geometry, becomes [65]

$$Q_t = -2\pi rLD_m\frac{dC}{dr} \tag{60}$$

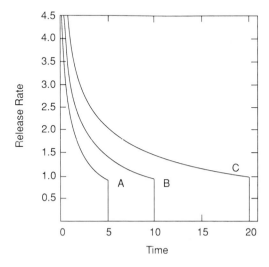

Figure 9 Release rate profile of agent from a slab containing different loading doses of dispersed solid agent ($\ell^2/8D_m = 1$, $2C_TC_M/\ell = 1$). Key: A, $C_T/C_M = 5$; B, $C_T/C_M = 10$; C, $C_T/C_M = 20$. (From Ref. 49.)

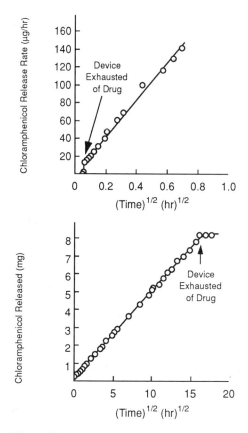

Figure 10 Release rate and cumulative drug
release profiles for chloramphenicol from a slab
device using the Higuchi model. (From Ref.
49.)

subjected to the following boundary conditions:

$$C = C_b K \qquad \text{at } r = R_0 \tag{61}$$

and

$$C = C_M \qquad \text{at } r = R(t) \tag{62}$$

Integration of Eq. (60), using Eqs. (61) and (62), gives [65]

$$\frac{dM_t}{dt} = Q_t = \frac{2\pi L D_m (C_M - C_b K)}{\ln[R_0/R(t)]} \tag{63}$$

Note that with the pseudo-steady-state assumption, as Eq. (63) shows, the concentration
profiles are no longer linear with respect to the radius. This phenomenon is somewhat
different from the slab device.

The original mass of drug in the diffusion region is $\pi L C_T [R_0^2 - R^2(t)]$. The conservation of mass becomes more difficult than the planar case unless one assumes that
$C_T \gg C_M$, and thus

$$\frac{dM_t}{dt} = -2\pi L C_T R(t) \frac{dR(t)}{dt} \tag{64}$$

Similar to the derivation of Eq. (56), equating the right-hand side of Eqs. (63) and (64), followed by simplification, yields

$$\{\ln(R_0) - \ln[R(t)]\}\, R(t)\, dR(t) = -\frac{D_m(C_M - C_bK)\, dt}{C_T} \tag{65}$$

In order to integrate Eq. (65), the following relationship should be used:

$$d[y^2 \ln(y)] = 2y \ln(y)\, dy + y\, dy \tag{66}$$

Thus the integration of Eq. (65) gives rise to the following implicit equation for R(t) as a function of t:

$$\frac{R^2(t)}{2} \ln \frac{R(t)}{R_0} + \frac{1}{4}[R_0{}^2 - R^2(t)] = \frac{D_m(C_M - C_bK)t}{C_T} \tag{67}$$

Equation (63) establishes the relationship between dM_t/dt and R(t), whereas Eq. (67) is an implicit solution for R(t) in terms of t. Thus, in order to obtain the release rate dM_t/dt at a given time, it would be necessary to solve Eq. (67) for R(t) for a given time using a trial-and-error method.

The experimental evidence of the pseudo-steady-state model for a cylindrical device is presented in Fig. 11 [65]. The photographs show that the release of medroxyprogesterone acetate from a silicone rubber cylinder forms a well-defined, time-dependent depletion zone.

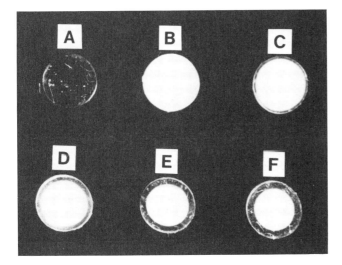

Figure 11 Photograph of cross-sectioned slices of medroxyprogesterone acetate-silicone rubber cylinders depicting zones of depletion (clear region) as a function of time of dissolution. Key: A, placebo; B, drug-filled, initial; C, 1 week; D, 2 weeks; E, 3 weeks; F, 4 weeks. (Reproduced with permission of the copyright owner, the American Pharmaceutical Association [65].)

Spherical geometry. Using a similar approach, Fick's first law, when applied to the spherical geometry, becomes [49,78]

$$Q_t = -4\pi r^2 D_m \frac{dC}{dr} \tag{68}$$

subjected to the following boundary conditions:

$$C = C_b K \qquad \text{at } r = R_0 \tag{69}$$

and

$$C = C_M \qquad \text{at } r = R(t) \tag{70}$$

Integration of Eq. (68), using Eqs. (69) and (70), and assuming that the diffusion rate is constant across the entire diffusion region at any time (pseudo-steady-state), gives

$$Q_t = -\frac{4\pi D_m (C_M - C_b K)}{[1/R(t)] - (1/R_0)} \tag{71}$$

Similar to the cylindrical geometry, as Eq. (71) shows, the concentration profiles for the pseudo-steady-state assumption are no longer linear with respect to the radius.

The amount of drug released, for $C_T \gg C_M$, is equal to $4/3\pi C_T[R_0^3 - R^3(t)]$. The conservation of mass in the diffusion region for the case of $C_T \gg C_M$ gives

$$\frac{dM_t}{dt} = -4\pi C_T R^2(t) \frac{dR(t)}{dt} \tag{72}$$

Similar to the derivation of Eq. (56), equating the right-hand sides of Eqs. (71) and (72), followed by simplification, yields

$$\left[R(t) - \frac{R^2(t)}{R_0} \right] dR(t) = \frac{D_m (C_M - C_b K) \, dt}{C_T} \tag{73}$$

Integration of Eq. (73) yields the implicit expression for R(t) as a function of time,

$$t = \frac{C_T}{D_m (C_M - C_b K)} \left(\frac{R^3(t) - R_0^3}{3R_0} - \frac{R^2(t) - R_0^2}{2} \right) \tag{74}$$

The rates and fractions released for the three geometries are plotted against time, t, as shown in Fig. 8. Figure 8 shows that none of these geometries gives zero-order release for the following reasons. For slab geometry, the area remains constant, yet the distance of receding boundary increases with time. The resulting release rate decreases with time. For the sphere and cylinder the situation is worse, since the area decreases with time, whereas the distance of receding boundary increases with time. These two factors magnify the decline of the release rate.

Special geometries that give zero-order or near-zero-order release. From the analyses described above, it is apparent that one potential disadvantage associated with diffusion-controlled matrix devices is their inability to achieve zero-order release kinetics. For the case of planar geometry, the release rate is inversely proportional to the distance of the receding front, which increases with time as the release rate decreases. Since the release rate is directly proportional to the area, the general approach to achieve zero-order release is to increase the cross-sectional area of the drug-receding front when drug is depleted, so that the reduction of the release rate due to the increase in diffusion distance can be compensated. One approach to achieving this goal is to release the drug from an orifice of the matrix. Two special geometries are discussed in this section, the hemispherical and the pie-shaped cylindrical matrices.

Hemispherical matrix releasing from a small cavity: This approach was first proposed by Rhine et al. [78] and Hsieh et al. [79]. The conceptual design of this device is depicted in Fig. 12. The hemispherical device is coated with an impermeable layer, except for a small cavity cut into the center of the flat surface. Assuming that $C_b = 0$, the diffusion equation for this system is the same as Eq. (68), except that the coefficient "-4" in Eq. (68) is replaced by "2." The pertinent boundary conditions for solving this diffusion equation become

$$C = 0 \qquad \text{at } r = a_i \tag{75}$$

$$C = C_M \qquad \text{at } r = R(t) \tag{76}$$

The solutions for release rate, cumulative amount of release, and the distance of receding front can be obtained following a similar procedure described earlier (pseudo-steady-state approach), as given by

$$C(r, t) = C_M \left(\frac{R(t)}{R(t) - a_i} \right) \left(1 - \frac{a_i}{r} \right) \tag{77}$$

The rate of release can be expressed by

$$\frac{dM_t}{dt} = 2\pi C_M D_m a_i \left(\frac{R(t)}{R(t) - a_i} \right) \tag{78}$$

When $R(t) >> a_i$, Eq. (78) is reduced to

$$\frac{dM_t}{dt} = 2\pi C_M D_m a_i \tag{79}$$

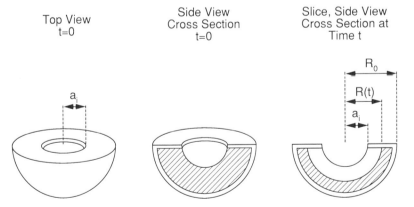

Top View
t=0

Side View
Cross Section
t=0

Slice, Side View
Cross Section at
Time t

Figure 12 Schematic diagram of drug release from the cavity of a hemispherical device: a_i, radius of the cavity; $R(t)$, radius of receding boundary; R_o, outer radius of the hemisphere. Outermost surface consists of insulated laminate. (Reproduced with permission of the copyright owner, the American Pharmaceutical Association [79].)

Equation (79) shows that the release rate is independent of the time, t, and therefore a zero-order release kinetics is attained. The cumulative amount of drug released is obtained by integrating Eq. (79) and gives a straight line passing through the origin with a slope of $2\pi C_M D_m a_i$.

The release kinetics for any value of a_i can also be derived by the following procedure. The cumulative amount of drug released at time t in the diffusion region, assuming that $C_T >> C_M$, is expressed by

$$M_t = \frac{2}{3} C_T \pi [R_0^3 - R^3(t) - a_i^3]$$ (80)

Following the same procedure described earlier, the implicit solution for R(t) can be derived as

$$\left(\frac{D_m C_M}{C_T a_i^2}\right) t = \frac{1}{3}\left(\frac{R(t)}{a_i}\right)^3 - \frac{1}{2}\left(\frac{R(t)}{a_i}\right)^2 + \frac{1}{6}$$ (81)

Plotted in Fig. 13 is the release profile of serum albumin in this device [78]. The solid curve represents the theoretical values of M_t, as predicted by Eq. (80), and the dot symbols are the experimental data. As seen, the device provided a zero-order release for over 50 days.

Pie-shaped cylindrical matrix releasing from an orifice: This geometry was first proposed by Brooke and Washkuhn [80] and Lipper and Higuchi [81]. The schematic diagram of this device is depicted in Fig. 14. Similar to the hemispherical device described earlier, the pie-shaped cylindrical device is coated with an impermeable layer, except for a small cavity cut into the center of the pie. Assuming that the drug concentration in the cavity is C_b and the partition coefficient is equal to unity, the diffusion equation for this system is the same as Eq. (60), except that the coefficient "-2π" in Eq. (60) is replaced by "2θ," where θ is the half-angle of the pie. The boundary conditions for solving this diffusion equation become

$$C = C_b \qquad \text{at } r = a_i$$ (82)

$$C = C_M \qquad \text{at } r = R(t)$$ (83)

The rate of release can be derived as

$$\frac{dM_t}{dt} = \frac{2\theta L D_m C_M}{(h/a_i) + \ln[R(t)/a_i]}$$ (84)

where a_i is the radius of the cavity and h is the boundary-layer thickness. Unlike Eq. (78) for the hemispherical device, Eq. (84) may not be further simplifed to give an equation to demonstrate the near-zero-order release kinetics.

The cumulative amount of drug released can be obtained by integrating Eq. (84) along with the mass balance equation to give

$$M_t = [R^2(t) - a_i^2][L\theta C_T - L\theta C_M + \frac{L\theta C_M}{2\{(h/a_i) + \ln[R(t)/a_i]\}}$$
$$- \frac{L\theta C_M a_i^2}{(h/a_i) + \ln[R(t)/a_i]} \ln \frac{R(t)}{a_i}$$ (85)

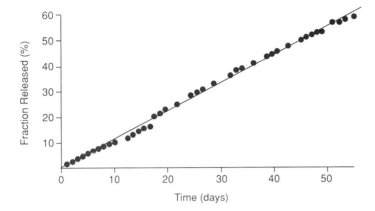

Figure 13 Cumulative release of bovine serum albumin versus time from the cavity of the hemispherical device. The matrix was made of ethylene-vinylacetate copolymer and bovine serum albumin. Standard error of the mean of the cumulative release at each time point was within 12%. (Reproduced with permission of the copyright owner, the American Pharmaceutical Association [79].)

The implicit expression for the radius of receding front, R(t), as a function of t can be derived as

$$t = \frac{[(h/a_i) - (1/2)][R^2(t) - a_i^2]C_T + R^2(t)C_T \ln[R(t)/a_i]}{2D_m C_M} \tag{86}$$

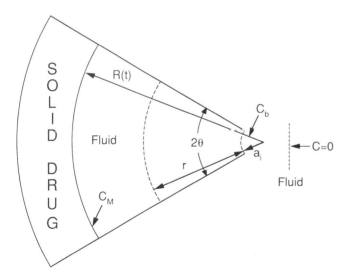

Figure 14 Schematic diagram to study the release kinetics of drugs from the cavity of a pie-shaped cylindrical device. (Reproduced with permission of the copyright owner, the American Pharmaceutical Association [81].)

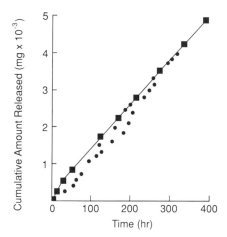

Figure 15 Cumulative mass of stearic acid
release from the cavity of a pie-shaped cylindrical device versus time. Key: ■, by
Eqs. (75) and (76); ●, experimental data.
(Reproduced with permission of the copyright owner, the American Pharmaceutical
Association [81]).

The theoretical and experimental profiles for stearic acid release in this device are plotted
in Fig. 15 [80]. The solid curve represents the theoretical values of M_t, computed by Eq.
(85), and are compared with the experimental values in closed-dot symbols. As seen from
this graph, a zero-order release is sustained for over 400 h.

The approach of releasing the drug from a small cavity of the matrix can be extended
to other geometries which are able to produce an increasing area to compensate for the
increasing diffusion distance. For instance, Kuu and Yalkowsky [82] modified a planar
matrix device by placing an impermeable membrane containing multiple holes at the
surface of the device so that the drug in the matrix is released only from the holes. After
this modification, near-zero-order release kinetics was achieved.

Exact Analysis

The pseudo-steady-state assumption greatly simplified the complexity of the moving-
boundary diffusion problem and thereby avoided solving complicated differential equations. For cases where the pseudo-steady-state assumption is no longer valid, a more
rigorous derivation is necessary. The exact analysis was proposed by Paul and McSpadden
[83], and the schematic concentration profiles of the dissolved and undissolved drug in
the matrix and the boundary layer are illustrated in Fig. 16. Fick's second law of diffusion
was used to establish the diffusion/dissolution problem. Their results permit evaluation
of the error associated with the pseudo-steady-state approximation.

The assumptions for this analysis are summarized below:

The concentration of drug in the bulk is constant, equal to C_b in the entire period of
release.

The drug-loading dose in the matrix is much greater than C_M.

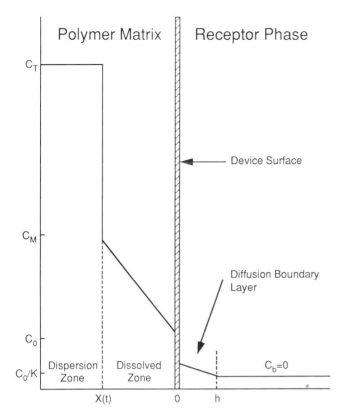

Figure 16 Concentration profile of drug in the polymeric matrix and in the receptor compartment based on the pseudo-steady-state assumption. (Reproduced with permission of the copyright owner, the American Pharmaceutical Association [89].)

Drug permeation through the polymeric matrix or membrane is governed solely by Fickian diffusion.

The diffusion coefficient of the drugs in the polymers, as well as in the solvent, is constant.

There is no interaction, e.g., chemical reaction or ion exchange, between the drug molecules and the matrix/membrane.

The conservation of mass is expressed by Fick's second law of diffusion:

$$\frac{\partial C}{\partial t} = D_m \frac{\partial^2 C}{\partial x^2} \tag{87}$$

The appropriate boundary conditions for solving Eq. (87) in the diffusion region $0 < x < X(t)$, assuming that the boundary-layer diffusional resistance is negligible, are given by

$$C = 0 \qquad \text{at } x = 0 \tag{88}$$

$$C = C_M \qquad \text{at } x = X(t) \tag{89}$$

and

$$(C_T - C_M) \frac{dX(t)}{dt} = D_m \frac{\partial C}{\partial x} \qquad \text{at } x = X(t) \tag{90}$$

The cumulative amount of release can be derived as

$$M_t = \frac{2AC_M}{erf(f)} \left(\frac{D_m t}{\pi} \right)^{1/2} \tag{91}$$

where $erf(f)$ is the value of an error function with argument f, defined by [68]

$$erf(f) = \frac{2}{\pi^{1/2}} \int_0^f e^{-\eta^2} d\eta \tag{92}$$

and f can be obtained by

$$\sqrt{\pi} \, f \exp(f^2) \, erf(f) = \frac{C_M}{C_T - C_M} \tag{93}$$

For given values of C_M and C_T, the corresponding value of f can be obtained from Eq. (93) by a trial-and-error method.

As indicated by the investigators, the maximum improvement in predicting M_t by the foregoing exact analysis relative to the pseudo-steady-state assumption is about 11.3%.

The solution for the concentration profile can be expressed by

$$C = C_M \frac{erf[x/(2(D_m t)^{1/2})]}{erf(f)} \tag{94}$$

Equation (94) gives a nonlinear concentration profile, in contrast to the linear form obtained using the pseudo-steady-state assumption.

The asymptotic properties of Eq. (91) can be examined for the following two cases. When $C_T \rightarrow C_M$, Eq. (93) indicates that f becomes very large and $erf(f)$ approaches unity. The equation becomes

$$M_t = 2AC_M \left(\frac{D_m t}{\pi} \right)^{1/2} \tag{95}$$

which is the same as the equation derived for the loading dose less than the solubility. On the other hand, for the case of $C_T \gg C_M$, Eq. (93) shows that f is small, and a series expansion permits Eq. (93) to give $erf(f)$ explicitly as follows:

$$erf(f) = \left(\frac{2C_M}{\pi(C_T - C_M)} \right)^{1/2} \tag{96}$$

which can be combined with Eq. (91) to give

$$M_t \approx A[2D_m C_M (C_T - C_M) t]^{1/2} \tag{97}$$

It can be seen that when the drug-loading dose is much greater than the solubility, Eq. (97) approaches the pseudo-steady-state release kinetics, that is, Eq. (59), in which C_M can be omitted using the condition $C_T \gg C_M$.

Problem 3. The solubility of the red organic dye Sudan III in water at 25°C is equal to 0.274 g/liter, and its diffusion coefficient in silicone rubber is 2.68×10^{-6} cm²/s. Use the exact analysis to determine the M_t-versus-t profiles for the following conditions:

C_T/C_M = 1.1, 2.6, 5.6, and 10.9

Also compare the profiles to those using the pseudo-steady-state assumption.

Solution. For each value of C_T/C_M, the value of f is obtained by trial and error using Eq. (93). The resulting values of f and erf(f) are listed below:

C_T/C_M	1.1	2.6	5.6	10.9
f	1.258	0.51	0.32	0.22
erf(f)	0.9228	0.5292	0.3491	0.2443

The predicted and experimental values of M_t as a function of time, using Eqs. (91) and (97) for the pseudo-steady-state assumption, are plotted in Fig. 17. As seen from the comparison, the error of the pseudo-steady-state assumption decreases with increasing values of C_T/C_M.

Amount of Drug Loading Lower Than Solubility

Fick's first law is useful only when the diffusion rate is constant and the steady-state and linear concentration gradients are maintained in the diffusion region. For the cases when the amount of drug loading is smaller than the drug solubility, the diffusion rate is not necessarily constant. Fick's second law should be used to describe the conservation of mass in the diffusion region. Because of the complexity of the mathematical manipulations, only planar geometry with constant diffusion coefficients will be discussed.

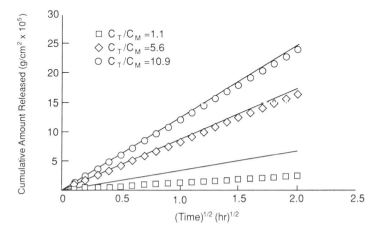

Figure 17 Comparison of the exact solution and the pseudo-steady-state approximation for the release of the red organic dye Sudan III from a planar silicone rubber monolithic device. The parameter C_T/C_M indicates the ratio of the solute loading dose to the solute solubility in the matrix. (From Ref. 83.)

For a membrane or matrix with thickness 2a, the region $-a < x < a$ being initially at zero concentration, the diffusion equation is

$$\frac{\partial C}{\partial t} = D_m \frac{\partial^2 C}{\partial x^2} \tag{98}$$

The initial condition and the two boundary conditions for solving Eq. (98) are given by

$$C = 0 \qquad -a < x < a, t = 0 \tag{99}$$

$$C = 0 \qquad \text{at } x = -a \tag{100}$$

and

$$C = 0 \qquad \text{at } x = a \tag{101}$$

Equations (100) and (101) indicate that the surface concentration is equal to zero at any time. The solutions to Eqs. (98)–(101) are given by Crank [66]. If M_t denotes the total amount of drug which was released from the sheet at time t, and M_∞ is the corresponding amount of drug after infinite time, then the fraction of drug release, M_t/M_∞, can be expressed by [66,84]

$$\frac{M_t}{M_\infty} = 1 - \sum_{n=0}^{\infty} \frac{8}{(2n+1)^2 \pi^2} \exp\left(\frac{-D_m (2n+1)^2 \pi^2 t}{4a^2}\right) \tag{102}$$

The corresponding solution which is useful for small times can be expressed by [66,84]

$$\frac{M_t}{M_\infty} = \left(\frac{D_m t}{a^2}\right)^{1/2} \left(\pi^{-1/2} + 2\sum_{n=1}^{\infty} (-1)^n \text{ ierfc } \frac{2na}{(D_m t)^{1/2}}\right) \tag{103}$$

where ierfc is the integrated complimentary function defined by [67–68]

$$\text{ierfc}(x) = \pi^{-1/2} \exp(-x^2) - x\,\text{erfc}(x) \tag{104}$$

In Eq. (104), erfc is the complementary error function, which relates to the error function erf by the following equation:

$$\text{erfc}(x) = 1 - \text{erf}(x) \tag{105}$$

For the early time approximation, which holds over the initial portion of the curve, Eq. (103) reduces to

$$\frac{M_t}{M_\infty} = 2\left(\frac{D_m t}{\pi a^2}\right)^{1/2} \qquad \text{for } 0 \le \frac{M_t}{M_\infty} \le 0.6 \tag{106}$$

For the later time approximation, which holds for the final portion of release, Eq. (102) reduces to

$$\frac{M_t}{M_\infty} = 1 - \frac{8}{\pi^2} \exp\left(\frac{-\pi^2 D_m t}{4a^2}\right) \qquad \text{for } 0.4 \le \frac{M_t}{M_\infty} \le 1.0 \tag{107}$$

Equations (106) and (107) are plotted against time in Fig. 18. As seen, the fraction of release increases with time, but it does not reach a plateau at the later time, indicating that a significant amount of drug remains in the membrane or matrix.

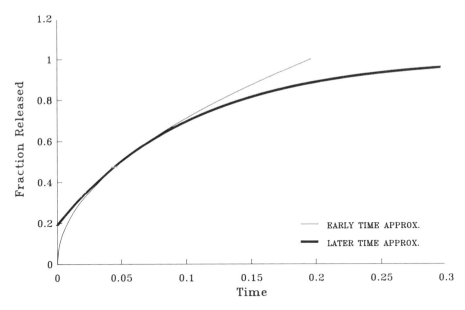

Figure 18 Fraction of drug release from a planar device as a function of time using the early and late time approximations. ($D_m/a^2 = 1$) (From Ref. 84, replotted.)

The release-rate profile can be obtained by differentiating Eqs. (106) and (107) with respect to time, giving

$$Q_t = M_\infty \left(\frac{D_m}{\pi a^2 t} \right)^{1/2} \tag{108}$$

for the early time approximation and

$$Q_t = \frac{2D_m M_\infty}{a^2} \exp\left(-\frac{\pi^2 D_m t}{4a^2} \right) \tag{109}$$

for the later time approximation.

Equations (108) and (109) are plotted against time in Fig. 19, indicating that the release rate falls rapidly at the onset of release and continuously decreases with time. It also indicates that no portion of the profile can be approximated by zero-order release.

RECIPIENT ENVIRONMENT

In the in-vitro release evaluation of a controlled-release device, the hydrodynamics often affects the overall transport of drug through membrane- and matrix-type devices. At steady state, the diffusion rate of a therapeutic agent in the matrix or membrane system is inversely proportional to the sum of individual diffusional resistances, including membrane, matrix, and boundary layer. Boundary-layer diffusion is no longer negligible when its magnitude becomes comparable to other diffusional resistances. The quantitative effect of the recipient environment on the release kinetics of either a matrix or a membrane system can be conveniently studied by the two-cell permeation apparatus depicted in Fig. 5. The boundary-

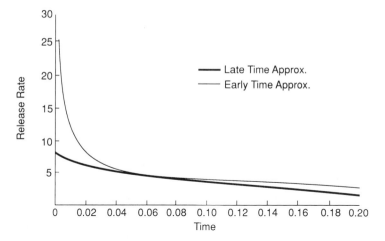

Figure 19   Release rate of drug from a planar device as a function of time using the early and late time approximations. ($D_m/a^2 = 1$) (From Ref. 84, replotted.)

layer diffusional resistance can be eliminated by increasing the rotational speed of the stirrer. Depicted in Fig. 20 are two typical profiles of the cumulative amount of drug permeated through a thin layer of membrane or matrix. Curve A represents the release under a well-stirred condition (intrinsic transport rate), while Curve B indicates release without sufficient stirring (apparent transport rate). As seen in the figure, the steady-state slope and the lag time of these two profiles are significantly different. Illustrated in Fig. 21 is the system having boundary layers on both sides of the membrane. For the simplified

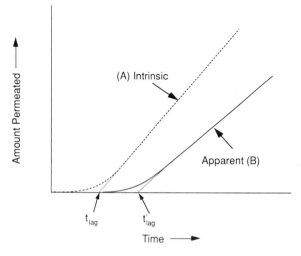

Figure 20   Effect of boundary-layer diffusion on the cumulative amount of release. Keys: (———), apparent profile; (----) intrinsic profile. (From Ref. 85.)

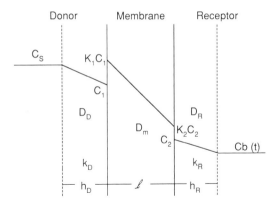

Figure 21 Concentration profile of drug in the polymeric membrane and in the boundary layers of the receptor and donor sides.

conditions $D_D = D_R = D$, $h_D = h_R = h$, and $K_1 = K_2 = K$, the apparent lag time, t_{lag}', can be derived [5,85,165] as

$$t_{lag}' = \frac{(D/k_m^2)[(4/3) + (\ell k_m K/2D_m)] + (\ell^2/D_m)(1 + \ell k_m K/6D_m) + \ell/k_m K}{2 + k_m K \ell/D_m}$$

(110)

In Fig. 21 and Eq. (110),

D_D = drug diffusion coefficient in the donor solution

D_R = drug diffusion coefficient in the receptor solution

D_m = drug diffusion coefficient in the polymeric membrane

k_D = mass transfer coefficient of the donor side, which is defined as D_D/h_D

k_R = mass transfer coefficient of the receptor side, which is defined as D_R/h_R

ℓ = thickness of membrane

K_1, K_2, K = partition coefficients, $C_{membrane}/C_{solution}$

If the resistance to the drug transport across the boundary layer is negligible, i.e., when $k_D \to \infty$ and $k_R \to \infty$, Eq. (110) reduces to

$$t_{lag} = \frac{\ell^2}{6D_m}$$

(111)

which is the same as Eq. (29) described earlier. The purpose of this section is to quantitatively determine the effect of recipient environment on the overall permeation rate for each individual controlled-release system.

Effect of Stirring on Boundary-Layer Diffusion

In the in-vitro evaluation of a delivery system, it is often necessary to know the thickness of the diffusion boundary layer of a specific hydrodynamic system, so that the importance of boundary-layer diffusion can be assessed. This is usually one of the most difficult tasks, since stirring closely relates to the flow pattern of the hydrodynamic system. And this needs to be investigated for each individual case. Instead of obtaining a theoretical solution, an empirical correlation is the common approach used to obtain this information. It is necessary to recall that the diffusion boundary-layer thickness, h, relates to the mass transfer coefficient, k_m, by the relationship $k_m = D/h$. In general, for a stirred system, the mass transfer coefficient can be correlated with the three dimensionless groups, the Sherwood-Reynolds-Schmidt (Sh-Re-Sc) number relationship [60,61,86,87]:

$$Sh = f\, Re^m\, Sc^n \tag{112}$$

where

$$Sh = \text{Sherwood number} = (k_m d)/D \tag{113}$$

$$Re = \text{Reynolds number} = (dU)/\nu \tag{114}$$

$$Sc = \text{Schmidt number} = \nu/D \tag{115}$$

In Eqs. (112)–(115), f is a proportionality constant, d is the characteristic length of the stirred system to be defined for each case, U is the effective velocity in cm/s to be determined for each case, D is the diffusion coefficient of drug in the boundary layer in cm^2/s, and ν is the kinematic viscosity of the solution in cm^2/s. The constants f, m, and n need to be determined by experimental data, depending on the permeation system.

Magnetically Stirred Two-Cell Permeation System

The experimental apparatus of the two-cell permeation system is shown in Fig. 5. Tojo et al. [60] have investigated the hydrodynamics of the magnetically stirred two-cell permeation system employing the approach presented by Colton [61] and using benzoic acid as the model drug. Benzoic acid has often been used as the model drug by researchers for calibration of dissolution and diffusion systems [52]. To study the hydrodynamics of the cell, a thin disk of benzoic acid of a suitable size is fabricated and mounted to the membrane permeation system [60,61]. A certain volume of solvent is added to the receptor cell, while the donor cell is kept empty. The stirring bar is activated to a desirable rotational speed. The temperature of the system is maintained constant by a water bath. A predetermined time intervals, a certain volume of the receptor solution is withdrawn and quickly replaced with an equal volume of solvent. The concentrations of the drug in the withdrawn samples are then assayed. The mass transfer coefficient for this system can be obtained using the following analyses.

Since the donor-side cell is empty, the dissolution of benzoic acid takes place only in the receptor cell. The mass balance equation for the drug in the receptor side can be written as

$$V \frac{dC_b(t)}{dt} = \frac{D_R}{h_R} A[C_s - C_b(t)] = k_R A[C_s - C_b(t)] \tag{116}$$

where V is the volume of the receptor compartment, $C_b(t)$ is the concentration of benzoic acid in the receptor solution at time t, A is the effective surface area for mass transfer, D_R is the diffusion coefficient of the drug in the receptor fluid, h_R is the thickness of the boundary layer in the receptor, C_s is the solubility of drug in the bulk solution, and k_R refers to the mass transfer coefficient on the receptor side. The solution of Eq. (116) using the initial condition, $C = C_{b0}$ at $t = 0$, gives

$$\ln\left(\frac{C_s - C_{b0}}{C_s - C_b(t)}\right) = \frac{k_R A t}{V} \qquad (117)$$

The characteristic length for the two-cell permeation system is defined as the length of the stirring bar, and the effective velocity for this system is defined as

$$U = \omega d \qquad (118)$$

where ω is the rotational speed in rad/s [60].

Equation (117) indicates that the mass transfer coefficient of the receptor cell, k_R, can be obtained from the slope, $(k_R A/V)$, in the plot of $\ln\{(C_s - C_{b0})/[C_s - C_b(t)]\}$ versus t. When the values of k_R are obtained at various rotational speeds, the values of constants f, m, and n can be determined by Eq. (112) using a nonlinear least-squares routine.

Problem 4. In order to determine the correlation equation for the two-cell permeation system, 170 ml of pure water or a solution containing 40% PEG 400 (polyethylene glycol 400) was added to the receptor cell, while the donor compartment remained empty. The rotational speed of the magnetic stirrer was maintained at a value ranging from 125 to 900 rpm. The physical properties of water and the PEG 400 solution at 37°C and the experimental parameters are listed in Table 6. Listed in Table 7 are the measured concentrations of benzoic acid in the receptor cell at various times and rotational speeds with 40% PEG 400 in water as the dissolution medium. Determine the correlation equation for the mass transfer coefficients and the corresponding thickness of the diffusion boundary layer.

Solution. For each rotational speed, the value of $\ln\{(C_s - C_{b0})/[C_s - C_b(t)]\}$ at each time point was computed and listed in Table 8. The resulting values in each column were plotted against time, as shown in Fig. 22. The slopes of these data at each rotational speed were determined using a linear regression analysis. The value of k_R at each rotational speed was then computed as (slope · V/A), according to Eq. (117). The values of the Reynolds, Schmidt, and Sherwood numbers were then calculated using Eqs. (113)–(115) and are listed in Table 9. A nonlinear least-squares method [88] was finally employed to determine the best-fit values of f, m, and n by minimizing the following equation:

$$SSQ = \sum_{i=1}^{P} [Sh(i) - f\,Re^m\,Sc^n]^2 \qquad (119)$$

Table 6 Physical Properties of Polyethylene Glycol Solutions at 37°C and Experimental Conditions for the Determination of Mass Transfer Coefficients

Physical properties or experimental parameters	Water	40% PEG400 in water
Density (g/ml)	0.9934	1.047
Viscosity (g/cm-s)	0.0069	0.0460
Solubility, C_s (mg/ml)	4.39	54.2
Diffusion coefficient (cm²/s)	1.45×10^{-5}	0.216×10^{-5}
Characteristic length (cm)	2.54	2.54
Volume of receptor (ml)	170	170
Surface area of membrane (cm²)	13.9	13.9

Table 7 Concentration, $C_b(t)$, of Benzoic Acid in the Receptor Cell at Various Times and Rotational Speeds

I. Solvent: water
    Temperature: 37°C
    $C_s = 4.39$ mg/ml
    $D = 1.45 \times 10^{-5}$ cm²/s
    $C_{b0} = 0$

t (min)	125 rpm $C_b(t)$ (mg/ml)	250 rpm $C_b(t)$ (mg/ml)	425 rpm $C_b(t)$ (mg/ml)	900 rpm $C_b(t)$ (mg/ml)
0				
5	0.0681	0.102	0.168	0.298
10	0.127	0.193	0.298	0.539
20	0.258	0.377	0.561	0.964
30	0.362	0.561	0.807	1.349
40	0.478	0.722	1.044	1.722

II. Solvent: 40% PEG 400
    Temperature: 37°C
    $C_s = 54.2$ mg/ml
    $D = 0.216 \times 10^{-5}$ cm²/s
    $C_{b0} = 0$
    $V = 170$ ml

t (min)	125 rpm $C_b(t)$ (mg/ml)	250 rpm $C_b(t)$ (mg/ml)	425 rpm $C_b(t)$ (mg/ml)	900 rpm $C_b(t)$ (mg/ml)
0				
5	0.170	0.212	0.423	0.697
10	0.318	0.487	0.718	1.177
20	0.613	0.927	1.343	2.166
30	0.948	1.385	2.003	3.296
40	1.281	1.797	2.714	4.246

where SSQ is the sum of the squares of residuals, Sh(i) is the observed values of the Sherwood number, and the second term in the bracket is the theoretical Sherwood number computed by Eq. (112) for given values of the Reynolds number and the Schmidt number. The resulting best-fit values of the parameters f, m, and n are $f = 0.01766$, m = 0.6819, and n = 0.3747. Therefore the final correlation equation becomes

$$Sh = 0.01766 \, Re^{0.6819} \, Sc^{0.3747} \tag{120}$$

It is necessary to emphasize that in order to express Re as a dimensionless group, the value of ω in Eq. (118) should be in rad/s rather than rps.

The values of the thickness of the diffusion boundary layer h_R as a function of the Reynolds and Schmidt numbers are obtained from the mass transfer coefficient k_R, which is defined as D_R/h_R. The resulting values of h are tabulated in the last column of Table 9.

Table 8   Computed Values of $\ln\{(C_s - C_{b0})/[C_s - C_b(t)]\}$ for Benzoic Acid at Various Times and Rotational Speeds

I.   Solvent: water
     Temperature: 37°C
     $C_s = 4.39$ mg/ml
     $D = 1.45 \times 10^{-5}$ cm^2/s
     $C_{b0} = 0$
     $V = 170$ ml

t (min)	125 rpm $\ln\{(C_s - C_{b0})/ [C_s - C_b(t)]\}$	250 rpm $\ln\{(C_s - C_{b0})/ [C_s - C_b(t)]\}$	425 rpm $\ln\{(C_s - C_{b0})/ [C_s - C_b(t)]\}$	900 rpm $\ln\{(C_s - C_{b0})/ [C_s - C_b(t)]\}$
0	0.0	0.0	0.0	0.0
5	0.0156	0.0234	0.0391	0.0703
10	0.0293	0.0449	0.0703	0.1309
20	0.0605	0.0898	0.1367	0.2480
30	0.0859	0.1367	0.2031	0.3672
40	0.1152	0.1797	0.2715	0.4980

II. Solvent: 40% PEG 400
     Temperature: 37°C
     $C_s = 54.2$ mg/ml
     $D = 0.216 \times 10^{-5}$ cm^2/s
     $C_{b0} = 0$
     $V = 170$ ml

t (min)	125 rpm $\ln\{(C_s - C_{b0})/ [C_s - C_b(t)]\}$	250 rpm $\ln\{(C_s - C_{b0})/ [C_s - C_b(t)]\}$	425 rpm $\ln\{(C_s - C_{b0})/ [C_s - C_b(t)]\}$	900 rpm $\ln\{(C_s - C_{b0})/ [C_s - C_b(t)]\}$
0	0.0	0.0	0.0	0.0
5	0.00314	0.00392	0.00784	0.0129
10	0.00588	0.00902	0.01333	1.0220
20	0.01137	0.01726	0.02510	0.0408
30	0.01765	0.02588	0.03765	0.0628
40	0.02392	0.03373	0.05137	0.0816

Source: Data obtained from Fig. 4 of Ref. [85] with permission.

Impeller-Stirred Bath System

The second type of correlation was proposed by Colton [61], who studied the mass transfer in an impeller-stirred bath system. The procedure for determining the constants f, m, and n of the correlation equation for this system is similar to that described in the magnetically stirred two-cell system. The correlation equations obtained have been used by several investigators [86,87], as given below:

Laminar boundary layer over the membrane surfaces:

$$\frac{k_m r}{D} = 0.285 \left(\frac{\omega r^2}{\nu}\right)^{0.55} \left(\frac{\nu}{D}\right)^{0.35} \qquad \text{where } 8000 < \frac{\omega r^2}{\nu} < 32{,}000 \qquad (121)$$

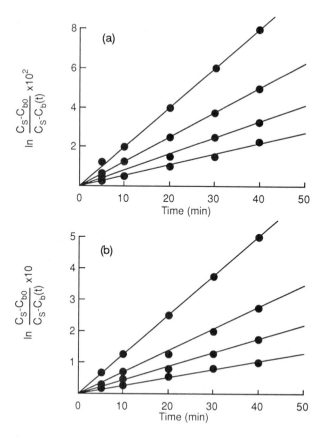

Figure 22 Dissolution profile of benzoic acid disk in the two-cell diffusion system: (a) 40% PEG 400 solution; (b) pure water; $V = 170$ ml. Numbers on the profiles are values of rotational speed of the magnetic stirring bar, ω (rpm). (From Ref. 60.)

Turbulent boundary layer over the membrane surfaces:

$$\frac{k_m r}{D} = 0.0443 \left(\frac{\omega r^2}{\nu}\right)^{0.55} \left(\frac{\nu}{D}\right)^{0.35} \qquad \text{where } 32{,}000 < \frac{\omega r^2}{\nu} < 82{,}000 \qquad (122)$$

where k_m is the mass transfer coefficient. In these equations, the characteristic length is defined as the cell radius r, and the effective velocity is defined as ωr.

The correlation equations for the magnetically stirred system, Eq. (120), and for the impeller-stirred system, Eqs. (121) and (122), conveniently express the mass transfer coefficient k_m or k_R in terms of the Reynolds number and the Schmidt number. However, it is important to emphasize that each equation was determined for a specific experimental condition. Thus, the equation obtained may not be used for other experimental conditions, even different temperatures, since as the temperature changes the viscosity also changes.

Effect of Recipient Environment on Release Kinetics

After presenting the effect of hydrodynamics on mass transfer or the boundary-layer diffusion, it is now appropriate to shift the subject to the effect of boundary-layer diffusion

Table 9 Summarized Values of Sherwood-Reynolds-Schmidt Numbers

ω (rpm)	ω (rad/s)	Percent PEG 400	$k_m = k_R$ (cm/s $\times$ 10^3)	Re	Sc	Sh	$h = h_R$ (cm $\times$ 10^3)
125	13.1	0	1.838	12,151	479	101	7.89
250	26.2	0	2.907	24,305	479	160	4.99
425	44.5	0	4.300	41,318	479	237	3.37
900	94.2	0	7.853	87,497	479	433	1.85
125	13.1	40	0.3842	1,921	20,340	142	37.7
250	26.2	40	0.5479	3,842	20,340	203	26.5
425	44.5	40	0.8028	6,532	20,340	297	18.1
900	94.2	40	1.282	13,833	20,340	475	11.3

on the overall release kinetics of a controlled-release device. When a controlled-release device is associated with a moving solid drug boundary and an appreciable boundary-layer diffusional resistance, the solution to the diffusion equation becomes very complicated. This is especially true when the release rate follows non-steady-state kinetics [83]. For this reason, the discussions here focus only on the two simpler mechanisms for drug release from devices, the steady-state conditions for the membrane and the pseudo-steady-state conditions for the homogeneous monolithic matrix.

Membrane-Type Device

A membrane permeation system using a two-cell reservoir to study transport kinetics, Fig. 5, has been used by a number of investigators owing to the convenience of the design. As Fig. 5 shows, the system consists of two cylindrical half-cells. The solution of each cell is stirred by a matched pair of magnetic stirring bars at a synchronous speed. The donor and receptor cells are enclosed inside a water jacket to maintain constant temperature. The rotational speed of the stirring bars can be controlled at a constant value by an external synchronous driving unit. The hydrodynamics of this system was investigated by Colton [61], and Tojo et al. [60], as described earlier.

If pseudo-steady-state can be assumed, the permeation rate, Q_t, of drug through a unilayer membrane under a perfect sink condition ($c_b = 0$) can be expressed by

$$Q_t = \frac{C_s}{(1/k_D) + (\ell/K_D D_m) + (K_R/K_D k_R)} \tag{123}$$

where

C_s = drug solubility in the bulk solution

D_m = drug diffusion coefficient in the membrane

D_m = drug diffusion coefficient in the donor-side boundary layer

D_R = drug diffusion coefficient in the receptor-side boundary layer

K_D = partition coefficient of drug between the membrane and the liquid in the donor side

K_R = partition coefficient of drug between the membrane and the liquid in the receptor side

k_D = mass transfer coefficient in the donor-side boundary layer, which is equal to D_D/h_D

k_R = mass transfer coefficient in the receptor-side boundary layer, which is equal to D_R/h_R

ℓ = thickness of the membrane

In Eq. (123), the denominator of the right-hand side consists of the sum of the three permeation resistances. For vigorous stirring, the diffusional resistance in two boundary layers may be negligible and Eq. (123) reduces to

$$Q_t = \frac{C_s}{\ell/(K_D D_m)} \tag{124}$$

The correction factor γ for the calculation of the intrinsic permeation rate from experimental data under nonideal stirring conditions is defined [60] as

$$\gamma = \frac{Q_t}{Q_\infty} = \left[1 + (\alpha + \beta) \frac{K_D d D_m}{Sh_R h_m D_R} \right]^{-1} \tag{125}$$

where $\alpha = k_R/k_D$ and $\beta = K_R/K_D$.

The value of γ calculated by Eq. (125) is plotted as a function of ω, the stirring speed, reflecting the effect of stirring on the release profiles.

Homogeneous Monolithic Device

The effect of boundary-layer diffusion on the overall permeation rate of a matrix-type device has been investigated by Tojo [89]. The physical model is depicted in Fig. 16. Assuming that the amount of drug loading is much greater than the solubility, in the pseudo-steady-state [76] approach, the conservation of mass in the diffusion region of the matrix can be expressed by

$$\frac{dM_t}{dt} = AD_m \frac{C_M - C_0(t)}{X(t)} \tag{126}$$

where C_M is the drug solubility in the matrix, $C_0(t)$ is the time-dependent drug concentration in the polymer phase adjacent to the bulk solution, and $X(t)$ is the distance of the receding boundary. All other parameters have been defined earlier.

The cumulative amount of drug released from the depletion zone at time t is given as

$$M_t = A\left\{ C_T - \frac{1}{2} [C_M + C_0(t)] \right\} X(t) \tag{127}$$

The derivative of M_t with respect to time can be expressed by

$$\frac{dM_t}{dt} = A\left(C_T - \frac{C_M + C_0(t)}{2} \right) \frac{dX(t)}{dt} + AX(t) \frac{d}{dt} \left(C_T - \frac{C_M + C_0(t)}{2} \right) \tag{128}$$

For the case where the amount of drug loading is much greater than the solubility, the last term of Eq. (128) can be omitted. Therefore, the relationship among the matrix diffusion, the conservation of mass in the diffusion region, and the boundary layer diffusion can be written as

$$\frac{dM_t}{dt} = AD_m \frac{C_M - C_0(t)}{X(t)} = A\left(C_T - \frac{C_M + C_0(t)}{2} \right) \frac{dX(t)}{dt} = \frac{Ak_m C_0(t)}{K} \tag{129}$$

where K is the partition coefficient between the matrix phase and the boundary-layer phase.

The length of the receding front $X(t)$ can be solved from Eq. (129) as

$$X(t) = \left[\left(\frac{bD_m}{a} \right)^2 + \frac{2C_M D_m t}{C_T - C_M/2} \right]^{1/2} - \frac{bD_m}{a} \tag{130}$$

where $a = k_m/K$ and

$$b = \frac{C_T - C_M}{C_T \quad C_M/2} \tag{131}$$

The time-dependent surface concentration $C_0(t)$ is obtained by solving the equality of the second and last terms of Eq. (129), as

$$C_0(t) = \frac{C_M}{1 + aX(t)/D_m} \tag{132}$$

Substituting Eqs. (130) and (132) for $C_0(t)$ and $x(t)$ in Eq. (129) with rearrangement gives

$$\frac{dM_t}{dt} = AD_m \frac{C_M - C_0(t)}{X(t)} \tag{133}$$

$$= \frac{AC_M}{\{(b/a)^2 + [2C_M t/D_m(C_T - C_M/2)]\}^{1/2} + (1 - b)/a} \tag{134}$$

Integration of Eq. (134) gives

$$M_t = \frac{2AaC_M}{c_1} \{(b^2 + c_1 t)^{1/2} + (b - 1) \ln[(b^2 + c_1 t)^{1/2} + 1 - b] - b\} \tag{135}$$

where

$$c_1 = \frac{2a^2 C_M}{D_m(C_T - C_M/2)} \tag{136}$$

For the case where the drug-loading dose, C_T, is much greater than the solubility, C_M, Eq. (135) can be simplified as

$$M_t = A\left[\left(\frac{D_m C_T K}{k_m} \right)^2 + 2C_T C_M D_m t \right]^{1/2} - \frac{AD_m C_T K}{k_m} \tag{137}$$

Equations (135) and (137) are termed the apparent release profile because of the effect of the boundary-layer diffusion. The typical plot of M_t versus $t^{1/2}$ in Eq. (137) is depicted as the solid curve in Fig. 23. The equation for the intrinsic release profile (the profile without boundary-layer diffusional resistance) can be obtained by letting $k_m \to \infty$. Equation (137) can be simplified as

$$M_{it} = A(2C_T C_M D_m t)^{1/2} \tag{138}$$

where M_{it} denotes the cumulative amount of drug release of the intrinsic kinetics. Equation (138) is essentially the same as Eq. (59) when the concentration in the bulk, C_b, in Eq. (59) is equal to zero.

The following procedure will demonstrate how to determine the intrinsic release profile

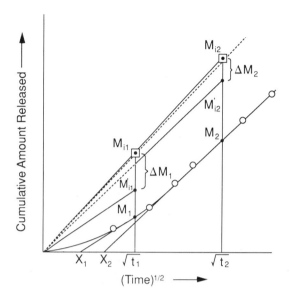

Figure 23 Graphic evaluation of the intrinsic release profile. Key: ○, apparent rate; □, intrinsic rate. (Reproduced with permission of the copyright owner, the American Pharmaceutical Association [89].)

based on the apparent release profile, Eq. (137). First, the ratio of M_t to M_{it}, denoted as ϕ, can be obtained by dividing Eq. (137) by Eq. (138), followed by simplifications, giving

$$\phi = \left[1 + \left(\frac{M_{it}K}{2C_Mk_mt}\right)^2\right]^{1/2} - \frac{M_{it}K}{2C_Mk_mt} \tag{139}$$

The derivatives of M_t and M_{it} versus $t^{1/2}$ are then obtained from Eqs. (137) and (138). These derivatives are equivalent to the tangents of the two profiles in Fig. 23. The ratio of these two derivatives, denoted as Φ, can be derived from Eqs. (137) and (138) as given by

$$\Phi = \frac{dM_t/dt^{1/2}}{dM_{it}/dt^{1/2}} = \left[1 + \left(\frac{M_{it}K}{2C_Mk_mt}\right)^2\right]^{-1/2} \tag{140}$$

Equations (139) and (140) contain the common term $M_{it}K/2C_Mk_mt$, which can be eliminated to give the relationship between ϕ and Φ as

$$\Phi = \frac{2\phi}{\phi^2 + 1} \tag{141}$$

The ratio of the derivatives, Φ, is also illustrated graphically in Fig. 23. For example, an arbitrary point on the horizontal axis in Fig. 23 is chosen as $t_1^{1/2}$. The corresponding values of the apparent and intrinsic profiles on the vertical axis are M_1 and M_{i1}, respectively. The tangent of the apparent profile at $t_i^{1/2}$ is then drawn and its intercept on the horizontal axis is x_1. A straight line is then drawn from the origin, parallel to the tangent x_1M_1. The value of this line on the vertical axis at $t_1^{1/2}$ is denoted as M_{i1}' and the difference between

M_{i1}' and M_{i1} is denoted as ΔM_1. Thus Φ, defined by Eq. (140), can be further expressed as

$$\Phi = \frac{dM_t/dt^{1/2}}{dM_{it}/dt^{1/2}} = \frac{M_{it}'}{M_{it}} \tag{142}$$

Equation (141) can be rewritten as

$$\Phi = \frac{M_{it}'}{M_{it}} = \frac{2M_t/M_{it}}{(M_t/M_{it})^2 + 1} \tag{143}$$

Letting $\Delta M_t = M_{it} - M_{it}'$, Eq. (143) can be expressed in terms of ΔM_t and M_t as

$$\frac{\Delta M_t}{M_t} = \left(\frac{\xi}{2 - \xi} \right)^{1/2} - \xi \tag{144}$$

where $\xi = M_{it}'/M_t$, which is a measurable quantity. Thus, for a given value of M_t, the value of ξ is determined, and the value ΔM_t can be evaluated by Eq. (144) from the obtained value of ξ. The corresponding intrinsic amount released is finally obtained as $M_{it}' + \Delta M_t$, as shown in Fig. 23.

The foregoing procedure for the determination of the intrinsic release profile from an apparent release profile is summarized below. The apparent release profile, Eq. (137), is plotted versus $t^{1/2}$. For any data point of interest, for instance, $t_1^{1/2}$ and $t_2^{1/2}$ in Fig. 23, two tangent lines are drawn and the intercepts on the horizontal axis are x_1 and x_2, respectively. Two straight lines are then drawn from the origin, parallel to these two tangent lines, and the values of these two straight lines on the vertical axis, at $t_1^{1/2}$ and $t_2^{1/2}$, are determined to be M_{i1}' and M_{i2}', respectively. The values of ΔM_t at these two points, denoted as ΔM_1 and ΔM_2, respectively, are computed by Eq. (144). The intrinsic amounts of release at these two points are finally calculated as $M_{i1}' + \Delta M_1$ and $M_{i2}' + \Delta M_2$, respectively. Other data points can be generated by the same procedure. The entire intrinsic release profile can be obtained by connecting these data points. The final curve would give a straight line if the theory is applicable.

This method has been applied to the release of chlormadinone acetate from a silicone elastomer by Tojo [89], as shown in Fig. 24. The intrinsic release obtained is in agreement with the theoretical linear relationship.

Method to Justify Well-Stirred Condition

In the foregoing discussions regarding the determination of the diffusion coefficients of solutes in polymeric matrices or membranes, it is always necessary to perform the experiment under a well-stirred condition so that the boundary-layer diffusion is negligible. The thickness of the boundary layer is dependent on the stirring parameter, i.e., the rpm, or flow rate. In general, the boundary layer is thinner at higher stirring speeds or flow rates. The appropriate method for determining the critical stirring parameter at which the boundary-layer diffusion is negligible is to vary the parameter until the fraction of absorption-versus-time profile becomes constant. This approach has been used by a number of investigators in immobilized biocatalyst studies. The example illustrated in Fig. 25 [90] is an experimental demonstration of increasing flow rate to increase the observed activity of the immobilized enzyme. As seen, when the flow rate is higher than 12 ml/min, the observed activity is no longer affected by the flow rate, indicating that the boundary-layer thickness is negligible at this critical flow rate.

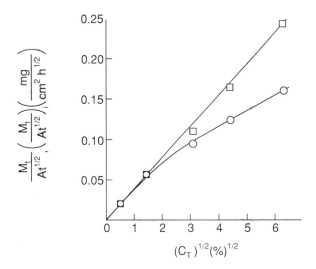

Figure 24 Correction of the apparent release kinetics to obtain the intrinsic release kinetics for the release of chlormadinone acetate from the solicone elastomer, using the method proposed in the text. On the vertical axis, $M_t/At^{1/2}$ and $(M_t/At^{1/2})_i$, respectively, represent the cumulative amount of release per unit area per square root of time of the apparent and the intrinsic release profiles. Key: $\bigcirc$, apparent; $\square$, intrinsic. (Reproduced with permission of the copyright owner, the American Pharmaceutical Association [89].)

PARTITION COEFFICIENT

Permeation is a consequence not only of diffusivity, but also of the distribution of the solute between the polymer and the recipient fluid surrounding it. The permeability coefficient of a drug in a polymer, K, is conventionally defined as the drug diffusion coefficient in the polymer, D_m, multiplied by the partition coefficient between the polymer and the recipient fluid, K, as given by

$$P = D_m K \tag{145}$$

The drug diffusion coefficient is a kinetic or nonequilibrium transport parameter, while the partition coefficient is an equilibrium thermodynamic property and is defined by

$$K = \frac{C_b{}^*}{C_b} \tag{146}$$

where C_b in the solute concentration in the bulk and $C_b{}^*$ is that in the polymer phase adjacent to the interface between the polymer and the bulk. For the case where the partition coefficient is independent of the drug concentrations, Eq. (146) can be written as

$$K = \frac{C_M}{C_s} \tag{147}$$

where C_M and C_s are the drug solubilities in the polymer and the bulk solution, respectively.

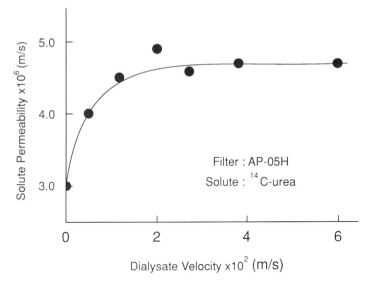

Figure 25 Experimental method to justify the well-stirred conditions.
(From Ref. 90.)

Whereas the diffusion coefficient reflects the mobility of the diffusing species in the
matrix and is dependent primarily on the molecular volume of the permeant and the
segmental mobility of the polymer, the partition coefficient depends largely on the chemistry
of the polymer-permeant pair. Leo et al. [91] have shown that for aqueous and nonaqueous
systems, the partition coefficient is an additive property of the functional groups present
in a molecule. There is evidence which suggests that this is true for polymer-permeant
pairs as well. For instance, the addition of a single hydroxyl group to progesterone forming
17α hydroxyprogesterone results in a 50-fold reduction in the silicone rubber-water partition
coefficient [92].

Table 10 [93] shows that the variation in the magnitude of the partition coefficient
is striking for a series of steroids where a several-thousand-fold difference is noted.

Table 10 Partition Coefficient (K) for
Steriods in a Silicone Rubber/Water
System at $37°C^a$

Steriod	K
Cortisol	5.5×10^{-3}
Estradiol	8×10^{-1}
Melengstiol acetate	18.87
Norethindrone	1.22
Norgestrel	3.19
19-Norprogesterone	33.3
Megestrol acetate	35.7
Mestranol	100
Progesterone	22.7
Testosterone	4.31

[a] Reported by Sundaram and Kincl [93].

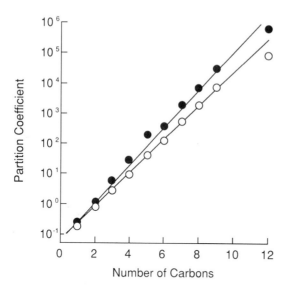

Figure 26 Semilogarithmic relationship between the partition coefficients of alkyl-*p*-aminobenzoates and the length of the alkyl chain. Key: ●, hexane-water; ○, silicone oil-water. (Reproduced with permission of the copyright owner, the American Pharmaceutical Association [19].)

Yalkowsky et al. [19] studied the partitioning of a homologous series of alkyl-*p*-aminobenzoates from water to liquid silicone polymer as a function of chain length (Fig. 26). It was reported that the exponential linearity between the partition coefficient and the length of alkyl chain can be expressed by the relationship

$$\log K_n = \log K_0 + n\pi \tag{148}$$

where K_n is the partition coefficient for the compound with a chain length of n carbons, K_0 is the partition coefficient for the reference homolog, and π is a constant, characteristic of the membrane and eluting solvent.

The partition coefficient plays a critical role in determining the rate-limiting release mechanism of a solute from a membrane or across a membrane [65,94–99]. For example, the cumulative amount of release of solute from a monolith of unit area into an eluting medium according to Fig. 16 for the case of $C_T \gg C_M$ can be described by simplifying Eq. (135) as

$$M_t = \frac{-AD_mhKC_T}{D} + A\left[\left(\frac{D_mhKC_T}{D}\right)^2 + 2C_TD_mC_Mt\right]^{1/2} \tag{149}$$

where C_T is the total concentration of drug in the polymer, D_e is the effective solute diffusion coefficient in the polymer matrix, D is the solute diffusion coefficient in the eluting medium, h is the thickness of the boundary layer, which is equal to D/k_m, and k_m is the mass transfer coefficient [92].

For very small values of K, Eq. (149) reduces to

$$M_t = A(2C_TC_MD_mt)^{1/2} \tag{150}$$

and therefore solute release becomes a diffusion-controlled process.

For very large values of K, however, it can be shown [92] that Eq. (149) reduces to

$$M_t = \frac{AC_sD_m}{h}t \tag{151}$$

Roseman and Yalkowsky [97] have demonstrated that the overall release kinetics of steroids and a homologous series of esters of p-aminobenzoic acid from monoliths of dispersed solute are biphasic, with the early portion following zero-order kinetics and the latter being dependent on $t^{1/2}$. A homologous series of esters of p-aminobenzoic serves as an excellent model to illustrate the dual nature of the profile. Considering a variation of n from one (methyl/p-aminobemzoate) to seven (heptyl p-aminobenzoate), for example, results in an increase in K by about a factor of 4. Plots of M_t versus $t^{1/2}$ are shown in Fig. 27. For small values of n, a linear relationship is observed. As n increases to above four (i.e., butyl), the duration of the nonlinear period increases dramatically. In fact, the hexyl and heptyl esters show a linear dependence of M_t on t.

Determination of Partition Coefficient

There are several methods of measuring the partition coefficient experimentally. Perhaps the simplest method is to plot the drug concentration in the polymer with respect to the equilibrium drug concentration in the external solution. The slope of such a sorption isotherm yields the partition coefficient. Care should be taken in making such a measurement, since in some instances the partition coefficient may be concentration-dependent [100]. As an example, the sorption isotherm of dexamethasone in water-swollen poly(2-

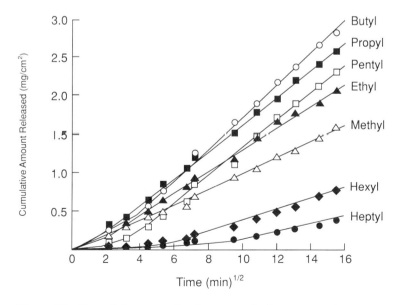

Figure 27 Plots of the accumulated amount of release versus the square root of time for esters of p-aminobenzoic acid from silicone rubber monoliths containing 5% dispersed solute. (From Ref. 97.)

hydroxyethylmethacrylate), PHEMA, is illustrated in Fig. 6 [69], where the slope of the curve determines the partition coefficient. In this case, an approach to saturation accompanied by a diminishing partition coefficient is observed at higher drug concentrations. Many of the partition coefficients reported in the literature were measured in saturated drug solutions and subsequently used in situations where the drug concentration may deviate considerably from saturation. Depending on the nonlinearity involved in the sorption isotherm, such practice can lead to appreciable error in the determination of the permeability parameters. Therefore, it is preferable to perform the experiment over the entire concentration range of interest in order that the concentration dependence of the partition coefficient can be established.

Another simple method of determining the partition coefficient is to perform a steady-state permeation experiment. This experiment can be carried out either on a free-standing membrane using a diffusion cell or on a fabricated membrane-reservoir device by a direct release measurement. Both methods require constant upstream and downstream drug concentrations, which can easily be achieved by having a saturated drug reservoir with excess solid on one side of the membrane and a large volume of drug-free aqueous phase on the other to maintain a true steady-state condition. Well-established steady-state flux equations [66] can then be used to estimate the permeability coefficient from the slope of amount of drug released-versus-time plot. If the diffusion coefficient can be determined independently, then the partition coefficient is known using Eq. (145). For example, for a membrane having planar geometry, the permeation rate can be expressed by

$$\frac{dM_t}{dt} = \frac{(C_{b1} - C_{b2})AD_mK}{\ell} \tag{152}$$

where M_t is the total amount of drug released at time t, dM_t/dt is the steady-state permeation rate at time t, C_{b1} and C_{b2} are the equilibrium drug concentrations in the solutions adjacent to the membrane, A is the membrane surface area, ℓ is the membrane thickness, and D_mK is equal to the permeability coefficient according to Eq. (145).

For cylindrical geometry, the working equation is

$$\frac{dM_t}{dt} = \frac{2\pi LD_mK(C_{b1} - C_{b2})}{\ln(R_o/R_i)} \tag{153}$$

where L is the height of the cylinder and R_i is the inner radius.

For a spherical geometry, the working equation is

$$\frac{dM_t}{dt} = \frac{4\pi D_mK(C_{b1} - C_{b2})R_iR_o}{R_o - R_i} \tag{154}$$

where R_i is the inner radius and R_0 is the outer radius.

Another method of determining the partition coefficient involves using a single permeation experiment as described above, but instead of considering only the steady-state permeation, the initial non-steady-state segment of the permeation profile must be characterized as well. Depending on the prior history of the membrane system, an initial release rate higher or lower than the steady-state value will be observed. By extrapolating the linear steady-state portion to the time axis, intercepts corresponding to either the time lag or burst effect can be readily obtained. The drug diffusion coefficient of the permeant in the membrane can then be determined from established time lag or burst effect equations, which are dependent on the geometry of the membrane. The permeability coefficient can be determined in the same experiment from steady-state permeation data, and therefore the partition coefficient is known from Eq. (145).

For example, in the case of planar geometry, the steady-state equation for the time-lag case is given by Eq. (202), and similarly for the burst effect system, by Eq. (205). These equations will be subsequently discussed in more detail in the section on temporal idiosyncracies.

Another efficient method of determining the partition coefficient is to follow the concentration change in a constant, finite volume as the drug is sorbed or desorbed by the membrane material. This method has the distinct advantage that both the drug diffusion coefficient and partition coefficient can be obtained accurately from just one experiment with a simple apparatus and a wide range of accessible experimental conditions. The exact solution to this diffusion system is well established, but the solutions require laborious numerical data-fitting routines. Lee [69] has derived approximate solutions for different geometries which allow an estimation of the diffusion coefficient and ultimately the partition coefficient from algebraic equations based on sorption data, as described by Eqs. (44)–(50).

Therefore, in a finite-volume sorption experiment, the monotonic decrease of drug concentration external to the membrane material is measured as a function of time. The left-hand side of Eq. (44) is then calculated and plotted versus time. From the slope of such a linear plot, the diffusion coefficient can be evaluated. The partition coefficient of the drug in the polymer can then be calculated from the same set of data using Eq. (145).

Tables 10 and 11 present partition coefficients of solutes for various polymer/water systems.

Problem 5.   A plot of amount of solute released from a polymeric monolithic device of unit area was found to be linear with time at steady state with a slope of 1.05×10^{-8} g/s-cm^2. If the thickness of the boundary layer is 2×10^{-3} cm and the aqueous diffusion coefficient of the drug is 7×10^{-7} cm^2/s, estimate the solute's polymer/solution partition coefficient. The solute's solubility in the polymer is 3.24×10^{-3} g/cm^3.

Solution.   The linear dependence of solute release on time suggests a partition-controlled release mechanism. The aqueous solubility of the drug can be determined from the slope of Eq. (151), as

$$C_s = \frac{h(1.05 \times 10^{-8} \text{ g/s-cm}^2)}{D}$$

$$= \frac{(2 \times 10^{-3} \text{ cm}) (1.05 \times 10^{-8} \text{ g/s-cm}^2)}{7 \times 10^{-7} \text{ cm}^2/\text{s}}$$

$$= 3 \times 10^{-5} \text{ g/ml}$$

The partition coefficient can then be evaluated by Eq. (147), as

$$K = \frac{3.24 \times 10^{-3}}{3 \times 10^{-5}}$$

$$= 108$$

DRUG SOLUBILITY IN POLYMER

Drug solubility in a polymeric system is an important consideration from two points of view. First, the manufacture of some drug delivery systems is very much dependent on

Table 11 Partition Coefficients of Solutes in Various Polymers

Solute	Polymer	Partition coefficient (polymer/water)	Ref.
Acetophenone[a]	Polyethylene	3.16	92
Amobarbital	Polydimethylsiloxane	0.0936	100
Amobarital	Polyethylene-co-vinyl acetate	18.7	100
Amobarbital	Poly-t-caprolactam-co-t-caprolactone	97.6	100
Amobarbital	Poly-t-caprolactone	27.5	100
Androstenedione	Silicone rubber	7.4	160
Androst-4-ene-3,17-dione	Polydimethylsiloxane	11.1	100
Androst-4-ene-3,17-dione	Polyethylene	2.30	100
Androst-4-ene-3,17-dione	Polyethylene-co-vinyl acetate	152	100
Androst-4-ene-3,17-dione	Poly(2-hydroxylethyl methacrylate)	62.7	161
Androst-4-ene-3,17-dione	Poly-t-caprolactam-co-t-caprolactone	230	100
Androst-4-ene-3,17-dione	Poly-t-caprolactone	89.9	100
Benzaldehyde[a]	Polyethylene	3.74	92
Benzoic acid[a]	Polyethylene	6.25	92
Chlormadinone acetate	Silicone rubber	82	153
Codeine	Polydimethylsiloxane	0.0472	100
Codeine	Polyethylene-co-vinyl acetate	2.00	100
Codeine	Poly(2-hydroxylethyl metacrylate)	11.9	161
Codeine	Poly-t-caprolactam-co-t-caprolactone	26.1	100
Codeine	Poly-t-caprolactone	3.26	100
Corticosterone	Polydimethylsiloxane	0.569	100
Corticosterone	Polyethylene-co-vinyl acetate	9.79	100
Corticosterone	Poly-t-caprolactam-co-t-caprolactone	75.9	100
Corticosterone	Poly-t-caprolactone	21.3	100
Cortisone	Polyethylene-co-vinyl acetate	1.71	100
Cortisone	Poly-t-caprolactam-co-t-caprolactone	37.9	100
Cortisone	Poly-t-caprolactone	6.85	100
Delmadinone acetate	Silicone rubber	140	104
Estra-3,17β-diol	Poly(2-hydroxylethyl methacrylate)	177	101
Estriol[b]	Polyurethane ether	133	92
Estrone	Silicone rubber	8.0	160
Ethylnodiol diacetate	Silicone rubber	108	92
Ethyl-p-aminobenzoate	Silicone rubber	0.966	92
Hydrocortisone	Poly(2-hydroxylethyl methacrylate)	27	100
Hydrocortisone	Silicone rubber	0.05	92
17-α-Hydroxyprogesterone	Poly(2-hydroxylethyl methacrylate)	83	101
17-α-Hydroxyprogesterone	Silicone rubber	0.89	95
Ketodesogestrel	Polyethylene-co-vinyl acetate	506	100
Ketodesogestrel	Poly-t-caprolactone	226	100
L-α-Acetylmethadol	Polyethylene	208	100
L-Methadone	Polydimethylsiloxane	1920	100
L-Methadone	Polyethylene	524	100
L-Methadone	Polyethylene-co-vinyl acetate	7800	100
L-Methadone	Poly-t-caprolactam-co-t-caprolactone	7860	100
L-Methadone	Poly-t-caprolactone	4360	100
Maltrexone	Polydimethylsiloxane	0.236	100
Maltrexone	Poly(2-hydroxylethyl methacrylate)	25.5	161

Table 11   Continued

Solute	Polymer	Partition coefficient (polymer/water)	Ref.
Medroxyprogesterone acetate	Silicone rubber	26.9	95
Meperidine	Polydimethylsiloxane	16.8	100
Meperidine	Polyethylene	5.83	100
Meperidine	Polyethylene-co-vinyl acetate	95.5	100
Meperidine	Poly-t-caprolactam-co-t-caprolactone	140	100
Meperidine	Poly-t-caprolactone	48.5	100
4-Methylacetophenone[a]	Polyethylene	12.5	92
4-Methylacetophenone[a]	Polyethylene	12.5	92
4-Methylbenzaldehyde[a]	Polyethylene	25.6	92
Naltrexone	Polyethylene-co-vinyl acetate	18.3	100
Naltrexone	Poly-t-caprolactam-co-t-caprolactone	88.2	100
Naltrexone	Poly-t-caprolactone	38.4	100
Norethindrone	Poly(2-hydroxylethyl methacrylate)	70	101
Norprogesterone	Silicone rubber	22.4	160
p-Aminoacetophenone	Silicone rubber	0.0321	150
Progesterone	Polydimethylsiloxane	133	100
Progesterone	Polyethylene	52.3	100
Progesterone	Polyethylene-co-vinyl acetate	1620	100
Progesterone	Poly(2-hydroxylethyl methacrylate)	129	101
Progesterone	Poly-t-caprolactam-co-t-caprolactone	1650	100
Progesterone	Poly-t-caprolactone	1250	100
Progesterone	Silicone rubber	45.0	92
Testosterone	Polydimethylsiloxane	49.7	100
Testosterone	Polyethylene	2.00	100
Testosterone	Polyethylene-co-vinyl acetate	165	100
Testosterone	Poly(2-hydroxylethyl methacrylate)	77.7	100
Testosterone	Poly-t-caprolactam-co-t-caprolactone	490	100
Testosterone	Poly-t-caprolactone	135	100

Note: Temperature is 37°C unless stated otherwise.
[a] Measured at 40°C.
[b] Measured at 30°C.
Source: Partially adapted from Ref. 92.

the drug solubility in the polymer. This is particularly true for monolithic systems and laminated structures. In the case of monolithic systems, with the drug loading dose less than the drug solubility, the active agent is physically blended with the polymer powder and then fused together by compression molding, injection molding, screw extrusion, calendering, or casting. In these processes, the drug is dissolved until appropriate concentration is reached. The detailed release kinetics of the monolithic systems as a function of loading dose, e.g., $C_T \gg C_s$ and $C_T \le C_s$, have been described earlier.

Second, drug solubility in the polymer plays an important role in solute release kinetics and mechanisms. In reservoir systems having a rate-controlling membrane or in nonporous matrix systems in which the drug is dispersed, drug dissolves into the surrounding polymer structure, diffuses through it, and finally partitions into the elution medium.

The solubilization of a drug crystal in a polymer consists of two consecutive steps: (1) the dissociation of drug molecules from their crystal lattice and (2) the solvation of the dissociated drug molecules by the polymer structure. Therefore, drug solubility in polymers will be dependent on the enthalpy and entropy of fusion in addition to the differences in the cohesive energy density between the solute and polymer. Michaels et al. [101] derived an expression based on Hildebrand's theory of microsolutes and Flory-Huggins' theory which allows an estimation of solute solubility in polymers. The expression is

$$\ln C_M = \ln \rho - (1 + \sigma) - \frac{\Delta S_f(T_M/T - 1)}{R} \tag{155}$$

where ΔS_f is the entropy of fusion, ρ is the density, T_M is the solute melting temperature, σ is the solute-polymer interaction constant, and

$$\sigma = \frac{V_A(\delta_A - \delta_p)^2}{RT} \tag{156}$$

where V_A is the molar volume of the solute, and δ_A and δ_p are the solubility parameters of the solute and polymer, respectively.

The solubility parameter is defined as the square root of the cohesive energy density (CED), where

$$CED = \frac{E_v}{V_1} \tag{157}$$

where E_v is the molar energy of vaporization and V_1 is the molar volume of the liquid.

Equations (155)–(157) indicate, therefore, that the smaller the difference between the solubility parameters or cohesive energy densities of the solute and polymer, the greater the solute solubility. Solubility parameters for various polymer systems are shown in Table 12.

If the polymer-solute system is considered as a solid solution, then the mole fraction solubility, $\bar{C}_M$, can be expressed as the sum of two terms: the solubility in an ideal solution and the logarithm of the activity coefficient of the solute,

$$\log \bar{C}_M = -\log \gamma - \frac{\Delta H_f}{2.303RT} \frac{T_M - T}{T_M} \tag{158}$$

where γ is the activity coefficient of the drug solute in the polymer and ΔH_f is the molar heat of fusion.

Chien [102] has suggested that the energy term in Eq. (158) be separated as follows:

$$\log \bar{C}_M = -\log \gamma + \frac{\Delta H_f^*}{2.303R} \frac{1}{T_M} - \frac{\Delta H_f^{**}}{2.303R} \frac{1}{T} \tag{159}$$

where ΔH_f^* is the energy required for dissociation of the drug molecules from their crystal lattice and ΔH_f^{**} is the energy required for solvation of the drug molecules into the polymer structure. Chien [102] has shown that if variations in $\log \gamma$ and ΔH_f^* are small among a homogeneous series of drug analogs, then Eq. (159) may be simplified to

$$\log \bar{C}_M = \text{constant} + \frac{\Delta H_f^*}{2.303R} \frac{1}{T_M} \tag{160}$$

Table 12 Solubility Parameters
of Polymers [162,163]

Polymer	δ_p $(cal/cm^3)^{1/2}$
Alkyd, med.-oil length	9.4
Cellulose derivatives:	
C. dinitrate	10.6
C. lacquer nitrate	11.5
C. diacetate	10.0
Ethyl cellulose	10.3
Ester gum	9.0
Ethylene-propylene	8.0
Nylon 66	13.6
Polyacrylonitrile	15.4
Polybutadiene	8.6
Poly(dimethylsiloxane)	7.3
Polyethylene	7.9
Poly(ethylene terephthaalate)	10.7
Polyisobutylene	8.1
Poly(methyl methacrylate)	9.5
Polypropylene	8.1
Polystyrene	9.1
Polysulfides	9.0–9.4
Poly(tetrafluorethylene)	6.2
Poly(vinyl acetate)	9.4
Poly(vinyl chloride)	9.7
Poly(vinylidene chloride)	12.2
Rubbers:	
Butadiene-acrylonitrile (70:30)	9.4
Butadiene-styrene (71.5:28.5)	8.1
cis-Polyisoprene	8.3
Polychloroprene	9.2

Chien [103] has measured the solubilities of steroids in liquid silicone and has shown the linear semilogarithmic dependence of the mole-fraction solubility of a drug on the reciprocal of its melting-point temperature (Fig. 28). If the aqueous solubility C_s of the drug is known, then C_M can be derived from the polymer-water partition coefficient, as given by Eq. (147).

The substitution of solubilities for thermodynamic activities is an approximation, for it assumes that the activity coefficients of the solute in each phase deviate from unity to the same degree so that K is concentration-independent. However, the use of K to determine C_M eliminates the need to consider heat-of-fusion terms, since these are embodied in the value of C_s. Pitt et al. [104] have addressed the feasibility of deriving K from the octanol-water partition coefficient (K_{oct}) of the drug using an extra thermodynamic relationship such as

$$\log K = a \log K_{oct} + b \tag{161}$$

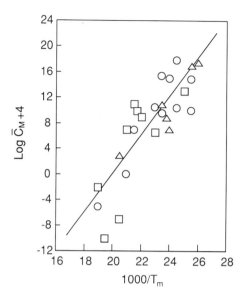

Figure 28 Relationship between the mole
fraction solubility of steroids and melting point,
T_M. Key: $\bigcirc$, testosterone derivatives; $\square$, progesterone derivatives; $\triangle$, estradiol derivatives.
(From Ref. 103.)

The form of Eq. (161) has already been shown to apply to a variety of low-molecular-
weight solute-solvent pairs [91]. An advantage of this method of estimating drug solubility
in polymers is that C_s and log K_{oct} together with the melting point are commonly reported
drug properties [105]. There is a large literature database of experimental K_{oct} values
[106]. Additionally, methods of calculating the K_{oct} of new structures are available, for
example, by the additive contributions of substituent groups or fragment constants [107,
108], by semiempirical calculations [109–111], or by the use of HPLC retention times
[112–114]. Even for drugs where C_s is not known, it can be accurately estimated from
the melting point and/or K_{oct} using one or more semiempirical relationships developed by
Hansch et al. [115] for liquids and by Yalkowsky et al. [116] for solids.

Drug solubility in a polymer system plays an important role in determining release
kinetics. For example, Michaels et al. [101] have shown that steroid permeability across
membranes is a function of melting point, as described by the following equation

$$\ln[J_{max}\ell \exp(1 + \sigma)] = \frac{-\Delta S_f}{R}\left[\left(\frac{T_M}{T}\right) - 1\right] + \ln \rho_m D_m \tag{162}$$

where J_{max} is the permeation flux from a saturated solution across a membrane of thickness
ℓ, ΔS_f is the entropy of fusion, σ is the solute-polymer interaction constant, ρ_m is the
density of the drug, and D_m is the drug diffusion coefficient in the polymer. The melting
point T_M in Eq. (162) in turn relates quantitatively to the drug's solubility in the membrane
as described previously. These researchers have demonstrated (Fig. 29) that plots of $\ln[J_{max}\ell$
$\exp(1 + \sigma)]$ versus $[(T_M/T) - 1]$ are linear for a class of steroids in a particular polymer
if ΔS_f, D_m, and δ_A are constant.

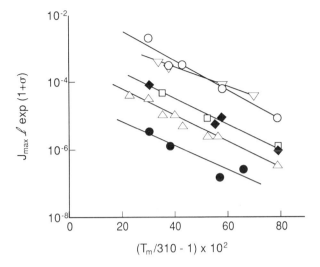

Figure 29 Dependence of steroid permeability in various polymers as a function of melting point, T_M. Key: $\bigcirc$, polydimethylsiloxane; $\bullet$, low density polyethylene; $\triangle$, poly(co-ethylenevinyl acetate) 9% w/w vinyl acetage; $\square$, poly(co-ethylenevinyl acetate) 18% w/w vinyl acetate; $\bigtriangledown$, poly(co-ethylenevinyl acetate) 40% w/w vinyl acetate; $\blacksquare$, poly(co-polytetramethylene ether glycol-diphenylmethane di-isocyanate. (From Ref. 101.)

Some solute solubilities in various polymer systems are shown in Table 13.

Problem 6. Estimate the solute solubility in a polymer at room temperature (25°C) for which the difference between the solute and polymer solubility parameters is 1.4 (cal/cm³)$^{1/2}$. The entropy of fusion for solute is 0.16 cal/g-K, the melting point is 353K, and the molar volume is 100 cm³/mol. Assume a solute density of 1 g/cm³.

Solution. The solute-polymer interaction constant is evaluated from Eq. (156) as

$$\sigma = \frac{V_A(\delta_A - \delta_p)^2}{RT}$$

$$= \frac{100 \text{ cm}^3/\text{mol} \, (1.4 \text{ cal/cm}^3)^2}{(1.987 \text{ cal/mol-K})(298 \text{ K})}$$

$$= 0.331$$

Then, according to Eq. (155), the polymer solubility can be calculated as

$$\ln C_M = \ln \rho - (1 + \sigma) - \frac{\Delta S_f(T_M/T - 1)}{R}$$

$$= \ln(1 \text{ g/cm}^3) - (1 + 0.331) - 0.16 \text{ cal/g-K}$$
$$(353 \text{ K}/298 \text{ K} - 1)/1.987 \text{ cal/mol-K}$$

$$C_M = 0.260 \text{ g/cm}^3$$

Table 13 Solubilities in Various Polymers at 37°C

Solute	Polymer	Solubility (mg/cm^3)	Ref.
Adrenosterone	Poly(etherurethane) 2000E	5.0	151
Adrenosterone	Poly(ethylene vinyl acetate)	3.5	151
Androstenedione	Poly(etherurethane) 2000E	10.0	151
Androstenedione	Poly(ethylene vinyl acetate)	5.6	151
Androstenedione	Silicone rubber	0.365	150
Androst-4-ene-3,17-dione	Poly-t-caprolactone	4.4	151
Androst-4-ene-3,17-dione	Polyethylene-co-vinyl acetate	7.4	151
Chlormadinone acetate	Silicone rubber	0.082	160
Codeine	Poly-t-caprolactone	29.3	151
Codeine	Polyethylene-co-vinyl acetate	18.0	151
Corticosterone	Poly(etherurethane) 2000E	11	151
Corticosterone	Poly(ethylene vinyl acetate)	3.9	151
Cortisone	Poly(etherurethane) 2000E	4.9	151
Cortisone	Poly(ethylene vinyl acetate)	1.8	151
Delmadinone acetate	Silicone rubber	0.850	153
11-Dehydrocortico-sterone	Poly(etherurethane) 2000E	18	151
11-Dehydrocortico-sterone	Poly(ethylene vinyl acetate)	4.9	151
11-Deoxycortico-sterone	Poly(ethylene vinyl acetate)	11	151
11-Deoxycortico-sterone	Poly(etherurethane) 2000E	25	151
Deoxycorticosterone acetate	Silicone rubber	0.105	154
11-Deoxy-17-hydroxy-corticosterone	Poly(ethylene vinyl acetate)	3.1	151
11-Deoxy-17-hydroxy-corticosterone	Poly(etherurethane) 2000E	7.8	151
Estriol	Polyurethane	2[a]	155
Estrone	Silicone rubber	0.324	150
Ethylnodiol diacetate	Silicone rubber	1.48	157
Ethyl-p-aminobenzoate	Silicone rubber	1.68	149
Hydrocortisone	Poly(ethylene vinyl acetate)	1.4	151
Hydrocortisone	Poly(etherurethane)	4.9	
Hydrocortisone	Silicone rubber	0.014	94
11-Hydroxyandro-stenedione	Poly(ethylene vinyl acetate)	9.2	151
11-Hydroxyandro-tenedione	Poly(etherurethane) 2000E	23	151
17α-Hydroxyprogesterone	Silicone rubber	0.0072	95
L-Methadone	Polyethylene-t-caprolactone	30.5	151
L-Methadone	Polyethylene-co-vinyl acetate	36.6	151
Maltrexone	Poly-t-caprolactone	16.9	151
Maltrexone	Polyethylene-co-vinyl acetate	8.1	151
Medroxyprogesterone acetate	Silicone rubber	0.0874	104
Norprogesterone	Silicone rubber	0.631	150
p-Aminoacetophenone	Silicone rubber	0.317	149
Progesterone	Silicone rubber	0.513	104
Progesterone	Polyethylene-t-caprolactone	17.6	151
Progesterone	Polyethylene-co-vinyl acetate	22.8	151

[a] Measured at 30°C.
Source: Partially adapted from Ref. 92.

VISCOSITY AND CONCENTRATION

Constant Viscosity in the Boundary Layer

The primary reason that the viscosity of the recipient environment plays an important role in the overall release kinetics can be seen from the Stokes-Einstein equation, Eq. (1), Eq. (4), or the Wilke-Chang correlation, Eq. (5). The diffusion coefficients in these equations are expressed inversely proportional to the viscosity of the fluid in the boundary layer. The laboratory observation of the effect of viscosity on the in-vivo absorption was given by Levy and Jusko [117], who reported that the rate of drug absorption by rats from viscous solutions decreases with increasing viscosity. A similar phenomenon was discovered by Nelson and Shah [118,119] in their in-vitro experiment. They found that the dissolution rate of a nondisintegrating pellet decreases in a polymeric solution of increasing viscosity.

Quantitative investigations regarding the effect of viscosity on drug dissolution were also performed by Nelson and Shah [120,121]. Their work indicates that the functional relationship between the viscosity and the dissolution rate depends on the hydrodynamic system used in the experiment. For instance, the dissolution rate of a drug, Q_t, in a flow cell having a shear rate α at the interface between the solid drug and the fluid stream can be expressed by

$$Q_t = 2.16D^{2/3}C_s\alpha^{1/3}r^{5/3} \tag{163}$$

where

$$\alpha = \frac{6F}{H^2W} \tag{164}$$

In Eqs. (163) and (164), r is the radius of the exposed circular surface, F is the flow rate, and H and W are the height and width of the flow cell, respectively. In Eq. (163), the dissolution rate Q_t is proportional to $D^{2/3}$. Since the diffusion coefficient is inversely proportional to the viscosity η, the resulting Q_t would be proportional to $\eta^{-2/3}$.

Variable Viscosity and Concentration in the Diffusion Boundary Layer

Variable viscosity in the boundary layer often occurs in the dissolution of highly soluble drugs, and the associated diffusion problem becomes very complicated. This problem has been investigated by Kuu et al. [22] using rotating-disk methodology. The effect of solution viscosity as a result of high drug concentration causes the drug diffusion coefficient to be variable in the boundary layer and can be expressed as

$$\frac{\eta}{\eta_0} = 1 + C^a \exp(bC) \tag{165}$$

where a and b are constants, η_0 is the viscosity of the dissolution medium, and η is the viscosity of the drug solution. The values of a and b for each drug, in general, need to be measured. For example, for sodium ampicillin, a = 0.956, b = 6.76, and for sodium salicylate, a = 1.16, b = 4.41. Plotted in Figs. 30a and 30b are the experimental and regressional viscosities of sodium ampicillin and sodium salicylate as a function of the concentration. As shown in the figures, Eq. (165) gives an excellent fit to the experimental data. Figures 30a and 30b represent drugs with extremely high and moderate viscosities,

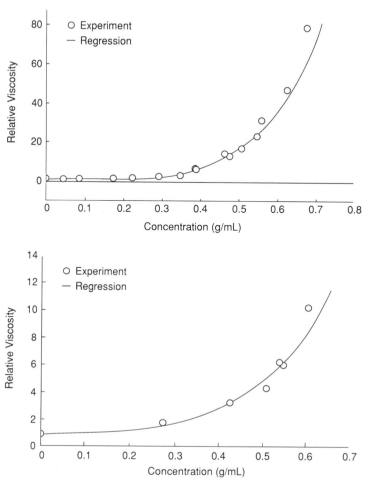

Figure 30   Relative viscosity (η/η_0)-concentration profiles of (a) sodium ampicillin and (b) sodium salicylate. Key: $\bigcirc$, experiment; —, regression. (From Ref. 22.)

and therefore should encompass most of the highly soluble drugs. Equation (165) will be used to establish the relationship between the viscosity and diffusion coefficients, as described below.

The Diffusion Coefficient of Highly Soluble Drugs

Generally, most highly soluble drugs are ionized. The diffusion coefficients of these drugs in the liquid phase have been found to be strongly dependent on the concentration and viscosity of the resulting solution [22]. The direct measurement of a variable diffusion coefficient as a function of concentration is not a trivial matter, because of the considerable experimental difficulties. In this section, correlation equations reported in the literature will be used to estimate their approximate values. Leffler and Cullinan [122] derived a correlation equation for predicting the diffusion coefficient of nonelectrolytes in terms of the concentration and viscosity of the mixture. Likewise, for concentrated electrolyte solutions, such as sodium ampicillin and sodium salicylate, the equation derived by Gordon [123] may be used:

$$D = D_0 \frac{\eta_0}{\eta} \frac{1}{C_B'V_B} \left(1 + \frac{m \, \partial \ln \gamma_\pm}{\partial m}\right) \tag{166}$$

where

C_B' = molar density of solvent, g-mol/cm^3

V_B = partial molar volume of solvent, cm^3/g-mol

η_0 = viscosity of solvent, cP

η = viscosity of solution, cP

m = molality of solute, g-mol/1000 g solvent

$\lambda_\pm$ = mean ionic activity coefficient of solute

Note that Eq. (166) indicates that the diffusion coefficient of the compound is inversely proportional to the viscosity of the mixture. Thus, the effect of viscosity on the diffusion coefficient of highly soluble drugs can dominate the effect of concentration at high solution viscosities.

In Eq. (166), the mean activity coefficient of the concentrated electrolyte may be estimated by the following equation [124]:

$$\ln \gamma_\pm = -\frac{1.172 Z_+ Z_- m^{1/2}}{1 + m^{1/2}} \tag{167}$$

where Z_+ and Z_- are the valences of cations and anions, respectively, and m is the molal concentration of drug.

The diffusion of electrolytes at low concentrations, D_0, is given by the Nernst-Haskell equation [1], as described earlier by Eq. (6). The evaluations for the other parameters in Eq. (166) are described below.

Evaluation of $C_B'V_B$.   As indicated by Kuu et al. [22], the value of $C_B'V_B$ can be determined from the concentration-density data. The partial molar volumes, V_B, of water were determined by the graphical method of tangent intercepts, as described by Lewis and Randall [125]. For the case of sodium ampicillin and sodium salicylate, the data of $C_B'V_B$ versus concentration can be approximated by a straight line, as given by

$$C_B'V_B = d_1 + d_2 C \tag{168}$$

The values of d_1 for sodium ampicillin and sodium salicylate are 0.997 and 0.995, respectively, whereas the values of d_2 are equal to 0.317 and 0.372, respectively.

Evaluation of $1 + m \, \partial \ln \gamma_\pm / \partial m$.   The experimental density-concentration data of highly soluble drugs, such as sodium ampicillin and sodium salicylate, exhibits a straight-line relationship [22] given by

$$\rho = a_0 + a_1 C \tag{169}$$

where ρ is the density in g/ml, and C is the concentration in g/ml. The respective values of a_0 and a_1 are 1.00 and 0.319 for sodium ampicillin, and 1.00 and 0.374 for sodium salicylate. Thus, the molal concentration, m, in Eqs. (166) and (167) can be converted to the concentration C in g/ml by the following equation:

$$m = \frac{1000 \, C}{(a_0 + a_1 C - C)M} \tag{170}$$

where M is the molecular weight of the drug.

The first derivative of Eq. (167) with respect to m can be expressed by

$$\frac{m \, \partial \ln \gamma_{\pm}}{\partial m} = 0.5862 m Z_+ Z_- \left(\frac{1}{(1 + \sqrt{m})^2} - \frac{1}{m + \sqrt{m}} \right) \tag{171}$$

Experiment shows that the values of the left-hand sides of Eq. (171) for sodium ampicillin and sodium salicylate are nearly constant. Thus Eq. (171) can be simplified as

$$1 + \frac{m \, \partial \ln \gamma_{\pm}}{\partial m} = 1 + b_0 \tag{172}$$

where the values of b_0 are equal to -0.137 for sodium ampicillin and -0.133 for sodium salicylate. These values imply that for viscous drug solutions, the effect of viscosity on the drug diffusion coefficient is much stronger than the effect of drug concentration.

Thus the combined effect of viscosity and concentration can be summarized by substituting Eqs. (165), (168), (169), and (172) into Eq. (166), which yields

$$D(C) = \frac{D_0 (1 + b_0)}{(d_1 + d_2 C)[1 + C^a \exp(bC)]} \tag{173}$$

Equation (173) shows that the diffusion coefficient $D(C)$ is a nonlinear function of drug concentration. Equation (173) is a desirable expression for the variable diffusion coefficient, since it is a function of drug concentration only.

The Diffusion Equation for a Rotating Disk

Kuu et al. [22] incorporated Eq. (173) into the convective diffusion equation of the rotating disk. Because of the high nonlinearity of the resulting diffusion equation, it is very difficult to solve using a conventional finite-difference method. Instead of solving the diffusion equation directly, it was transformed to the following integral equation:

$$Y = 1 - \frac{\int_0^x [1/G(C_s Y)] \exp \left[\int_0^t (f_1 \xi^2 / G(C_s Y) v_0^{1/2}) \, d\xi \right] dt}{\int_0^\infty [1/G(C_s Y)] \exp \left\{ \int_0^t [f_1 \xi^2 / G(C_s Y) v_0^{1/2}] \, d\xi \right\} dt} \tag{174}$$

where $G(C)$ is a dimensionless variable defined by

$$G(C) = \frac{D(C)}{D_0} \tag{175}$$

and the parameter f_1 is defined as

$$f_1 = \frac{-0.51 \omega^{3/2} h^3}{D_0} \tag{176}$$

whereas the dimensionless variables X and Y are defined as

$$X = \frac{z}{h} \tag{177}$$

$$Y = \frac{C}{C_s} \tag{178}$$

In Eq. (177), h is the diffusion boundary-layer thickness assuming a constant diffusion coefficient and a constant viscosity, defined by Levich [56] as

$$h = 0.51 \left(\frac{D}{\nu_0}\right)^{1/3} h_0 \tag{179}$$

where ν_0 is the kinematic viscosity of the dissolution medium in the bulk, and h_0 is the hydrodynamic boundary-layer thickness for constant viscosity, approximated by [56]

$$h_0 = 3.6 \left(\frac{\nu_0}{\omega}\right)^{1/2} \tag{180}$$

The introduction of Eqs. (177) and (178) conveniently confines the diffusion region X in the range of 0 to the vicinity of 1.0, corresponding to the values of Y from 0 to 1.0. The iterative computation procedure for solving Eq. (174) includes the following steps:

1. Select an appropriate value of the upper integration limit.
2. Select appropriate values of Y at the selected intervals of X.
3. Evaluate $G(C_sY)$ from Eqs. (173) and (175).
4. Compute the new values of Y for all selected values of X by Eq. (174).
5. Repeat steps 3 and 4 starting from the new values of Y.
6. Convergence is attained when two successive approximations yield the same values of Y everywhere to the order of accuracy required.
7. Verify that the value of Y reaches zero at the selected upper limit of integration. If not, the range of the upper integration limit is adjusted until convergence is attained.

A modified Levich equation was then derived by Kuu et al. [22], as expressed by

$$Q_t = 0.556AD_0^{2/3}\nu_0^{-1/6}G(C_s)S_0C_s\,\omega^{1/2} \tag{181}$$

where the initial slope in the X-Y profile, S_0, is obtained by taking the first derivative of Eq. (174) with respect to X. When $X = 0$, the following equation results:

$$S_0 = -\frac{1}{G(C_sY)\displaystyle\int_0^\infty [1/G(C_sY)]\exp\left\{\int_0^t [f_1\xi^2/G(C_sY)\nu_0^{1/2}]\,d\xi\right\}dt} \tag{182}$$

It is noted that the value of S_0 in Eq. (182) is independent of ω, since f_1 is independent of ω as described earlier. The denominator of Eq. (182) is obtained by the iterative computation procedure described earlier.

Equation (181) indicates that the plot of Q_t versus $\omega^{1/2}$ gives a straight line passing through the origin with a slope equal to $0.556AD_0^{2/3}\nu_0^{-1/6}G(C_s)S_0C_s$. This result is similar to the following conventional Levich equation [56]:

$$Q_t = 0.62AD^{2/3}\nu_0^{-1/6}\omega^{1/2}C_s \tag{183}$$

Equation (183) indicates that the plot of the dissolution rate against $\omega^{1/2}$ gives a straight line passing through the origin with a slope equal to $0.62AD^{2/3}\nu_0^{-1/6}C_s$. By comparing Eqs. (181) and (183), it can be seen that the slope in Eq. (181) is no longer independent of the diffusion coefficient. It depends on $G(C_s)$ and S_0, which need to be determined for each drug of interest.

The foregoing studies establish the effect of variable viscosity of the boundary layer on the diffusion rate of a rotating disk. For other types of dissolution systems, such as the convective diffusion systems studied by Nelson and Shah [118–121], a similar procedure can be used for the derivations.

The following problem is used to study the dissolution of sodium ampicillin and sodium salicylate from the rotating disk using the foregoing theory.

Problem 7. Determine the diffusion coefficient of sodium ampicillin and sodium salicylate and predict the dissolution rate of sodium ampicillin and sodium salicylate from a rotating disk using the above theory.

Solution. The values of D_0 for sodium ampicillin and sodium salicylate have been determined earlier in Problem 1, as 7.92×10^{-6} cm^2/s and 10.7×10^{-6} cm^2/s, respectively.

The numerical integration of Eq. (174) was conducted by a FORTRAN program developed based on the trapezoidal rule (Appendix). Prior to using the computer program to solve the variable-diffusion-coefficient problem, it was first tested by solving the simpler case of a constant-diffusion problem, by replacing $G(C_s Y)$ in Eq. (174) by a constant. The resulting values of $1 - Y$ are plotted as curve A in Fig. 31, which is identical to the analytical solution obtained by Levich [56] and Riddiford [126]. The validated procedure was then used to solve the dimensionless concentration profiles of sodium ampicillin and sodium salicylate by Eq. (174). The results are plotted as curve B and curve C in Fig. 31. The thickness of the diffusion boundary layer, defined as the boundary at which the drug concentration drops to 1% of C_s, estimated from Fig. 31, are 1.15h for sodium ampicillin, 1.20h for sodium salicylate, and 1.40h for the case of constant diffusion coefficients. The results imply that the viscous boundary layer is significantly thinner than the one with dilute concentration. The values of $G(C_s)$ and S_0 computed by Eqs. (175) and (176) are 0.0112 and -33.1 for sodium ampicillin, and 0.0595 and -8.54 for sodium salicylate.

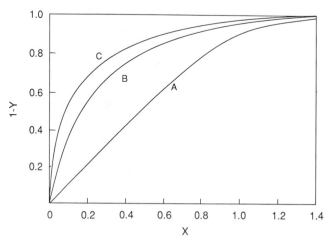

Figure 31 Plots of the dimensionless concentration $1 - Y$ versus the dimensionless distance x. Key: A, constant diffusion coefficient; B, sodium salicylate; C, sodium ampicillin. (From Ref. 22.)

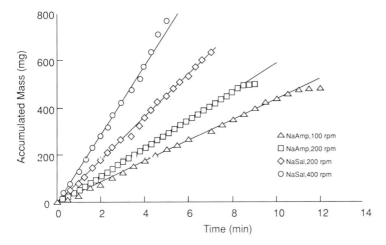

Figure 32 Typical plots of the accumulated mass-versus-time profiles of the dissolved sodium ampicillin and sodium salicylate from the rotating disk. Key: +, sodium ampicillin at 100 rpm; □, sodium ampicillin at 200 rpm; ⇕ sodium salicylate at 200 rpm; ○, sodium salicylate at 400 rpm; ———, regression. (From Ref. 22.)

The accumulated mass-versus-time profiles of sodium salicylate and sodium ampicillin at various rotational speeds yield straight lines for most part of the dissolution process, as indicated by the typical examples in Fig. 32. The rates of dissolution, obtained from the slopes of the straight-line portion of these curves, are plotted against the square root of the rotational speeds, as shown by the squares in Figs. 33a and 33b. The dissolution rates predicted by the conventional Levich equation, using D_0 as the drug diffusion coefficient, are plotted as curve I in Figs. 33a and 33b. The dissolution rates predicted by the modified Levich equation, Eq. (181), are then plotted as curve II in Figs. 33a and 33b, using the previously computed values of $G(C_s)$ and S_0, as well as the following data: $A = 1.645$ cm^2 and $\nu_0 = 0.8930 \times 10^{-2}$ cm^2/s. Note that, in Figs. 33a and 33b, the dissolution rate predicted by the Levich equation without correction of the diffusion coefficient shows significant discrepancy from the experimental data. The slope of curve I in Fig. 33a gives 176% error, while that in Fig. 33b shows 99% error. After correction, the errors in the slopes of curve II, in Figs. 33a and 33b, are reduced to 8% and 6%, respectively. The close agreement between the experimental values and the corrected dissolution rates indicates that the proposed theory has been suitably applied to the rotating-disk dissolution.

Other constants and parameters for these two drugs are listed in Table 14.

SOLUTE RELEASE FROM FLUID FLOW SYSTEMS

Currently, there are various means of delivering drugs by the intravenous route, such as infusion devices, syringe pumps, and metered burettes. Frequently, intravenous drug administration consists of a reconstituted admixed or premixed drug solution which is delivered via an administration set at an infusion rate which is controllable. However,

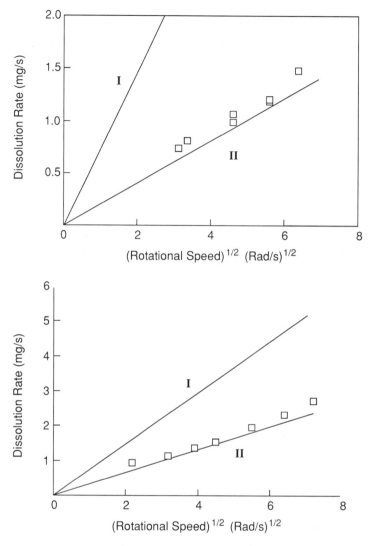

Figure 33 Dissolution rates of (a) sodium ampicillin and (b) sodium salicylate on the rotating disk. Key: □, experiment; I, conventional Levich equation; II, modified Levich equation. (From Ref. 22).

previously presented controlled-release or sustained-release concepts have application in intravenous drug delivery as well. A controlled-release device can be placed in the administration set so that the drug is released in a fluid flow stream to the patient. In fact, numerous patents in the last decade, such as Refs. 127 through 131, have been granted which describe this concept. One of the systems is depicted in Fig. 34, where the device with formulated medication is placed between the large-volume parenteral (LVP) and the patient. In order to efficiently control the flow of the diluent (the fluid in the LVP) for these types of in-line drug delivery, several components are also included as part of the i.v. set, such as the drip chamber, filter, and check valve. The drug-release mechanisms

Table 14 Values of Parameters Used in Problem 7

Sodium salicylate:
  a = 1.159
  a_0 = 1.0
  a_1 = 0.3739
  b = 4.411
  b_0 = -0.1332
  C_s = 0.65 g/ml
  D_0 = 10.0 × 10^{-6} cm²/s
  d_1 = 0.995
  d_2 = 0.3719
  $G(C_s)$ = 0.05947
  M = 160.1
  S_0 = -8.5363
  λ_- = 33.71 (cm²/Ω/g-equiv) ($^-$Salicy)
  Z_+ = 1
  Z_- = 1

Sodium ampicillin:
  a = 0.9562
  a_0 = 1.0
  a_1 = 0.3189
  b = 6.759
  b_0 = -0.1374
  C_s = 0.67 g/ml
  D_0 = 7.92 × 10^{-6} cm²/s
  d_1 = 0.9971
  d_2 = 0.3169
  $G(C_s)$ = 0.01118
  M = 371.4
  S_0 = -33.106
  λ_- = 21.16 (cm²/Ω/g-equiv) ($^-$Amp)
  Z_+ = 1
  Z_- = 1

Miscellaneous:
  A = 1.645 cm²
  C_s = 11.0 mg/ml (anhydrous ampicillin)
  Fa = 96,500 C/g-equiv
  ν_0 = 0.8930 × 10^{-2} cm²/s
  λ_+ = 50.1 (cm²/Ω/g-equiv)

Source: Some values were obtained from Ref. 22; the others were computed in the text.

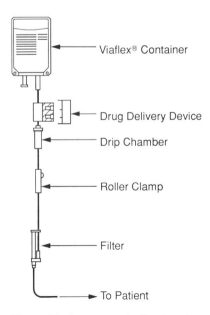

Figure 34 Intravenous in-line drug delivery system.

include diffusion control, ion exchange, iontophoresis, osmosis, etc. The advantages of in-line drug delivery systems are numerous, as listed below.

The device contains fail-safe features and is very convenient to use. The potential for human error in preparing the medication in the conventional i.v. administration can be lowered.

Unstable drugs can be conveniently formulated in solid dosage forms and placed in the device. The degradation of these drugs during the conventional dosage preparation can be minimized.

Since the large-volume parenteral (LVP) can be used as the diluent to deliver drugs, a potential cost reduction for the hospital with regard to preparing and delivering medication to patients can be achieved.

The amount of drug waste produced in the hospital's nursing activities due to medication changes can be reduced.

The device can accommodate a wide range of drugs and doses without altering the design.

In these systems, the concentration of the drug reaching the patient depends not only on the release rate from the device but also on the flow rate, the volume of fluid, and the geometry of the system. This is due to the fact that the released drug molecules from the device need to travel through a long fluid path prior to reaching the patients.

This phenomenon can be investigated using the concept of residence-time distribution (RTD), which is well known in the area of chemical reaction engineering [132,133]. This theory states that the traveling time of each drug molecule in the fluid path is, in general, not equal. Instead, a statistical distribution, such as a normal or Gaussian distribution, is usually observed. This phenomenon is termed the "dispersion" effect. The quantitative notation to express the RTD is E(t). The physical implication of E(t) is that it characterizes the degree of mixing of drug molecules in the fluid path. The RTD for a completely mixed vessel can be expressed by the following exponential function [132,133]:

$$E(t) = \frac{1}{\bar{t}} \exp\left(-\frac{t}{\bar{t}}\right) \tag{184}$$

Using $E(\tau) = \bar{t}E(t)$, the dimensionless RTD, $E(\tau)$, becomes

$$E(\tau) = \exp(-\tau) \tag{185}$$

where the dimensionless time, τ, is defined as

$$\tau = \frac{t}{\bar{t}} \tag{186}$$

where $\bar{t}$ is the mean residence time, equal to the ratio of the flow rate to the fluid volume in the entire system, denoted as F/V.

If the flow pattern of a vessel is laminar, its E(t) can be derived as

$$E(t) = \frac{\bar{t}^2}{2t^3} \tag{187}$$

in terms of the real time t, or

$$E(\tau) = \frac{1}{2\tau^3} \tag{188}$$

in terms of the dimensionless time τ.

In Eqs. (184) and (187), it can be seen that E(t) is a function of the known quantity $\bar{t}$. Thus, the RTD for these idealized systems can be determined without experiments.

For other nonideal flow systems, the degree of dispersion is usually determined by tracer experiments. The results of experiments can be interpreted by one of the following two models.

Tanks-in-Series Model

If the degree of dispersion is high, the vessel can be postulated as consisting of a number of equal-volume, well-mixed tanks. The RTD can then be expressed by the following tank-in-series model [132]:

$$E(t) = \frac{1}{\bar{t}_i} \left(\frac{t}{\bar{t}_i} \right)^{N-1} \frac{1}{(N-1)!} \exp\left(-\frac{t}{\bar{t}_i} \right) \tag{189}$$

in terms of the real time t, or

$$E(\tau) = \frac{N(N\tau)^{N-1}}{(N-1)!} \exp(-N\tau) \tag{190}$$

in terms of the dimensionless time τ. In Eqs. (189) and (190), $\bar{t}_i$ is the mean residence time of each vessel, N is the number of tanks, and $\tau = t/\bar{t} = t/N\bar{t}_i$.

Axial Dispersion Model

If the degree of dispersion is relatively low and the radial dispersion is negligible, the following axial dispersion model may be used [132]:

$$E(\tau) = \frac{1}{2(\pi\mathring{D}/uL)^{0.5}} \exp\left(-\frac{(1-\tau)^2}{4\mathring{D}/uL} \right) \tag{191}$$

for a small degree of dispersion in an open or closed vessel, and

$$E(\tau) = \frac{1}{2(\pi\tau\mathring{D}/uL)^{0.5}} \exp\left(-\frac{(1-\tau)^2}{4\tau\mathring{D}/uL} \right) \tag{192}$$

for a high degree of dispersion in an open vessel. Eqs. (191) and (192), $\mathring{D}$ is the axial dispersion coefficient, u is the velocity, L is the length of the tubing, and τ is the dimensionless time defined by Eq. (186).

The expression of RTD in terms of the real time, E(t), can be obtained by substituting Eq. (186) for τ in Eq. (191) or (192), followed by dividing the resulting E(τ) by $\bar{t}$. In Eqs. (191) and (192), the values of $\mathring{D}$ need to be determined by experiment.

If there are several components in the fluid path, for example, the device chamber, drip chamber, tubing, and filter, the degree of dispersion in each component and in the entire system will be different. The best approach to obtain the overall dispersion effect is to study the degree of dispersion in each component and combine them using the consecutive convolution integrals. For the case of three components, the overall RTD, denoted as $E_{all}(t)$, can be expressed by

$$E_{all}(t) = \int_0^t E_3(t-t') \left[\int_0^{t'} E_1(t'')E_2(t-t'') \, dt'' \right] dt' \tag{193}$$

It is important to note that since the RTD in each component is not equal unless the flow pattern is identical, it is necessary to express $E_{all}(t)$ in terms of the real time, t, in Eq. (193), instead of the dimensionless, τ, to avoid unnecessary confusions. Equation (175) can be written for any number of components for a particular system. The integration of Eq. (193) can be evaluated numerically if an analytical solution is not possible.

Once the RTD of a system is determined, the release rate profile reaching the patient can be computed by the following convolution integrals [68]:

$$C(t) = \int_0^t Con(t') \, E_{all}(t - t') \, dt' \tag{194}$$

where Con(t) is the concentration-versus-time profile of the drug immediately released out of the device (without dispersion). Equation (194) is virtually applicable to any concentration profile Con(t), and is not necessarily limited to zero-order or first-order kinetics. A FORTRAN source program designed for computing the concentration profile C(t) in Eq. (194) is listed in the Appendix. In this program, Con(t) is assumed to be a Weibull function, and the overall RTD, E_{all}, is assumed to be either Eq. (191) or (192).

The following problem is used to demonstrate how to use Eq. (175) to derive the overall RTD for a particular flow system.

Problem 8. Derive the overall RTD of an in-line system which consists of a device and the tubing, provided that the container is completely mixed and the flow pattern in the tubing is laminar.

Solution. The overall RTD of the device and the tubing can be evaluated by inserting Eq. (184), denoted as $E_1(t)$, and Eq. (187), denoted as $E_2(t)$, into Eq. (193), using the two-component convolution integral, giving

$$E_{all}(t) = \int_0^t E_1(t') E_2(t - t') \, dt' \tag{195}$$

Equation (195) can be evaluated numerically. To obtain a semianalytical solution for Eq. (195), the following procedure can be used.

First, it is more convenient to convert Eq. (188) to a double-exponential function using a nonlinear least-squares routine [88], as shown by the following equation, expressed in terms of the dimensionless time, τ_2:

$$E_2(\tau_2) = 1.103 \exp[-1.943(\tau_2 - 0.5)] + 3.0 \exp[-6.937(\tau_2 - 0.5)] \tag{196}$$

Second, Eq. (196) is then transformed to express in terms of the real time t using Eqs. (185)–(186):

$$E_2(t) = \frac{1.103}{\bar{t}_2} \exp\left[-\frac{1.943}{\bar{t}_2}(t - 0.5\bar{t}_2)\right] + \frac{3.0}{\bar{t}_2} \exp\left[-\frac{6.937}{\bar{t}_2}(t - 0.5\bar{t}_2)\right] \tag{197}$$

The convolution integral Eq. (195) is then performed and the obtained final solution is given below:

$$E_{all}(t) = f_8 \exp(-f_3 + f_4) - f_8 \exp(-f_1 t + f_9) \tag{198}$$
$$+ f_{10} \exp(-f_6 t + f_7) - f_{10} \exp(-f_1 t + f_{11})$$

where

$$f_1 = \frac{1}{\bar{t}_1}$$

$$f_2 = \frac{1.103}{\bar{t}_2}$$

$$f_3 = \frac{1.943}{\bar{t}_2}$$

$$f_4 = 0.9715$$

$$f_5 = \frac{3.0}{\bar{t}_2}$$

$$f_6 = \frac{6.937}{\bar{t}_2}$$

$$f_7 = 3.468$$

$$f_8 = \frac{f_1 f_2}{f_1 - f_3}$$

$$f_9 = 0.5\bar{t}_2(f_1 - f_3) + f_4$$

$$f_{10} = \frac{f_1 f_5}{f_1 - f_6}$$

and

$$f_{11} = 0.5\bar{t}_2(f_1 - f_6) + f_7$$

The foregoing exercise demonstrates that when the number of components increase, the analytical solution for $E_{all}(t)$ becomes very tedious and a numerical approach becomes more convenient.

TEMPORAL IDIOSYNCRASIES

Although a controlled-release device is generally designed to achieve zero-order kinetics, the initial release rates of an actual device are often higher or lower than steady-state value, depending on the history of the device. These phenomena are termed temporal idiosyncrasies. The two common phenomena associated with the storage history are time lag and burst effect.

The mathematical expression for the burst effect and the time lag can be derived from the release kinetics. In this section, studies will focus on a reservoir device having a membrane of thickness ℓ. The typical permeation conditions to be studied in this section are that one face of the membrane at $x = 0$ contacts with a bulk solution of constant concentration C_{b1}, and the other face at $x = \ell$ contacts with another bulk solution of constant concentration C_{b2}. Provided that the membrane is initially at a uniform concentration KC_{b0}, the cumulative amount of diffusant M_t which has passed through the membrane at time t is expressed by the following polynomial [66]:

$$M_t = AD_m(KC_{b1} - KC_{b2})\frac{t}{\ell} + \frac{2A\ell}{\pi^2}\sum_{n=1}^{\infty}\frac{KC_{b1}\cos n\pi - KC_{b2}}{n^2}$$
$$\left[1 - \exp\left(-\frac{D_m n^2 \pi t}{\ell^2}\right)\right] + \frac{4AKC_{b0}\ell}{\pi^2}\sum_{n=0}^{\infty}\frac{1}{(2n+1)} \qquad (199)$$
$$\left[1 - \exp\left(-\frac{D_m(2n+1)^2\pi^2 t}{\ell^2}\right)\right]$$

The mathematical expressions for the time lag and burst effects can be deduced from Eq. (199), as described below.

Time Lag

When a membrane-type controlled-release device is used immediately after it is manufactured, the membrane is theoretically free of drug. It will require some time to establish the concentration gradient within the membrane before the drug molecules can emerge from the surface of the device. It can be conceived that both C_{b0} and C_{b2} are essentially zero. In this case, Eq. (199) reduces to

$$\frac{M_t}{A\ell KC_{b1}} = \frac{D_m t}{\ell^2} - \frac{1}{6} - \frac{2}{\pi^2} \sum_{n=1}^{\infty} \frac{(-1)^n}{n^2} \exp\left(-\frac{D_m n^2 \pi^2 t}{\ell^2}\right) \tag{200}$$

where the following relationship has been used in the derivation:

$$\sum_{n=1}^{\infty} \frac{\cos n\pi}{n^2} = -\frac{\pi^2}{12} \tag{201}$$

The asymptotic line for Eq. (200) can be obtained by letting $t \to \infty$, as given by [84]:

$$M_t = \frac{AD_m KC_{b1}}{\ell} \left(t - \frac{\ell^2}{6D_m}\right) \tag{202}$$

The intercept of Eq. (202) on the time axis is equal to $\ell^2/6D_m$, which is termed the lag time.

Burst Effect

For the case of the burst effect, the device has been stored for some time before use, during which the agent gradually permeates through the membrane and eventually saturates the entire membrane. When it is placed in a desorbing environment, some of the agent in the membrane is released at a high initial rate. In this case, C_{b0} is initially equal to C_{b1}, whereas C_{b2} is equal to zero. Equation (199) can be simplified as

$$\frac{M_t}{A\ell KC_{b1}} = \frac{D_m t}{\ell^2} - \frac{1}{6} - \frac{2}{\pi^2} \sum_{n=1}^{\infty} \frac{(-1)^n}{n^2} \exp\left(-\frac{D_m n^2 \pi^2 t}{\ell^2}\right)$$
$$+ \frac{1}{2} - \frac{4}{\pi^2} \sum_{n=0}^{\infty} \exp\left(-\frac{D_m (2n+1)^2 \pi^2 t}{\ell^2}\right) \tag{203}$$

where the following relationship has been used for the derivation:

$$\sum_{n=0}^{\infty} \frac{1}{(2n+1)^2} = \frac{\pi^2}{8} \tag{204}$$

The asymptotic line for Eq. (203) can be obtained by letting $t \to \infty$, giving [84]

$$M_t = \frac{AD_m KC_{b1}}{\ell} \left(t + \frac{\ell^2}{3D_m}\right) \tag{205}$$

Similar sets of equations for spherical and cylindrical membrane geometries are well established [66].

Relationship Between Time Lag and Burst Effect

The magnitudes of the time lag and burst effect are dictated by the diffusion coefficient of the agent in the membrane and by the thickness of the membrane. If M_t is plotted against time using Eq. (200), the intercept of the linear portion of the plot on the time axis is denoted as $\ell^2/6D_m$, which is the definition of the time lag as indicated by Eq. (202). Likewise, if M_t is plotted against time using Eq. (205), the intercept of the linear portion of the plot on the time axis is denoted as $-\ell^2/3D_m$, which gives the definition of the burst effect. The results of these plots are illustrated in Fig. 35.

The release rate as a function of time, Q_t, for a controlled-release membrane containing a time-lag effect can be obtained by differentiating Eq. (200) with respect to t. The resulting relative release rate can be expressed by

$$\frac{Q_t}{Q_\infty} = 1 + 2 \sum_{n=1}^{\infty} (-1)^n \exp\left(-\frac{D_m n^2 \pi^2 t}{\ell^2}\right) \tag{206}$$

where Q_∞ is the steady-state release rate.

Likewise, the relative release rate for a membrane exhibiting a burst effect can be derived from Eq. (203) as

$$\frac{Q_t}{Q_\infty} = 1 + 2 \sum_{n=1}^{\infty} (-1)^n \exp\left(-\frac{D_m n^2 \pi^2 t}{\ell^2}\right) \\ + 4 \sum_{n=1}^{\infty} (2n + 1)^2 \exp\left(-\frac{D_m (2n + 1)^2 \pi^2 t}{\ell^2}\right) \tag{207}$$

Typical profiles described by Eqs. (206) and (207) are plotted in Fig. 36. Also indicated in Fig. 36 are some multiples of the time lag for each effect. It can be seen that the release rate is within 99% of the steady-state values after three lag times, which is defined as $\ell^2/$

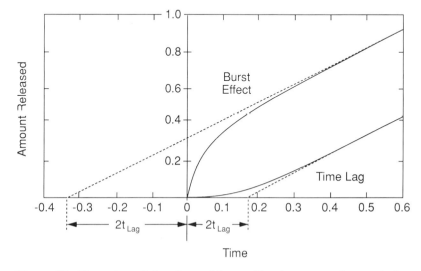

Figure 35 Illustration of time lag and burst effect in a reservoir-type device using the amount of agent released-versus-time profiles ($D_m K C_{bl}/\ell = 1$ and $\ell^2/D_m = 1$). (From Ref. 84.)

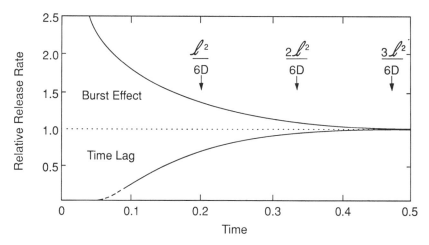

Figure 36 Illustration of time lag and burst effect in a reservoir-type device using the relative release rate-versus-time profiles ($\ell^2/D_m = 1$). (From Ref. 84.)

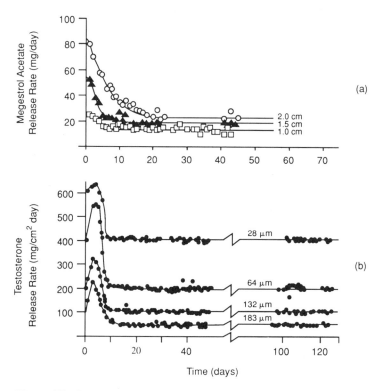

Figure 37 Burst effect profiles for megestrol acetate and testosterone cylinders made of Silastic, Dow Corning's medical-grade silicone rubber. (a) Effect of capsule length on the release rates of megestrol acetate. (From Ref. 134.) (b) Effect of membrane thickness on the release rates of testosterone. (From Ref. 135.)

$6D_m$. It is also interesting to note from Eqs. (206) and (207) that the time required to reach steady state depends only on D_m and ℓ, not on the cumulative amount of release.

Illustrated in Fig. 37a is the burst effect phenomena of the release of megestrol acetate from cylindrical polydimethylsiloxane capsules of varying length [134]. The high initial release rate, which persists for up to 18 days, is due to the steroid accumulation in the membrane.

If the agent reservoir is only partially equilibrated with the membrane during storage, or if the device is stored at one temperature and then used at another, the release profile becomes more complex, as illustrated in Fig. 37b for the release profile of testosterone through silicone rubber [135].

CONCLUSIONS: APPLICATIONS OF RELEASE KINETICS TO DESIGN AND OPTIMIZATION OF CONTROLLED-RELEASE SYSTEMS

This chapter critically reviews the concepts and factors that influence the kinetics of solute release using a fundamental, physical model approach. During the early stage of development, applications of mathematical modeling and optimization using the previously described concepts serve as powerful tools in aiding the design and optimization of controlled-release systems. Based on this approach, drug candidates can be rapidly screened from a database that stores the physical and chemical properties of the drugs. Moreover, modeling permits the rapid determination of the rate-controlling mechanism for drug release. With the aid of modeling, unnecessary in-vitro experiments can also be avoided.

In the later stage of research and development activities, modeling can also be applied to predict the in-vivo performance of the device, thus providing valuable information for clinical trials, such as experimental design, assessment of therapeutic blood levels and in-vitro/in-vivo correlations. Described below are four types of strategies that are frequently encountered in the design and optimization of a controlled-release system.

Type I Situation

The type I situation is concerned with the prediction of the in-vitro performance for a controlled-release system for given designs. The in-vitro performance includes the release rate-versus-time profiles and the duration of the zero-order or non-zero-order release. This issue appears to be a trivial computation problem, since the only task is to perform computations based on the release kinetic equations derived in this chapter. Yet some precautions need to be taken, as the actual delivery conditions are often considerably different from the idealized ones. For instance, the pseudo-steady-state assumption employed to derive the release kinetics from the monolithic device requires that the amount of drug loading, C_T, is much greater than the drug solubility in the polymer, C_M. In reality, this criterion may not be attainable. To fill the gap between the actual and idealized conditions, the following suggestions may be worth considering.

1. Identify a physical model that is closest to the actual conditions of the device.
2. Determine the rate-controlling step, by evaluating the relative diffusional resistance of each step. Discard any negligible diffusional resistances so that the problem can be simplified.
3. If boundary-layer diffusion resistance is not negligible, the appropriate value of the mass transfer coefficient may be obtainable from correlation equations.

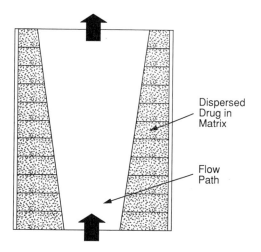

Figure 38 Approximation method for modeling of a controlled-release device with irregular geometry.

4. The release kinetics derived in this chapter are based on the three regular geometries, planar, cylindrical, or spherical. For cases where the geometry of the device is not regular, an approximation approach may be necessary to simplify the solution. If, unfortunately, an approximation is not possible, the problem becomes very complex, and one may have to consult a finite-element numerical method (a numerical approach to solving irregular geometry problems), e.g., the computer software package FIDAP [136]. For instance, depicted in Fig. 38 is an irregular geometry device in which drug is released from a conical matrix into the flow path of varying cross-sectional areas. To simplify the problem, the irregular geometry may be approximated by a number of flat cylinders. The release rate of each element is evaluated as a normal cylinder. The overall release rate, Q_t, can then be obtained by summing the individual release rates of each element, as

$$Q_t = \sum_{i=1}^{N} Q_t(i) \tag{208}$$

where $Q_t(i)$ is the release rate of each element and N is the number of elements.

5. If the physical model is too complicated for direct evaluation of the release kinetics, modeling may still be useful to obtain the upper and lower limits of the release rate. The task can be regarded as successful if the range between the two limits is sufficiently small.

Type II Situation

The type II situation relates to determining the optimum design of a controlled-release system using in-vitro release requirements. The approach to achieving this objective can be viewed as the opposite of that of the type I situation. The necessary considerations for modeling this situation have been described in the type I situation, and in addition, a parameter estimation technique is required. To obtain the optimum design, the parameters of the release kinetics need to be searched so that the device can achieve the release rate that is closest to the desirable target. For a simple problem, especially the one with one parameter and linear release kinetics, the parameter may be determined by algebraic calculations or a linear least-squares algorithm, as will be demonstrated in Problem 9a (pp. 130–132). For complicated problems, usually ones with multiple parameters and highly nonlinear release kinetics, a special parameter estimation technique should be used. There are several nonlinear parameter estimation techniques that have been conventionally used; among them the nonlinear least-squares algorithm is the best known. To use this method, the desirable release rate or rate profile is treated as the hypothetical experimental data and the values computed from the theoretical release kinetics as the predicted data. The nonlinear least-squares algorithm is then performed by minimizing the following sum of squares, SSQ:

$$SSQ = \sum_{i=1}^{N} W(i)[Q_t(i) - \hat{Q}_t]^2 \tag{209}$$

if the dependent variable is the release rate, Q_t, or

$$SSQ = \sum_{i=1}^{N} W(i)[M_t(i) - \hat{M}_t]^2 \tag{210}$$

if the dependent variable is the cumulative amount of release, M_t. In Eqs. (209) and (210), N is the number of hypothetical data points, $Q_t(i)$ is the release rate of data point i, $\hat{Q}$ is the estimated release rate, $M_t(i)$ is the cumulative amount of release of data point i, $\hat{M}_t$ is the estimated amount of release, and $W(i)$ is the nonnegative weighting factor for data point i. The choice of $W(i)$ depends on the significance or importance of each data point; in many cases, $W(i) = 1$. Since the in-vitro kinetics are often in an implicit form or sometimes cannot be expressed by algebraic equations, the choice of an appropriate nonlinear least-squares algorithm is critical. One of the most versatile and efficient algorithms is given by Powell, namely, the SSQMIN algorithm [137,138], and the complete FORTRAN subroutine developed using Powell's SSQMIN algorithm is available in Ref. 88. An example using this algorithm to determine the parameters of a cylindrical monolithic device will be given in Problem 9b (pp. 130–134).

Type III Situation

The pattern of the type III situation is similar to that of the type I situation except that the prediction includes the in-vivo response of a controlled-release system. The situation is aimed at determining the plasma drug concentration-versus-time profiles from which the therapeutic effect can be assessed. For drugs with short biological half-lives, the plasma level of the drug declines rapidly. Thus both in-vitro and in-vivo kinetics need to be considered in order to obtain a proper control of the therapeutic effect. This situation is a combination of in-vitro kinetics and the pharmacokinetics of the drug. The rationale of combining in-vitro and in-vivo kinetics is exemplified by the following simple example.

A drug is given to a patient at a constant rate via an intravenous infusion (a zero-order releasing device). If the drug follows a two-compartment pharmacokinetic model, the plasma drug levels can be expressed by

$$c(t) = \frac{k_0(K_{21} - \alpha)(e^{\lambda_1 T} - 1)}{V_c\lambda_1(\lambda_2 - \lambda_1)} e^{-\lambda_1 t} + \frac{k_0(K_{21} - \beta)(e^{\lambda_2 T} - 1)}{V_c\lambda_2(\lambda_1 - \lambda_2)} e^{-\lambda_2 t} \qquad (211)$$

where k_0 is the infusion rate; V_c is the apparent volume of distribution; T is the infusion time; and K_{21}, α, β, λ_1, and λ_2 are pharmacokinetic constants [139]. Equation (211) can be regarded as the in-vivo response of a zero-order controlled-release device with complete absorption. Equation (211) is actually a combination of the constant-rate infusion and the pharmacokinetics of the drug.

For release kinetics that cannot be expressed by a zero-order or first-order form, the plasma levels may not be easily expressed by a simple equation, e.g., Eq. (211). For these cases, the following numerical convolution integral is useful in combining in-vitro release kinetics and the pharmacokinetics, provided that the drug follows linear pharmacokinetics:

$$C(t) = \frac{dose}{V_c} \int_0^t Rate(t)\, R(t - t')\, dt' \qquad (212)$$

where t' is a dummy variable and $R(t)$ is the normalized pharmacokinetics of the drug. The FORTRAN source program developed to perform the convolution integral, Eq. (212), is listed in the Appendix.

Type IV Situation

The pattern of the type IV situation is similar to that of the type II situation except that the in-vivo response of the drug in the controlled-release system needs to be considered in design optimization. This situation is logically the opposite of the type III situation and is especially applicable to drugs with narrow therapeutic ranges. In this case, the desirable plasma drug profiles are usually given as the input information for the design and optimization of the device. The technique for solution of this situation is similar to that of the type II situation, except that the pharmacokinetics of the drug need to be included in the computation scheme.

Problem 9. The following problem is employed to demonstrate the application of the optimization technique to determine the best-fit parameters associated with the type I and type II situations.

The following information is given for a cylindrical monolithic device:

$C_M = 100$ mg/ml.
$C_T = 500$ mg/ml.
$D_M = 5.0 \times 10^{-7}$ cm^2/s $= 1.8 \times 10^{-3}$ cm^2/h.

(a) Determine the release rate at 40.0 h, provided that the length of the device is equal to 1.375 cm and the radius is 1.6 cm.

(b) Determine the optimum values of the length and the radius of the device to satisfy the following release-rate profile:

Time (h)	Rate, Q_t (mg/h)
1.0	100.0
2.0	65.0
3.0	53.0
4.0	45.0
5.0	39.0
10.0	27.0
15.0	22.0
20.0	19.0
25.0	17.0
30.0	16.0
40.0	15.0
50.0	14.0

It is assumed that all release rates in the above table are equally important.

Solution. As described by Eq. (67), the radius of the receding boundary of a cylindrical monolithic device is an implicit function of the time t. Equation (67) can be rewritten as

$$FF = \frac{R^2(t)}{2} \ln \frac{R(t)}{R_o} + \frac{R_o^2 - R^2(t)}{4} - \frac{D_m C_M t}{C_T} \tag{213}$$

where FF is a function. For a given value of R_o, the solution for R(t) for a value of t should satisfy the condition FF = 0. The release rate for a cylindrical device is given by Eq. (63) and is reexpressed as

$$Q_t = \frac{2\pi L D_m C_M}{\ln[R_o/R(t)]} \tag{214}$$

(a) This case relates to the type I situation. The release rate Q_t cannot be readily calculated by Eq. (214), since R(t) is an implicit function of the time t, as indicated by Eq. (213). A trial-and-error method should be used to evaluate R(t) from Eq. (213). A value of R(t) is first tried and substituted into Eq. (213) along with the given values of R_o, D_m, C_M, C_T, and t. The functional value of FF is then evaluated by Eq. (213). If FF is different from zero, a new value of R(t) is tried. This procedure is repeated until FF is approximately equal to zero and the solution of R(t) is obtained. The trial-and-error procedure is listed in the following table:

Trial no.	R(t) cm	FF
1	1.4	0.042128
2	1.5	0.014696
3	1.592	−0.01200
4	1.54	0.003269
5	1.562	−0.00313
6	1.552	−0.00021 ≈ 0.0

Thus, the final value for R(t) is 1.552 cm. Equation (214) is then used to calculate the release rate, Q_t, giving

$$Q_t = \frac{2\pi(1.375)(1.8 \times 10^{-3})(100)}{\ln(1.6/1.552)}$$

$$= 51.05 \text{ mg/h}$$

(b) This case can be classified as the type II situation. This is not a simple release-rate calculation problem like case (a), since the target release rate is not a constant, but is a function of time. A nonlinear parameter estimation method should be used to determine the two parameters R_o and L. The values of R_o and L are first guessed for every time point t(i). The corresponding value of R(t) should be evaluated by a trial-and-error method as indicated in case (a). In order to incorporate this trial-and-error procedure into the nonlinear parameter estimation routine, an automatic computation program must be established. This searching procedure is automatically performed by the half-interval search routine [140], which was developed as a FORTRAN subroutine HALF and is listed in the Appendix. The computational scheme is illustrated in Fig. 39, in which a pair of radii, R_{L1} and R_{R1}, are chosen between 0 and R_o, and the corresponding values of the function FF are evaluated by Eq. (213), denoted as FF(R_{L1}) and FF(R_{R1}). Here "1" in R_{L1} and R_{R1} indicates the first step of iteration. The value of FF at the halfway point between R_{L1} and R_{R1}, denoted as $FF1_{1/2}$, is then evaluated again by Eq. (213). $FF1_{1/2}$ will be either zero or have the sign of FF(R_{L1}) or FF(R_{R1}). If the value is not zero, a second pair of R_{L2} and R_{R2} can be chosen from the three values, R_{L1}, R_{R1}, and $(R_{L1} + R_{R1})/2$, so that FF(R_{L2}) and FF(R_{R2}) are opposite in sign, while

$$|R_{L2} - R_{R2}| = \frac{1}{2}|R_{L1} - R_{R1}| \tag{215}$$

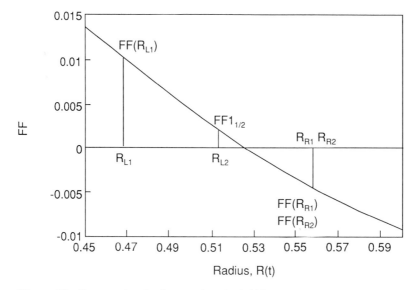

Figure 39 Computational scheme using the half-interval method to determine the radius of the receding boundary for a cylindrical monolithic device.

Here, "2" in R_{L2} and R_{R2} indicates the second step of iteration. Continuing in this manner, there is always a point z in the interval $[R_{Lk}, R_{Rk}]$ for which $FF(z) \approx 0$, where "k" denotes the k-th step of iteration.

Since the given values of the dependent variable are Q_t, Eq. (209) is used to design the nonlinear least-squares computations. And since all release rates in the table are equally important, the weighting factor should be set equal to unity. The resulting values of R(t), the guessed values R_o and L, and the given values of D_m and C_M are then substituted into Eq. (214) to evaluate Q_t. The computations can be performed by any commercially available nonlinear least-squares package, such as SAS, BMDP, or NONLIN. For the convenience of using a personal computer to perform the nonlinear least-squares computations, a FORTRAN source code, CYLINDER.FOR, of the main program was developed and is listed in the Appendix. This program was designed to use Powell's nonlinear least-squares algorithm, called SSQMIN, as described in Ref. 88.

The entire computational scheme is summarized in the flow diagram in Fig. 40. The final parameters obtained are

$L = 3.316 \text{ cm}$

$R_o = 0.704 \text{ cm}$

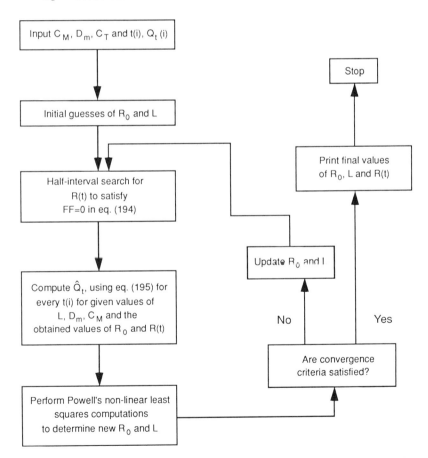

Figure 40 Flow diagram for the parameter estimation of the cylindrical monlithic device.

Table 15 Best-Fit Values of R(t) and $\hat{Q}_t$ Obtained
by the Nonlinear Least-Squares Computations

i	t(i) (h)	$Q_t(i)$ (mg/h)	$\hat{Q}_t(i)$ (mg/h)	R(t) (cm)
1	1.0	100.00	95.89	0.6772
2	2.0	65.00	67.05	0.6659
3	3.0	53.00	54.28	0.6572
4	4.0	45.00	46.66	0.6498
5	5.0	39.00	41.46	0.6433
6	10.0	27.00	28.54	0.6175
7	15.0	22.00	22.81	0.5974
8	20.0	19.00	19.38	0.5803
9	25.0	17.00	17.04	0.5651
10	30.0	16.00	15.31	0.5512
11	40.0	15.00	12.88	0.5262
12	50.0	14.00	11.21	0.5039

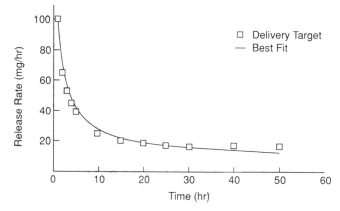

Figure 41 Comparison of the target and best-fit release-rate profiles.

The estimated values of R(t) and the rates for all time points are listed in Table 15. Plotted in Fig. 41 are the target and fitted release-rate profiles, indicating that an excellent fit of parameters has been obtained.

Summary

This chapter has established the theoretical framework for the release kinetics of solutes and therapeutic agents from a variety of polymeric controlled-release systems. Release rates depend on the physicochemical properties of the diffusing species and the geometric designs of various polymeric devices. These have been described with the appropriate limiting mathematical cases. The four situations that were previously described represent the realistic circumstances that exist when the researcher sets out to optimize drug delivery and eventually the plasma concentration of a particular therapeutic agent. The interface between the device and the surrounding biological tissue, fluid dynamics, pharmacokinetics, and metabolism of the drug will all come into play. Each combination of the drug and the delivery system should be carefully designed to meet the clinical application. This chapter is not intended to give an exhaustive review of the literature, but it is directed toward providing the reader with a grasp of the fundamental principles that govern the transport of molecules through and from various polymeric drug delivery systems. It is hoped that readers will consider these concepts along with the examples that are provided in the other chapters of the text in their quest to optimize drug release. Delivery systems such as transdermal, ocular, intranasal, parenteral, extended-release oral, bioerodible, osmotic, and pesticide, which are described in this text, represent the value of specialized therapeutic delivery systems, as well as the applications of some of the concepts presented here.

APPENDIXES: FORTRAN SOURCE PROGRAMS

FORTRAN Source Program for Computing the Variable-Diffusion-Coefficient Problem

```
C
C The source program was written using FORTRAN 77.
C Numerical integration for eq. (174) and (182) using
C the trapezoidal rule
C
      IMPLICIT REAL*8(A-H, O-Z)
      DIMENSION X(100),Y(100),YY(100),YYY(100),YY1(100),
     & Y2(100),G(100),Y1(100),Y3(100)
C
C  ROTATE2.I is the input file
C  ROTATE2.O is the output file
C  ROTATE2A.PRN and ROTATE2B.PRN are the output files for
C   graphic purpose.
C  SUM11.O is the output file for the integrated values
C  SUM11.
C
      OPEN(5,FILE='ROTATE2.I')
      OPEN(6,FILE='ROTATE2.O')
      OPEN(7,FILE='ROTATE2A.PRN')
      OPEN(0,FILE='ROTATE2B.PRN')
      OPEN(9,FILE='SUM11.O')
C
C NT is the number of integration points, ITER is
C number of iteration
C
      READ(5,*) NT,ITER
      WRITE(6,1) NT,ITER
      WRITE(0,1) NT,ITER
    1 FORMAT(/,' TOTAL NUMBER OF POINTS =',I5,/,
     &        ' NUMBER OF ITERATION =',I5)
C
      READ(5,*) A1,A2,A3,A4,A5,D0,CS,B0
      WRITE(6,4) A1,A2,A3,A4,A5,D0,CS,B0
      WRITE(0,4) A1,A2,A3,A4,A5,D0,CS,B0
    4 FORMAT(/,' A1 =',D12.4,/,' A2 =',D12.4,/,' A3=',D12.4,
     &/,' A4 =',D12.4,/,' A5 =',D12.4,/,' D0 =', D12.4,/,
     &' CS =',F10.5,/,' B0 =',F10.5,//)
```

```
C
C  ANUO is the kinematic viscosity at infinite dilution
C
      READ(5,*) ANUO,W,B1,B2
      WRITE(6,21) ANUO,W,B1,B2
      WRITE(0,21) ANUO,W,B1,B2
   21 FORMAT(/,' NUO =',D12.4,/,' W =',F8.2,/,' B1 =',F10.5,
     & /,' B2 =',F10.5 )
      READ(5,*,END=999) (X(I), I=1,NT)
      WRITE(6,5)
      WRITE(0,5)
    5 FORMAT(/,' X(I), I=1, NT:',/)
C
      WRITE(6,2) (X(I), I=1,NT)
      WRITE(0,2) (X(I), I=1,NT)
    2 FORMAT(6F10.5)
C
C Initial guess of Y(I) at selected X(I)
C
      READ(5,*,END=999) (Y(I), I=1,NT)
      WRITE(6,6)
      WRITE(0,6)
    6 FORMAT(/,' Y(I), I=1, NT:',/)
C
      WRITE(6,3) (Y(I), I=1,NT)
      WRITE(0,3) (Y(I), I=1,NT)
    3 FORMAT(6F10.5)
C
C  W is the rotation speed in radian; ANUO the kinematic
C  viscosity;
C  DELTA is diffusion boundary layer thickness, F1 is
C  eq. (176);
C  DELO is hydrodynamic boundary layer thickness, eq. (180);
C  DO is the diffusion coefficient of the in the bulk.
C
      W = W*2.*3.14/60.0
      DELO = 3.6*DSQRT(ANUO/W)
      DELTA = 0.51*(DO/ANUO)**0.33333D0*DELO
      F1 = -0.51*DSQRT(W**3/ANUO)*DELTA**3/DO
C
      INTO = 0
C
 1000 CONTINUE
C
C  G(I) is D(C)/Do in eq. (175).  D(C) is expressed by
C  eq. (173), and Y=C/CS in eq. (178).
C
C     WRITE(6,8)
C     WRITE(0,8)
C   8 FORMAT(/,' G(I) =',/)
C
      DO 111 I = 1, NT
      S1 = CS*Y(I)
      E1 = 1./( 1. + S1**A1*DEXP(A2*S1) )
      E2 = 1./(B1 + B2*CS*Y(I))
      G(I) = (1.-BO)*E1*E2
C
C     WRITE(6,9) (G(I), I=1, NT)
C     WRITE(0,9) (G(I), I=1, NT)
C   9 FORMAT(7D11.3)
```

```
C
 111   CONTINUE
C
-----------------------------------------------------------------
C       Compute the denominator of eq. (174)
C
        DO 550 J = 1, NT
C
C DO 200 is used to perform the first integration.
C The integration is performed for each value of J in DO 550
C (i.e. for each value of X in eq. (174)).
C
        DO 200 I = 1, J
        IF(X(I) .EQ. 0.D0) THEN
           YY(I) = 0.0
        ELSE
           YY(I) = F1*X(I)**2.D0/G(I)
        ENDIF
C
C       WRITE(6,27) YY(I),X(I)
C       WRITE(0,27) YY(I),X(I)
C 27    FORMAT(' YY(I)=',D12.4,' X(I)=',D12.4)
  200   CONTINUE
C
C perform the first integration by trapezoidal rule.
C SUM1X is the integrated value.
C
        CALL TRAPEZ(J,X,YY,SUM1X)
C
C       WRITE(6,48) J,SUM1X
C       WRITE(0,48) J,SUM1X
C 48    FORMAT(' J =',I3,' SUM1X=',D12.4)
C
C YYY(J) is the integrand of the second integration.
C
        YYY(J) = DEXP(SUM1X)/G(J)
C
C       WRITE(6,49) YYY(J),X(J)
C       WRITE(0,49) YYY(J),X(J)
C 49    FORMAT(' DO 550, YYY(J)=',D12.4,' X(J)=',D12.4)
C
   550 CONTINUE
C
C Perform the second integration by trapezoidal rule.
C SUM11 is the integrated value of the denominator.
C
C ### Note in the following functional CALL, the first
C ### argument is NT, which indicates the second integration
C ### of the numerator is from 0 to the infinite.
C
        CALL TRAPEZ(NT,X,YYY,SUM11)
C
C       WRITE(6,53) SUM11
C       WRITE(0,53) SUM11
C 53    FORMAT(/,' SUM11 =',D12.4,/)
C
=================================================================
C Perform integration of the numerator of eq. (174).
C
C       For each x
```

```
C
      DO 2000 II = 1, NT
      DO 50 J = 1, II
C
C DO 100 is used to perform the first integration.
C
      DO 100 I = 1, J
      YY(I) = F1*X(I)**2/G(I)
C
C     WRITE(6,31) YY(I),X(I)
C     WRITE(0,31) YY(I),X(I)
C 31  FORMAT(' YY(I)=',D12.4,' X(I)=',D12.4)
 100  CONTINUE
C
C SUMX is the integrated value of the first integration.
C
      CALL TRAPEZ(J,X,YY,SUMX)
C
C     WRITE(6,33) J,SUMX
C     WRITE(0,33) J,SUMX
C 33  FORMAT(' J =',I3,' SUMX=',D12.4)
C
C YYY(J) is the value of the integrand of the second
C integration.
C
      YYY(J) = DEXP(SUMX)/G(J)
C
C     WRITE(6,29) YYY(J),X(J)
C     WRITE(0,29) YYY(J),X(J)
C 29  FORMAT(' DO 50, YYY(J)=',D12.4,' X(J)=',D12.4)
C
  50  CONTINUE
C
C SUMXX is the value of the second integration
C ### Note in the following functional CALL, the first
C     argument
C ### TRAPEZ is II, which indicates the second integration
C ### of the numerator is from 0 to X.
C
      CALL TRAPEZ(II,X,YYY,SUMXX)
C
C     WRITE(6,11) II,SUMXX
C     WRITE(0,11) II,SUMXX
C 11  FORMAT(' II =',I3,' SUMXX =',D12.4)
C
      Y2(II) = 1.0 - SUMXX/SUM11
      Y3(II) = SUMXX/SUM11
C
      WRITE(9,56) INTO,SUM11
  56  FORMAT(' INTO =',I4,'  SUM11 =',D14.5)
 2000 CONTINUE
C
===============================================================
      RES = 0.0
      DO 28 I = 1, NT
      DEL = DABS(Y2(I)-Y(I))
      RES = RES + DEL**2
      Y(I) = Y2(I)
  28  CONTINUE
      WRITE(6,35) INTO,RES
      WRITE(0,35) INTO,RES
```

```
   35   FORMAT(/,' INTO =',I4,' RES =',D12.4,' Y2(I) =',/)
        WRITE(6,34) (Y2(I), I=1, NT)
        WRITE(0,34) (Y2(I), I=1, NT)
   34   FORMAT(6D12.4)
C
C Check convergence.  If it does'nt converge, replace the
C old values of Y(I) by the new values and perform the
C second iteration.
C
        INTO = INTO + 1
        IF(INTO .GT. ITER) GOTO 36
        IF(RES .GT. 1.D-16) GOTO 1000
   36   CONTINUE
C
C   Compute eq. (182).
C
        S0 = -1./G(1)/SUM11
        WRITE(6,55) G(1),SUM11,S0
        WRITE(0,55) G(1),SUM11,S0
   55   FORMAT(/,' The value of K(Cs),  G(1) =',D14.5,/,
       & ' The value of the denominator',/,
       & ' of eq. (174), SUM11 =',D14.5,//,' The initial rate,
       & slope of',/,' eq. (174), S0 =',D14.5,/)
C
        WRITE(6,15) INTO,RES
        WRITE(0,15) INTO,RES
   15   FORMAT(//,' CONVERGENCE IN ITERATION NO.',I4,/,
       &  ' WITH RESIDUAL =',D12.4,//,
       &  ' FINAL SOLUTION:',//,
       & '         X                Y2',/,60('*'))
C
        DO 85 I = 1, NT
        WRITE(6,37) X(I),Y2(I)
        WRITE(0,37) X(I),Y2(I)
        WRITE(7,37) X(I),Y2(I)
   37   FORMAT(1X,F8.4,5X,F10.6)
   85   CONTINUE
C
        WRITE(6,39)
   39   FORMAT(//,'          X                Y3',/,60('*'))
        DO 185 I = 1, NT
        WRITE(6,38) X(I),Y3(I)
        WRITE(0,38) X(I),Y3(I)
        WRITE(8,38) X(I),Y3(I)
   38   FORMAT(1X,F8.4,5X,F10.6)
  185   CONTINUE
C
  999   STOP
        END
C ************************************************************
C  Integration using the trapezoidal rule
C
        SUBROUTINE TRAPEZ(N,X,Y,SUM1)
        IMPLICIT REAL*8(A-H, O-Z)
        DIMENSION X(100),Y(100)
C
C       WRITE(6,5)
C       WRITE(0,5)
C   5   FORMAT(/,' IN TRAPEZ, X(I), I=1, N:',/)
C
C       WRITE(6,2) (X(I), I=1,N)
```

```
C       WRITE(0,2) (X(I), I=1,N)
C 2     FORMAT(6D12.4)
C
C       WRITE(6,6)
C       WRITE(0,6)
C 6     FORMAT(/,' IN TRAPEZ, Y(I), I=1, N:',/)
C
C       WRITE(6,3) (Y(I), I=1,N)
C       WRITE(0,3) (Y(I), I=1,N)
C 3     FORMAT(6D12.4)
C
        SUM1 = 0.
C
        DO 100 I = 1, N-1
C
        H = X(I+1)-X(I)
        YH = Y(I) + Y(I+1)
C
C       WRITE(6,10) H,YH
C       WRITE(0,10) H,YH
C 10    FORMAT(' IN TRAPEZ, H=',D12.4,' YH=',D12.4)
C
        SUM1 = SUM1 + YH/2. * H
C
 100    CONTINUE
C
        RETURN
        END
C
*************************************************************
```

The input data are listed below:

```
 36    30
0.95619  6.7588    0.0    0.0     0.0    7.92D-6   0.67  0.1327
0.89304D-2   100.0  0.9971    0.3169
0.0  0.005  0.01  0.02  0.03  0.04  0.05  0.06  0.07  0.08
0.09  0.1   0.15  0.2   0.25  0.3   0.35  0.4   0.45  0.5
0.55  0.6   0.65  0.7   0.75  0.8   0.85  0.9   0.95  1.0
1.05  1.1   1.2   1.3   1.4   1.5
0.5   0.5   0.5   0.5   0.5   0.5   0.5   0.5    0.5   0.5
0.5   0.5   0.5   0.5   0.5   0.5   0.5   0.5    0.5   0.5
0.5   0.5   0.5   0.5   0.5   0.5   0.5   0.5    0.5   0.5
0.5   0.5   0.5   0.5   0.5   0.5
```

FORTRAN Source Program for Computing the Convolution Integral

```
C       CONVOLUTION INTEGRAL FOR WEIBULL FUNCTION AND
C       DISPERSION MODEL
C## The lag-time in the Weibull function is excluded ##
C   The TRAPEZOIDAL rule is used to conpute the integration.
C
C   The WEIBULL fucntion is expressed by:
C
C       %Del = f3 - f3*exp(-f2*t**f1)     (1)
C
C   where
```

```
C    t  = time in min
C    f3 ~ 100
C    f1 and f2 are constants
C
C The rate versus time profile is the derivative of eq. (1):
C                          f1-1          f1
C          %Rate = f1 f2 f3 t      exp(-f2 t   )    (2)
C
C The input parameters for this computer program are B1,B2
C and B3
C   B1=f1, B2=f2, B3=f3/100, therefore, eq. (2) becomes
C   a probability density function, with total area under
C   the curve equal to 1.0.  The input function in the
C   subroutine is:
C
C   Rate = Dose/V * (B1 B2 B3 * T**(B1-1.D0)*DEXP(-B2*T**B1)
C
C      TRAPEZOIDAL RULE TO COMPUTE AREA UNDER THE CURVE
C
       IMPLICIT REAL*8(A-H,O-Z)
       CHARACTER*16 DISPERS
       COMMON/BLOCK1/B1,B2,B3
       COMMON/BLOCK2/Q,TBAR,V,DUL
       COMMON/BLOCK3/DISPERS
C
       OPEN(5,FILE='CONVL1.I')
       OPEN(6,FILE='CONVL1.O')
       OPEN(7,FILE='CONVL1.O2')
       OPEN(8,FILE='CONVL1.O3')
       OPEN(9,FILE='CONVL1.O4')
       OPEN(10,FILE='CONVL1.O5')
C
       WRITE(6,10)
  10   FORMAT(/,' CONVOLUTION INTEGRAL FOR RTD MODEL',/ )
       WRITE(0,10)
C
C  Input the degree of dispersion, LARGE or SMALL
C
       READ(5,111) DISPERS
 111   FORMAT(A)
       WRITE(6,112) DISPERS
       WRITE(0,112) DISPERS
 112   FORMAT(/,' DISPERSION =',A,/)
C
C      B1,B2,B3 ARE THE COEFFICIENTS OF THE WEIBULL FUNCTION
C
          READ(5,*) B1,B2,B3
          WRITE(6,3) B1,B2,B3
   3      FORMAT(/,' B1 =',E13.5,/,' B2 =',E13.5,/,' B3 =',
      & E13.5,/ )
C
C      Q, TBAR, V, D ARE RTD CONSTANTS
C
       READ(5,*) Q,TBAR,V,DUL
       WRITE(6,7) Q,TBAR,V,DUL
   7   FORMAT(' Q =',F7.2,' mL/min',/,' TBAR =',F7.2,'min',/,
      &' V =',F7.2,' mL',/,' DUL =',F7.4,/ )
C
          WRITE(0,4)
          WRITE(6,4)
   4      FORMAT(//,'  T   RATE.(%/min)     AREA',/,/,60('*') )
```

```
C
C  A IS THE LOWER LIMIT OF CONVOLUTION INTEGRAL
C  AREA IS AREA UNDER THE CURVE, TT1 IS BEGINNING POINT OF
C  TRAPEZOID, TT2 IS ENDING POINT OF TRAPEZOID
C  HITE1 AND HITE2 ARE THE HEIGHTS OF TRAPEZOID
C
       AREA = 0.0
       TT1 = 0.0
       HITE1 = 0.0
C
 100   READ(5,*,END=999) N,A,T
       TT2 = T
         CALL CONVOL(N,A,T,CONC)
C
       HITE2 = CONC
       AREA = AREA + (HITE1+HITE2) * (TT2-TT1)/2.0
C
         WRITE(0,25) T,CONC,AREA
         WRITE(6,25) T,CONC,AREA
  25   FORMAT(F7.2,2F12.5)
C
       TT1 = T
       HITE1 = CONC
       GOTO 100
C
 999   STOP
       END
C
C  ***********************************************************
       SUBROUTINE CONVOL(N,A,T1,CONC)
       IMPLICIT REAL*8(A-H,O-Z)
C
       B = T1
       H = (B - A)/N
C
  20   CONTINUE
C
       SUM = 0.0
       VALUE1 = 0.0
C
       DO 100 K = 1, N+1
C
C  Y IS t'
C
       Y = A + REAL(K-1)*H
C
       CALL INPUT(Y,AMY)
C      WRITE(9,4) Y,AMY,T1
C  4   FORMAT(' Y=',E12.4,2X,'AMY=',E12.4,2X,'T1=',E12.4)
C
       CALL RTD(T1,Y,EFUNY)
       VALUE2 = AMY*EFUNY
C
       SUM = SUM + (VALUE1+VALUE2)*H/2.0
       VALUE1 = AMY*EFUNY
 100   CONTINUE
C
       CONC = SUM
       RETURN
       END
```

```
C
C  *********************************************************
C ... THE CONVOLUTION INTEGRAL OF EQ. (194)
C
      SUBROUTINE INPUT(Y,CON)
      IMPLICIT REAL*8(A-H,O-Z)
C
      COMMON/BLOCK1/B1,B2,B3
C
C Y is t'; B1, B2, B3 are constants.
C CON is CON(t') in eq. (194), expressed by the following
C    Weibull function.
C
       CON = B1*B2*B3*(Y)**(B1-1.D0)*DEXP(-B2*(Y)**B1)
C
      RETURN
      END
C
*********************************************************
      SUBROUTINE RTD(T1,Y,EALL)
      IMPLICIT REAL*8(A-H,O-Z)
      CHARACTER*16 DISPERS
      COMMON/BLOCK2/Q,TBAR,V,DUL
      COMMON/BLOCK3/DISPERS
C
C  EALL is Eall(t-t') in eq. (194), T1 is t, and  Y is t'.
C    AL is the length of tubing.
C    U is the velocity.
C    A is the cross-sectional area of tubing.
C    V is the total volume of the tubing.
C    D is the dispersion coefficient.
C
      TH = (T1-Y)/TBAR
C
      IF(DISPERS .EQ. 'SMALL') GOTO 100
C
C  MODEL FOR LARGE DISPERSION, eq. (192)
C
      IF(TH .LE. 0.D0) THEN
           EALL = 0.0
      ELSE
           S1 = (1.0-TH)*(1.0-TH)/4.D0/TH/DUL
           IF(S1 .GE. 700.D0) S1 = 700.D0
           EALL = 1.0/TBAR/2./DSQRT(3.14D0*TH*DUL)
     &           * DEXP(-S1)
      ENDIF
C
C          WRITE(0,3) TH,S1
C  3       FORMAT(' TH =',D12.4,'  S1 =',D12.4)
C          WRITE(0,4) S1
C  4       FORMAT(' S1 =',D12.4)
C
      GOTO 200
C
C  MODEL FOR SMALL DISPERSION, eq. (191).
C
 100  CONTINUE
      S1 = (1.0-TH)*(1.0-TH)/4.D0/DUL
      IF(S1 .GE. 700.D0) S1 = 700.D0
      EALL = 1.0/TBAR/2./DSQRT(3.14D0*DUL)
     &          * DEXP(-S1)
```

```
C
C       WRITE(8,1) T1,EALL,S1
C       WRITE(0,1) T1,EALL,S1
C 1     FORMAT(' T1=',F7.2,'  E=',D11.3,'  S1=',D11.3 )
C       WRITE(7,2) Q,V,TBAR,D
C 2     FORMAT(' Q=',F7.2,' V=',D11.3,' TBAR=',D11.3,'D=',
C       & F7.4 )
C
  200   CONTINUE
        RETURN
        END
C ******************************************************
```

The typical input data are listed below:


```
 LARGE
   1.512      0.1469     98.51
   4.0   2.75    11.0   0.1427
              500         0.0         0.01
              500         0.0         0.03
              500         0.0         0.13
              500         0.0         0.18
              500         0.0          0.2
              500         0.0         0.23
              500         0.0         0.33
              500         0.0         0.43
              500         0.0         0.53
              500         0.0         0.63
              500         0.0         0.93
              500         0.0         1.63
              500         0.0         2.02
              500         0.0         2.03
              500         0.0         2.08
              500         0.0         2.13
              500         0.0         2.23
              500         0.0         2.33
              500         0.0         2.43
              500         0.0         2.53
              500         0.0         2.63
              500         0.0         3.63
              500         0.0         4.63
              500         0.0         5.63
              500         0.0         6.63
              500         0.0         7.63
              500         0.0         8.63
              500         0.0         9.63
              500         0.0        10.63
              500         0.0        11.63
              500         0.0        12.63
              500         0.0        13.63
              500         0.0        14.63
              500         0.0        15.63
              500         0.0        16.63
              500         0.0        17.63
              500         0.0        18.63
              500         0.0        19.63
              500         0.0        20.63
              500         0.0        21.63
              500         0.0        22.63
```

FORTRAN Source Program for Parameter Estimation
of the Cylindrical Monolithic Device

```
C
C NOTE: This program needs to be used along with the
C nonlinear least squares subroutine SSQMIN listed in
C reference 88, with the following modifications:
C      SUBROUTINE SSQMIN(M,N,F,X,E,ESCALE,IPRINT,MAXFUN,FF,
C     & NI,NO)
C in reference 88 is replaced by
C      SUBROUTINE SSQMIN(M,N,F,X,E,ESCALE,IPRINT,MAXFUN,FF,
C     & NI,NO,R,QTEST)
C and
C      CALL CALFUN(NI,NO,M,N,F,X)
C in reference 88 is replaced by
C      CALL CALFUN(NI,NO,M,N,F,X,R,QTEST)
C
C
       IMPLICIT REAL*8(A-H,O-Z)
       DIMENSION F(200),X(8),E(8),T(200),QT(200),R(200),
      &          QTEST(200)
       COMMON /BLOCK1/ T,QT
       COMMON/BLOCK2/ DM,CM,CT
C
C      THE DIMENSION FOR W SHOULD BE GREATER THAN:
C           (N + (M+2*N) * (N+1))
C
       OPEN(5,FILE='CYLINDER.I')
       OPEN(6,FILE='CYLINDER.O')
C
       NI = 5
       NO = 6
       READ(5,*) DM,CM,CT
       WRITE(6,1) DM,CM,CT
    1  FORMAT(/,' Diffusion coefficient, DM =',D12.4,
      & ' cm**2/hr',/,' Polymer solubility =',F10.4,
      & ' mg/mL',/,' Drug loading dise, CT =',F10.4,
      & ' mg/mL',/ )
C
       READ(5,*) M,N,MAXFUN,IPRINT
       READ(5,*) ESCALE
       DO 100 I = 1, N
       READ(5,*) X(I),E(I)
  100  CONTINUE
C
       DO 300 K = 1, M
       READ(5,*,END=999) T(K),QT(K)
  300  CONTINUE
C
       WRITE(6,13)
   13  FORMAT(/,27HPOWELL REGRESSION ALGORITHM )
C
C.............................................................
       CALL SSQMIN(M,N,F,X,E,ESCALE,IPRINT,MAXFUN,FF,
      &            NI,NO,R,QTEST)
C.............................................................
       WRITE(6,4) FF
    4  FORMAT(/,2X,32HTHE SUM-OF-SQUARES DIFFERENCE =
      & ,1PD16.8 )
       WRITE(0,4) FF
```

```
      WRITE(6,15)
      WRITE(0,15)
  15  FORMAT(/,2X,'FINAL COEFFICIENT VALUES:',5X,'J',
     & 16X,'X(J)',/,/,70('*') )
C
      DO 200 J = 1, N
      WRITE(6,6) J, X(J)
C
      WRITE(0,6) J, X(J)
   6  FORMAT(I5,10X,D16.8)
 200  CONTINUE
C
      WRITE(6,8)
   8  FORMAT(
     &' NO.      T          Qt,(mg/hr)    Qest(t)        R(t) ',
     &  /,70('*'))
C
      DO 400 J = 1, M
      WRITE(6,9) J,T(J),QT(J),QTEST(J),R(J)
   9  FORMAT(I5,3X,F7.1,2F12.2,F12.4)
 400  CONTINUE

C
 999  STOP
      END
C ************************************************************
      SUBROUTINE CALFUN(NI,NO,M,N,F,X,R,QTEST)
      IMPLICIT REAL*8(A-H,O-Z)
      DIMENSION F(200),T(200),QT(200),QTEST(200),X(8),R(200)
      COMMON /BLOCK1/ T,QT
      COMMON/BLOCK2/ DM,CM,CT
C
C   TSEC is time in sec
C   QTSEC is the rate in g/sec
C
      AL = X(1)
      R0 = X(2)
      DO 10 I = 1, M
      T1 = T(I)
      CALL HALF(T1,R0,RT)
      R(I) = RT
      QTEST(I) = 2.*3.14159*AL*DM*CM/DLOG(R0/RT)
      F(I) = QT(I) - QTEST(I)
  10  CONTINUE
      RETURN
      END
C **********************************************
C  The half-interval search scheme
C
      SUBROUTINE HALF(T,R0,RT)
      IMPLICIT REAL*8(A-H,O-Z)
      COMMON/BLOCK2/ DM,CM,CT
C
      R1 = 0.001*R0
      R2 = R0
  10  CONTINUE
      RM = (R1+R2)/2.
      Y1 = 0.5*R1**2*DLOG(R1/R0) + 0.25*(R0**2 - R1**2)
     &    - DM*CM*T/CT
      Y2 = 0.5*R2**2*DLOG(R2/R0) + 0.25*(R0**2 - R2**2)
     &    - DM*CM*T/CT
      YM = 0.5*RM**2*DLOG(RM/R0) + 0.25*(R0**2 - RM**2)
```

```
      &       - DM*CM*T/CT
C
C     WRITE(0,1) T,R0,R1,R2,RM,YM
C  1  FORMAT(' T=',F6.2,' R0=',F6.3,' R1=',F6.3,' R2=',F6.3,
C     &  ' RM=',F7.4,' YM=',D10.3)
C
      IF(YM .EQ. 0.D0) GOTO 100
      IF(DABS(R2-R1) .LE. 1.0D-7) GOTO 100
      IF(YM .LT. 0.D0) THEN
         R2 = RM
         GOTO 10
      ENDIF
      IF(YM .GT. 0.D0) THEN
         R1 = RM
         GOTO 10
      ENDIF
 100  CONTINUE
      RT = RM
      RETURN
      END
******************************************************************
```

The input data are listed below:

```
   1.8D-3  100.0  500.0
     12          2          300          1
   1.0D03
   4.9         1.0D-3
   0.4         1.0D-3
          1.0     100.0
          2.0      65.0
          3.0      53.0
          4.0      45.0
          5.0      39.0
         10.0      27.0
         15.0      22.0
         20.0      19.0
         25.0      17.0
         30.0      16.0
         40.0      15.0
         50.0      14.0
```

REFERENCES

1. R. Reid, J. M. Prausnitz, and B. E. Poling, *The Properties of Gases and Liquids*, 4th ed., McGraw-Hill, New York (1987).
2. C. O. Bennett and J. E. Myers, *Momentum, Heat, and Mass Transfer*, McGraw-Hill, New York (1962).
3. A. H. P. Skelland, *Diffusional Mass Transfer*, John Wiley, New York (1974).
4. J. T. Edward, Molecular volumes and the Stokes-Einstein equation, *J. Chem. Educ.*, 47:261 (1970).
5. G. L. Flynn, S. H. Yalkowsky, and T. J. Roseman, Mass transport phenomena and models: Theoretical concepts, *J. Pharm. Sci.*, 63:479–510 (1974).
6. R. B. Bird, W. E. Stewart, and E. N. Lightfoot, *Transport Phenomena*, John Wiley, New York (1960).
7. A. L. Hines and R. N. Maddox, *Mass Transfer, Fundamentals and Applications*, Prentice-Hall, Englewood Cliffs, N.J. (1985).

8. C. Tanford, *Physical Chemistry of Macromolecules*, John Wiley, New York, chap. 6 (1961).

9. S. Bretsznajder, *Prediction of transport and other physical properties of fluids*, Permegon Press, Oxford, chap. 8 (1971).

10. C. R. Wilke, Estimation of liquid diffusion coefficients, *Chem. Eng. Progr.*, *45*:218–224 (1949).

11. L. G. Longsworth, Diffusion measurements, at 1°, of aqueous solutions of amino acids, peptides, and sugars, *J. Am. Chem. Soc.*, *74*:4155–4159 (1952).

12. L. G. Longsworth, Diffusion measurements, at 25°, of aqueous solutions of amino acids, peptides, and sugars, *J. Am. Chem. Soc.*, *75*:5705–5709 (1953).

13. J. H. Hildebrand, Motions of molecules in liquids: viscosity and diffusivity, *Science*, *174*:490–493 (1971).

14. J. C. M. Li and P. Chang, Self-diffusion coefficient and viscosity in liquids, *J. Chem. Phys.*, *23*:518–520 (1955).

15. R. E. Beck and J. S. Schultz, Hindrance of solute diffusion within membranes as measured with microporous membranes of known pore geometry, *Biochim. Biophys. Acta*, *255*:273–303 (1972).

16. D. A. Goldstein and A. K. Solomon, Determination of equivalent pore radius for human red cells by osmotic pressure measurement, *J. Gen. Physiol.*, *44*:1–17 (1960).

17. S. H. Yalkowsky and G. Zografi, Micellar properties of long-chain acylcarnitines, *J. Colloid Sci.*, *34*(4):525–533 (1970).

18. S. H. Yalkowsky and G. Zografi, Potentiometric titration of monomeric and micellar acyl-carnitines, *J. Pharm. Sci.*, *59*(6):798–802 (1970).

19. S. H. Yalkowsky, G. L. Flynn, and T. G. Slunick, Importance of chain length on physicochemical and crystalline properties of organic homologs, *J. Pharm. Sci.*, *61*:852 (1972).

20. J. Dainty and C. R. House, Unstirred layers in frog skin, *J. Physiol.*, *182*:66–78 (1966).

21. G. Le Bas, *The Molecular Volumes of Liquid Chemical Compounds*, Longmans, Green, New York (1915), cited by Reid et al. [4].

22. W. Y. Kuu, M. R. Prisco, R. W. Wood, and T. J. Roseman, Studies of dissolution behavior of highly soluble drugs using a rotating disk, *Int. J. Pharm.*, *55*:77–89 (1989).

23. J. M. Padfield and I. W. Kellaway, The Diffusion of penicillin G and ampicillin through phospholipid Sols, *J. Pharm. Pharmacol.*, *27*:348–352 (1975).

24. S. J. Desai, P. Singh, A. P. Simonelli, and W. I. Higuchi, Investigation of factors influencing release of solid drug dispersed in inert matrices II, quantitation of procedures, *J. Pharm. Sci.*, *55*:1224–1229 (1966).

25. N. A. Peppas and D. L. Meadows, Macromolecular structure and solute diffusion in membrane: An overview of recent theories,'' *J. Membrane Sci.*, *16*:361–377 (1983).

26. N. A. Peppas and S. R. Lustig, Solute diffusion in hydrophilic network structures, in *Hydrogels in Medicine and Pharmacy*, *Vol. I*, *Fundamentals* (N. A. Peppas, Ed.), CRC Press, Boca Raton, Fla. (1986).

27. C. T. Reinhart, R. W. Korsmeyer, and N. A. Peppas, Macromolecular network structure and its effects on drug and protein diffusion, *Int. J. Pharm. Tech. & Prod. Mfr.*, *2*:9–16 (1981).

28. J. Crank and G. S. Park, Eds., *Diffusion in Polymers*, Academic Press, New York (1968).

29. C. N. Satterfield, C. K. Colton, and W. H. Pitcher, Restrictive diffusion in liquids within fine pores, *AIChE J.*, *19*(3):628–635 (1973).

30. H. Faxén, Die Bewegung einer starren kugel langs der achse eines mit zahrer flüssigkeit gefüllten rohres, *Ark. Mat. Astron. Fys.*, *17*(27):1 (1922), cited by Renkin.

31. E. Renkin, Filtration, diffusion, and molecular sieving through porous cellulose membranes, *J. Gen. Physiol.*, *38*:225–243 (1954).

32. W. D. Stein, *The Movement of Molecules Across Cell Membranes*, Academic Press, New York, p. 112 (1967).

33. K. Lakshminarayanaiah, *Transport Phenomena in Membranes*, Academic Press, New York, p. 329 (1969).

34. B. K. Davis, Diffusion in polymer gel implants, *Proc. Natl. Acad. Sci.*, *71*(8):3120–3123 (1974).

35. J. Comyn, Ed., *Polymer Permeability*, Elsevier Applied Science, New York (1985).

36. C. K. Colton, C. N. Satterfield, and C. J. Lai, Diffusion and partitioning of macromolecules within finely porous glass, *AIChE J.*, *21*(2):289–298 (1975).

37. N. A. Peppas and C. T. Reinhart, Solute diffusion in swollen membranes. Part I. A new theory, *J. Membrane Sci.*, *15*:275–287 (1983).

38. P. G. deGennes, *Scaling Concepts in Polymer Physics*, Cornell University Press, Ithaca, N.Y., 1979.

39. S. Wisniewski and S. W. Kim, Permeation of water-soluble solutes through poly(2-hydroxyethyl methacrylate) and poly(2-hydroxyethyl methacrylate) cross-linked with ethylene glycol dimethacylate, *J. Membrane Sci.*, *6*:299–308 (1980).

40. H. Yasuda and C. E. Lamaze, Permselectivity of solutes in homogeneous water-swollen polymer membranes, *J. Macromol. Sci. Phys.*, *B5*:111 (1971).

41. N. A. Peppas and H. J. Moynihan, Solute diffusion in swollen membranes. IV. Theories for moderately swollen networks, *J. Appl. Polym. Sci.*, *30*:2589–2606 (1985).

42. H. J. Moynihan, M. S. Honey, and N. A. Peppas, Solute diffusion in swollen membranes. Part V: Solute diffusion in poly(2-hydroxyethyl methacrylate), *Polym. Eng. Sci.*, *26*(17):1180–1185 (1986).

43. A. Peterlin, Transport of small molecules in polymers, in *Controlled Drug Delivery*, Vol. *I*, *Basic Concepts* (S. D. Bruck, Ed.), CRC Press, Boca Raton, Fla. (1983).

44. R. S. Harland and N. A. Peppas, Solute diffusion in swollen membranes. VI. A model for diffusion in heterogeneous media, *Polym. Bull.*, *18*:553–556 (1987).

45. R. S. Harland and N. A. Peppas, Solute diffusion in swollen membranes. VII. Diffusion in semicrystalline networks, *Colloid. Polym. Sci.*, *267*:218–225 (1989).

46. P. Rempp and J. E. Herz, ''Model networks: Synthesis and structure,'' *Angew. Makromol. Chemie*, *76*:373 (1979).

47. C. G. Pitt, A. L. Andrady, Y. T. Bao, and N. K. P. Samuel, Estimation of rates of drug diffusion in polymers, *ACS Symp. Ser.*, 348 (Controlled Release Technol.: Pharm. Appl.): 49–70 (1987).

48. D. W. McCall, E. W. Anderson, and C. M. Huggins, Self-diffusion in linear dimethylsiloxanes, *J. Phys. Chem.*, *34*:804–808 (1961).

49. R. Baker, *Controlled Release of Biologically Active Agents*, John Wiley, New York (1987).

50. H. Nogami, T. Nagai, and A. Suzuki, Studies on powdered preparations XVII. Dissolution rate of sulfonamides by rotating disk method, *Chem. Pharm. Bull.*, *14*:329–338 (1966).

51. H. Nogami, T. Nagai, and T. Yotsuyanagi, Dissolution phenomena of organic medicinals involving simultaneous phase changes, *Chem. Pharm. Bull.*, *17*:499–509 (1969).

52. S. Prakongpan, W. I. Higuchi, K. H. Kwan, and A. M. Molokhia, Dissolution rate studies of cholesterol monohydrate in bile acid-lecithin solutions using the rotating-disk method, *J. Pharm. Sci.*, *65*:685–689 (1976).

53. M. S. Wu, W. I. Higuchi, J. L. Fox, and M. Friedman, Kinetics and mechanism of hydroxyapatite crystal dissolution in weak acid buffers using rotating disk method, *J. Dent. Res.*, *55*:496–505 (1976).

54. K. G. Mooney, M. A. Mintun, K. J. Himmelstein, and V. J. Stella, Dissolution kinetics of carboxylic acids I: Effect of pH under unbuffered conditions, *J. Pharm. Sci.*, *70*:13–22 (1981).

55. H. Grijseels, D. J. A. Crommelin, and C. J. De Blaey, Hydrodynamic approach to dissolution rate, *Pharm. Weekbl. Sci. Ed.*, *3*:129–144 (1981).

56. V. G. Levich, *Physicochemical Hydrodynamics*, Prentice-Hall, Englewood Cliffs, N.J., pp. 60–78 (1962).

57. J. W. McBain and C. R. Dawson, The diffusion of potassium chloride in aqueous solution, *Proc. Roy. Soc.*, London, Ser. A., *148*:32 (1935).

58. G. Taylor, Dispersion of soluble matter in solvent flowing slowly through a tube, *Proc. Roy. Soc.*, *A219*:186–203 (1953).

59. G. Taylor, Conditions under which dispersion of a solute in a stream of solvent can be used to measure molecular diffusion, *Proc. Roy. Soc.*, *A225*:473–477 (1954).

60. K. Tojo, Y. Sun, M. M. Ghannam, and Y. W. Chien, Characterization of a membrane permeation system for controlled drug delivery system, *AIChE J.*, *31*:741–746 (1985).

61. C. K. Colton, Permeability and transport studies in batch and flow dialyzers with applications to hemodialysis, Sc.D. thesis, Department of Chemical Engineering, M.I.T., Cambridge, Mass. (1969).

62. K. A. Smith, C. K. Colton, E. W. Merrill, and L. B. Evans, Convective transport in a batch dialyzer: Determination of true membrane permeability from a single measurement, *Chem. Eng. Progr. Symp. Ser.*, *64*(84):45–58 (1968).

63. J. H. Petropoulos and D. G. Tsimboukis, Boundary layer effects on membrane permeation rates, *J. Membrane Sci.*, *27*:359–361 (1986).

64. R. W. Wood and M. J. Mulski, Methodology for the determination of water vapor transport across plastic films, *Int. J. Pharm.*, *50*:61–66 (1989).

65. T. J. Roseman and W. I. Higuchi, Release of medroxyprogesterone acetate from a silicone polymer, *J. Pharm. Sci.*, *59*:353–357 (1970).

66. Crank, J., *The Mathematics of Diffusion*, 2nd ed., Clarendon Press, Oxford (1975).

67. H. S. Carslaw and J. C. Jaeger, *Conduction of Heat in Solids*, Oxford University Press, Oxford (1959).

68. E. Kreyszig, *Advanced Engineering Mathematics*, John Wiley, New York, pp. 716–725 (1967).

69. P. I. Lee, Determination of diffusion coefficients by sorption from a constant, finite volume, in *Controlled Release of Bioactive Materials*, R. Baker, Ed., Academic Press, New York, pp. 135–153 (1980).

70. P. I. Lee, Diffusional release of a solute from a polymeric matrix—approximate analytical solutions, *J. Membrane Sci.*, *7*:255–275 (1980).

71. P. C. Carman and A. W. Haul, Measurement of diffusion coefficients, *Proc. Roy. Soc.*, *222A*:109 (1954).

72. B. Berner and D. Kivelson, Paramagnetically enhanced relaxation of nuclear spins. Measurement of diffusion, *J. Physical Chem.*, *83*:1401–1405 (1979).

73. N. Muramatsu and A. P. Minton, An automated method for rapid determination of diffusion coefficients via measurements of boundary spreading, *Anal. Biochem.*, *168*:345–351 (1988).

74. G. Convath, B. Leclerc, F. Falson-Rieg, J. P. Devissaguet, and G. Couarraze, *In situ* determination of the diffusion coefficient of a solute in a gel system using a radiotracer, *J. Controlled Release*, *9*:159–168 (1989).

75. S. N. B. Thomas, H. W. Blanch, and D. S. Soane, A novel optical method for the measurement of biomolecular diffusion in polymer matrices, *Biotech. Progress*, *5*:126–131 (1989).

76. T. Higuchi, Mechanism of sustained-action medication, *J. Pharm. Sci.*, *52*(12):1145–1149 (1963).

77. R. W. Baker, H. K. Lonsdale, and R. M. Gale, Membrane-controlled delivery systems, in *Controlled Release Pesticide Symposium*, Sept. 16–18 (N. Cardarelli, Ed.), University of Akron, Akron, Ohio (1974).

78. W. Rhine, V. Sukhatme, D. S. T. Hsieh, and R. Langer, A new approach to achieve zero-order release kinetics from diffusion-controlled polymer matrix systems, in *Controlled Release of Bioactive Materials* (R. Baker, Ed.), Academic Press, New York, pp. 177–188 (1980).

79. D. S. Hsieh, W. D. Rhine, and R. Langer, Zero-order controlled-release polymer matrices for micro- and macromolecules, *J. Pharm. Sci.*, *72*:17–22 (1983).

80. D. Brooke and R. J. Washkuhn, Zero-order drug delivery system: Theory and preliminary testing, *J. Pharm. Sci.*, *66*:159–162 (1977).

81. R. A. Lipper and W. I. Higuchi, Analysis of theoretical behavior of a proposed zero-order drug delivery system, *J. Pharm. Sci.*, *66*:163–164 (1977).

82. W. Y. Kuu and S. H. Yalkowsky, Multiple-hole approach to zero-order release, *J. Pharm. Sci.*, *74*:926–933 (1985).

83. D. R. Paul and S. K. McSpadden, Diffusional release of a solute from a polymer matrix, *J. Membrane Sci.*, *1*:33–48 (1976).

84. R. W. Baker and H. K. Lonsdale, Controlled-release: Mechanisms and rates, in *Advances in Experimental Medicine and Biology*, Vol. 47 (A. C. Tanquary and R. E. Lacey, Eds.), Plenum Press, New York (1974).

85. K. Tojo, Y. Sun, M. Ghannam, and Y. Chien, Simple evaluation method of intrinsic diffusivity for membrane-moderated controlled release, *Drug Development Ind. Pharm.*, *11*(6&7):1363–1371 (1985).

86. D. D. Do and A. A. Elhassadi, A theory of limiting flux in a stirred batch cell, *J. Membrane Sci.*, *25*:113–132 (1985).

87. F. Bellucci and E. Drioli, Protein ultrafiltration: An experimental study, *J. Appl. Polymer Sci.*, *19*:1639–1647 (1975).

88. J. L. Kuester and J. H. Mize, *Optimization Techniques with FORTRAN*, McGraw-Hill, New York, pp. 251–271 (1973).

89. K. Tojo, Intrinsic release rate from matrix-type drug delivery systems, *J. Pharm. Sci.*, *74*(6):685–687 (1985).

90. K. Sakai, K. Ozawa, R. Mimura, and H. Ohashi, Comparison of methods for characterizing microporous membranes for plasma separation, *J. Membrane Sci.*, *32*:3–17 (1987).

91. A. Leo, C. Hansch, and D. Elkin, Partition coefficients and their uses, *Chem. Rev.*, *71*:525–616 (1971).

92. T. J. Roseman and N. F. Cardarelli, Monolithic polymer devices, in *Controlled Release Technologies: Methods, Theory and Applications*, Vol. I (A. F. Kydonieus, Ed.), CRC Press, Boca Raton, Fla., pp. 21–54 (1980).

93. K. Sundaram and F. A. Kincl, Sustained release hormonal preparations. 2. Factors controlling the diffusion of steroids through dimethyl polysiloxane membranes, *Steroids*, *12*:517 (1968).

94. G. L. Flynn, N. F. H. Ho, S. Hwang, E. Owada, A. Molokhia, C. R. Behl, W. I. Higuchi, T. Yotsuyanagi, Y. Shah, and J. Park, Interfacing matrix release and membrane absorption— analysis of steroid absorption from a vaginal device in the rabbit doe, in *Controlled Release Polymeric Formulations* (D. R. Paul and F. W. Harris, Eds.), *ACS Symp. Ser.*, *33*:87 (1976).

95. T. J. Roseman, Release of steroids from a silicone polymer, *J. Pharm. Sci.*, *61*:46 (1972).

96. Y. W. Chien and H. J. Lambert, Controlled drug release from polymeric delivery devices. II. Differentiation between partition controlled and matrix controlled drug release mechanisms, *J. Pharm. Sci.*, *63*:515 (1974).

97. T. J. Roseman and S. H. Yalkowsky, Importance of solute partitioning on the kinetics of drug release from matrix systems, in *Controlled Release Polymeric Formulations* (D. R. Paul and F. W. Harris, Eds.), *ACS Symp. Ser.*, *33*:33 (1976).

98. T. J. Roseman and S. H. Yalkowsky, Influence of solute properties on release of *p*-aminobenzoic acid esters from silicone rubber: Theoretical considerations, *J. Pharm. Sci.*, *63*:1639 (1974).

99. Y. W. Chien, H. J. Lambert, and T. K. Lin, Solution-solubility dependency of controlled release of drug from polymer matrix: Mathematical analysis, *J. Pharm. Sci.*, *64*:1643 (1975).

100. W. R. Good and P. I. Lee, Membrane-controlled reservoir drug delivery systems, in *Medical Applications of Controlled Release*, Vol. I (R. S. Langer and D. L. Wise, Eds.), CRC Press, Boca Raton, Fla., pp. 1–39 (1984).

101. A. S. Michaels, P. S. L. Wong, R. Prather, and R. M. Gale, Thermodynamic method of predicting the transport of steroids in polymer matrices, *AIChE J.*, *21*(6):1073–80 (1975).

102. Y. W. Chien, Methods to achieve sustained drug delivery–The physical approach: Implants, in *Sustained and Controlled Release Drug Delivery Systems* (J. R. Robinson, Ed.), Marcel Dekker, New York, p. 243 (1978).

103. Y. W. Chien, Thermodynamics of controlled release from polymeric delivery devices, presented at American Chemical Society Symposium on Controlled Release Polymeric Formulations, New York, April 4–9, 1979.

104. C. G. Pitt, Y. T. Boa, A. L. Andrady, and P. N. K. Samuel, The correlation of polymer-water and octanol-water partition coefficients: Estimation of drug solubilities in polymers, *Int. J. Pharm.*, *45*:1–11 (1988).

105. S. H. Yalkowsky and W. A. Morozowich, A physical chemical basis for prodrugs in drug design, in *Drug Design*, Vol. 9 (E. J. Ariens, Ed.), Academic Press, New York, pp. 122–185 (1980).

106. C. Hansch and A. Leo, *The Pomona College Medicinal Chemistry Project*, Pomona College, Claremont, Calif. (1988).

107. C. Hansch and A. Leo, Substituent constants for correlation analysis, in *Chemistry and Biology*, John Wiley, New York (1979).

108. W. T. Nanta and R. F. Rekker, *The Hydrophobic Fragmentation Constant*, Elsevier, New York (1977).

109. R. W. Taft, M. H. Abraham, G. R. Famini, R. M. Doherty, J.-L. M. Abboud, and M. J. Kamlet, Solubility properties in polymers and biological media 5: An analysis of the physicochemical properties which influence octanol-water partition coefficients of aliphatic and aromatic solutes, *J. Pharm. Sci.*, *74*:807–814 (1985).

110. G. Klopman and L. D. Iroff, Calculation of partition coefficients by the charge density method, *J. Comput. Chem.*, *2*:157–160 (1981).

111. D. E. Leahy, Intrinisic molecular volume as a measure of the cavity term in linear solvation energy relationships: Octanol-water partition coefficients and aqueous solubilities, *J. Pharm. Sci.*, *75*:629–636 (1986).

112. M. S. Mirrlees, S. J. Moulton, C. T. Murphy, and P. J. Taylor, Direct measurement of octanol-water partition-coefficients by high-pressure liquid chromatography, *J. Med. Chem.*, *19*:615–619 (1976).

113. S. R. Unger, J. R. Cook, and J. S. Hollenberg, Simple procedure for determining octanol-aqueous partition, distribution and ionization coefficients by reversed-phase high-pressure liquid chromatography, *J. Pharm. Sci.*, *67*:1364–1367 (1978).

114. J. C. Caron and B. Schroot, Determination of partition coefficients of glucocorticosteroids by high-performance liquid chromatography, *J. Pharm. Sci.*, *73*:1703–1706 (1984).

115. C. Hansch, J. E. Quinlan, and G. L. Lawrence, The linear free-energy relationship between partition coefficients and the aqueous solubility of organic liquids, *J. Org. Chem.*, *33*:347–350 (1968).

116. S. H. Yalkowsky, C. C. Valvani, and T. J. Roseman, Solubility and partitioning VI: Octanol solubility and octanol-water partition coefficients, *J. Pharm. Sci.*, *72*:866–870 (1983).

117. G. Levy and W. J. Jusko, Effect of viscosity on drug absorption, *J. Pharm. Sci.*, *54*:219–224 (1965).

118. K. G. Nelson and A. C. Shah, Convective diffusion model for a transport-controlled dissolution rate process, *J. Pharm. Sci.*, *64*:610–614 (1975).

119. A. C. Shah and K. G. Nelson, Evaluation of a convective diffusion drug dissolution rate model, *J. Pharm. Sci.*, *64*:1518 (1975).

120. K. G. Nelson and A. C. Shah, Mass transport in dissolution kinetics I: Convective diffusion to assess the role of fluid viscosity under forced flow conditions, *J. Pharm. Sci.*, *76*:799 (1987).

121. K. G. Nelson and A. C. Shah, Mass transport in dissolution kinetics II: Convective diffusion to assess the role of fluid viscosity under conditions of gravitational flow, *J. Pharm. Sci.*, *76*:910 (1987).

122. J. Leffler and H. T. Cullinan, Jr., Variation of liquid diffusion coefficients with composition, *Ind. Eng. Chem. Fundam.*, *9*:84–88 (1970).

123. A. R. Gordon, Diffusion constant of an electrolyte and its relation to concentration, *J. Chem. Phys.*, 5:522–526 (1937).

124. G. M. Barrow, *Physical Chemistry*, 3rd ed., McGraw-Hill, New York (1973).

125. G. N. Lewis and M. Randall, *Thermodynamics and the Free Energy of Chemical Substance*, McGraw-Hill, New York (1923).

126. A. C. Riddiford, The rotating disk system, in *Advances in Electrochemistry and Electrochemical Engineering*, Vol. 4 (P. Delahay, Ed.), Interscience, New York (1966).

127. F. Theeuwes, U.S. Patent 4,511,352, "Parenteral delivery system with in-line container," assigned to ALZA Co., Palo Alto, Calif. (April 16, 1985).

128. A. Wolfe, J. M. Davenport, F. Theeuwes, and S. I. Yum, U.S. Patent 4,533,348, "In-line drug dispenser for use in intravenous therapy," assigned to ALZA Co., Palo Alto, Calif. (August 6, 1985).

129. F. Theeuwes, U.S. Patent 4,552,555, "System for intravenous delivery of a beneficial agent," assigned to ALZA Co., Palo Alto, Calif. (November 12, 1985).

130. F. Theeuwes, U.S. Patent 4,586,922, "Intravenous system for delivering a beneficial agent," assigned to ALZA Co., Palo Alto, Calif. (May 6, 1986).

131. L. H. Tran, U.S. Patent 4,715,850, "Therapeutic agent delivery system and method," assigned to Controlled Release Technologies, Inc., Batavia, Ill. (December 29, 1987).

132. O. Levenspiel, *Chemical Reaction Engineering*, 2nd ed., John Wiley, New York, chap. 9 (1972).

133. C. Y. Wen and L. T. Fan, *Models for Flow Systems and Chemical Reactors*, Marcel Dekker, New York (1975).

134. F. A. Kincl and H. W. Rudel, Sustained release hormonal preparations, *Acta Endocrinol. Suppl.* 5 (1970).

135. R. L. Shippy, S. T. Hwong, and R. G. Bunge, Controlled release of testosterone using silicone rubber, *J. Biomed. Mater. Res.*, 7:55 (1973).

136. FIDAP, *General Introduction Manual*, Fluid Dynamics International, Evanston, Ill. (1990).

137. M. J. D. Powell, A method for minimizing a sum of squares of nonlinear functions without calculating derivatives, *Computer J.*, 7:303–307 (1965).

138. D. M. Himmelblau, *Applied Nonlinear Programming*, McGraw-Hill, New York, p. 167 (1972).

139. M. Gibaldi and D. Perrier, *Pharmacokinetics*, 2nd ed., Marcel Dekker, New York (1982).

140. B. Carnahan, H. A. Luther, and J. O. Wilkes, *Applied Numerical Methods*, John Wiley, New York (1969).

141. P. A. Witherspoon and D. N. Saraf, Diffusion of methane, ethane, propane, and butane in water from 25 to 43°C, *J. Phys. Chem.*, 69:3752–3755 (1965).

142. L. G. Longsworth, Diffusion in liquids, in *American Institute of Physics Handbook*, 2nd ed. (D. E. Gray, Coord. Ed.), McGraw-Hill, New York, pp. 2–205 to 2–212 (1963).

143. C. Gary-Bobo and H. W. Weber, Diffusion of alcohols and amides in water from 4 to 37°, *J. Phys. Chem.*, 73:1155–1156 (1969).

144. W. D. Stein and S. Nir, Spontaneous and induced changes in the membrane potential and resistance of Acetabularia mediterranea, *J. Membrane Biol.*, 5:246–249 (1971).

145. W. J. Albery, A. R. Greenwood, and R. F. Kibble, Diffusion coefficients of carboxylic acids, *Trans. Faraday Soc.*, 63:360–368 (1967).

146. R. G. Stehle and W. I. Higuchi, *In vitro* model for transport of solutes in three-phase system II: Experimental considerations, *J. Pharm. Sci.*, 61:1931 (1972).

147. M. Ihnat and D. A. I. Goring, Shape of the cellodextrins in aqueous solution at 25°, *Can. J. Chem.*, 45:2353–2361 (1967).

148. M. A. Gonzales, J. Nematollaji, W. L. Guess, and J. Autian, Diffusion, permeation, and solubility of selected agents in and through polyethlene, *J. Pharm. Sci.*, 56:1288 (1967).

149. G. L. Flynn and T. J. Roseman, Membrane diffusion. II. Influence of physical adsorption

on molecular flux through heterogeneous dimethylpolysiloxane barriers, *J. Pharm. Sci.*, *60*:1788 (1971).

150. R. E. Lacey and D. R. Cousar, Factors affecting the release of steroids from silicones, in *Advances in Experimental Medicine and Biology*, Vol. 47 (A. C. Tanquary and R. E. Lacey, Eds.), Plenum Press, New York, p. 117 (1974).

151. E. K. L. Lee, H. K. Lonsdale, R. W. Baker, E. Driolli, and P. A. Bresnahan, Transport of steriods in poly(etherurethane) and poly(ethylene vinyl acetate) membranes, *J. Membrane Sci.*, *24*:125–243 (1985).

152. J. Halebliab, R. Runkel, N. Muller, J. Christopherson, and K. Ng, Steroid release from silicone elastomer containing excess drug in suspension, *J. Pharm. Sci.*, *60*:541 (1971).

153. J. S. Kent, Controlled release of delmadinone acetate from silicone rubber tubing: In vitro–in vivo correlations to diffusion model, in *Controlled Release Polymeric Formulations*, *ACS Symposium Series* 33 (D. R. Paul and F. W. Harris, Eds.), American Chemical Society, Washington, D.C., p. 157 (1976).

154. Y. W. Chien, H. J. Lambert, and L. F. Rozek, Controlled release of deoxycorticosterone acetate from matrix-type silicone devices: *In vitro–in vivo* correlation and prolonged hypertension animal model for cardiovascular studies, in *Controlled Release Polymeric Formulations*, ACS Symposium Series 33 (D. R. Paul and F. W. Harris, Eds.), American Chemical Society, Washington, D.C., p. 72 (1976).

155. R. W. Baker, M. E. Tuttle, H. K. Lonsdale, and J. W. Ayres, Development of an estriol-releasing intrauterine device, *J. Pharm. Sci.*, *68*:20 (1979).

156. C. F. Most, Jr., Some filler effects on diffusion in silicone rubber, *J. Appl. Polym. Sci.*, *14*:1019 (1970).

157. Y. W. Chien, H. J. Lambert, and D. E. Grant, Controlled drug release from polymeric devices. I. Technique for rapid in vitro release studies, *J. Pharm. Sci.*, *63*:365 (1974).

158. W. R. Vezin and A. T. Florence, Diffusion of small molecules in amorphous polymers, *J. Pharm. Pharmacol.*, *29*(suppl.):44 (1977).

159. J. C. Fu, C. Hagemeir, D. L. Moyer, and E. W. Ng, A unified mathematical model for diffusion from drug-polymer composite tablets, *J. Biomed. Mater. Res.*, *10*:743 (1976).

160. J. Haleblian, R. Runkel, N. Mueller, J. Christopherson, and K. Ng, Steroid release from silicone elastaner containing excess drug in suspension, *J. Pharm. Sci.*, *60*:541 (1971).

161. G. M. Zentner, J. R. Cardinal, J. Feijen, and S.-Z. Song, Progestin permeation through polymer membranes IV: Mechanism of steroid permeation and functional group contributions to diffusion through hydrogel films, *J. Pharm. Sci.*, *68*:970–975 (1979).

162. H. Burrell, Solubility parameters, *Interchem Rev.*, *14*(1):3 (1955).

163. A. Beerbower, L. A. Kaye, and D. A. Pattison, Picking the right elastomer to fit your fluids, *Chem. Eng.*, *74*:118 (1967).

164. C. M. Gary-Bobo, R. DiPolo, and A. K. Solomon, Role of hydrogen-bonding in nonelectrolyte diffusion through dense artificial membranes, *J. Gen. Physiol.*, *54*:369–382 (1969).

165. J. A. Barrie, J. D. Levine, A. S. Michaels, and P. Wong, Diffusion and solution of gases in composite rubber membranes, *Trans. Farady Soc.*, *59*:869–78 (1963).

3
Diffusion-Controlled Matrix Systems

Ping I. Lee *University of Toronto, Toronto, Ontario, Canada*

INTRODUCTION

Historically, the most popular drug delivery system has been the matrix system containing uniformly dissolved or dispersed drug, such as tablets and granules, because of its low cost and ease of fabrication. However, as shown in Fig. 1, the release behavior is inherently first-order in nature, with continuously diminishing release rate for all three standard geometries: slab, cylinder, and sphere. This is the result of an increase in diffusional resistance and a decrease in effective area at the diffusion front as drug release proceeds. With the growing awareness of the need to optimize therapy, matrix systems providing programmable rates of drug delivery (in most cases other than the typical first-order delivery) are becoming more important. One example is the use of zero-order (or constant-rate) delivery systems to maintain and prolong optimum therapeutic drug concentration in the blood for drugs with short biological half-lives [1].

Methods of altering the kinetics of drug release from the inherent first-order behavior, especially to achieve a constant rate of drug release from matrix devices, have involved the use of geometry factors, erosion/dissolution control and swelling control mechanisms, nonuniform drug loading, and matrix-membrane combinations. The fundamentals of diffusion phenomena and the effects of geometry and rate-controlling membranes on the kinetics of drug release have been described in detail in the previous chapter. In this chapter we examine additional methodologies for modifying drug release from matrix devices. Special emphasis is placed on the mechanism and analysis of drug-release kinetics, and examples relating to the application of these approaches are provided wherever possible.

SYSTEMS INVOLVING A DIFFUSIONAL MOVING BOUNDARY

Polymeric materials are often used to fabricate matrix-type drug delivery systems via a variety of processes, e.g., casting, compaction, extrusion, and lamination. In these sys-

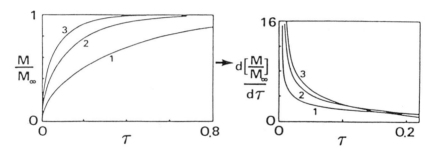

Figure 1 Inherent first-order drug release behavior in matrix systems: (left) cumulative release; (right) release rate. Key: 1, slab; 2, cylinder; 3, sphere.

tems, the drug is usually dissolved or dispersed uniformly throughout the device. Drug release from such matrix systems is governed mainly by the diffusion process, although other processes, such as swelling and erosion, may also take place simultaneously, depending on the nature of the drug and the matrix material. One universal feature during the release of a dissolved or dispersed drug from a matrix device involves the presence of a moving diffusional front separating an undissolved core from a partially extracted region (Fig. 2a). In the case of an erodible or swellable matrix, the release kinetics are further complicated by the presence of a second moving boundary, namely, the eroding or swelling polymer front (Fig. 2b).

To gain insight into factors governing the release kinetics, one often has to carry out a careful theoretical analysis of the associated transport problem. The mathematical descriptions of diffusion problems involving moving boundaries mentioned above are known as moving-boundary problems, free-boundary problems, or simply Stefan problems. Except in some special cases, only a few exact solutions are known because of the nonlinearity introduced by the moving boundary. This difficulty can be overcome by introducing approximate analytical solutions to specific diffusion problems. The results of these approaches and their relevance to the interpretation of release kinetics are illustrated below.

The general diffusion equation describing the kinetics of drug release from a polymer matrix of dimension a is

$$\frac{\partial C}{\partial t} = x^{-\alpha} \frac{\partial}{\partial x}\left(x^{\alpha} D \frac{\partial C}{\partial x}\right) \tag{1}$$

where D is the drug diffusion coefficient in the matrix, a is the half-thickness or radius ($x = 0$ at the center and $x = a$ at the surface), and $\alpha = 0$, 1, and 2 for slab (or sheet), cylinder, and sphere, respectively. The boundary conditions needed to solve Eq. (1) vary depending on the type of system under consideration.

Rigid Matrix: Perfect Sink Condition

The "rigid" matrix considered here involves essentially negligible or no movement of the device surface due to swelling or erosion. The drug concentration in the external release medium is maintained under sink condition and assumed not to be affected by the drug release. This is the simplest moving-boundary problem, as it involves only one moving diffusion front.

The "rigid" matrix assumption is generally valid for nonswellable hydrophobic

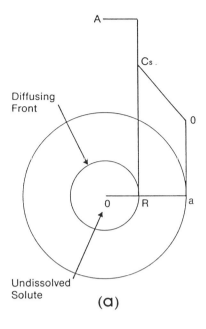

(a)

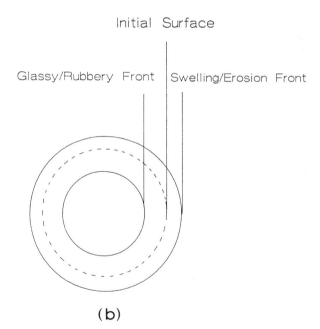

(b)

Figure 2 Moving boundaries involved in matrix devices during drug release; (a) rigid matrix—single diffusional front; (b) erodible/swellable matrix—multiple fronts.

polymer matrices, such as PVC or EVAc, and for hydrophilic polymer matrices swollen prior to the release study. The sink condition is normally approximated by using a large volume of releasing medium with adequate stirring to minimize the hydrodynamic boundary-layer effect. Under these conditions, the initial and boundary conditions for Eq. (1) are

$$R(t) = a \qquad \text{at } t = 0 \tag{2a}$$

$$C(x,t) = 0 \qquad \text{at } x = a \tag{2b}$$

$$C(x,t) = C_s \qquad \text{at } x = R(t) \tag{2c}$$

$$D\frac{\partial C}{\partial x} = (A - C_s)\frac{dR}{dt} \qquad \text{at } x = R(t) \tag{2d}$$

where A is the drug loading per unit volume, C_s is the drug solubility in the matrix, and R(t) is the time-dependent position of the moving diffusion front, the only moving boundary in this problem. The sharp concentration discontinuity at the front [Fig. 2a and Eq. (2d)] implies that transfer of the drug from dispersed to dissolved state is rapid compared to the diffusion process. In other words, the diffusion process is assumed to be rate-limiting.

The mathematical solution to this moving-boundary problem is very important for describing the drug release kinetics from matrix devices with drug loadings greater than the solubility of the drug in the matrix. Although the exact and pseudo-steady-state solutions to this problem have been described in the previous chapter, they are briefly repeated below for a comparison with another, more accurate approximate analytical solution.

The only exact solution for Eqs. (1) − (2), originally described by Carslaw and Jager [2] and later adopted by Paul and McSpadden [3], is the one available for the semi-infinite slab geometry:

$$M = \frac{2}{\sqrt{\pi}\,\text{erf}(\xi)}\,[C_s\sqrt{Dt}] \tag{3}$$

with

$$\sqrt{\pi}\xi\exp(\xi^2)\,\text{erf}(\xi) = \frac{C_s}{(A - C_s)} \tag{3a}$$

where M is the drug released per unit area.

Since the exact solution involves a transcendental expression that is cumbersome for routine usage, several useful approximate analytical solutions were developed and reported in the literature. A familiar example is Higuchi's equation, based on a pseudo-steady-state approximation [4]:

$$M = [C_s(2A - C_s)Dt]^{1/2} \tag{4}$$

However, because of the pseudo-steady-state assumption involved, which assumes a linear drug concentration profile in the matrix, Higuchi's equation is valid only when the drug loading is in great excess of the drug solubility ($A \gg C_s$). At the limit of $A \rightarrow C_s$, Higuchi's equation gives a result 11.3% smaller than the exact solution.

Based on a refined heat balance integral method, Lee [5] presented a simple analytical solution for this problem, which has been shown t be uniformly valid for all A/C_s values:

$$M = C_s\left[\frac{1 + H}{\sqrt{3H}}\right](Dt)^{1/2} \tag{5}$$

where

$$H = 5\left(\frac{A}{C_s}\right) + \left[\left(\frac{A}{C_s}\right)^2 - 1\right]^{1/2} - 4$$

when Eq. (5) is applied to the analysis of drug release, the deviations from the exact results are consistently one order of magnitude smaller than those of Higuchi's equation. At $A/C_s > 1.04$, the accuracy of Eq. (5) is within 1% of the exact result obtained from Eq. (3).

Therefore, Eqs. (3), (4), and (5) all predict a square-root-of-time dependence for the drug release. However, Eq. (5) is much easier to use routinely than the exact solution, Eq. (3), and is much more accurate than the pseudo-steady-state result, Eq. (4), particularly at low A/C_s values. The latter case occurs quite often in delivery systems involving hydrophilic polymers and drugs of high water solubilities.

Both Higuchi [6] and Lee [5] have derived results for the release of dispersed drug from spherical matrices based on the respective approximations stated above. The results are all expressed in parametric form, since it is not possible to obtain the drug release as a direct function of time.

Since both results are correlated through the reduced diffusion front position, δ, and are more complex than Eqs. (4) and (5), only the more accurate results of Lee are given here in terms of the fractional release:

$$\frac{M}{M_\infty} = [1 - (1 - \delta)^3]\left(1 - \frac{C_s}{A}\right) + 3\delta\left(\frac{C_s}{A}\right)$$
$$\left[\left(a_1 + \frac{a_2}{2} + \frac{a_3}{3}\right) - \left(\frac{a_1}{2} + \frac{a_2}{3} + \frac{a_3}{4}\right)\delta\right] \quad (6a)$$

where M_∞ is the total amount of drug released, with

$$\frac{Dt}{a^2} = \frac{1}{12}\left[6\left(\frac{A}{C_s}\right) - 4 - a_3\right] - \frac{1}{3}\left(\frac{A}{C_s} - 1\right)\delta^3 \quad (6b)$$

where

$$a_1 = 1 \qquad a_2 = -a_3 - 1$$

and

$$a_3 = 1 - \left(1 - \frac{A}{C_s}\right)(1 - \delta) - \left(\left[1 - \left(1 - \frac{A}{C_s}\right)(1 - \delta)\right]^2 - 1\right)^{1/2}$$

Interested readers should refer to the original references for details. It should be noted that the limitations on (A/C_s) values described for the pseudo-stead-state solutions in slab geometry applications also apply to Higuchi's results on spherical matrices.

Example. The following fractional release data were measured for the release of oxprenolol HCl from a spherical hydrogel bead (diameter = 1 mm) at 25°C:

t (min)	0	6	23	49.2	83	121.5	163	205
M/M_∞	0	0.17	0.32	0.45	0.56	0.65	0.72	0.77

If the drug loading-to-solubility ratio A/C_s is 1.5 for this spherical hydrogel system and the bead was swollen prior to the release study, determine the oxprenolol diffusion co-efficient in this hydrogel. Assume that the change in sample dimension during the drug release is negligible.

Solution. From Eqs. (6a) and (6b), we first calculate M/M_∞ and Dt/a^2 by setting δ at different values:

δ	0	0.1	0.2	0.3	0.4	0.5	0.6	0.7
M/M_∞	0.17	0.32	0.45	0.46	0.56	0.65	0.72	0.77
Dt/a^2	0	0.00366	0.01393	0.0297	0.04974	0.07292	0.0979	0.12323

By plotting Dt/a^2 versus t, one obtains a straight line. Since a = $\frac{1}{2}$ mm = 0.05 cm, the diffusion coefficient for oxprenolol HCl calculated from the slope ($= D/a^2$) of the straight line is 2.5×10^{-8} cm^2/s.

Rigid Matrix: Release into a Constant, Finite Volume

The release of a dissolved or dispersed drug from a rigid matrix into a constant, finite volume is particularly relevant to the situation of a body cavity of limited volume containing a drug delivery device whose release rate is in excess of the removal rate of the drug. Intuitively, the external drug concentration (C_b) will be building up monotonically with time due to the condition of a constant, finite volume. The assumption of a perfect sink is no longer realistic here; however, the assumption of a negligible hydrodynamic boundary-layer effect is still reasonable if the diffusion rate of the drug in the external aqueous environment is much larger than that in the device.

For the sake of simplicity, we will illustrate only the case of dissolved drug, that is, $A \leq C_s$. For this problem, the boundary conditions for Eq. (1) are as follows:

$$C(x, t) = sC_b \qquad \text{at } x = a \tag{7a}$$

$$C(x, t) = A \qquad \text{at } x = R(t) \tag{7b}$$

$$D\frac{\partial C}{\partial x} = -\left(\frac{\lambda a}{a + 1}\right)\frac{\partial C}{\partial t} \qquad \text{at } x = a \tag{7c}$$

$$\frac{\partial C}{\partial x} = 0 \qquad \text{at } x = R(t) \tag{7d}$$

$$R(t) = a \qquad \text{at } t = 0 \tag{7e}$$

where λ is the effective volume ratio, defined as

$$\lambda = \frac{V}{2s\sigma a} \tag{7f}$$

for a slab,

$$\frac{V}{s\pi a^2 l}$$

for a cylinder, and

$$\frac{3V}{4s\pi a^3}$$

for a sphere, where V is the total liquid volume, a is the half thickness of a slab or the radius of a cylinder or sphere, s is the distribution coefficient at the surface, σ is the surface area of each side of the slab, and l is the length of the cylinder.

Again, utilizing a refined heat balance integral approach, accurate approximate solutions for different geometries, which closely approximate the exact solutions, have been given by Lee [7]:

$$Slab \qquad \frac{M}{M_\infty} = (1 + \lambda)(1 - \theta_b) \tag{8a}$$

where θ_b is evaluated from

$$\frac{Dt}{a^2} = \frac{3\lambda^2}{4}\left[\ln \theta_b + \frac{1}{2\theta_b^2} - \frac{1}{2}\right] \tag{8b}$$

$$Cylinder \qquad \frac{M}{M_\infty} = (1 + \lambda)(1 - \theta_b) \tag{9a}$$

where θ_b is defined by

$$\frac{Dt}{a^2} = \frac{3}{40}\lambda^3\left(\frac{1}{\theta_b} - 1\right)^3 + \frac{3(3\lambda + 5)}{160}\lambda^2\left(\frac{1}{\theta_b} - 1\right)^2$$
$$+ \frac{3\lambda^2(5 - 3\lambda)}{80}\left(\frac{1}{\theta_b} - 1\right) - \frac{3\lambda^2(5 - 3\lambda)}{80}\ln\left(\frac{1}{\theta_b}\right) \tag{9b}$$

$$Sphere \qquad \frac{M}{M_\infty} = (1 + \lambda)\left[1 - \frac{4\lambda}{4(1 + \lambda) - (2 - \delta)^2}\right] \tag{10a}$$

where δ is evaluated from

$$\frac{Dt}{a^2} = -\frac{\delta}{3} - \left(\frac{\lambda + 2}{3}\right)\ln\left[\frac{4\lambda + 4 - (\delta - 2)^2}{4\lambda}\right] + \frac{2}{3}$$
$$(1 + \lambda)^{1/2}\ln\left|\frac{[2(1 + \lambda)^{1/2} + (\delta - 2)][(1 + \lambda)^{1/2} + 1]}{[2(1 + \lambda)^{1/2} - (\delta - 2)][(1 + \lambda)^{1/2} - 1]}\right| \tag{10b}$$

As shown in Figs. 3a, 3b, and 3c for all three geometries, these results closely approximate the exact solutions for the major part of the drug release. The deviation starts to appear only at the tail end of the release. For λ values greater than 10, which corresponds to the experimental condition of negligible external concentration buildup, the results become indistinguishable from those predicted for perfect sink conditions. It is interesting to observe from Figs. 3a–3c that the results predict a faster completion of drug release with smaller effective volume ratios (λ), although the actual amount released will be smaller in this case due to the finite-volume effect. This corresponds to either a small volume of releasing medium or a large drug partition coefficient for a given delivery system. In addition, the nonlinear square-root-of-time plots in Figs. 3a–3c suggest a shift from $t^{0.5}$ dependence for larger λ's to n^n dependence with $n < 0.5$ for smaller λ's. Expressions similar to Eqs. (8)–(10) can also be used to determine diffusion coefficients from sorption experiments carried out in a constant, finite volume. This aspect and a relevant example have been described in detail in the previous chapter.

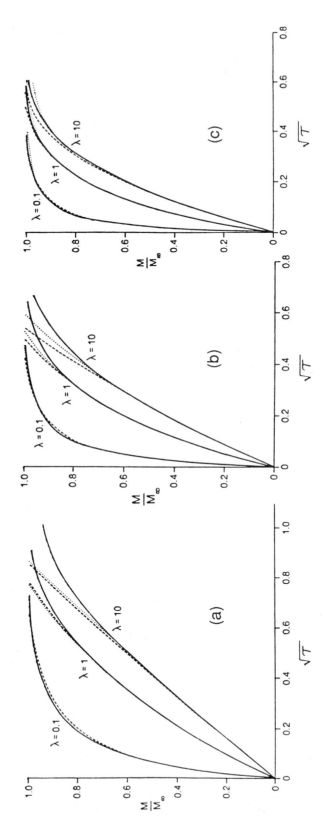

Figure 3 Predicted fractional drug releases ($A \leq C_s$) in a constant, finite volume from (a) slab, (b) cylinder, and (c) sphere: ———, exact; ········, short time solution; ‑‑‑‑‑‑, Eqs. (8) through (10); $\tau = Dt/a^2$. (From Ref. 7.)

Effect of Porosity

When well-defined pores exist throughout the matrix structure, the kinetics of drug release from a rigid matrix can still be described by Eqs. (3)–(10), provided that an effective drug diffusion coefficient D_{eff} and effective distribution coefficient s_{eff} are utilized in place of D and s. When drug diffusion occurs only through the solvent-filled porous network, the effective drug diffusion coefficient can be related to the matrix structure by

$$D_{eff} = \frac{D\epsilon}{\tau^*}$$

and the effective distribution coefficient is expressed as

$$s_{eff} = \epsilon$$

where ϵ is the porosity expressed as the volume fraction of void space in the membrane, and τ^* is the tortuosity factor expressed as the ratio of the effective average path length in the porous medium to the shortest distance in the direction of mass flow. Since the ratio ϵ/τ^* represents essentially the fractional area available for drug release, an increase in porosity or a decrease in tortuosity will increase D_{eff}.

The porosity of a polymer matrix can be measured by various methods involving optical, inhibition, mercury intrusion, gas expansion, and density measurements [49]. On the other hand, the tortuosity factor is difficult to estimate and quite dependent on the model (in general, $1 \leq \tau^* \leq 3$). The release behavior can be further complicated by drug diffusion through both the solvent-filled pores and the polymer matrix. Recently, Chang and Himmelstein [42] have successfully utilized a concentration- and position-dependent diffusion coefficient to take into account the effect of increased porosity, due to dissolution of the drug and excipient, on overall drug release.

SYSTEMS INVOLVING POLYMER EROSION/DISSOLUTION

In some systems the polymer matrix either erodes or dissolves, thereby releasing the entrapped drug to the surrounding environment. The term "erosion" is generally reserved for polymers containing hydrolytically or enzymatically labile bonds that undergo hydrolysis or enzymatic cleavage in the environment of use, whereas the term "dissolution" generally describes the result of physical disentanglement of polymer chains by a swelling solvent without involving any chemical reaction. Since the mathematical analyses of both the polymer erosion and dissolution are similar and the moving boundaries involved are the same, both processes will be considered together here.

The Erosion Process

As pointed out by Heller [8], polymer erosion can be controlled by the following three types of mechanisms: (1) water-soluble polymers insolubilized by hydrolytically unstable crosslinks; (2) water-insoluble polymers solubilized by hydrolysis, ionization, or protonation of pendant groups; (3) hydrophobic polymers solubilized by backbone cleavage to small water-soluble molecules. These mechanisms represent extreme cases, since the actual erosion may occur by a combination of mechanisms. Details of the physical and chemical properties of various erodible polymers can be found in the next chapter, on erodible systems.

The release of a dissolved or dispersed drug from an erodible polymer matrix can be controlled by either a bulk-erosion or a surface-erosion mechanism. The situation in which polymer erodes by a pure surface-erosion mechanism is of special interest because the rate of drug release from such devices having constant geometry (slab) will be constant. However, the corresponding release rates from both the cylindrical and spherical geometries will decrease with time. In cases where a diffusional contribution is present in addition to surface erosion, exact analytical solutions are not available owing to the inherent nonlinearity. Lee [5] has presented approximate analytical solutions for the drug release from a surface-erodible polymer slab. When the eroding front moves at a constant velocity, the following solutions have been obtained for the case $A > C_s$:

$$\frac{M}{M_\infty} = \left[1 - \frac{1}{2}\left(\frac{C_s}{A}\right)\right]\delta + \left(\frac{Ba}{D}\right)\left(\frac{Dt}{a^2}\right) \tag{11a}$$

where δ is evaluated from

$$\frac{Dt}{a^2} = \frac{1}{6h}\left[3\left(\frac{A}{C_s}\right) - 2\right]\left[\delta - \frac{1}{2h}\ln(1 + 2\delta h)\right] \tag{11b}$$

and

$$h = \frac{1}{2}\left(1 - \frac{A}{C_s}\right)\left(\frac{Ba}{D}\right) \tag{11c}$$

with B defined as the surface-erosion rate constant having the dimension of velocity; and δ is the relative separation between the diffusion and eroding fronts defined as $\delta = (S - R)/a$, where S is the time-dependent position of the erosion front defined as $S = a - Bt$, R being the time-dependent position of the diffusion front, and a the half-thickness of the slab. It can easily be shown that the rate of movement of diffusion front is

$$-\frac{a}{D}\frac{dR}{dt} = -\frac{1}{\delta[A/C_s - 1]} \tag{11d}$$

In the limit of $A \rightarrow C_s$, or a dissolved drug system, Lee's analysis gives:

$$\frac{M}{M_\infty} = \left(\frac{4Dt}{3a^2}\right)^{1/2} + \left(\frac{Ba}{D}\right)\left(\frac{Dt}{a^2}\right) \tag{12a}$$

and the rate of movement of the diffusion front is

$$-\frac{a}{D}\frac{dR}{dt} = 6\left(\frac{a^2}{Dt}\right)^{1/2} + \left(\frac{Ba}{D}\right) \tag{12b}$$

The parameter Ba/D is the erosion rate-to-matrix permeability ratio, which measures the relative contribution of erosion and diffusion processes. As shown in Fig. 4, where the fractional release is plotted as a function of time and Ba/D, the drug release generally starts with typical first-order kinetics and shifts toward zero-order kinetics with a limiting slope of Ba/D. When the erosion process dominates the diffusion process, i.e., large Ba/D, almost complete zero-order release can result. This apparent zero-order release region has been attributed to the *identical rate* of movement of the diffusing and eroding fronts at a large time—in other words, a synchronization of front velocities. This synchronization of front velocities is an important phenomenon common to all delivery systems involving polymer surface erosion and dissolution. In general, the drug release from such

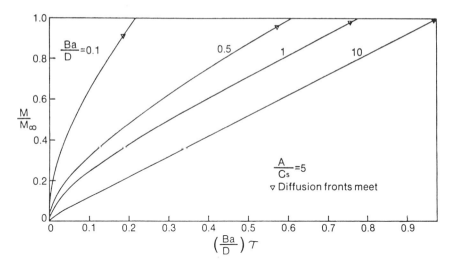

Figure 4   Fractional drug release from an erodible polymer slab ($A/C_s = 5$) as a function of the ratio of erosion rate to matrix permeability. $\tau = Dt/a^2$. (From Ref. 5.)

systems will be zero-order if the parameter Ba/D for the system is large enough for synchronization of front velocities to occur early in time. This synchronization of front velocities is best illustrated in Fig. 5 for the same example considered in Fig. 4. It is clear that the larger the value of Ba/D, the sooner the front synchronization will be reached. This can usually be achieved with a fast-eroding device (large B), a tight polymer matrix (small D), or a thick sample (large a).

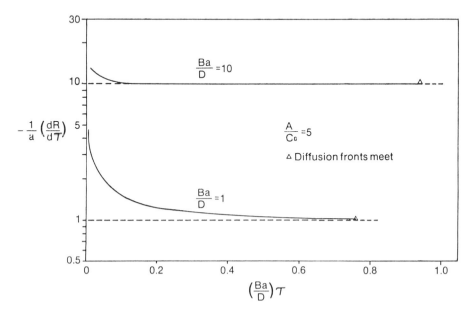

Figure 5   Time dependence of diffusion front velocity in an erodible polymer slab during drug release as a function of the erosion rate-to-matrix permeability ratio. $\tau = Dt/a^2$. (From Ref. 5.)

Such synchronization behavior can also be realized for an erodible system containing dissolved drugs, by utilizing Eqs. (12a) and (12b), in which the fractional release is expressed as a combination of $\sqrt{t}$ and t dependencies, and the diffusion front velocities as the combination of a constant Ba/D and a $1\sqrt{t}$-dependent term. At small t values, the first-order behavior in the fractional release and diffusion-front velocity becomes important. However, at large t values, the linear dependence will dominate the fractional release, and a constant diffusion front velocity will prevail.

Although the mathematical analysis described above does not consider the swelling effect explicitly, similar approaches can be utilized to treat systems with swelling [29]. In general, Eqs. (11) and (12) describe the inward movement of the polymer surface and the associated drug release behavior. The surface erosion rate constant B characterizes the steady-state inward movement of the polymer surface. Therefore, inherently, the analysis does not consider transient processes such as volume swelling or induction time, which may stall or cause outward movement of the polymer surface. Nevertheless, Eqs. (11) and (12) should describe the increase in gel thickness and the dissolution behavior of a swellable polymer system, particularly after the initial outward movement of the swelling polymer surface has been transformed into an inward movement of the dissolving polymer surface. This aspect will be examined in more detail below.

The Dissolution Process

The process of polymer dissolution is a key step during drug release from many non-cross-linked hydrophilic polymer matrices involving controlled-release tablets or granules. These systems are generally prepared either by compressing granulated hydrophilic polymers such as hydroxypropyl or hydroxymethyl cellulose with dispersed drug or by casting and drying of a polymer solution containing a suitable amount of dissolved or dispersed drug. The hydrophilic polymers utilized here are generally glassy in the de-hydrated state. Therefore, when such systems are in contact with an aqueous dissolution medium such as the gastric fluid, two distinctive processes—namely, swelling and true dissolution—occur, leading to the overall dissolution of these systems. In the early stage of the process, the polymer swelling starts because of the transition from glassy to rubbery state due to water penetration, and the matrix thickness increases. Subsequently, when the water concentration at the polymer surface exceeds a critical concentration of macromolecular disentanglement, the true dissolution process occurs.

The schematic diagram of the front positions observed during the overall dissolution is presented in Fig. 6, where R is the glassy/rubbery swelling front and S is the interface between the swollen polymer and the dissolution medium. During the initial swelling step before the start of true dissolution, front R moves inward and front S moves outward. Once the true dissolution has started, front R continues to move toward the center of the device, and front S now moves inward as well. After the disappearance of the glassy core, front R vanishes, and only front S continues to moves inward. It has to be noted that when front S starts to moves inward in the same direction as R, a synchronization of these front movements develops, and the rubbery gel thickness becomes independent of time (see Fig. 7). In essence, this is the same synchronization of front velocities described earlier for systems involving polymer erosion, since the dissolution and erosion of polymers involve the same physical phenomenon. Therefore, the mathematical analysis presented above for systems with polymer erosion should also describe the general behavior of front synchronization and associated drug release. For example, Eq. (11b), which describes

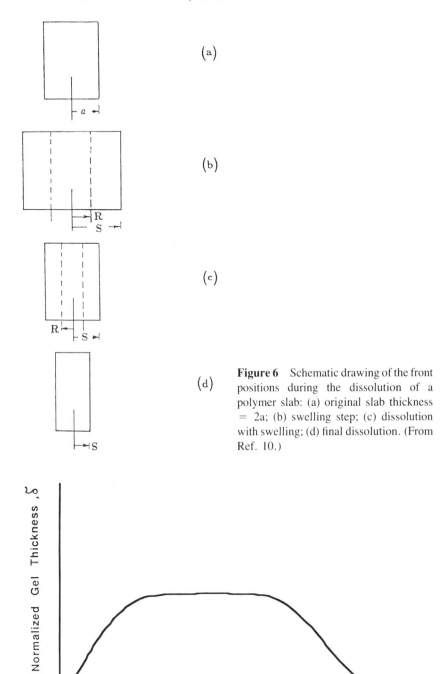

Figure 6 Schematic drawing of the front positions during the dissolution of a polymer slab: (a) original slab thickness = 2a; (b) swelling step; (c) dissolution with swelling; (d) final dissolution. (From Ref. 10.)

Figure 7 Dependence of rubbery gel thickness (normalized $\delta = (S - R)/a$) on dissolution time during drug release from a polymer matrix. (From Ref. 11.)

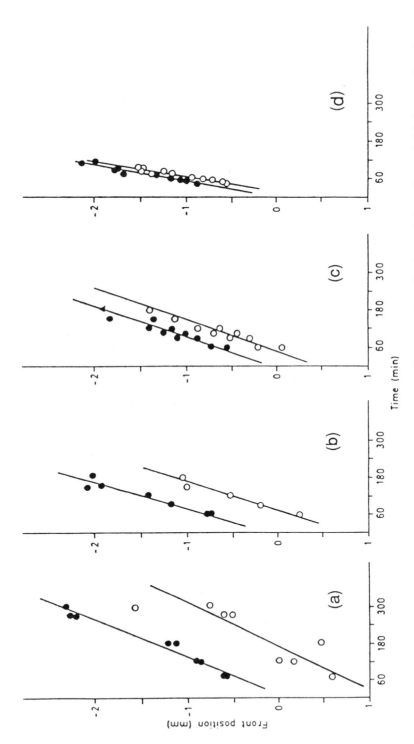

Figure 8 Experimental verification of the synchronization of swelling (○) and eroding (●) fronts during release of diclofenac sodium from erodible polyvinyl alcohol tablets: (a) 25 rpm; (b) 50 rpm; (c) 100 rpm; (d) 200 rpm. (From Ref. 9.)

the time-dependent change of the relative separation between diffusion and eroding fronts, can now be used to describe the normalized gel thickness as a function of time for the present dissolution system. Similarly, Eq. (11a) can now be used to describe the fractional drug release due to polymer dissolution. In fact, Eq. (11b) predicts an initial increase in the normalized gel thickness with a square-root-of-time dependence before reaching a plateau. By differentiating Eq. (11b) with respect to time and letting the derivative $d\delta/dt = 0$, one can calculate the maximum normalized gel thickness in the plateau region of Fig. 7 where the synchronization of fronts will occur:

$$\delta_{max} = \left(\left[4 \frac{A}{C_s} - 1 \right] \frac{Ba}{D} \right)^{-1} \tag{11e}$$

where B now stands for the rate constant of polymer dissolution.

Similar to the erosion results predicted from Eq. (11a) and shown in Fig. 4, the fractional drug release due to polymer dissolution also starts with a typical first-order kinetics and shifts toward zero-order kinetics with a limiting slope of Ba/D. Experimental evidence presented by Colombo et al. [9] on the swelling and dissolution of polyvinyl alcohol matrix tablets has unequivocally confirmed the existence of front synchronization and its link to zero-order drug release (Fig. 8).

More recently, in an attempt to arrive at a predictive model, Lee and Peppas [10] have presented mathematical analyses for the overall dissolution process in the absence as well as in the presence of drugs [11]. These models take into account such polymer parameters as the critical solvent volume fraction for the glass/rubber transition (C^*), disentanglement polymer volume fraction at the dissolving polymer surface (C_d), and polymer density (ρ_p), which are either known or can be measured or calculated. They have formulated the associated moving-boundary problem in the exact form, which requires a numerical scheme for the solutions. To simplify the approach, a pseudo-steady-state approximation was introduced to obtain the desired results. Their models predict:

Without drug $\quad -\dfrac{D\delta}{aKC_d} - \dfrac{D^2(C^* - C_d)(2 - C^*)}{aKC_d(1 - C^*)}$

$$\times \ln \left[1 - \frac{aKC_d(1 - C^*)\delta}{D(2 - C^*)(C^* - C_d)} \right] = \frac{Dt}{a^2} \tag{13a}$$

Since

$$0 \le \frac{aKC_d(1 - C^*)\delta}{D(2 - C^*)(C^* - C_d)} \le 1$$

Eq. (13a) can be simplified to

$$\delta \simeq \left[\frac{2(2 - C^*)(C^* - C_d)Dt}{(1 - C^*)a^2} \right]^{1/2} \tag{13b}$$

where K is the mass transfer coefficient of the dissolved polymer at surface S, and δ is the normalized gel thickness defined similar to Eq. (11b).

With drug $\quad \dfrac{a\delta}{K_dC_d} + \dfrac{[D_s(2 - C^* - C_s)(C^* + C_s - C_d - C_o) + D(C_s - C_o)(C^* + C_s)]}{(1 - C^* - C_s)K_d^2C_d^2}$

$$\times \ln\left[1 - \frac{(1 - C^* - C_s)K_dC_da\delta}{\begin{array}{c}D_s(2 - C^* - C_s)(C^* + C_s - C_d - C_o)\\ + D(C_s - C_o)(C^* + C_s)\end{array}}\right] + t = 0 \quad (14a)$$

When

$$\frac{(1 - C^* - C_s)K_dC_da\delta}{D_s(2 - C^* - C_s)(C^* + C_s - D_d - C_o) + D(C_s - C_o)(C^* + C_s)} \leq 1$$

Eq. (14a) reduces to

$$\delta \approx \left[2\left(\frac{\begin{array}{c}D_s(2 - C_s - C^*)(C^* + C_s - C_d - C_b)\\ + D(C_s - C_o)(C^* + C_s)\end{array}}{1 - C^* - C_s}\right)\left(\frac{t}{a^2}\right)\right]^{1/2} \quad (14b)$$

and the corresponding fractional drug release is

$$\frac{M}{M_\infty} = \frac{C_o}{C_{cd}a}\left\{\left[\frac{2[-D_s(C^* + C_s - C_d - C_o) + D(C_s - C_o)}{\begin{array}{c}+ (D/C_o)(C_s - C_o)]^2(1 - C^* - C_s)t\\ [D_s(2 - C^* - C_s)(C^* + C_s - C_d - C_o)\\ + D(C^* + C_s)(C_s - C_o)]\end{array}}\right]^{0.5} + K_dC_dt\right\} \quad (14c)$$

where

C_s = drug solubility in volume fraction at the drug front R

C_0 = drug volume fraction at the gel/solution interface S

C^* = polymer volume fraction at R

C_d = polymer volume fraction at S

C_{cd} = drug volume fraction in the glassy core

D = drug diffusion coefficient in the swollen polymer

D_s = solvent diffusion coefficient in the polymer matrix

K_d = mass transfer coefficient of the dissolved drug at S

Most of these drug concentration terms can either be measured or estimated from known solubility data. Other parameters such as K, K_d, and C_o can be determined from the slope of the linear drug release portion as well as the normalized gel thickness at the maximum or plateau region. The threshold polymer disentanglement concentration C_d has to be estimated from knowledge of the molecular characteristics of the dissolution process. Recently, a simple rheological method has been reported for the estimation of C_d for several commercially available hydrophilic polymers [12]. The polymer volume fraction for the glass/rubber transition C^* can be calculated from polymer viscoelastic considerations [13]:

$$C^* = \frac{1}{\rho_p}\left(\frac{1}{\rho_p} + \frac{T_g - T}{(\beta/\alpha_f)} \times \frac{1}{\rho_s}\right) \quad (15)$$

where T_g is the glass transition temperature of the polymer, T is the experimental temperature, α_f is the linear expansion coefficient, β is the expansion coefficient contribution of the solvent to the polymer, and ρ_p and ρ_s are the densities of the dry polymer and solvent, respectively.

The predictive power of Eqs. (14a)–(14c) for systems containing drugs has been demonstrated recently for drug release from a polyvinyl alcohol matrix [11]. The measured drug release and normalized gel thickness have been shown to follow the anticipated time dependence. In this case, more component volume fractions and diffusion coefficients have to be determined or estimated. Interested readers should refer to Ref. 11 for additional details. In cases when the volume fractions and diffusion coefficients are not known, the drug release behavior from soluble polymers can still be described by an equation similar in form to that of Eq. (12a), where the fractional drug release is related to a diffusion term (with $t^{1/2}$ dependence) and an erosion/dissolution term (with t dependence). At small t values, the $t^{1/2}$ term will be important. At large t values, however, the linear t dependence will dominate the fractional release as a result of the front synchronization.

Example. When Eqs. (13) and (15) are applied to the dissolution of a PMMA (poly-methylmethacrylate) slab (0.22 cm thick) in MEK (methyl ethyl ketone) at 26°C, we have $T_g = 105°C$, $T = 26°C$, $\alpha_f \simeq 3.7 \times 10^{-4}$ K according to Ferry [13], $\beta = 0.20$ according to Fujita and Kishimoto [14], $\rho_p = 1.95$ g/cm^3, and $\rho_s = 0.785$ g/cm^3. Thus, the value of C* was evaluated as 0.74. D for MEK in PMMA was taken as 5.2×10^{-7} cm^2/s according to Hwang and Cohen [50], and the half-thickness of the slab is a = 0.11 cm. The experimental results have been reported by Parsonage et al. [34] and are reproduced in Fig. 9. It is clear that the normalized gel thickness has a square-root-of-time dependence as predicted by Eq. (13b). From the slope of the square-root-

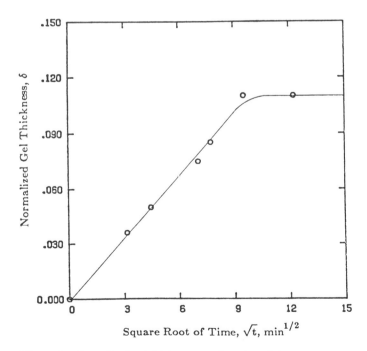

Figure 9 Normalized gel thickness as a function of the square-root-of-dissolution time for the dissolution of PMMA in MEK at 26°C. (From Ref. 34.)

of-time plot, the disentanglement polymer volume fraction C_d was calculated to be 0.74. Therefore, within experimental error, the glass/rubbery transition and the disentanglement volume fractions for the MEK-PMMA system are close to each other. This indicates that PMMA chains disentangle and dissolve shortly after their transition from the glassy to the rubbery state.

SWELLING-CONTROLLED SYSTEMS

Kinetic Considerations

Swelling phenomena are generally encountered in both the hydrophilic and hydrophobic polymer matrices during the release of entrapped water-soluble drugs in an aqueous environment. When the polymer is cross-linked, either chemically through covalent bonding or physically through entanglement or crystallite formation, the swelling will continue to an equilibrium state at which the elastic and osmotic swelling forces balance each other. Depending on the relative magnitude of the rate of polymer swelling to the rate of drug diffusion, various release profiles may be possible. The situation where the polymer structural rearrangement takes place rapidly in response to the swelling solvent as compared to the rate of drug diffusion generally leads to Fickian diffusion, or the so-called first-order release, characterized by a square-root-of-time dependence in both the amount released and the penetrating diffusion front position in a slab geometry.

The case of particular interest is the glassy hydrogel system which, out of necessity, is usually stored in the dehydrated glassy state before usage because of drug stability considerations. The release of water-soluble drugs from such initially dehydrated hydrogels generally involves the simultaneous absorption of water and desorption of drug via a swelling-controlled diffusion mechanism. Similar to the transport of organic penetrant in glassy polymers, such diffusion and swelling in glassy hydrogels generally do not follow a Fickian diffusion mechanism. The slow reorientation of polymer chains in order to accommodate the penetrating solvent molecules often leads to a variety of anomalous sorption behaviors, particularly when the experimental temperatures are near or below the glass-transition temperature of the hydrogel. In cases where the sorption process is completely governed by the rate of polymer relaxation, the so-called Case II transport, characterized by a linear time dependence in both the amount diffused and the penetrating swelling front position, results. In most systems, the intermediate situation, which is often termed non-Fickian or anomalous diffusion, will prevail whenever the rates of diffusion and polymer relaxation are comparable.

Phenomenologically, one can express the fraction released M/M_∞ as a power function of time t for the short-time period.

$$\frac{M}{M_\infty} = k't^n \tag{16}$$

where k' is a constant characteristic of the system and n is an exponent characteristic of the mode of transport. For $n = 0.5$, the solvent diffusion or drug release follows the well-known Fickian diffusion mechanism. For $n > 0.5$, non-Fickian or anomalous diffusion behavior is generally observed. The special case of $n = 1$ describes a Case II transport mechanism, which is particularly interesting since the drug release from such devices having constant geometry will be of constant rate (zero-order). When the fractional drug release from an initially dehydrated hydrogel sheet is plotted as a function of the square

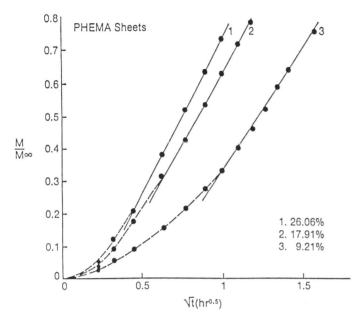

Figure 10 Effect of loading level on thiamine HCl release from glassy PHEMA sheets at 37.5°C. (From Ref. 32.)

root of time, as shown in Fig. 10 for thiamine HCl release from a poly(2-hydroxyethyl methacrylate) (PHEMA) sheet, linearity in the plot is observed only at large times. This reflects the non-Fickian and time-dependent nature of the initial swelling period. As water penetrates a glassy hydrogel matrix containing dissolved or dispersed drug, it requires a finite amount of time for the polymer chains to rearrange to an equilibrium state in order to accommodate the penetrating solvent. Once the hydrogel matrix is significantly hydrated, drug release becomes Fickian, giving rise to the linearity in the square-root-of-time plot (Fig. 10).

To examine the relative rates of diffusion and polymer relaxation, a criteria based on the diffusion Deborah number [15] can be defined:

$$(DEB)_D = \frac{\lambda_m}{\theta_D} \tag{17}$$

where λ_m is a mean relaxation time of the polymer/solvent system relating closely to its shear relaxation modulus, and θ_D is a characteristic diffusion time defined by a^2/D with a the sample half-thickness or radius and D the diffusion coefficient. Since the sample dimension as well as concentration and temperature will affect $(DEB)_D$ in any given system, it is understood that a single value of the Deborah number should at least provide an average value over the range of these experimental parameters. For moderate to large Deborah numbers, $(DEB)_D \approx 1$ or > 1, non-Fickian (anomalous) diffusion, including the limiting Case II transport, can be expected depending on whether the rate of rearrangement of polymer chains is comparable to or smaller than the diffusion rate. On the other hand, classical Fickian diffusion in either the rubbery or glassy state can be expected in the limit of either very small or very large Deborah numbers, i.e., $(DEB)_D \ll 1$ or $(DEB)_D \gg 1$.

Another parameter, the swelling interface number S_W, defined as

$$S_W = v\frac{\delta(t)}{D} \tag{18}$$

has also been proposed to determine the mechanism of drug release from swellable polymer matrices [16], where v is the velocity of the penetrating swelling front, $\delta(t)$ is the time-dependent thickness of the swollen phase, and D is the diffusion coefficient in the swollen phase. S_W is a pseudo-Peclet number that compares the relative mobilities of the penetrating solvent and the drug in the presence of macromolecular relaxation in the polymer. For $S_W \ll 1$, the rate of drug diffusion through the swollen region is much faster than the rate at which the glassy/rubbery front advances, and zero-order release kinetics for the drug are expected. Hopfenberg and Hsu [17] were able to demonstrate this phenomena experimentally in glassy polymer systems undergoing Case II swelling. In the case of $S_W \gg 1$, the swelling front advances faster than the diffusion of the drug, and therefore Fickian diffusion will be observed. For values of $S_W \approx 1$, non-Fickian drug release behavior is anticipated. The dependence of the drug release behavior on various criteria for determining diffusion and release mechanisms is summarized in Table 1. It should be noted that both dimensionless parameters $(DEB)_D$ and S_W are valuable in the conceptual realization of various diffusion mechanisms. However, $(DEB)_D$ has recently been shown to be more general than S_W in predicting drug release behavior in hydrogels [18].

Most of the existing theories on diffusion in glassy polymers consider the transport of a single penetrant, the solvent. The interpretation of various observed anomalous sorption kinetics has been based on the following three approaches: (1) diffusion with convection model, where a constant-swelling-front velocity due to Case II diffusion is incorporated either into the boundary condition or into the diffusion equation as a convective term [19,20]; (2) differential-swelling stress model, where the swelling front velocity is related to the swelling stress exerted by the penetrating solvent on the glassy matrix at the moving front [21,22]; and (3) molecular relaxation model, where the penetrant-induced polymer molecular relaxation is taken into account through the use of a variable surface concentration, a time-dependent diffusion coefficient, or a time-dependent solubility [23–26].

For the case of drug-loaded hydrogels, the release kinetics and swelling behavior are further complicated by the presence of an additional component: the water-soluble drug [27]. Unlike the situation with a single penetrant, the presence of a water-soluble drug alters both the swelling osmotic pressure and the associated time-dependent relaxation of the polymer network during the simultaneous absorption of water and release of drug. Only a few attempts, with limited success, have been made to model these swelling-controlled systems. For example, Good [28] employed a time-dependent drug diffusion coefficient that was arbitrarily set to be proportional to the fractional solvent absorption. The results were used to fit experimental drug release data from initially dehydrated

Table 1 Characteristics of Diffusion in Polymers

Type	$M/M_\infty = k't^n$	$(DEB)_D$	S_W
Fickian diffusion	$n = 0.5$	$\ll 1$ or $\gg 1$	$\gg 1$
Anomalous (non-Fickian diffusion)	$n \neq 0.5$	≈ 1	≈ 1
Case II transport	$n = 1$	> 1	$\ll 1$

hydrogels. Lee [5,29] analyzed drug release from polymers involving moving boundaries, such as the swelling and erosion fronts; and as noted at the beginning of this chapter, accurate approximate analytical solutions for various geometries were obtained. Korsmeyer and Peppas [30] developed mathematical models based on a drug diffusion coefficient that depends on the concentration of the swelling solvent in a functional form consistent with the free-volume theory. Recently, Lee [31,32] demonstrated that by incorporating a time dependence explicitly into the drug diffusion coefficient to reflect the polymer relaxation process, various release behaviors from hydrogels ranging from Case II to Fickian can be described by the analytical solutions to the corresponding moving-boundary problem formulated for a swellable dispersed system. The predicted release behavior is consistent with physical observations and the Deborah number concept.

Time-Dependent Diffusion Coefficient Approach

The importance of polymer relaxation on drug release from swelling-controlled hydrogels has been recognized for some time; however, it has not been taken into account appropriately in the equation governing diffusion. Crank [33] first introduced a time-dependent (or history-dependent) diffusion coefficient to interpret observed anomalous sorption kinetics of penetrant in polymers. The diffusion coefficient was assumed to be affected partly by an instantaneous response attributable to fast local movements of individual molecular groups or small chain segments, and partly by a slow drifting response that resulted from the relatively slow uncoiling and rearrangement of large segments of the polymer chains. Good agreement between the model and experimental results for a single-penetrant system was demonstrated by Crank.

By adopting a similar time-dependent diffusion coefficient, Lee [31,32] demonstrated that various drug release behaviors from glassy hydrogels can also be consistently described. The rationale of employing a time-dependent diffusion coefficient is quite evident, since the various observed anomalous diffusion behaviors share a common physical origin, i.e., the slow penetrant-induced polymer molecular relaxation. Furthermore, one can show that the drug diffusion coefficient defined in a polymer-fixed frame of reference is proportional to the square of the polymer volume fraction. This frame of reference is often used to describe systems with considerable volume swelling. Since the polymer volume fraction is a strong function of time during the swelling process, a significant time dependence is therefore expected in the drug diffusion coefficient.

To examine the effect of time dependent diffusion coefficient on the release behavior from a swellable polymer system containing dissolved or dispersed drug, Lee defined the following time-dependent drug diffusion coefficient:

$$D(t) = D_i + (D_\infty - D_i) [1 - \exp(-kt)] \tag{19}$$

where D_i is the instantaneous part of the drug diffusion coefficient, D_∞ is the drug diffusion coefficient at swelling equilibrium, and k is the average relaxation constant controlling the approach to swelling equilibrium in the specific polymer-drug-solvent combination. Equation (19) is equivalent to the history-dependent diffusion coefficient introduced by Crank [33]; however, the concentration dependence is neglected for the sake of simplicity. By defining a new time variable,

$$dT = D(t) dt \tag{20a}$$

where

$$T = D_\infty \left\{ t - \left(1 - \frac{D_i}{D_\infty} \right) \frac{1}{k} [1 - \exp(-kt)] \right\} \tag{20b}$$

the transient diffusion equation, Eq. (1) for slab geometry, reduces to a form similar to that of a constant diffusion coefficient, where t is now replaced by T. The analytical solutions for both the dissolved and dispersed systems have been given by Lee [32]:

Dissolved systems ($A \leq C_s$). The exact solution for this swellable, dissolved system is

$$\frac{M}{M_\infty} = 1 - \sum_{n=0}^{\infty} \frac{8}{(2n+1)^2 \pi^2} \exp \left\{ - (n + 0.5)^2 \pi^2 \right.$$
$$\left[\tau - \left(1 - \frac{D_i}{D_\infty} \right) \left(\frac{D_\infty}{ka^2} \right) \right] \left[1 - \exp \left(- \frac{ka^2}{D_\infty} \right) \right] \right\} \tag{21}$$

where

$$\tau = \frac{D_\infty t}{a^2}$$

and other parameters such as A, C_s, and a have been defined previously.

For small times, Eq. (21) can be simplified to

$$\frac{M}{M_\infty} = \frac{4}{\sqrt{\pi}} \left\{ \tau - \left(1 - \frac{D_i}{D_\infty} \right) \frac{D_\infty}{ka^2} \left[1 - \exp \left(- \frac{ka^2}{D_\infty} \tau \right) \right] \right\}^{1/2} \tag{22}$$

Dispersed system ($A > C_s$). The exact solution for this swellable, dispersed system is

$$\frac{M}{M_\infty} = \frac{1}{(A/C_s)\,\text{erf}(\eta)} \frac{2}{\sqrt{\pi}} \left[\tau - \left(1 - \frac{D_i}{D_\infty} \right) \frac{D_\infty}{ka^2} \left[1 - \exp \left(- \frac{ka^2}{D_\infty} \tau \right) \right] \right]^{1/2} \tag{23}$$

with

$$\pi^{1/2} \eta \exp(\eta^2)\, \text{erf}(\eta) = \frac{C_s}{A - C_s} \tag{24}$$

The relative penetration of the diffusion front that separates the dissolved region from the undissolved core is expressed as

$$\frac{\xi}{a} = 2\eta \left\{ \tau - \left(1 - \frac{D_i}{D_\infty} \right) \frac{D_\infty}{ka^2} \left[1 - \exp \left(- \frac{ka^2}{D_\infty} \tau \right) \right] \right\}^{1/2} \tag{25}$$

where ξ is the penetration distance and η is evaluated from Eq. (24).

Similar to Eq. (17), the parameter D_∞/ka^2 in the above equations is essentially the Deborah number for the release system, defined as

$$(\text{DEB})_R = \frac{D_\infty}{ka^2}$$

which describes the relative magnitude of polymer relaxation time (k^{-1}) to the characteristic diffusion time (a^2/D_∞). Based on the Deborah number concept discussed earlier, it is now possible to describe various observed release behaviors in glassy hydrogels. As illustrated qualitatively in Fig. 11, when $(\text{DEB})_R$ is very small, the diffusion coefficient D will

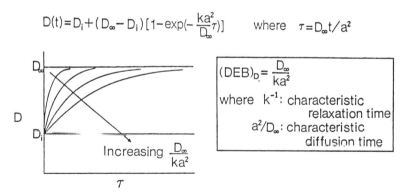

$$D(t) = D_i + (D_\infty - D_i)[1 - \exp(-\frac{ka^2}{D_\infty}\tau)] \quad \text{where} \quad \tau = D_\infty t/a^2$$

$$(DEB)_D = \frac{D_\infty}{ka^2}$$

where k^{-1}: characteristic relaxation time
a^2/D_∞: characteristic diffusion time

Figure 11 Effect of Deborah number for the release system $(DEB)_R$ on the time-dependent drug diffusion coefficient. (From Ref. 32.)

approach the constant diffusion coefficient D_∞ rapidly, giving rise to a Fickian diffusion behavior. At the other limit, when $(DEB)_R$ is very large, the diffusion coefficient remains essentially constant at D_i, resulting in a slower Fickian diffusion. At intermediate values of $(DEB)_R$, where it takes a finite amount of time for the diffusion coefficient to approach its equilibrium value, a time-dependent anomalous diffusion behavior will result. Therefore, the smaller the Deborah number, the faster a constant equilibrium diffusion coefficient will be approached and the earlier a Fickian diffusion behavior will prevail.

The fractional drug release predicted from Eqs. (21)–(24) is shown in Figs. 12 and 13 as a function of the square root of reduced time for $D_i/D_\infty = 0.05$. It is clear that the release behaviors for both the dissolved and dispersed systems are Fickian for $(DEB)_R$

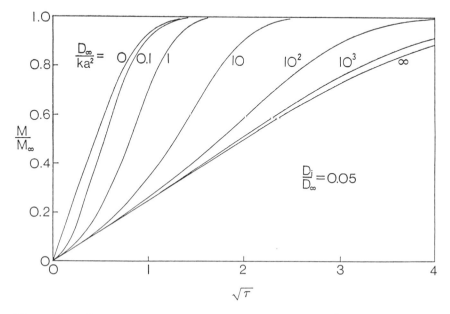

Figure 12 Effect of release Deborah number, $(DEB)_R = D_\infty/ka^2$, on fractional drug release from a swellable polymer slab containing dissolved drug. (From Ref. 32.)

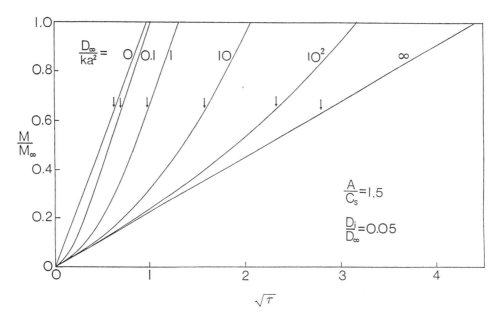

Figure 13 Effect of release Deborah number, $(DEB)_R = D_\infty/ka^2$, on fractional drug release from a swellable polymer slab containing dispersed drug; $\downarrow$ indicates where solvent fronts meet. (From Ref. 32.)

$\rightarrow 0$ as characterized by the linear dependence in the square-root-of-time plot. This is the so-called rubbery-state Fickian diffusion, where the polymer molecular relaxation process is fast compared to the diffusive-transport process. For $(DEB)_R \approx 1$ or > 1, the release behavior shifts to anomalous diffusion (including Case II), where the molecular relaxation process is occurring at a comparable or slightly slower time scale than that of diffusion. In fact, Figs. 12 and 13 show that at $D_i/D_\infty \ll 1$, the drug release will be constant-rate (or zero-order) for $(DEB)_r \geq 10$. This aspect has recently been confirmed by Vyavahare et al. [18] for the release of benzoic acid derivatives from glassy poly-HEMA. When $(DEB)_R \gg 1$, there is effectively no time variation of the polymer structure during diffusion, and the release behavior appears to be Fickian, as characterized again by the linear dependence in the square-root-of-time plot. This approaches the so-called glassy-state Fickian diffusion, which apparently is governed by the constant instantaneous portion of the diffusion coefficient D_i. These predictions are consistent with literature results [15–22].

By treating the dispersed system $(A/C_s > 1)$ as a moving-boundary problem, the relative penetration of the diffusion front is described by Eq. (25) and shown in Fig. 14. It is evident that a full spectrum of front penetration behavior ranging from Fickian to Case II is predicted. Again, its dependence on $(DEB)_R$ or D_∞/ka^2 is consistent with the diffusion characteristics described in Table 1.

Example. To demonstrate the utility of the time-dependent diffusion coefficient approach, data from Fig. 10 for the thiamine HCl release from initially dehydrated poly-HEMA sheets as a function of drug loading have been analyzed with Eqs. (23) and (24). The results are shown in Fig. 15 and Table 2. Rather than pure curve fitting, the equilibrium diffusion coefficient D_∞ is first calculated from the latter part of the experimental release curve using a large-time approximation of the Fickian diffusion results. This is a reasonable

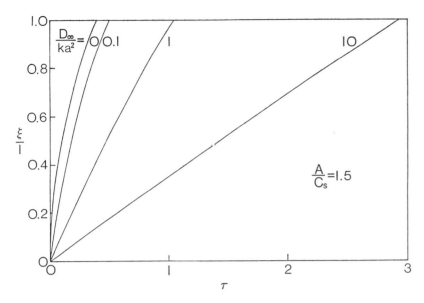

Figure 14 Effect of release Deborah number, $(DEB)_R = D_\infty/ka^2$, on relative penetration of the diffusion front in a swellable polymer slab containing dispersed drug. (From Ref. 32.)

approach, since at large times the hydrogel matrix is already fully swollen by the solvent while the drug diffusion is still taking place. The obtained experimental D_∞ is then used in conjunction with other experimental data, such as A/C_s values and the release curves (Fig. 10), to calculate the corresponding release Deborah number $(DEB)_R$ and the polymer

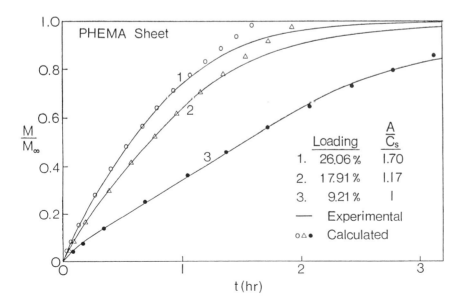

Figure 15 Effect of loading level on thiamine HCl release from glassy PHEMA sheets at 37.5°C. Data points calculated from Eqs. (23) and (24). (From Ref. 32.)

Table 2 Characteristics of Thiamine HCl Release
from Poly-HEMA Sheets

A/C_s	D_∞ $(10^{-7}$ cm^2/s)	$(DEB)_R$ or D_∞/ka^2	k $(10^{-4}$ s$^{-1})$
1	2.13	0.33	2.40
1.17	3.84	0.20	7.21
1.70	5.55	0.17	12.51

$a = 0.0516$ cm^2, $D_i/D_\infty = 0.01$.

relaxation constant k from Eqs. (23) and (24). Indeed, as shown in Table 2, the calculated k increases with drug loading, suggesting a faster polymer relaxation rate or a rate-limiting Fickian diffusion step. This is in agreement with the observation that the release of thiamine HCl from an initially dehydrated poly-HEMA matrix becomes more Fickian with increasing thiamine HCl loading levels.

Dimensional Changes During Drug Release

The change of sample dimension as a function of time during the dynamic swelling of glassy hydrogels often reflects the overall effect of the relaxation and diffusion processes. It has been shown [27] that the swelling behavior and drug-release kinetics in glassy hydrogels can be significantly affected by the local drug concentration. The presence of a water-soluble drug alters both the swelling osmotic pressure and the associated time-dependent relaxation of the hydrogel network during the simultaneous absorption of water and desorption of drugs. Detailed studies on the dimensional changes during the entire course of the simultaneous water penetration and drug release from hydrogel beads have been reported by Lee [27,35].

A typical example illustrating the effect of drug loading on the fractional release from glassy PHEMA beads is shown in Fig. 16 for thiamine HCl. The corresponding initial

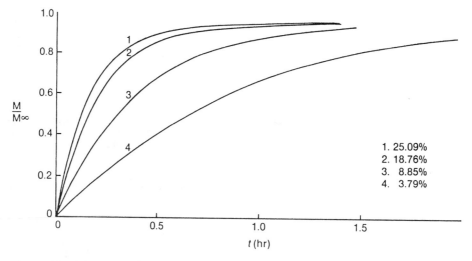

Figure 16 Effect of loading level on thiamine HCl release from glassy PHEMA beads at 37.5°C. (From Ref. 27.)

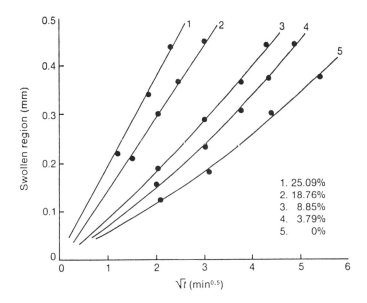

Figure 17 Effect of loading level on swelling front penetration in glassy PHEMA beads during thiamine HCl release at 37.5°C. (From Ref. 27.)

solvent front penetration is presented as a square-root-of-time plot in Fig. 17. Both the fractional release and solvent front penetration are observed to behave more Fickian with higher thiamine HCl loadings. Similar observations have recently been reported for oxprenolol HCl in PHEMA beads [35]. Such a transition can be considered as a change of relative importance of the diffusion process versus the polymer relaxation as a function of drug loading. As shown by the example in the previous section, this observed trend

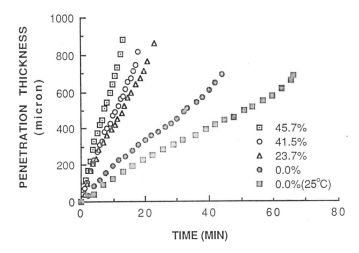

Figure 18 Effect of loading level on swelling front penetration during release of oxprenolol HCl from glassy PHEMA beads at 37°C. (From Ref. 35.)

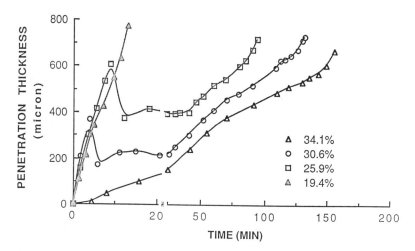

Figure 19 Effect of loading level on swelling and drug dissolution front penetration during release of diclofenac sodium from PHEMA beads at 37°C. (From Ref. 35.)

can also be related to the larger polymer relaxation rate constant k at higher drug loading levels.

The swelling front penetration in hydrogels is also affected by the drug solubility. Figures 18 and 19 show the penetrating front movements in PHEMA beads loaded, respectively, with oxprenolol HCl and diclofenac sodium. For oxprenolol HCl, a drug with high water solubility (~77%), only one swelling front is observed at all loading levels, and the swelling front penetrates faster with increasing drug loading (Fig. 18). On the other hand, two moving penetrating fronts are observed for diclofenac sodium, a drug with low water solubility (~2.65%), in a drug loading range of about 25–30% (Fig. 19). Below this range, a single fast-moving swelling front is observed, whereas above this range, a single slow-moving drug dissolution front separating an opaque drug core from a clear swollen region appears. With 25–30% diclofenac sodium loading, a fast-moving swellling front is initially visible; however, a precipitation of diclofenac sodium starts to appear after a short penetration period, which eventually develops into a slow-moving drug dissolution front. Such a transition of fronts is evident from Fig. 19, where the apparent drop in penetration thickness reflects the transition from a fast-moving swelling front to a slow-moving drug dissolution front. The general schemes of observed swelling boundaries and dimensional changes during drug release in relation to drug solubilities are summarized in Fig. 20.

The transient swelling front penetration, as shown in Figs. 18 and 19, exhibits an initially nonlinear region (ranging from Fickian to non-Fickian behavior) followed by a short linear transition region before accelerating toward the core. It can be shown [36] that the apparent acceleration of penetrating front near the core is a natural outcome of the spherical geometry and not a super-Case II transport behavior as suggested by some authors. A typical theoretical plot of penetration front movement in different geometries is shown in Fig. 21 for a loading-to-solubility ratio (A/C$_s$) of 3. It is clear that the initial part of the moving-front movement is characteristically Fickian and an acceleration of penetrating front near the core is predicted in radically symmetric geometries (cylinders and spheres). Such an acceleration is not predicted for the planar geometry.

(a)

(b)

(c)

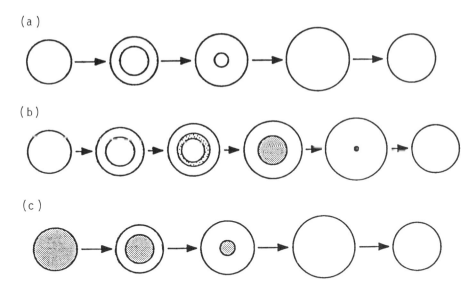

Figure 20   Schematic drawing of observed swelling boundaries and dimensional changes during drug release in relation to drug solubilities. (a) High drug solubility, all loading levels; low drug solubility, low loading levels. (b) Low drug solubility, intermediate loading levels. (c) Low drug solubility, high loading levels. (From Ref. 35.)

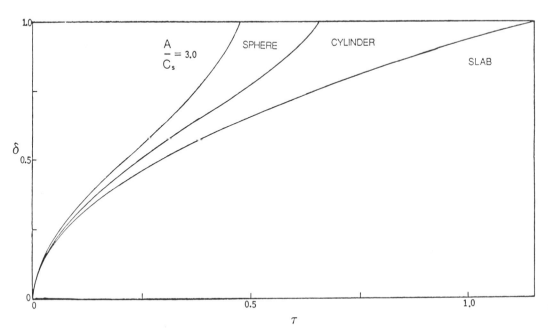

Figure 21   Predicted effect of geometry on moving-boundary position as a function of time ($\tau = Dt/a^2$) at $A/C_s = 3$. (From Ref. 36.)

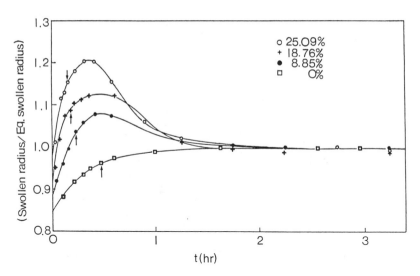

Figure 22   Effect of loading level on transient dimensional changes during release of thiamine HCl from glassy PHEMA beads at 37.5°C; ↑ indicates where solvent fronts meet. (Replotted based on data from Ref. 27.)

The transient dimensional changes during the simultaneous solvent penetration and drug release from PHEMA beads are shown in Fig. 22 for thiamine HCl. The spherical geometry eliminates the anisotropy and edge effect normally associated with dimensional measurements in glassy polymer sheets. As shown in Fig. 22, the radius of an unloaded bead increases monotonically toward the equilibrium radius, whereas that of drug-loaded beads goes through a maximum followed by a gradual approach to an equilibrium value. Since the reference state in Fig. 22 is the final equilibrium swollen radius, the plots converge to the same equilibrium value of 1. The presence of homogeneously dissolved or dispersed drug in PHEMA beads generates an additional osmotic driving force to that due to the polymer alone. This alters both the total swelling osmotic pressure and the associated time-dependent relaxation of the hydrogel network during the simultaneous sorption of water and desorption of drug. Intuitively, the swelling by water tends to increase, whereas the release of drug tends to decrease the dimension of the hydrogel bead. A combination of these two competing processes should result in a maximum in the transient dimensional changes for drug-loaded hydrogel beads, qualitatively agreeing with experimental observations. Since the osmotic contribution from the drug has not been considered in the above scheme, the difference between the experimentally obtained and the calculated transient dimensional changes has been used to estimate the osmotic contribution to the overall dimensional increases [35].

SYSTEMS WITH NONUNIFORM INITIAL CONCENTRATION PROFILES

Diffusion- and erosion-controlled matrix systems where the drug is uniformly dissolved or dispersed in a polymer generally exhibit release rates that continuously diminish with time. This is a result of the increasing diffusional resistance and decreasing area at the diffusion/erosion front due to geometry limitations. In order to overcome such inherent drawbacks, especially to achieve a constant rate of drug release, methods involving the use of geometry factors and the surface erosion-controlled and swelling-controlled mechanisms discussed earlier have been utilized. While the first mechanism may be of limited practicability, the applicability of the latter two mechanisms for constant-rate drug release is restricted by the need to maintain a constant surface area at the diffusion or erosion front.

The potential of an important area that has not been fully explored involves the use of a nonuniform initial concentration distribution as the mechanism for regulating drug release from matrix devices. The idea involves the use of an increasing drug loading toward the core of the device to compensate for the inherent decrease in release rate due to increasing diffusional resistance and decreasing area at the diffusion/erosion front. However, criteria for selecting a specific drug distribution to achieve a desired release rate profile are more complex and not immediately obvious. Several empirical matrix systems of this nature were initially reported in the literature [37,38], but with no physical rationale or analysis given. Lee [39,41] first provided a theoretical basis for such an approach and reported a novel method for achieving a constant rate of drug release from glassy hydrogel beads via an immobilized nonuniform initial drug distribution. Recently, Chang and Himmelstein [42] have extended the analysis of such systems by considering both the dissolution and diffusion processes and by using a concentration- and position-dependent diffusion coefficient to take into account the effect of increased void volume due to dissolution. Such principles have also been applied to the design and study of a coated "gradient matrix system" for drug release by van Bommel et al. [43].

In this section, the effect of nonuniform initial drug concentration distribution on the release kinetics will be reviewed for both the diffusion-controlled and surface erosion-controlled matrix systems. Various methods for achieving nonuniform drug loading will also be examined.

Kinetics of Drug Release—Diffusion-Controlled Systems

For a diffusion-controlled matrix (planar sheet, cylinder, or sphere) with radius or half-thickness a and an initial drug concentration distribution $f(x)$, the fractional drug releases can be obtained from solutions to Eq. (1) assuming dissolved drug, constant diffusion coefficient, and perfect sink conditions [39,41]:

$$\text{Planar sheet} \quad \frac{M}{M_\infty} = 1 - \left\{ \sum_{n=0}^{\infty} \frac{(-1)^{n+1}}{2n+1} I_1(n) \right. \tag{26}$$
$$\left. \exp\left[-\frac{(2n+1)^2\pi^2\tau}{4} \right] \middle/ \sum_{n=0}^{\infty} \frac{(-1)^{n+1}}{2n+1} I_1(n) \right\}$$

with

$$I_1(n) = \int_0^1 f(\xi) \cos\left[(n+0.5)\pi\xi\right] d\xi$$

where

$$\xi = \frac{x}{a}$$

and

$$\tau = \frac{Dt}{a^2}$$

Cylinder $\quad \dfrac{M}{M_\infty} = 1 - \left[\displaystyle\sum_{n=1}^{\infty} \dfrac{I_2(n)}{\beta_n J_1(\beta_n)} \exp(-\beta_n^2 \tau) \middle/ \displaystyle\sum_{n=1}^{\infty} \dfrac{I_2(n)}{\beta_n J_1(\beta_n)} \right]$ (27)

with

$$I_2(n) = \int_0^1 \xi f(\xi) J_0(\beta_n \xi)\, d\xi$$

where J_0 and J_1 are Bessel functions of the first kind of zero and first order, respectively, and the β_n's are the roots of $J_0(\beta) = 0$.

Sphere $\quad \dfrac{M}{M_\infty} = 1 - \left[\displaystyle\sum_{n=1}^{\infty} \dfrac{(-1)^{n+1}}{n} I_3(n) \right.$

$$\left. \exp(-n^2\pi^2\tau) \middle/ \displaystyle\sum_{n=1}^{\infty} \dfrac{(-1)^{n+1}}{n} I_3(n) \right]$$ (28)

with

$$I_3(n) = \int_0^1 \xi f(\xi) \sin(n\pi\xi)\, d\xi$$

The assumption of a constant diffusion coefficient is generally valid for diffusion from polymer matrices without appreciable swelling or when the time scale for solvent penetration and swelling is much shorter than that for drug release. For most swellable polymers, however, a constant diffusion coefficient may not rigorously characterize the initial swelling period because of the existence of a time-dependent polymer relaxation in addition to diffusion. The diffusion from such a swelling system with nonuniform drug distribution can be readily analyzed by incorporating the same time-dependent diffusion coefficient as defined in Eqs. (20a) and (20b). Such an approach has been shown to be successful in analyzing the kinetics of drug release from swellable polymers with *both* uniform and nonuniform initial drug concentration distributions [32,44]. In this case, the resulting fractional drug releases have the same form as in Eqs. (26)–(28), except that τ is replaced by the new time variable T defined by Eq. (20b).

Based on Eqs. (26)–(28), the predicted characteristics of drug release from diffusion-controlled matrix systems as a function of initial drug concentration distribution are summarized in Fig. 23 for all geometries. Both the uniform and parabolic initial drug distribution show first-order release behavior with continuously diminishing rate; however, the initial rate of release is drastically reduced for the parabolic distribution. In contrast, an initially sigmoidal drug distribution is shown to be capable of introducing a characteristic inflection point, and therefore, the cumulative release curve becomes considerably more linear. As a result, a prolonged constant rate of drug release and a characteristic time lag similar to those of a membrane-reservoir system are obtained. This concept has been demonstrated in a novel approach to constant-rate drug release from glassy hydrogel beads via immobilized nonuniform drug concentration distributions [40,41].

As is also evident from Fig. 23, an initial drug distribution following an ascending staircase function inevitably results in a first-order release behavior and, as expected, there is a sizable increase in the initial release rate over that of the uniform distribution. On the other hand, a descending staircase function approximating the sigmoidal profile in a stepwise fashion introduces a considerable constant-rate region into the cumulative-release

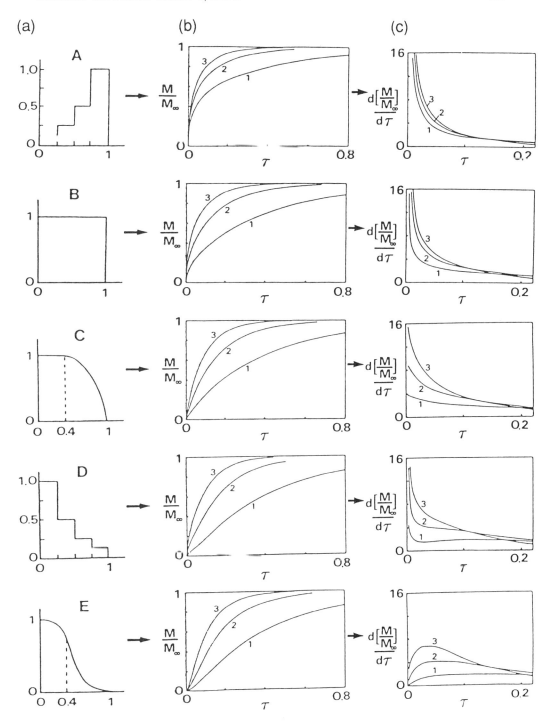

Figure 23 Predicted characteristics of drug release from diffusion-controlled matrix system as a function of initial drug concentration distribution: (a) concentration; (b) cumulative release; (c) release rate. Key: 1, planar sheet; 2, cylinder; 3, sphere. (From Ref. 39.)

curve for the planar geometry. The corresponding constant-rate region diminishes in the cylindrical and spherical geometries. Although the sigmoidal drug distribution appears to provide the largest reduction in initial release rate and the longest constant-rate period, the discontinuous concentration profile in the form of a staircase function is the one most amenable to large-scale fabrication, since it can readily be obtained by coating, compaction, and lamination techniques.

Kinetics of Drug Release—Surface Erosion-Controlled Systems

Polymer erosion can be controlled by a variety of mechanisms, ranging from backbone cleavage to swelling and dissolution. This aspect will be covered in detail in Chapter 4. For surface-erodible systems with uniform drug distribution, the effect of diffusional contributions on the release kinetics has been examined above. Here, only the situation in which polymers erode by a purely surface erosion mechanism with no diffusional contribution will be considered for systems with nonuniform drug distribution.

For a surface-erodible polymer matrix containing dissolved or dispersed drug with initial radius or half-thickness a, a time-dependent surface erosion front position S, and an initial drug distribution f(x), the fractional release has been obtained by appropriate mass balance taking into consideration the moving erosion front [39]:

$$\frac{M}{M_\infty} = \frac{\int_S^a f(x)x^n\, dX}{\int_0^a f(x)x^n\, dX} \tag{29}$$

with n = 0, 1, and 2 for planar sheet, cylinder, and sphere, respectively. All other notations have been defined previously.

For most polymers, the surface erosion can be characterized by a constant rate process after an initial induction period. In this case, the erosion front position is described by S = a − Bt, where, again, B is the surface erosion rate constant having the same dimension as a velocity. Equation (29) can then be rewritten as

$$\frac{M}{M_\infty} = \frac{\int_{1-Bt/a}^1 f(\xi)\xi^n\, d\xi}{\int_0^1 f(\xi)\xi^n\, d\xi} \tag{30}$$

Based on Eq. (30), the predicted characteristics of drug release from surface erosion-controlled matrix systems as a function of initial drug concentration distribution are shown in Fig. 24 for all three geometries. The uniform initial drug distribution can produce a constant rate of drug release only from the planar geometry, because of its constant area at the eroding front (neglecting edge effect). The corresponding releases from both the cylindrical and spherical geometries all exhibit decreasing release rate with time. On the other hand, both the parabolic and sigmoidal distributions result in zero-order release characteristics from all three geometries; however, the duration of constant rate of drug release is longer for planar geometry than that for cylindrical and spherical geometries.

When the initial drug distribution follows an ascending staircase function, the surface-erosion mechanism results in a first-order release behavior for all three geometries, but with discontinuities in the release rates. In contrast, the initial drug distribution following a descending staircase function results in an oscillatory release rate from the cylindrical

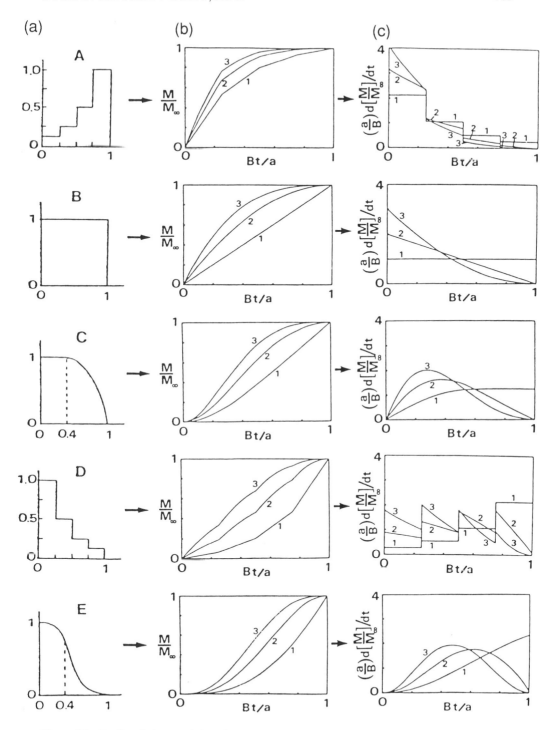

Figure 24 Predicted characteristics of drug release from surface erosion-controlled matrix systems as a function of initial drug concentration distribution: (a) concentration; (b) cumulative release; (c) release rate. Key: 1, planar sheet; 2, cylinder; 3, sphere. (From Ref. 39.)

and spherical geometries, whereas the corresponding release rate from a planar matrix increases in a stepwise fashion.

The analyses presented in Eqs. (26)–(30) will be useful in assessing the effect of an arbitrary initial drug concentration distribution on the release behavior from diffusion-controlled and surface erosion-controlled matrix systems. For surface erosion-controlled matrices, the presence of diffusional contribution can result in drug release behavior intermediate between that of the diffusion- and surface erosion-controlled systems. One example is that for the initial drug distribution following a descending staircase function in which presence of the diffusional contribution tends to smooth out the drug distribution at early times, giving rise to an extended constant-rate releasing period.

In principle, one can calculate the desired initial drug distribution from Eqs. (26)–(30) with respect to any given release profile. However, this may involve cumbersome numerical deconvolutions to evaluate f(x) for diffusion-controlled systems. The corresponding mathematical operations in surface erosion-controlled systems are easier to carry out because of the simplicity of Eqs. (29) and (30). One special case of interest is evident from Eq. (30), where a complete zero-order (or constant-rate) drug release can be obtained with an initial drug distribution f(x) given by

$$f(x) = \frac{1}{x^n} \tag{31}$$

provided that the surface erosion is characterized by a constant-rate process. Here n = 0, 1, and 2 are for planar sheet, cylinder, and sphere, respectively.

Methods to Achieve Nonuniform Drug Loading

Controlled-Extraction Process

Despite the theoretical prospect of having a totally relaxation-controlled (Case II) situation, thereby achieving zero-order release, swelling-controlled systems still have the same drawback suffered by all other matrix systems, a continuously diminishing release rate. This is the result of the increasing diffusional resistance and decreasing area at the penetrating diffusion front, especially for nonplanar geometries. To complicate matters further, local drug loading has been shown to cause deviations from Case II behavior in hydrogels, as discussed in the previous section on swelling-controlled systems.

Based on the realization that a sigmoidal type of drug distribution is needed to achieve a constant rate of drug release and the fact that the sigmoidal distribution is characteristic of glassy polymers partially penetrated by a swelling solvent undergoing non-Fickian diffusion, Lee [40,41] developed a controlled-extraction process to partially penetrate and extract drug-loaded hydrogel beads with a swelling solvent. The non-Fickian diffusion behavior allows the development of a sigmoidal concentration profile for both the drug and solvent as depicted in Fig. 25. Immediately after separating the extracting solvent, the controlled-extracted beads are freeze-dried under vacuum to remove the swelling solvent and to immobilize a nonuniform, sigmoidal drug distribution. A schematic diagram of the concentration profiles and experimental steps involved in such a controlled-extraction process is shown in Fig. 26. This process has been applied to cross-linked hydrogel beads based on HEMA co-polymers, using a very water-soluble drug, oxprenolol HCl (~77% soluble in water). With such a high water solubility, it is usually very difficult to achieve a zero-order release from a conventional membrane-reservoir system.

The in-vitro oxprenolol HCl release from such controlled-extracted beads, compared with that of unextracted control, is shown in Fig. 27. As predicted earlier, a release time

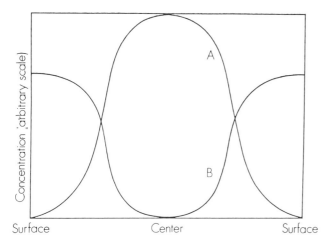

Figure 25 Schematic diagram of (A) drug and (B) solvent distributions during controlled extraction of a drug-loaded glassy hydrogel matrix.

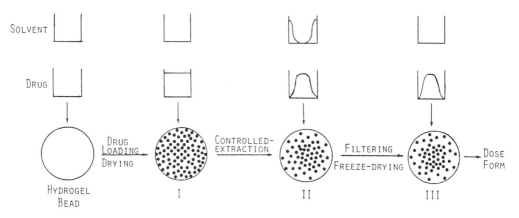

Figure 26 Schematic diagram of the concentration profiles and experimental steps involved in the controlled-extraction process.

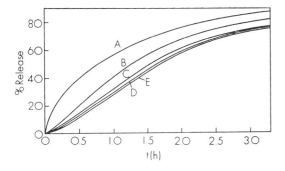

Figure 27 Effect of controlled extraction time in water on the release of oxprenolol HCl from hydrogel beads: A, 0 min; B, 5 min; C, 15 min; D, 20 min; E, 30 min. (From Ref. 41.)

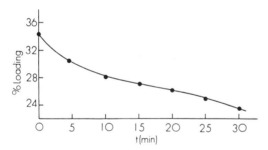

Figure 28 Oxprenolol HCl loading as a function
of controlled extraction time in water. (From Ref.
41.)

lag and constant-release region similar to that of membrane-reservoir devices have been
introduced by the process. With an increase in the extraction time, the constant-rate release
region is extended, and the corresponding release rate is progressively decreased. Inevitably, a certain amount of drug will be lost during the controlled-extraction process.
However, as shown in Fig. 28, only up to 10% of the drug loading is removed. Stability
tests on the system reveal that, in the absence of moisture, the constant-release characteristics can be preserved for an extended period of time. The diffusion of entrapped drug
does not occur until the hydrogel matrix is swollen at the time of usage.

The immobilization of a nonuniform drug concentration distribution and the maintenance of it in the absence of moisture do not violate thermodynamic laws. Although

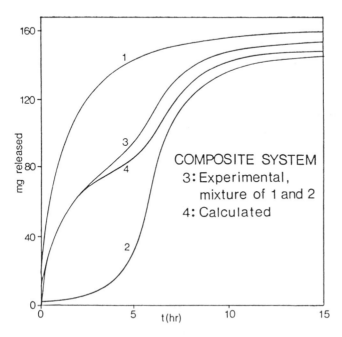

Figure 29 Predicted release characteristics of a composite dose
based on the combination of untreated and controlled-extracted
hydrogel beads.

thermodynamics may predict an eventual equilibrium in drug distribution, it does not specify the kinetic process leading to the equilibrium. In other words, it does not specify how long it takes to get from one state to the other. In a hydrogel matrix, the drug diffusion coefficients in the dehydrated glassy matrix are generally five to six orders of magnitude smaller than that in the swollen rubbery phase, thereby preserving the immobilized drug distribution virtually indefinitely.

In addition to being able to achieve a constant rate of drug release without a saturated reservoir and a rate-controlling membrane, this controlled-extraction process provides a rational and practical way of modifying the release kinetics from hydrogels. Recently, it was demonstrated through the use of a time-dependent diffusion coefficient [44] that the requirement of a sigmoidal drug distribution for the generation of a constant rate of drug release may be relaxed for swelling-controlled systems when the Deborah number for the release system $(DEB)_R$ is sufficiently large and the contribution from the instantaneous part of the diffusion coefficient in Eq. (19) is small. This aspect certainly will broaden the applicable drug distributions for achieving zero-order drug release from swelling-controlled systems.

Example. With an appropriate combination of hydrogel bead composition and extracting solvent, adjustable time lags can be generated by the controlled-extraction process [45,46], in addition to the constant rate of drug release already described above. Since the untreated control has a characteristic first-order release behavior, a combination of such untreated control and controlled-extracted hydrogel beads should enable one to achieve a release profile typical of a twice-a-day (t.i.d.) regimen from a *single* composite dose (Fig. 29).

Lamination Process

The desired sigmoidal drug distribution may be easy to achieve in glassy hydrogels; however, it may not be readily applicable to other matrix materials. As discussed above, a descending staircase function approximating the sigmoidal profile in a stepwise fashion may be the best alternative, since it generates a considerable constant-rate region in drug release and is much easier to achieve with conventional coating, compaction, and lamination techniques. Therefore, a multilayered tablet, a multilayered plastic laminate, or a pellet/ tablet core with multiple coatings can easily be engineered to contain decreasing drug loading toward the surface of the device via a specific descending staircase function [37,38,43].

As an example, the acetaminophen releases from a laminated matrix prepared from a blend of cellulose acetate phthalate/pluronic F68 (70:30) are presented in Figs. 30a and 30b as a function of the initial stepwise concentration profile [47,48]. As expected from predictions of Fig. 23, the sample with a smoother inflection in the stepwise drug distribution exhibits a constant rate of acetaminophen release right from the beginning (Fig. 30a), whereas the one with a more pronounced inflection in the descending staircase distribution also exhibits a sharp inflection in the release profile (Fig. 30b).

In view of the experimental flexibility in achieving essentially an unlimited number of nonuniform initial drug distribution patterns in polymer systems, initial concentration distribution as a mechanism for regulating drug release not only provides release characteristics not easily achievable from other delivery mechanisms (such as oscillatory or pulsatile release patterns), it allows one to program the drug release profile more effectively to meet specific temporal therapeutic requirements.

Example. Based on the results of Eqs. (29) and (30), design drug distribution profiles in surface-erodible systems that can give rise to pulsatile delivery patterns with the same

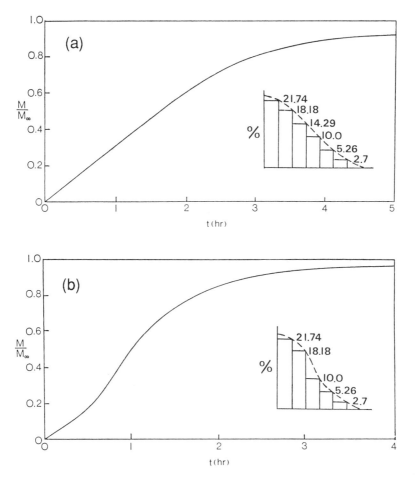

Figure 30 Characteristics of acetaminophen release from laminated matrices as a function of the initial stepwise drug concentration profile; (a) has a less pronounced inflection than (b).

frequency and peak rate for all three geometries. Also illustrate the necessary drug distribution profiles for zero-order release from surface-erodible systems.

Solution. The pulsatile delivery pattern can easily be obtained by alternating drug-loaded and drug-free layers in a laminate. As shown in Figs. 31a–31c, the drug loadings in the cylindrical and spherical laminates would have to be arranged in a decreasing manner toward the surface in order to maintain the same peak rate as compared with the slab geometry. The frequency of the pulses can be controlled by the thickness of the drug-free layer. For zero-order release, Fig. 31d shows the required drug profiles for all three geometries based on Eq. (31). Although a constant drug distribution is sufficient to produce a zero-order release from a surface-erodible slab (or sheet), a drug distribution increasing toward the center is needed for both the cylindrical and spherical geometries to compensate for the decreasing area at the surface of erosion.

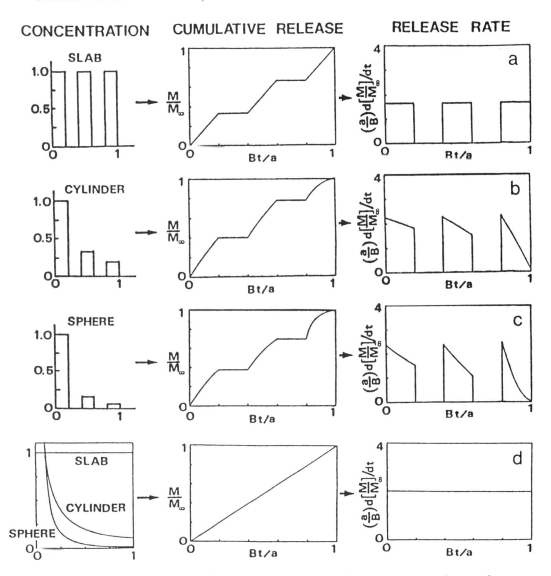

Figure 31 Drug-concentration profiles leading to programmable release patterns from surface erosion-controlled polymer laminates: (a–c) pulsatile delivery; (d) zero-order delivery.

REFERENCES

1. J. R. Robinson and V. H. L. Lee (Eds.), *Controlled Drug Delivery*, 2nd ed., Marcel Dekker, New York (1987).
2. H. S. Carslaw and J. C. Jaeger, *Conduction of Heat in Solids*, 2nd ed., Clarendon Press, Oxford (1959).
3. D. R. Paul and S. K. McSpadden, Diffusional release of a solute from a polymer matrix, *J. Membrane Sci.*, *1*:33–48 (1976).
4. T. Higuchi, Rate of release of medicaments from ointment bases containing drug in suspension, *J. Pharm. Sci.*, *50*:874–875 (1961).

5. P. I. Lee, Diffusional release of a solute from a polymeric matrix—Approximate analytical solutions, *J. Membrane Sci.*, 7:255–275 (1980).
6. T. Higuchi, Mechanism of sustained-action medication, *J. Pharm. Sci.*, 52:1145–1149 (1963).
7. P. I. Lee, Determination of diffusion coefficients by sorption from a constant finite volume, in *Controlled Release of Bioactive Materials* (R. Baker, Ed.), Academic Press, New York, pp. 135–153 (1980).
8. J. Heller, Controlled release of biologically active compounds from bioerodible polymers, *Biomaterials*, 1:51–57 (1980).
9. P. Colombo, A. Gazzaniga, C. Caramella, U. Conte, and A. LaManna, In vitro programmable zero-order release drug delivery system, *Acta Pharm. Technol.*, 33:15–20 (1987).
10. P. I. Lee and N. A. Peppas, Prediction of polymer dissolution in swellable controlled-release systems, *J. Control. Rel.*, 6:207–215 (1987).
11. R. S. Harland, A. Gazzaniga, M. E. Sangalli, P. Colombo, and N. A. Peppas, Drug/polymer matrix swelling and dissolution, *Pharm. Res.*, 5:488–494 (1988).
12. C. Caramella, F. Ferrari, M. C. Bonferoni, M. Ronchi, and P. Colombo, Rheological properties and diffusion dissolution behavior of hydrophilic polymers, *Boll. Chim. Farm.*, 128:298–302 (1989).
13. J. D. Ferry, *Viscoelastic Properties of Polymers*, John Wiley, New York, p. 277 (1980).
14. H. Fujita and A. Koshimoto, Diffusion-controlled stress relaxation in polymers II. Stress relaxation in swollen polymers, *J. Polym. Sci.*, 28:547–567 (1958).
15. J. S. Vrentas, C. M. Jarzebski, and J. L. Duda, A Deborah number for diffusion in polymer-solvent systems, *AICHE J.*, 21:894–901 (1975).
16. R. W. Korsmeyer and N. A. Peppas, Macromolecular and modelling aspects of swelling-controlled systems, in *Controlled Release Delivery Systems* (T. J. Roseman and S. Z. Mansdorf, Eds.), Marcel Dekker, New York, pp. 77–90 (1983).
17. H. B. Hopfenberg and K. C. Hsu, Swelling-controlled constant rate delivery systems, *Polym. Eng. Sci.*, 18:1186–1191 (1978).
18. N. R. Vyavahare, M. G. Kulkarni, and R. A. Mashelkar, Zero-order release from glassy hydrogels. I. Enigma of the swelling interface number, *J. Membrane Sci.*, 49:207–222 (1990).
19. T. T. Wang, T. K. Kwei, and H. L. Frisch, Diffusion in glassy polymers, III, *J. Polym. Sci.* A-2, 7:2019–2028 (1969).
20. A. Peterlin, Diffusion in a glassy polymer with discontinuous swelling. II. Concentration distribution of diffusion as a function of time, *Makromol. Chem.*, 124:136–142 (1969).
21. N. L. Thomas and A. H. Windle, A theory of Case II diffusion, *Polymer*, 23:529–542 (1983).
22. C. Gostoli and G. C. Sarti, Influence of rheological properties in mass-transfer phenomena: Super Case II sorption in glassy polymers, *Chem. Eng. Commun.*, 21:67–79 (1983).
23. J. Crank, Influence of molecular relaxation and internal stress on diffusion in polymers, *J. Polym. Sci.*, 11:151–163 (1953).
24. F. A. Long and D. Richman, Concentration gradients for diffusion of vapours in glassy polymers and their relaxation to time-dependent diffusion phenomena, *J. Am. Chem. Soc.*, 82:513–519 (1960).
25. J. H. Petropoulos and P. R. Roussis, Influence of transverse differential swelling stress on the kinetics of sorption of penetrants by polymer membranes, *J. Membrane Sci.*, 3:343–356 (1978).
26. J. H. Petropoulos, Interpretation of anomalous sorption kinetics in polymer-penetrant systems in terms of time-dependent solubility coefficient, *J. Polym. Sci., Polym. Phys. Ed.*, 22:1885–1900 (1984).
27. P. I. Lee, Dimensional changes during drug release from a glassy hydrogel matrix, *Polym. Commun.*, 24:45–47 (1983).
28. W. R. Good, Diffusion of water soluble drugs from initially dry hydrogels, in *Polymeric Delivery Systems* (R. Kostelnik, Ed.), Gordon & Breach, New York, pp. 139–153 (1976).
29. P. I. Lee, Controlled drug release from polymeric matrices involving moving boundaries, in

Controlled Release of Pesticides and Pharmaceuticals (D. H. Lewis, Ed.), Plenum, New York, pp. 39–48 (1981).

30. R. W. Korsmeyer and N. A. Peppas, Modelling drug release from swellable systems, *Proc. Int. Symp. Control. Rel. Bioactive Mater.*, *10*:141–142 (1983).

31. P. I. Lee, Kinetics of drug release from hydrogel matrices, *J. Control. Rel.*, *2*:277–288 (1985).

32. P. I. Lee, Interpretation of drug-release kinetics from hydrogel matrices in terms of time-dependent diffusion coefficients, in *Controlled-Release Technology: Pharmaceutical Applications* (P. I. Lee and W. R. Good, Eds.), ACS Symp. Ser. No. 348, American Chemical Society, Washington, D.C., pp. 71–83 (1987).

33. J. Crank, A theoretical investigation of the influence of molecular relaxation and internal stress on diffusion in polymers, *J. Polym. Sci.*, *11*:151–168 (1953).

34. E. E. Parsonage, N. A. Peppas, and P. I. Lee, Properties of positive resists. II. Dissolution characteristics of irradiated poly(methy methacrylate) and poly(methyl methacrylate-co-maleic anhydride), *J. Vacuum Sci. Technol.*, *Pt. B*, *5*:538-545 (1987).

35. P. I. Lee and C. J. Kim, Probing the mechanisms of drug release from hydrogels, *J. Control. Rel.* (accepted 1990).

36. P. I. Lee and C. J. Kim, Effect of geometry on the swelling front penetration in glassy polymers, submitted to *J. Membrane Sci.*

37. C. B. Bogentoft and C. H. Appelgren, Methods for preparing preparations having controlled release of an active component, U.S. Patent 4,289,795 (September 15, 1981).

38. S. B. Mitra, Oral sustained release drug delivery using polymer film composites, *Polym. Prep.*, *24*:51–52 (1983).

39. P. I. Lee, Initial concentration distribution as a mechanism for regulating drug release from diffusion controlled and surface erosion controlled matrix systems, *J. Control. Rel.*, *4*:1–7 (1986).

40. P. I. Lee, Novel approach to zero-order drug delivery via immobilized non-uniform drug distribution in glassy hydrogels, *J. Pharm. Sci.*, *73*:1344–1347 (1984).

41. P. I. Lee, Effect of non-uniform initial drug concentration distribution on the kinetics of drug release from glassy hydrogel matrices, *Polymer*, *25*:973–978 (1984).

42. N. J. Chang and K. J. Himmelstein, Dissolution-diffusion controlled constant-rate release from heterogeneously loaded drug-containing materials, *J. Control. Rel.*, *12*:201–212 (1990).

43. E. M. G. van Bommel, J. G. Fokkens, and D. J. A. Crommelin, A gradient matrix system as a controlled release device. Release from a slab model system, *J. Control. Rel.*, *10*:283–292 (1989).

44. P. I. Lee, Effects of time-dependent swelling on the kinetics of drug release from polymers with initially non-uniform concentration distribution, *Proc. Int. Symp. Control. Rel. Bioactive Mater.*, *15*.97–98 (1988).

45. P. I. Lee, Active agent containing hydrogel devices wherein the active agent concentration profile contains a sigmoidal concentration gradient for improved constant release, their manufacturers and use, U.S. Patent 4,624,848 (November 25, 1986).

46. K. F. Mueller, Release and delayed release of water-soluble drugs from polymer beads with low water swelling, in *Controlled-Release Technology: Pharmaceutical Applications* (P. I. Lee and W. R. Good, Eds.) ACS Symp. Ser. No. 348, American Chemical Society, Washington, D.C., pp. 139–157 (1987).

47. P. I. Lee, Membrane-forming polymeric systems, U.S. Patent 4,729,190 (March 8, 1988).

48. P. I. Lee, Programmable drug delivery from diffusion controlled and surface erosion controlled matrix systems, *Abstracts—AphA Acad. Pharmaceut. Sci.*, *15*(2):88 (1985).

49. F. A. L. Dullien, *Porous Media—Fluid Transport and Pore Structure*, Academic Press, New York (1979).

50. D. H. Hwang and C. Cohen, Diffusion and relaxation in polymer-solvent systems. 2. Poly(methyl methacrylate)/methyl ethyl ketone, *Macromolecules*, *17*:2890–2895 (1984).

4

Erodible Systems

Eyal Ron   *Genetics Institute, Cambridge, Massachusetts*

Robert Langer   *Massachusetts Institute of Technology, Cambridge, Massachusetts*

INTRODUCTION

Erodible polymers as vehicles for drug delivery are attractive because surgery to remove the implanted device (after depletion of the drug supply) is unnecessary. Polymers may be either natural (e.g., polypeptides, polysaccharides) or synthetic [e.g., poly(D,L-lactic acid), polyanhydride]. Drugs may be dissolved or dispersed uniformly throughout the matrix, incorporated within an erodible reservoir, or they may be an integral part of the polymeric chain.

Controlled-release polymeric systems based on chemically controlled drug release may be classified as follows:

1. Erodible systems
2. Drug-polymer conjugates

The drugs may be released either by degradation of the polymer matrix or by hydrolytic cleavage of the polymer-drug bonds. Terms that describe the process are erodible, degradable, and absorbable. In this chapter, the prefix ''bio-'' used with one of the foregoing terms indicates that the process occurs within a biological environment. (Some suggest that the prefix ''bio-'' should be reserved solely to describe a biological degradation process, e.g., via enzymes). In an erodible system, solubilization of the solid polymer occurs as a consequence of a chemical reaction, dissolution of a water-soluble polymer, ionization, or protonation. The polymer system must physically erode and therefore lose mass. In this process, the polymer chain length may or may not be shortened. In a degradable system, the polymer becomes chain-shortened. Finally, ''absorbable system'' describes a polymer that is broken down by a living system, either enzymatically, hydrolytically, phagocytically, or simply by solubilization, with the breakdown products being removed via normal metabolic means. Many polymers are referred to as erodible, degradable, and

OCH$_3$ CO$_2$H   OCH$_3$ CO$_2^-$

$\longrightarrow$

CO$_2$R   CO$_2$R

absorbable, these terms often being used interchangeably in the literature. Yet a polymer may have only one of the three characteristics without necessarily possessing the other two. In this chapter, we generally use these terms interchangeably.

ERODIBLE POLYMERS

Introduction

Many erodible systems (e.g., esters, anhydrides, etc.) have water-labile bonds incorporated into the polymeric chain. Under physiological conditions, the labile bonds are hydrolyzed, and the polymeric chain breaks down to oligomers that are easily removed either by the body or the environment. Other mechanisms for degradation are (1) incorporation of bonds that can be cleaved by proteolytic enzymes at the site of delivery, and (2) inclusion of polymeric chains with specific side groups that undergo chemical reaction, mainly hydrolysis, and subsequent dissolution of the initially insoluble polymer.

Polymer Erosion Mechanisms

The erosion mechanisms of polymers can be described both physically and chemically.

Chemical Erosion

There are three general chemical mechanisms that describe polymer erosion, as has been described by Heller [1] (Fig. 1).

Mechanism I.   Mechanism I describes the degradation of water-soluble macromolecules that are cross-linked to form a three-dimensional network. As long as the crosslinks remain intact, the network is insoluble. Degradation in these systems can occur either at the crosslinks to form soluble backbone polymeric chains (type IA) or at the main chain to form water-soluble fragments (type IB). Generally, degradation of type IA polymers provide high-molecular-weight, water-soluble fragments, while degradation of type IB polymers provide low-molecular-weight, water-soluble oligomers and monomers.

With the system in an aqueous environment, drug solubility is important. Highly water-soluble drugs generally leach out rapidly despite the erosion rate. To avoid this situation, slowly eroding systems should be used with somewhat insoluble molecules or with large hydrophilic/water-soluble macromolecules that become entangled in these systems and thus are unable to escape until a sufficient number of crosslinks have been cleaved.

An example of a type IA system is the release of bovine serum albumin (BSA) from hydrogels. Data on varying the amounts of the cross-linking comonomer in poly(N-vinyl pyrrolidone) hydrogel are shown in Fig. 2 [2]. At low cross-linking levels, the hydrogel swells extensively, and BSA is released rapidly. As crosslink concentrations increase, swelling is reduced, and release of BSA is slowed. The hydrogel does not degrade during the release period.

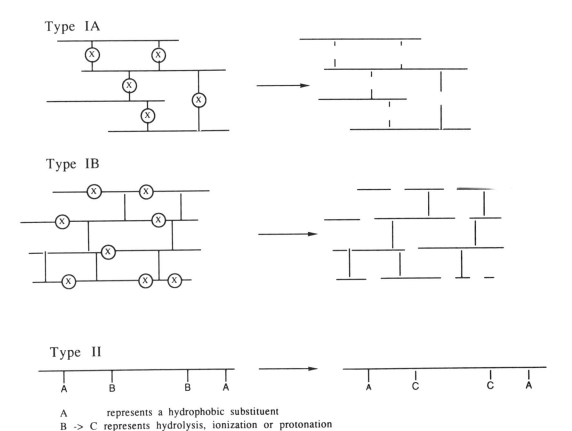

Figure 1 Schematic representation of degradation mechanisms. (From Ref. 1.)

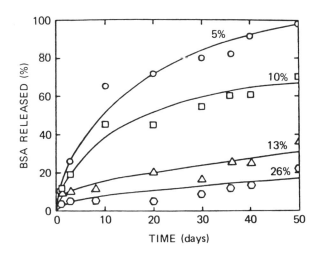

Figure 2 Bovine serum albumin cumulative release from N-vinylpyrrolidone hydrogels as function of crosslink density. (From Ref. 2.)

201

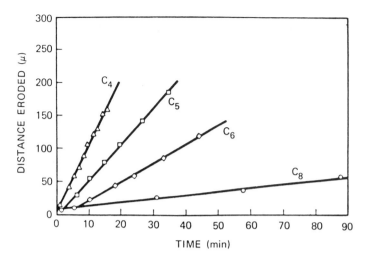

Figure 3   Effect of size of the ester group in half-esters of poly(methyl vinyl ether-*co*-maleic anhydride) on rate of erosion at pH 7.4. (From Ref. 5.)

Mechanism II.   Mechanism II describes the dissolution of water-insoluble polymers with side groups that are converted to water-soluble polymers as a result of ionization, protonation, or hydrolysis of the groups. With this mechanism the polymer does not degrade, and its molecular weight remains essentially unchanged. Materials displaying type II erosion include cellulose acetate derivatives [3] and partially esterified copolymers of maleic anhydride [4]. These polymers become soluble by ionization of the carboxyl group as shown by type II erosion in Fig. 1 and Scheme I. The number of carbon atoms in the ester side group will affect the dissolution rate of this polymer. With small ester groups, only a low degree of ionization is sufficient to solubilize the polymer. This could be accomplished at low pH (e.g., at a pH as low as 5). As the size of the alkyl group increases, so does the hydrophobicity, and progressively more ionization is needed to solubilize the polymer, resulting in an increasingly higher dissolution pH (Fig. 3) [5]. Similarly, a higher degree of esterification of the polymeric backbone results in a more hydrophobic polymer that will require a higher pH to degrade. The effect of pH on the rate of polymer erosion is illustrated by the release of hydrocortisone from partially esterified poly(methyl vinyl ether-*co*-maleic anhydride) (Fig. 4) [6].

Mechanism III.   Mechanism III describes the degradation of insoluble polymers with labile bonds. Hydrolysis of the labile bonds causes scission of the polymer backbone, thereby forming low-molecular-weight, water-soluble molecules.

   Polymers undergoing type III erosion (Fig. 1) include poly(lactic acid), poly(glycolic acid) and their copolymers, poly(ortho esters), polyamides, poly(alkyl 2-cyanoacrylates), and polyanhydrides.

   A more complete listing of synthetic and natural polymers that have been studied as type III bioerodible drug delivery systems is presented in Table 1 [6] along with pertinent references.

   The three mechanisms discussed are not mutually exclusive; combinations of them can occur. One may envision a cross-linked polymer (mechanism I) with water-labile bonds in its backbone (mechanism III). It is also possible to achieve two independent erosion rates. The first erosion rate may depend on the rate of formation of water-insoluble

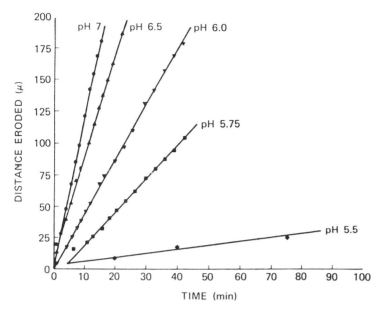

Figure 4 Effect of erosion medium pH on erosion rate of half-esters poly(methyl vinyl ether-*co*-maleic anhydride). (From Ref. 5.)

Table 1 Examples of Biodegradable Polymers

1.	Polyamides	
	a. Albumin	(7)
	b. Poly(hydroxylalkyl-*co*-L-glutamine)	(8)
	c. Collagen	(9)
	d. Poly(proline-*co*-glutamic acid)	(10)
	e. Poly(gelatin-*co*-lysine ester)	(11)
	f. Poly(L-glutamic acid-*co*-γ-ethyl-L-glutamate	(10)
2.	Polyesters	
	a. Poly(β-hydroxybutyrate)	(12)
	b. Poly(lactic acid)	[a]
	c. Poly(glycolic acid)	[a]
	d. Poly(ε-caprolactone)	(13)
	e. Poly(alklene oxalates)	(14)
	f. Polydioxanone	(15)
	g. Poly(alkylene diglycolates)	(16)
3.	Poly(ortho esters)	[a]
4.	Polyanhydrides	[a]
5.	Polyurethanes	(17)
6.	Polyacetals	(18)
7.	Polyiminocarbonates	(19)
8.	Polysaccharides	(20)
9.	Poly(vinyl pyrrolidone)	(3)
10.	Poly(alkyl 2-cyanoacrylates)	(21)
11.	Inorganic polymers	(22)
12.	Polyphosphazenes	(23)
13.	Poly(phosphate esters)	(24)

[a] Reported in this chapter.
Source: Adapted from Ref. 6.

203

polymer chains, while the second erosion rate may depend on the scission hydrolysis of the polymeric backbone.

Physical Erosion

The physical erosion mechanisms may be characterized as heterogeneous or homogeneous.

Heterogeneous erosion. In heterogeneous erosion, also called surface erosion, the polymer erodes only at its surface, and maintains its physical integrity (e.g., device shape, molecular weight) as it degrades. As a result, its drug release kinetics are predictable, and zero-order release kinetics will be obtained by applying the appropriate geometry (see p. 209 [25].

Homogeneous erosion. Few bioerodible polymers exhibit surface erosion as their predominant mechanism for drug release. Most polymers undergo homogeneous erosion; i.e., the hydrolysis occurs at an even rate throughout the polymer matrix. Generally, these polymers tend to be more hydrophilic than those exhibiting surface erosion. As a result, water penetrates the polymer matrix and increases the rate of drug diffusion. In homogeneous erosion, there is a loss in the integrity of the polymer matrix.

Both heterogeneous and homogeneous erosion represent extreme cases. Most erodible drug delivery systems are combinations of the two processes. In addition to hydrophobicity/hydrophilicity, morphology [26] and structural sequence [27] govern the erosion rate and type. Crystalline regions exclude water, with highly crystalline polymers tending to undergo heterogeneous erosion. Large hydrophobic blocks in block copolymers will similarly tend toward surface erosion. The addition of drugs also affects the above parameters, so that each drug/polymer system must be examined individually. Thus, each property is not exclusive, and all factors contribute to the type and rate of erosion.

Characteristics of Common Erodible Polymers

Highlighted below are characteristics of some of the more widely studied families of erodible polymers.

Polyamides

Polyamides generally degrade enzymatically. Accordingly, there is the potential for variability in the degradation rate from subject to subject and from site to site. However, the environment of the amide bond can be tailored for cleavage by a specific proteolytic enzyme. Thus, this specificity can provide a delivery system such that the polymer can degrade at the desired location by a site-specific enzyme. As an example, poly(γ-ethyl-L-glutamate-co-L-glutamic acid) (Scheme 2) displays two types of degradative mechanisms. Type II erosion occurs initially where the ethyl ester-side pendent chains are hydrolyzed to form the water-soluble copolymer. This polymer diffuses away and then it is degraded enzymatically (type III), in the liver and kidney, to naturally occurring L-glutamic acid. The release kinetics from devices made of this copolymer exhibit initially high diffusional release rates and then gradually level off [28].

Polyesters

Polyesters, especially poly(lactic acid) (PLA), poly(glycolic acid) (PGA), copolymers of those (PLGA), and poly(ε-caprolactone) are among the most extensively studied erodible polymers. The advantage of the first two is that they are already used clinically as absorbable suture material [29] and as erodible prosthesis parts [30]. Approval of these materials for human use by the Food and Drug Administration (FDA) make them attractive candidates for use in drug delivery devices. Also noteworthy is the fact that lactic acid is a naturally occurring product of glycolysis.

The degradation mechanism of the polyesters appears to be homogeneous. The relative erosion rate depends on the chemistry of the ester (i.e., bond energy) and the ratio of monomers. Thus, the $t_{1/2}$ of implanted pellets (ca. 5 mg in rats) quickly drops from 5 months for PGA and 6 months for PLA to about 1 week for the copolymer PLGA 50:50 [31].

One application of controlled-release drug delivery from a biodegradable device made of PLGA is for the release of the luteinizing-hormone releasing hormone (LHRH) and its more potent synthetic analogs. This native hormone causes the production of estrogen, progesterone, and testosterone and initiates the reproductive cycle. The synthetic analogs can initiate down-regulation to desensitize the LHRH receptors [32], an effect that can be utilized in conditions that require reduction in the level of estrogen, progesterone, and testosterone. In addition to contraception for both males and females, potential applications include treatment of cancer of the prostate gland, endometriosis, and mammary cancer. The first FDA-approved product to release an LHRH synthetic analog (leuprolide acetate) for the treatment of prostatic cancer is the Lupron-Depot from Takeda-Abbott [33]. The conventional treatment required once-daily injection of the drug over a long period. The controlled-release poly(D,L-lactide-co-glycolide) 75:25 microspheres require injection only once a month (Fig. 5). Similar products are currently in use or are in clinical trials at Imperial Chemical Industries, England—Zoladex [34], at Syntex, California—Nafarelin [35], and at Ibsen-Biotech, France—D-Trp-6-LHRH [36].

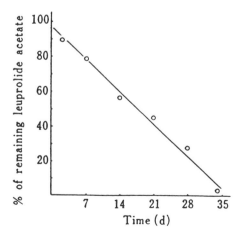

Figure 5 Release profile of leuprolide acetate from poly(lactide-co-glycolide) 75:25 microspheres. (From Ref. 33c.)

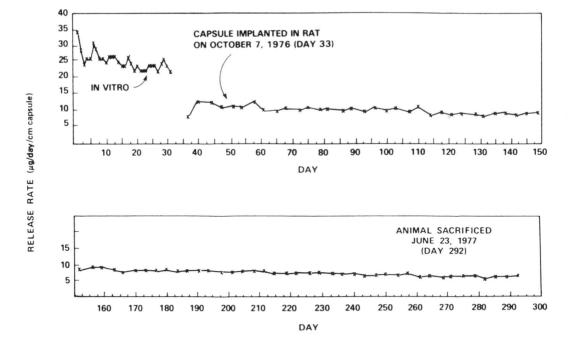

Figure 6 Daily release rate of norgestrel from poly(ε-caprolactone) capsule in rat after 32 days in vitro. Release rate was determined by measurement of radioactivity. (From Ref. 38.)

Another polyester, poly(ε-caprolactone), hydrolyzes to 6-hydroxyhexanoic acid (ε-hydroxycaproic acid) [37]. The polymer is permeable to lipophilic drugs and is therefore useful for the release of steroid drugs. In fact, the high permeability of poly(ε-caprolactone) and its copolymers has permitted the development of biodegradable reservoir devices that erode completely after the drug is depleted. Devices made of this polymer were capable of subdermal delivery of the contraceptive, levonorgestrel, at 50 μg/day for at least 1 year (Fig. 6) [38].

Poly(ortho esters)

Poly(ortho esters) were initially developed by Alza Corp. (under the trade names Chronomer and Alzamer) [39]. These polymers are more vulnerable to acidic than basic hydrolysis [40]. Chronomer is made of 2,2-dialkoxytetrahydrofurane, 1,6-hexanediol, and 1,4-cyclohexanedimethanol, and when hydrolysed it releases hydroxybutyric acid [41]. This erosion by-product, being acidic, further catalyzes the erosion and causes an increasing erosion rate with time.

Chronomer has been used by the World Health Organization as a biodegradable contraceptive implant [42]; however, further work was discontinued because local irritation at the implantation site was observed in clinical trials.

Another poly(ortho ester) system, based on the reaction of ketene acetals and polyols, was devised by Heller and co-workers and is currently being developed further by Merck, Sharp & Dohme [43]. Acidic excipients are incorporated into this system as catalysts to yield a surface-eroding delivery system [44]. This system is currently being explored for use in the release of the worm-killing agent, ivermectin, to protect dogs and cats.

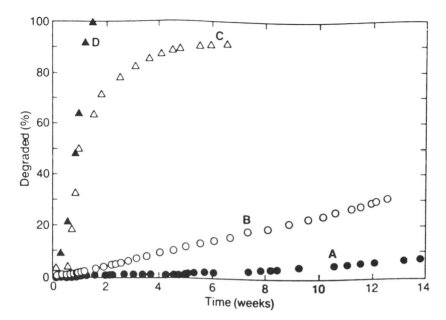

Figure 7 Degradation rate profiles of poly(1,3-bis(carboxyphenoxy)propane-*co*-sebacic anhydrides) at 37°C and pH 7.4. A, poly(CPP); B, poly(CPP-SA) 85:15; C, poly(CPP-SA) 45:55; D, poly(CPP-SA) 21:79. (From Ref. 47.)

Polyanhydrides

Polyanhydrides were first prepared as an alternative to polyester fibers for the garment industry [45]. However, owing to their hydrolytic instability, polyanhydrides were unsuitable as textile fibers. This same quality made polyanhydrides suitable as biodegradable polymers for drug delivery [46]. By varying the ratio of the hydrophobic moiety 1,3-bis(*p*-carboxyphenoxy)propane (CPP) and sebacic acid (SA), controlled degradation rates, from days to years, have been achieved (Fig. 7) [47].

One of the polyanhydrides, poly(CPP-SA) 20:80, is at present under Phase III clinical trials by NOVA Pharmaceutical Co. In these trials, nitrosoureas are administered as chemotherapeutic drugs following surgical removal of a brain tumor by applying the polymer/chemotherapeutic drug directly to the tumor site via a controlled-release biodegradable polyanhydride. This treatment provides a higher localized antitumor drug concentration, yet it minimizes the risk to other healthy tissues. Preliminary results showed good patient response to the treatment, and the quality of their life improved dramatically.

Polyanhydrides have also been used to release model polypeptidic drugs such as recombinant bovine somatotropin and insulin for more than 3 weeks (as measured by HPLC and radioimmunoassay) (Fig. 8) [48].

Drug Release Mechanisms

The release of drugs from the erodible polymers may occur by any of the mechanisms presented in Fig. 9 [49].

In mechanism A, the drug is attached to the polymeric backbone by a labile bond. This bond has a higher reactivity toward hydrolysis than the polymer reactivity to breakdown. In mechanism B, the drug is in a core surrounded by a bioerodible rate-controlling

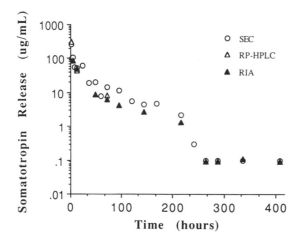

Figure 8 Release of recombinant bovine somatotropin from poly((carboxy-phenoxy)hexane) measured by different methods. (From Ref. 48.)

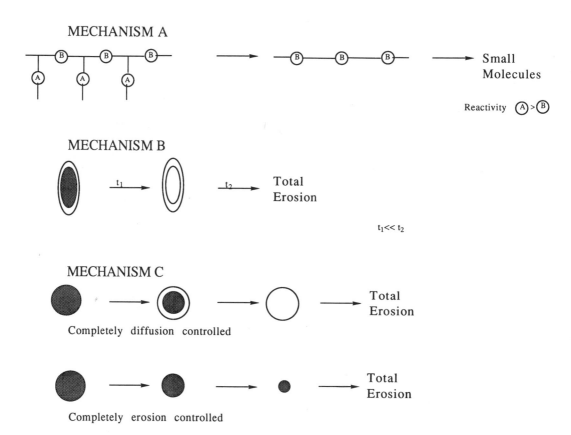

Figure 9 Possible release mechanisms from erodible polymers. (From Ref. 49.)

membrane. This is a reservoir-type device (see Chapter 3) that provides erodibility to eliminate surgical removal of the drug-depleted device.

Mechanism C describes a homogeneously dispersed drug in a bioerodible polymer. The drug is released by either erosion, diffusion, or a combination of both.

Mathematics of Erosion and Release

Heterogeneous Erosion

For an ideal heterogeneous eroding system (negligible edge effects), the release kinetics will equal the degradation rate of the polymer. As the degradation boundary moves into the bulk, the drug is released into the surrounding media.

For a simple slab of area A containing a uniform dispersed drug (C_0), the amount of drug released at any time t is

$$\frac{dM_t}{dt} = BC_0A \tag{1}$$

where B is the surface degradation rate (cm/s). Integrating Eq. (1), the following mass balance expression is obtained:

$$M_\infty = ABC_0t_\infty \tag{2}$$

where t_∞ is the time for total erosion. The expression for fractional drug release is

$$\frac{M_t}{M_\infty} = \frac{t}{t_\infty} \tag{3}$$

For a sphere with radius r the equivalent expression is

$$\frac{M_t}{M_\infty} = 1 - \left[1 - \left(\frac{t}{t_\infty} \right) \right]^3 \tag{4}$$

Homogeneous Erosion

The release of drug from a homogeneous eroded matrix is more complicated. The release in such cases is a combination of both diffusion and erosion. Diffusional drug release without erosion is described in Chapter 3. (The kinetics of drug release from a slab (flat matrix) device containing uniformly dissolved or dispersed drug generally follows a square-root-of-time relationship.)

The Higuchi model [50] (which assumes drug loading $> C_s$) describes the dissolution of the drug initially from the surface of the slab and progressively through a thicker drug-depleted layer. Using this model for a slab, the following expression for drug release has been derived [51]:

$$\frac{dM_t}{dt} = \frac{A}{2} \left(\frac{2PC_0}{t} \right)^{1/2} \tag{5}$$

where M_t is drug released at time t, P is the permeability of the polymer (see Chapter 3), A is the surface area of both sides of the slab (cm^2), and C_0 is the initial concentration of the drug in the slab (mg/cm^3). For an eroding system the situation is more complex, as the permeability P increases with time. This increase in permeability is attributed to the gradual loosening of the matrix by hydrolytic cleavage.

The change of permeability with time is difficult to estimate, but the following simplification can be made:

$$\frac{P}{P_0} = \frac{N_0}{(N_0 - Z)} \tag{6}$$

where N_0 is the initial number of bonds, Z is the number of cleaved bonds, and P_0 is the original permeability. If we assume simple first-order kinetics, where the rate of bond cleavage dZ/dt is proportional to the number of cleaved bonds present, then

$$\frac{dZ}{dt} = K(N_0 - Z) \tag{7}$$

where K is the first-order rate constant. Then, by integration,

$$Kt = \ln\left[\frac{N_0}{(N_0 - Z)}\right] \tag{8}$$

Equations (6) and (8) can be combined to yield

$$P = P_0\, e^{Kt} \tag{9}$$

Substituting this expression into the Higuchi equation (5) yields

$$\frac{dM_t}{dt} = \frac{A}{2}\left[\frac{2P_0\, e^{Kt}C_0}{t}\right]^{1/2} \tag{10}$$

A plot of the integrated form of this equation along with diffusional release without erosion is shown in Fig. 10. Without erosion the drug release follows square-root of time

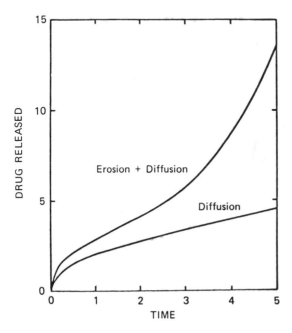

Figure 10 Theoretical curves of Eq. (10) along with diffusional release without erosion. (From Ref. 2.)

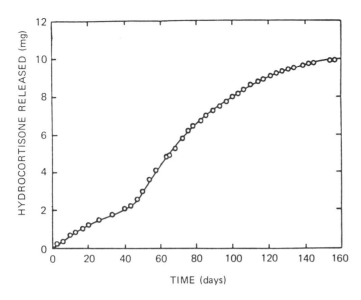

TIME (days)

Figure 11 Release of hydrocortisone from polyester spheres. (From Ref. 2.)

kinetics ($t^{1/2}$ kinetics—first-order). However, because of the polymer degradation the permeability rises, and the normal decline in release rate is slowed and then completely reversed (Fig. 11).

Similarly, mathematical models of homogeneous degradation and release rates of drugs from microspheres were developed [52]. In this analysis, the swelling of the matrix, the concentration dependence of the solute diffusion coefficient, and the external mass transfer resistance are neglected to simplify calculations. A cross section of a sphere undergoing erosion is presented in Fig. 12 at the region $a \geq r \geq R$ (where r is the average radius at time t). The kinetic expression describing release from a sphere is

$$\frac{dM_t}{dt} = k_0 4\pi R^2 \tag{11}$$

where k_0 is the erosion rate constant $\left(\dfrac{mg}{hr \cdot cm^2}\right)$. M_t, the amount of biologically active

material that is released from the device at time t, is defined as follows:

$$M_t = \left(\frac{4\pi}{3}\right) C_0(a^3 - R^3) \tag{12}$$

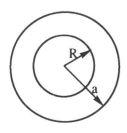

Figure 12 Cross section of a sphere that has eroded from "a" to "R."

Substituting the mass balance equation (12) into the kinetic expression (11) yields

$$\frac{d[(4\pi C_0/3)(a^3 - R^3)]}{dt} = k_0 4\pi R^2 \tag{13}$$

or

$$\frac{d(a^3 - R^3)}{dt} = \frac{3k_0}{C_0 R^2} \tag{14}$$

Differentiation and simplification of Eq. (14) yields

$$\frac{dR}{dt} = \frac{-k_0}{C_0} \tag{15}$$

The relationship for R as a function of time is

$$R = a - \left(\frac{k_0}{C_0}\right) t \tag{16}$$

Substituting Eq. (16) into Eq. (12) yields

$$M_t = \left(\frac{4\pi C_0}{3}\right)\left[a^3 - \left(a - \frac{k_0}{C_0 t}\right)^3\right] \tag{17}$$

and since

$$M_\infty = \frac{4\pi C_0 a^3}{3} \tag{18}$$

where M_∞ is the amount of drug released following exhaustion of the device, then

$$\frac{M_t}{M_\infty} = 1 - \left(1 - \frac{k_0 t}{C_0 a}\right)^3 \tag{19}$$

For heterogeneous degradation, the relationship M_t/M_∞ is directly proportional to time, while for homogenous degradation, the relationship between M_t/M_∞ is not linear with time.

More detailed analyses of devices undergoing either homogeneous or heterogeneous degradation are described in the literature [53]. These analyses discuss drug diffusion, permeability of the polymer to the drug, solute diffusion in the erodible device, and drug solubility in the matrix.

As the foregoing mathematical models indicate, for surface erosion a constant delivery rate is provided only when the device does not change its surface area as a function of time. A slab-shaped device provides the required geometry (neglecting edge effects). Spheres and cylinders display delivery rates that decrease with time. However, specialized geometries such as hollow cylinders and cloverleaf shapes may also be designed to produce zero-order release [54].

ERODIBLE POLYMERS CONTAINING PENDENT SUBSTITUENTS

Introduction

In another approach to development of erodible drug delivery devices, the therapeutic agent is chemically attached to the polymeric backbone. In such systems, the drug is

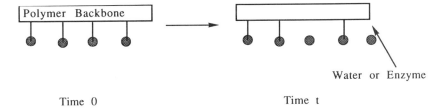

Figure 13   Idealized diagram of a chemically controlled pendent-chain drug delivery system.

released gradually from the polymer backbone via hydrolytic or enzymatic cleavage (Fig. 13) at a controlled rate. A significant advantage of the pendent-chain systems over other controlled-release systems is that more than 80% of the total delivery system can be the drug itself. Some systems contain 70–90% w/w of drug attached to inert polymer carrier. The polymeric backbone may be also modified by attaching a site-specific antibody to direct the drug to the target organ [55].

The ideal polymeric backbone in pendent-chain systems should degrade to products that are either cleared from the body or absorbed. Further, the polymer carriers should neither cause immunological reactions nor should the drug/polymer device function as a hapten and induce allergic reactions.

The drug-polymer attachment may be covalent or ionic, or function through weaker secondary molecular forces. The drug may be part of the polymeric backbone or attached to the side chain either directly or through a spacer group. A spacer group is usually included because it can be hydrolyzed enymatically. Enzymes at the targeted site then attack the spacer and release the drug. Similarly, the backbone may contain biostable segments linked with biodegradable segments to increase versatility.

Drug Attachment

Two synthetic routes have been employed in the preparation of polymers that contain bioactive pendent substituents [56]. In the first, the active agent is converted to a polymerizable derivative that subsequently polymerizes to afford the macromolecular combination (Fig. 14). In the second, bioactive agents are chemically bound to preformed synthetic or naturally occurring polymers (as spacer groups) as described in Fig. 15. The major advantage of the latter is the versatility of these processes. The copolymers can be prepared from several different monomers that hydrolyze under selected conditions. Also to be considered are the difficulties in synthesizing the polymer (Fig. 15, top) from the monomer-drug, which can lead to chemical changes in the drug due to side reactions or complete loss of the drug attributable to undesirable branching and/or crosslinking.

As an example of drug attachment, polypeptides containing pendent bioactive substituents were prepared by the polymerization of alcohol derivatives. Cholesterol was

Figure 14   Drug (D) is converted to a polymerized derivative (D-M) that subsequently is polymerized.

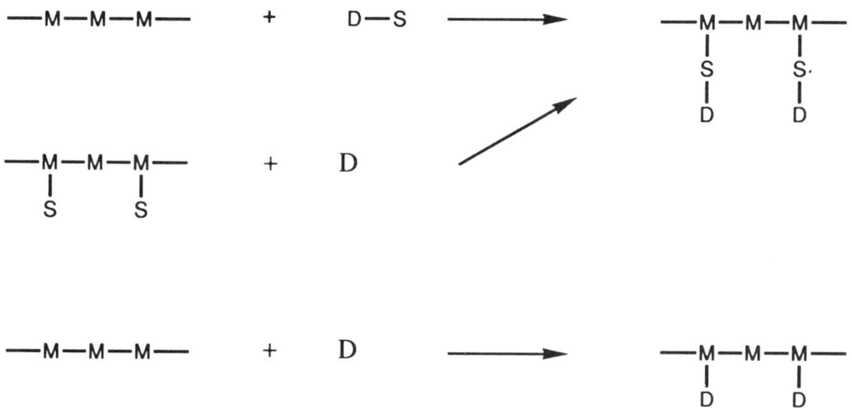

Figure 15 Synthesis and polymerization of drug derivatives. M, monomer; S, spacer; D, drug.

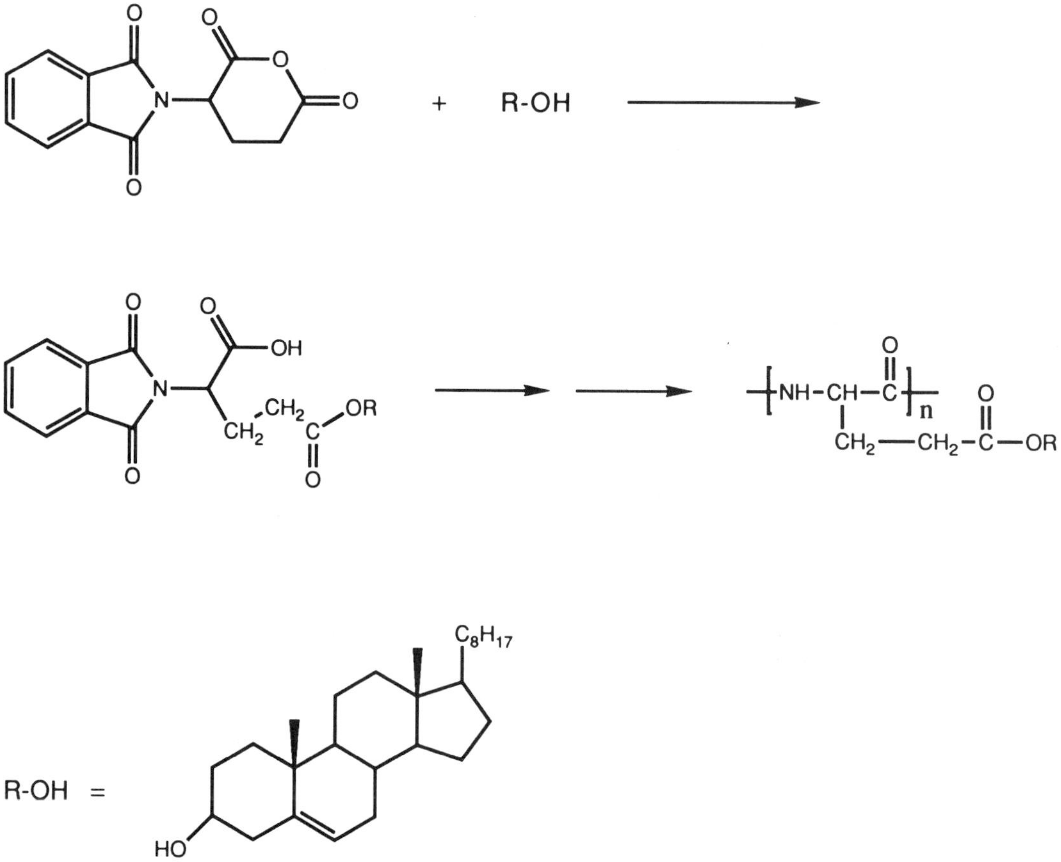

Figure 16 Cholesterol attachment to N-carboxyanhydride L-glutamic acid, and its polymerization.

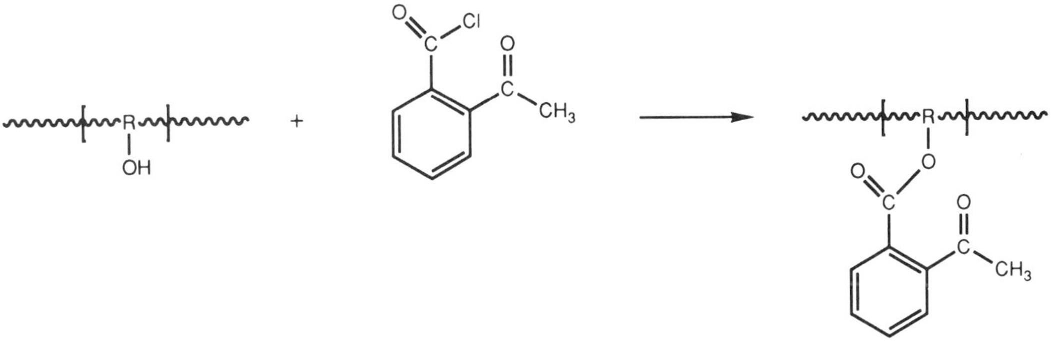

Figure 17 Attachment of acid chlorides to a polymeric chain backbone containing pendent hydroxyl groups.

attached to the N-carboxyanhydride of L-glutamic acid. The combination was polymerized to yield α-poly(Γ-cholesteryl L-glutamate) (Fig. 16) [57]. The reaction of bioactive agent with preformed polymers is presented in Fig. 17. The active agent reacts with the polymer containing either pendent hydroxyl or amino groups [58].

Rowland et al. [59] attached a *p*-phenylene-diamine mustard and an immunoglobulin (I_g) (Fig. 18) from rabbit antiserum against mouse lymphoma cells (EL-4) to poly(glutamic acid). This antitumor agent was injected into mice that were inoculated with EL-4 lymphoma cells. The highly toxic drug (LD_{50} of the free drug is 5 mg/kg) was raised to a less toxic level (LD_{50} of 200 mg/kg) for the pendent drug. In addition, survival of the mice bearing EL-4 lymphoma was 100 days for the pendent drug, but only 25 days for mice treated with the free drug.

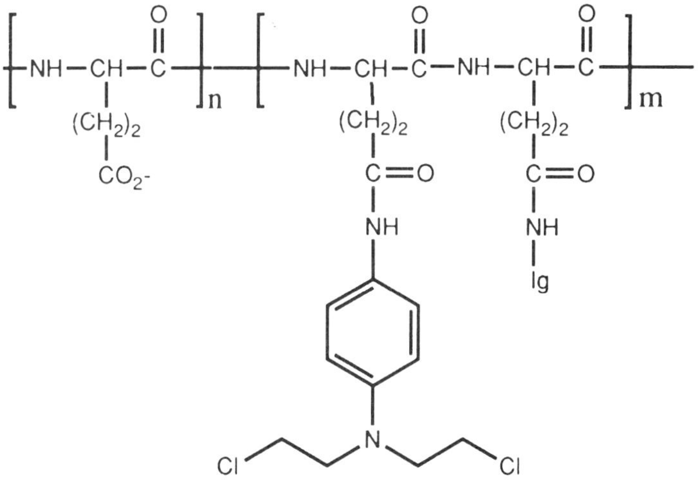

Figure 18 Product of the attachment of *p*-phenylenediamine mustard and immunoglobin (Ig) from rabbit antiserum against mouse lymphoma cells (EL4) to poly(glutamic acid).

Hydrolysis Kinetics

The release kinetics of these systems are more complex than those described above. To achieve zero-order release, cleavage of the drug from the polymer must be the rate-limiting step. Consider the cleavage of the pendent-chain system given by the equation

$$P\text{-}Dr + \text{cleaving agent} \xrightarrow{\ k\ } P + Dr \tag{20}$$

Then, if the rate of diffusion of the cleaving agent into the polymer matrix and the rate of the diffusion of the drug, Dr, through the polymer matrix is much faster than the rate of cleavage, the release rate is limited by the chemical reaction. Thus the release rate is constant only if the thermodynamic activity of the bound drug is constant with time, and the cleaving agent (e.g., acid, base, enzyme) is present in significant excess of P-Dr. The thermodynamic activity of the bound drug depends on the chemical environment and the amount of bound drug present in a certain volume.

The factors that govern the release kinetics are as follows:

1. Homogeneous or heterogeneous degradation, or a combination of the two
2. Rate of permeation of the water
3. Rate of diffusion of the liberated drug from the polymer matrix
4. Hydrolysis rate of the bound drug to produce a free drug

A water-soluble drug provides the simplest situation for modeling the hydrolysis kinetics. If the liberated drug diffuses rapidly through the polymer, the system will follow first-order kinetics (diffusion time is considered rapid if it is short compared to reaction time). Therefore, the release rate, R_r, can be defined as

$$R_r = R_h = kC_0 e^{-kt} \tag{21}$$

where R_h is the rate of hydrolysis, k is the hydrolysis rate constant, and C_0 is the initial concentration of drug-polymer linkages. For a purely heterogeneous reaction (i.e., no water penetration occurs), the geometry and the size of the system will affect the rate. The following expression will hold for an insoluble sphere [60]:

$$R_r = \frac{dM}{dt} = k4\pi r^2 C_0 \tag{22}$$

where r is the average radius at time t.

A different mathematical model has been suggested for a matrix that contains hydrophobic and hydrophilic zones [61]. In this model, the surface area is important. For this model, R_w = rate of water permeation and R_h = hydrolysis rate. Three possibilities exist:

1. $R_h \gg R_w$: The hydrolysis reaction is much faster than the water permeation. The hydrolysis occurs at the interface and zero-order release occurs until the depletion of the drug, then

$$R_r = R_w C_0 \tag{23}$$

2. $R_h \approx R_w$: The release rate is controlled by a combination of these two rates. If R_h is constant and independent of drug concentration, then

$$R_r = R_h R_w t \tag{24}$$

and

$$R_r = R_w C_0 \tag{25}$$

R_r increases with time until the drug concentration at the surface becomes zero. After that point, R_r becomes constant as in the case $R_h \gg R_w$. If R_h is first-order, then

$$R_r = R_w C_0 (1 - e^{-kt}) \tag{26}$$

3. $R_h \ll R_w$: In this case the water permeation is considerably faster than the rate of hydrolysis. If the diffusion of the free drug from the matrix is rapid, R_h determines the release rate. This case is analogous to homogeneous hydrolysis [Eq. (21)].

CONCLUSIONS

In this chapter the main elements of erodible controlled-release systems were outlined. The drugs contained in these systems may be released either by degradation of the polymer matrix or by hydrolytic cleavage of the polymer-drug bonds. Two main classes are identified as follows:

1. Erodible systems
2. Drug-polymer conjugates

For the first class, the polymers are initially insoluble and are subsequently hydrolyzed either by the body or the environment. Drug contained in these systems is released as the polymer erodes. In the second class, the therapeutic agent is chemically attached to the polymeric backbone. In such systems, the drug is detached from the polymer backbone via hydrolytic or enzymatic cleavage at a controlled rate.

Both systems have advantages and disadvantages. Ease of fabrication, preservation of drug viability, and loading are some of the factors that will determine which class will be used.

PROBLEMS

1. Assuming heterogeneous degradation, design a device that will yield an increasing release rate.
2. Calculate and graph the cumulative release of drug from 100 mg of 10%-loaded spheres (r = 50 μm) after 1, 2, 10, 30, and 60 days if the release persists more than 100 days (heterogeneous erosion).
3. For a slab, what type of release rate would you expect from combined heterogeneous and homogeneous erosion?
4. How will the molecular weight of polymers affect the release rate?
5. Name all factors that affect the erosion rate of a polymer matrix (nonenzymatic degradation). Explain what the effect of each of these factors is.
6. One of the reasons for the burst effect, in heterogeneous degradation, is drug particles that touch the outer surface and can diffuse immediately from the sphere. (a) What is the mathematical model for this phenomenon [particles (with radius r) randomly distributed throughout a polymeric sphere (with radius R)]? (b) For spheres with R = 1000, 500, and 100 μm, calculate the fraction of particles (r = 50 μm) that will burst.
7. The data given in Table 2 are for fraction erosion of spheres made of an erodible

Table 2   Fraction Polymer Eroded—P_t/P_∞ (Experimental)

time (h)	100 μm	660 μm	475 μm	250 μm
0.00	0.000	0.000	0.000	0.000
0.13	0.000	0.000	0.007	0.000
0.25	0.000	0.011	0.013	0.011
0.38	0.001	0.017	0.013	0.011
0.50	0.002	0.020	0.016	0.027
0.63	0.003	0.021	0.016	0.029
0.75	0.003	0.024	0.017	0.032
1.00	0.003	0.025	0.017	0.033
1.25	0.004	0.041	0.019	0.035
1.50	0.004	0.042	0.210	0.036
2.00	0.004	0.042	0.023	0.039
2.50	0.006	0.044	0.034	0.039
3.00	0.008	0.044	0.043	0.045
4.00	0.010	0.050	0.049	0.050
5.00	0.045	0.058	0.055	0.058
6.00	0.067	0.069	0.062	0.069
7.00	0.074	0.075	0.073	0.082
8.50	0.091	0.092	0.101	0.118
14.20	0.216	0.227	0.295	0.381
23.80	0.404	0.455	0.573	0.655
28.00	0.463	0.526	0.634	0.703
32.50	0.517	0.585	0.680	0.738
48.00	0.662	0.719	0.771	0.818
73.50	0.792	0.833	0.860	0.893
96.00	0.854	0.897	0.918	0.932
120.00	0.910	0.936	0.941	0.949
144.00	0.942	0.950	0.951	0.956
193.00	0.960	0.964	0.965	0.970
263.00	0.975	0.975	0.978	0.980
358.00	0.985	0.985	0.986	0.988
477.00	0.993	0.993	0.993	0.995
597.00	0.996	0.997	0.997	0.998
766.00	1.000	1.000	1.000	1.000

polyanhydride, poly(CPP-SA) 20:80. This fraction could be derived, similarly from Eq. (4), to yield $P_t/P_\infty = 1 - (1 - k_0 t/pa)^3$, where k_0 is defined as the erosion constant and p is the density of the sphere. (a) Plot the expected cumulative fraction of polymer eroded (P_t/P_∞) for spheres with initial diameters of 1000, 660, 475, and 250 μm from t = 0 to t = 100 h. Use a value of $k_0/p = 2.5$ μm/h. (b) Compare the experimental data with your results. Comment on the deviation of the experimental results from your theoretical results.

ANSWERS

1.   Devices whose exposed surface area increases with time and will lead to increasing release rate. An example is a cone composed of a degradable polymer coated with a nondegradable polymer cut off at the top.

$$\frac{dM_t}{dt} = BC_0A \qquad (Eq.\ 1)$$

B is the surface degradation rate (cm/s), A is the surface area (cm²), and c_0 is the initial drug concentration.

2. $\quad \dfrac{M_t}{M_\infty} = 1 - \left[1 - \left(\dfrac{t}{t_\infty} \right) \right]^3 \qquad (Eq.\ 4)$

Total drug loading − MM_∞ = 100 mg × 10% = 10 mg.
t_∞ = 100 days

$$\frac{M_1}{10} = 1 - \left[1 - \left(\frac{1}{100} \right) \right]^3 = 2.97\%$$

$$\frac{M_2}{10} = 1 - \left[1 - \left(\frac{2}{100} \right) \right]^3 = 5.88\%$$

$$\frac{M_{10}}{10} = 1 - \left[1 - \left(\frac{10}{100} \right) \right]^3 = 27.10\%$$

$$\frac{M_{30}}{10} = 1 - \left[1 - \left(\frac{30}{100} \right) \right]^3 = 65.70\%$$

$$\frac{M_{60}}{10} = 1 - \left[1 - \left(\frac{60}{100} \right) \right]^3 = 93.60\%$$

$$\frac{M_{100}}{10} = 1 - \left[1 - \left(\frac{100}{100} \right) \right]^3 = 100\%$$

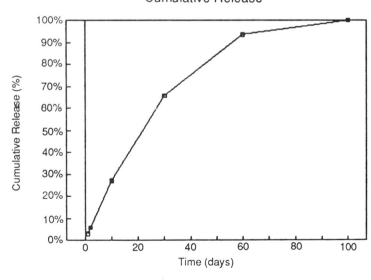

Cumulative Release

3. For heterogeneous erosion

$$\frac{dM_t}{dt} = BC_0A \qquad (Eq.\ 1)$$

For homogeneous erosion,

$$\frac{dM_t}{dt} = \frac{A}{2} \left[\frac{2P_0\,e^{Kt}C_0}{t} \right]^{1/2}$$ (Eq. 10)

Without erosion, the release follows first-order kinetics ($t^{1/2}$). With erosion, the initial release period will follow zero-order kinetics (t) that will accelerate with the diffusional release.

4. Assume that molecular weight is the only factor affecting erosion and that all factors that could effect erosion (e.g., crosslinking, crystallinity) are either unimportant or do not change as a function of molecular weight. Using this assumption, all molecular-weight polymers should have the same degradation rate. To allow both drug permeability and polymer solubility, the polymer has to reach a low-molecular-weight or monomeric stage. Therefore high-molecular-weight polymers will result in slower release rates.

5. Water has to come in contact with the polymer. Hydrophobicity of the polymer will affect the rate of water penetration into the bulk. Also, the degree of crystallinity of the polymer will affect the rate of erosion, as higher crystalline regions within the polymer will prevent water penetration. pH has effects on bond cleavage and therefore on the degradation rate. Morphological factors, such as porosity, may also affect the ease of water permeability. The bonds that make the polymer are also important, since some are more hydrolytically unstable.

6. (a) The volume of a spherical shell with width r/2 is

$$\frac{4}{3\pi R^3} - \frac{4}{3\pi(R - r/2)^3}$$ (a)

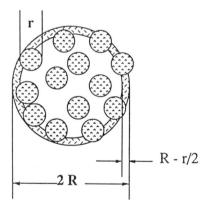

If the particles are randomly distributed throughout the sphere, any particle whose center occupies this volume [Eq. (a)] will *diffuse* from the sphere, as it will be in contact with the environment.

(b) The ratio of the above volume to the sphere is

$$\left[\frac{4}{3\pi R^3} - \frac{4}{3\pi(R - r/2)^3} \right] \Big/ \left(\frac{4}{3\pi R^3} \right) \quad \text{or} \quad 1 - \left(1 - \frac{r}{2R} \right)^3$$

R (μ)	r (μ)	Percent of particles within r/2
1000	50	0.6%
500	50	14.3%
100	50	57.8%

It is clear that as the ratio of particles to the sphere decreases, the burst will be more distinct.

7. Using the equation $P_0/P_\infty = 1 - (1 - k_0 t/pa)^3$, we get the following graph:

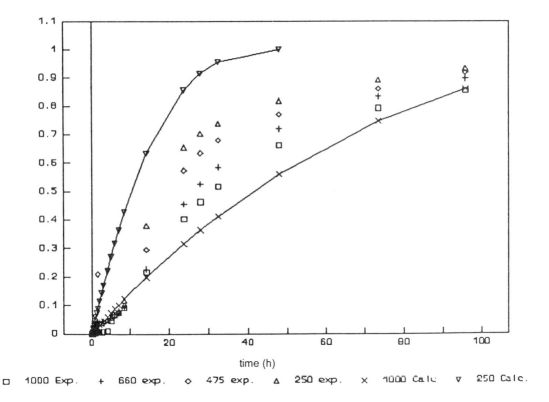

Pt/P

time (h)

□ 1000 Exp.   + 660 exp.   ◇ 475 exp.   △ 250 exp.   × 1000 Calc   ▽ 250 Calc.

Comparing the experimental and the calculated date (only 1000 and 250 μm are presented), we see that smaller spheres, i.e., those with higher surface area, erode faster. Also, the experimental data have a time lag in the erosion. The cause for this time lag include (1) only the monomers are soluble and therefore, experimentally, the polymer has to break down to monomers before erosion is detected; (2) water penetration is slow, but as the erosion progresses the water penetrates the bulk—deviating from heterogeneous degradation.

REFERENCES

1. (a) J. Heller, *Biomaterials*, *1*:51–57 (1980); (b) J. Heller, *CRC Crit. Rev. Ther. Drug Carrier Systems*, *1*:39 (1984).

2. J. Heller and R. W. Baker, in *Controlled Release of Bioactive Molecules* (R. Baker, Ed.), Academic Press, New York, pp. 1–17 (1980).

3. (a) J. G. Wagner and S. Long, *J. Pharm. Sci.*, *49*:121 (1960); (b) L. O. Wilken, Jr., M. M. Kochhar, D. P. Bennett, and E. P. Cosgrove, *J. Pharm. Sci.*, *51*:484 (1962); (c) G. D. Hiatt, J. W. Mench, and B. Fulkerson, *Ind. Eng. Chem. Prod. Res. Dev.*, *3*:295 (1964); (d) J. W. Mench and B. Fulkerson, *Ind. Eng. Chem. Prod. Res. Dev.*, *7*:2 (1968).

4. (a) L. C. Lappas and W. Mckeehan, *J. Pharm. Sci.*, *51*:808 (1965); (b) ibid., *54*:176 (1965); (c) ibid., *56*:1257 (1967).

5. J. Heller, R. W. Baker, R. M. Gale, and J. O. Rodin, *J. Appl. Polym. Sci.*, *22*:1991 (1978).

6. H. B. Rosen, J. Kohn, K. Leong, and R. Langer, in *Controlled Release Systems—Fabrication Technology, Vol. 11*, (D. S. Hsieh, Ed.), CRC Press, Boca Raton, Fla. pp. 83–110 and references within (1988).

7. T. K. Lee and T. D. Sokoloski, *Science*, *213*:233 (1981).

8. H. Rosen, Master's thesis, Massachusetts Institute of Technology, Cambridge, Mass. (1981).

9. W. G. Bradley and G. L. Wilkes, *Biomater. Med. Devices Artif. Organs*, *5*:159 (1977).

10. A. A. Randall, British Patent 1,049, 290 (1966).

11. K. R. Sidman, A. D. Schwope, W. D. Steber, S. E. Rudolph, and S. B. Poulin, *J. Membr. Sci.*, *7*:277 (1980).

12. J. N. Baptist, U.S. Patents 3,036,959 (1962); 3,044, 942 (1962); 3,225,766 (1965).

13. C. G. Pitt, T. A. Marks, and A. Schindler, in *Controlled Release of Bioactive Materials* (R. W. Baker, Ed.), Academic Press, New York, p. 19 (1980).

14. S. W. Shalaby and D. D. Jamiolkowsky, U.S. Patent 4,130,639 (1978).

15. N. Doddi, C. C. Versfelt, and D. Wasserman, U.S. Patent 4,052,988 (1977).

16. D. J. Casey and M. Epstein, U.S. Patent 4,048,256 (1977).

17. B. Master, P. Cefelin, T. E. Lipatova, L. A. Bakalo, and G. G. Lugovskya, *J. Polym. Sci. Polym. Symp.*, *66*:259 (1979).

18. J. Heller, D. W. H. Penhale, R. F. Helwing, B. K. Fritzinger, and R. W. Baker, *AIChE Symp. Ser.*, *206*:28 (1981).

19. (a) C. Li and J. Kohn, *Macromolecules*, *22*(5):2029 (1989); (b) J. Kohn and R. Langer, *Biomaterials*, *7*:176 (1986).

20. E. Schacht, L. Ruys, J. Vermeersch, and J. P. Remon, *J. Control. Rel.*, *1*:33 (1984).

21. W. R. Vezin and A. T. Florence *J. Biomed. Mater. Res.*, *14*:93 (1980).

22. (a) G. Daculi, N. Passsuti, S. Martin, C. Duedon, R. Z. Legeros, and S. Raher, *J. Biomed. Mat. Res.*, *24*:379 (1990); (b) K. De Groot, *Ceramics in Surgery*, 79 (1983).

23. (a) H. R. Allcock, *Proc. 16th Int. Symp. on Controlled Release of Bioactive Materials* (R. Perlman and J. A. Miller, Eds.), CRS, Chicago, pp. 18–19 (1989); (b) C. W. J. Grolleman, A. C. de Visser, J. G. C. Wolke, C. P. A. T. Klein, H. van der Goot, and H. Timmerman, *J. Control. Rel.* *4*:133 (1986); (c) C. W. J. Grolleman, A. C. De Visser, J. G. C. Wolke, H. van der H. Goot, and H. Timmerman, *J. Control. Rel.*, *4*:119 (1986).

24. N. H. Li, M. Richards, K. Brandt, and K. Leong, *Polym. Prepr.*, *30*:454 (1989).

25. H. B. Hopfenberg, in *Controlled Release Polymeric Formulations* (D. R. Paul and F. W. Harris, Eds.), Washington, D.C., pp. 26–32 (1976).

26. E. Mathiowitz, E. Ron, G. Mathiowitz, C. Amato, and R. Langer, *Macromolecules*, *23*:3212 (1990).

27. E. Ron, E. Mathiowitz, G. Mathiowitz, A. Domb, and R. Langer, *Macromolecules*, due April (1991).

28. K. R. Sidman, A. D. Schwope, W. D. Steber, S. E. Rudolph, and S. B. Poulin, *Polym. Prepr.*, *20*(2):27 (1979).

29. H. Laufman and T. Rubel, *Surg. Gynecol. Obstet.*, *145*:597 (1977).

30. (a) O. Bostman, E. Hirrvensalo, S. Vainiopaa, A. Makela, K. Vihtonen, P. Tormala, and P. Rokkanen, *Clin. Orthopaed. Related Res.*, 196 (1987); (b) O. Bostman, S. Vainiopaa, E.

Hirrvensalo, A. Makela, K. Vihtonen, P. Tormala, and P. Rokkanen, *J. Bone and Joint Surg.*, *69-b*:615 (1987).

31. R. A. Miller, J. M. Brady, and D. E. Cutright, *J. Biomed. Mater. Res.*, *11*:711 (1977).

32. B. H. Vickery, G. I. McRae, L. M. Sanders, J. S. Kent, and J. J. Nestor, Jr., in *Long Acting Contraceptive Delivery Systems* (G. I. Zatuchni and A. Goldsmith, Eds.), Harper & Row, Philadelphia, pp. 180–189 (1984).

33. (a) H. Okada, *Proc. Int. Symp. Control. Rel. Bioactive Mater.*, *16*:12 (1989); (b) Y. Ogawa, M. Yamamoto, H. Okada, T. Yashiki, and T. Shimamoto, *Chem. Pharm. Bull.*, *36*:1095 (1988); (c) Y. Ogawa, M. Yamamoto, S. Takada, H. Okada, and T. Shimamoto, *Chem. Pharm. Bull.*, *36*.1502 (1988), (d) Y. Ogawa, H. Okada, M. Yamamoto, and T. Shimamoto, *Chem. Pharm. Bull.*, *36*:2576 (1988); (e) H. Okada, T. Heya, Y. Ogawa, and T. Shimamoyo, *J. Pharm. Exp. Therap.*, *244*:744 (1988).

34. (a) F. G. Hutchinson and B. J. A. Furr, in *Drug Delivery Systems Fundamentals and Techniques* (P. Johnson and J. P. Lloyd-Jones, Ed.), Ellis Horwood, VCH Pubs., New York, pp. 106–119 and references within (1988); (b) F. G. Hutchinson, U.S. Patent 4,767,628 (1988).

35. (a) L. M. Sanders, B. A. Kell, G. I. McRae, and G. W. Whitehead, *J. Pharm. Sci.*, *75*:356 (1986); (b) L. M. Sanders, J. S. Kent, G. I. McRae, B. H. Vickery, T. R. Tice, and D. H. Lewis, *J. Pharm. Sci.*, *73*:1294 (1984); (c) L. M. Sanders, J. S. Kent, G. I. McRae, B. H. Vickery, T. R. Tice, and D. H. Lewis, *Arch. Andrology*, *9*:91 (1982).

36. (a) M. Parmar, et al., *Lancet*, *2*:1201 (1985); (b) J. Zorn, *Contr. Gynec. Obstet.*, *16*:254 (1987); (c) N. Lahlou, *J. Clin. Endo. Met.*, *65*:946 (1987).

37. C. G. Pitt, M. M. Gratzl, G. L. Kummel, J. Surles, and A. Schindler, *Biomaterials*, *2*:215 (1981).

38. C. G. Pitt, T. A. Marks, and A Schindler, in *Controlled Release of Bioactive Materials* (R. Baker, Ed.), Academic Press, New York, pp. 19–44 (1980).

39. N. S. Choi and J. Heller, U.S. Patents 4,093,709 (1978); 4,131, 648 (1978); 4,138,344 (1979); 4,180,646 (1979).

40. E. H. Cordes and H. G. Bull, *Chem. Rev.* *94*:581 (1974).

41. G. Benagiano, E. Schmitt, D. Wise, and M. Goodman, *J. Polym. Symp.*, *66*:129 (1979).

42. (a) *World Health Organization 11th Annual Report*, *p. 61* (1982); (b) *World Health Organization 10th Annual Report*, pp. 62–63 (1981).

43. (a) T. H. Nguyen, T. Higuchi, and K. J. Himmelstein, *J. Control. Rel.*, *5*:1 (1987); (b) R. V. Sparer, C. Shih, C. D. Ringeisen, and K. J. Himmelstein, *J. Control. Rel.*, *1*:23 (1984); and references within.

44. (a) J. Heller, *CRC Crit. Rev. in Therap. Drug Carrier Sys.*, *1*:39 (1984). (b) J. Heller, in *Polymers in Medicine II (E. Chielini, P. Giusti, C. Migliaresi, and L. Nicolais, Eds.), Plenum Press, New York pp. 357–368 (1986).*

45. J. Hill and W. H. Carothers, *J. Am. Chem. Soc.*, *54*:1569 (1932).

46. H. B. Rosen, J. Chang, G. E. Wnek, R. J. Linhardt, and R. Langer, *Biomaterials*, *4*:131 (1983); (b) K. W. Leong, B. C. Brott, and R. Langer, *J. Biomater. Res.*, *19*:941 (1985); (c) K. W. Leong, P. D. D'Amore, M. Marletta, and R. Langer, *J. Biomed. Mater. Res.*, *20*:51 (1986).

47. K. W. Leong, B. C. Brott, and R. Langer, *J. Biomed. Mater. Res.*, *19*:941 (1985).

48. E. Ron, T. Turek, E. Mathiowitz, M. Chasin, M. Hegnman, and R. Langer, *16 Int. Symp. Control. Rel. Bioactive Mater.*, Aug. 6–9, Chicago (1989).

49. J. Heller, *J. Control. Rel.* *2*:167 (1985).

50. T. Higuchi, *J. Pharm. Sci.*, *50*:874 (1961).

51. J. Heller, *Controlled Release of Bioactive Materials* (R. Baker, Ed.), Academic Press, New York pp. 1–17 (1980).

52. (a) H. B. Hopfenberg, in *Controlled Release Polymeric Formulations* (D. R. Paul, and F. W. Harris, Eds.), ACS Symp. Ser. No. 33, American Chemical Society, Washington, D.C.,

pp. 26–32 (1976); (b) P. I. Lee, *J. Membrane Sci.*, 7:255–275 (1980); (c) A. G. Thombre and K. J. Himmelstein, *AIChE J.*, *31*(5):759 (1985); (d) R. Baker, in *Controlled Release of Biologically Active Agents*, John Wiley, New York pp. 84–131 (1987).

53. (a) D. O. Cooney, *AIChE J.*, *18*:446 (1972); (b) R. W. Baker and H. K. Lonsdale, *Am. Chem. Soc. Div. Org. Coat. Plast. Chem. Prepr.*, *3*:229 (1976); (c) J. Heller and R. W. Baker, in *Controlled Release of Bioactive Materials* (R. W. Baker, Ed.), Academic Press, New York, p. 1 (1980); (d) P. I. Lee, *J. Membr. Sci.*, 7:225 (1980); (e) A. G. Thombre and K. J. Himmelstein, *Biomaterials*, 5:250 (1984).

54. D. O. Cooney, *AIChE J.*, *18*:446 (1972).

55. (a) H. Ringsdorf, *J. Polym. Sci. Polym.* Symp., *51*:135 (1975); (b) H. Ringsdorf, in *Polymeric Delivery Systems, Mildl. Macromol. Monogr.* No. 5 (R. J. Kostelnik, Ed.), Gordon & Breach, New York, p. 197 (1978); (c) A. Trouet, *ibid.*, p. 157; (d) E. P. Goldberg, *ibid.*, p. 197.

56. F. W. Harris, in *Medical Applications of Controlled Release* (R. S. Langer and D. L. Wise, Eds.), CRC Press, Boca Raton, Fla., pp. 103–128 (1984).

57. M. M. Dahr and K. L. Agarwal, *Steroids*, *3*:139 (1964).

58. K. Kratzel, E. Kaufmann, O. Kraupp, and H. Stormann, *Monatsh. Chem.*, *92*:379 (1961).

59. G. F. Rowland, G. J. O'Neill, and D. A. Davies, *Nature* (London), *255*:487 (1975).

60. A. N. Neogi and G. G. Allan, in *Advances in Experimental Medicine and Biology No. 47, Controlled Release of Biologically Active Agents* (A. C. Tanquary and R. E. Lacey, Eds.), Plenum Press, New York, p. 195 (1974).

61. N. Tani, M. VanDress, and J. M. Anderson, in *Controlled Release of Pesticides and Pharmaceuticals* (D. H. Lewis, Ed.), Plenum Press, New York, p. 79 (1981).

5

Osmotic Drug Delivery

Robert L. Jerzewski   *Bristol-Myers Squibb Pharmaceutical Research Institute, New Brunswick, New Jersey*

Yie W. Chien   *Rutgers—The State University of New Jersey, Piscataway, New Jersey*

INTRODUCTION

During the past three decades significant advances have been made in the area of controlled drug delivery. This was due, in part, to the evolving disciplines of biopharmaceutics, pharmacokinetics, and pharmacodynamics. In particular, pharmacokinetic analyses of drug absorption, distribution, and elimination, as a function of time, have yielded such useful concepts as the therapeutic index (TI) or ratio of toxic concentration to minimum effective concentration, plasma half-life ($t_{1/2}$), and dosing interval (DI). In a typical therapeutic regimen, the drug dose (D) and the dosing interval are optimized to maintain drug concentration within the therapeutic window, thus ensuring efficacy while minimizing toxic side effects. The relationship between dosing interval and therapeutic index was described by Theeuwes and Bayne as follows [1]:

$$DI < t_{1/2} \frac{\ln TI}{\ln 2} \tag{1}$$

The practical consequence of this relationship is that the smaller the therapeutic index, the shorter the dosing interval. Thus, dosing frequency is increased to maintain drug concentrations within the therapeutic window. Surveys indicated that dosing more than once or twice daily greatly reduces patient compliance. Hence, the primary objective for controlled drug delivery is to maintain drug concentration within the therapeutic window, improve patient compliance to the dosage regimen by decreasing dosing frequency, and improve drug efficacy while reducing toxic side effects. A diagrammatic illustration of controlled versus conventional dosage delivery is shown in Fig. 1.

Rate-controlled systemic medication may be accomplished through several routes of administration; e.g., transdermal, nasal, peroral, rectal, subcutaneous implantation, and intramuscular injection. Moreover, numerous approaches or technologies have been used

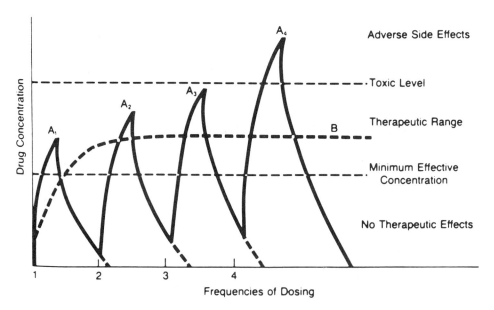

Figure 1 Simulation of blood concentration profiles resulting from multiple doses of conventional dosage form (A) as compared to a single dose of controlled release (zero-order) dosage form (B). (From Ref. 3.)

successfully to control the systemic delivery of drug. These approaches or technologies can be classified as follows [2]:

1. Preprogrammed drug delivery systems
2. Activation-controlled drug delivery systems
3. Feedback-regulated drug delivery systems

Most delivery systems marketed to date can be classified as preprogrammed drug delivery systems. For this class, the delivery system has been designed, or preprogrammed, to release drug at a therapeutically effective delivery rate. Generally, the drug activity or concentration gradient serves as the energy source that controls the release of drug from the delivery system based on Fick's laws of diffusion. Examples of this class include polymer membrane permeation-controlled drug delivery systems, polymer matrix diffusion-controlled drug delivery systems, and microreservoir dissolution-controlled drug delivery systems [3].

The second class of delivery systems uses a physical or chemical process to provide the necessary energy to activate the controlled release of drugs. One source of energy, namely, osmotic pressure, has been used extensively in the fabrication of drug delivery systems, and is the topic of this chapter.

THEORY OF OSMOTIC PRESSURE-ACTIVATED DRUG DELIVERY

Osmotic pressure, like vapor pressure and boiling point, is a colligative property of a solution in which a nonvolatile solute is dissolved in a volatile solvent. For a dilute, ideal solution of nonelectrolyte that follows Raoult's law, the vapor pressure (or escaping tendency) of the volatile solvent is reduced proportionally by the number of solute molecules present in the solution. In addition to being dependent on the concentration of solute, the osmotic pressure is also affected by the vapor pressure of the pure solvent. Given an ideal solution of nonelectrolyte, where attractive forces between solute and solvent and between solvent and solvent are equivalent, equations for osmotic pressure can be derived from solution thermodynamics in a manner that closely resembles the derivation of pressure equations for an ideal gas [4].

Osmotic pressure arises from at least two phases which are moving toward equilibrium. Thus, the Gibbs free energy of the system must be taken into consideration in the derivation of equations for osmotic pressure. Chemical potential (μ_i) is an important thermodynamic property as it describes the partial molar Gibbs free energy of component i at constant temperature and pressure:

$$\mu_i = \left[\frac{\partial G}{\partial n_i} \right]_{p,T,n_{j \neq i}} \tag{2}$$

Equation (2) indicates that the chemical potential is a differential quantity, which represents the slope of G as a function of n_i, at constant p and T. It follows that the chemical potential is equal to the Gibbs free energy per mole of substance. Generally, the chemical potential for a pure substance is regarded as the standard state, designated by the symbol, μ_i^0.

It should be pointed out that in an open system, the number of moles of a component may increase or decrease, and the Gibbs free energy for that system may be expressed as a sum of various contributions:

$$dG = \left[\frac{\partial G}{\partial T} \right]_{p,n} dT + \left[\frac{\partial G}{\partial p} \right]_{T,n} dp + \Sigma \left[\frac{\partial G}{\partial n_i} \right]_{p,T,n_j} dn_i \tag{3}$$

By substituting the familiar relationships for entropy,

$$-S = \left[\frac{\partial G}{\partial T} \right]_{p,n} \tag{4}$$

and volume,

$$V = \left[\frac{\partial G}{\partial p} \right]_{T,n} \tag{5}$$

the more familiar form of the equation is obtained:

$$dG = -S \, dT + V \, dp + \Sigma \, \mu_i \, dn_i \tag{6}$$

At constant temperature and pressure, Eq. (6) is reduced to

$$dG = \Sigma \, \mu_i \, dn_i \tag{7}$$

For a system which consists of several phases, the total Gibbs free energy is equal to the sum of the Gibbs free energies for all the phases $\alpha, \beta, \ldots$.

$$dG = dG_\alpha + dG_\beta \ldots \tag{8}$$

For equilibrium to be established, it is required that $dG = 0$; so Eq. (7) can be expressed as

$$\Sigma \, \mu_i \, dn_i = 0 \tag{9}$$

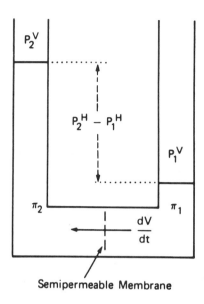

Figure 2 Osmosis cell. (From Ref. 6.)

If we restrict the argument to a system consisting of only two phases, the law of conservation of matter mandates that any changes in the amount of component in one phase must equal those in the second phase:

$$dn_{i\alpha} = -dn_{i\beta} \tag{10}$$

From the previous equations one can see that under the conditions of constant temperature, pressure, at equilibrium, the chemical potentials in these phases should be equal:

$$\mu_{i\alpha} = \mu_{i\beta} \tag{11}$$

The practical significance of Eq. (11) comes to light in a typical osmotic pressure experiment depicted in Fig. 2. One side of the U-shaped osmosis cell contains pure solvent, while the other contains a solution. The two sides are separated by a semipermeable membrane, which is selective to the transport of solvent only. In such an experiment, the solvent will travel from the solvent side to the solution side until such time as the hydrostatic pressure created by the solvent flux is sufficiently high to stop further flux.

An explanation of these results based on solution thermodynamics is as follows: From the previous arguments, it is known that, at equilibrium, the chemical potential on both sides of the membrane should be equal. Furthermore, on the side which contains pure solvent, μ_1 equals μ_1^0. To further describe the chemical potential on the solution side, the following equation which introduces the concept of activity is necessary:

$$\mu_1 = \mu_1^0 + RT \ln \alpha_1 \tag{12}$$

From this equation it should be obvious that as activity, or α_1, approaches unity, the chemical potential (μ_1) approaches that of the pure substance. Moreover, activity can be equated to partial pressure through the relationship between partial Gibbs free energy and partial molar volume, which results in the following equation:

$$\alpha_1 = \frac{p_1}{p_1^0} \tag{13}$$

Keeping in mind that the chemical potential has its greatest value for a pure substance and that the presence of a solute effectively lowers the activity, one can see that the presence of a positive pressure, such as the osmotic pressure (Π), is required to achieve a chemical potential equal to that of the pure solvent. This concept is represented by following equation [4]:

$$\mu_1 = \mu_1^0 + RT \ln \alpha_1 + \int_{P_1^0}^{P_1^0 + \Pi} V_1 \, dp \tag{14}$$

For the case of a dilute nonelectrolyte solution that obeys Raoult's law, integration and subsequent expansion and simplification of Eq. (14) results in the functional relationship between osmotic pressure and solute amount:

$$n_2 = \left[\frac{\Pi V_1}{RT} \right] \tag{15}$$

where n_2 is the number of moles of solute and V_1 is volume of solvent. This equation was derived by J. H. van't Hoff, who in 1901 received the Nobel Prize in Chemistry for his efforts. By simple rearrangement of Eq. (15), osmotic pressure is related to solute concentration (C):

$$\Pi = CRT \tag{16}$$

Example 1.   Ten grams of sucrose (MW = 342 g/mole) is dissolved in 0.1 liter of water at 25°C. Calculate the osmotic pressure of the resulting solution.

From Eq. (15),

$$n = \left[\frac{\Pi V}{RT} \right]$$

where

$$n = \frac{10.0 \text{ g}}{342 \text{ g/mol}} = 0.0292 \text{ mol}$$

$$R = 0.082 \text{ liter atm K}^{-1} \text{ mol}^{-1}$$

$$T - 298 \text{ K}$$

Thus,

$$\Pi = \frac{nRT}{V} = (0.0292 \text{ mol}) \times (0.082 \text{ liter atm K}^{-1} \text{ mol}^{-1}) \times \left(\frac{298 \text{ K}}{0.10 \text{ liter}} \right)$$

$$\Pi = 7.15 \text{ atm}$$

From the following discussion, the basis for osmotic pressure should be clear. It should also be apparent that osmotic pressure can be harnessed as an energy source, which can be utilized to release drug from a device of appropriate design. Referring once again to Fig. 2, one can envisage that volume flow of solvent (dV/dt) depends on several factors. These factors include membrane characteristics, differential osmotic pressure between half-cells, differential hydrostatic pressure between half-cells, and the difference between osmotic pressure and hydrostatic pressure as the mass (volume) transfer process approaches equilibrium. Membrane characteristics that affect dV/dt are membrane area (A), membrane thickness (l), permeability coefficient (Lp), and the reflection coefficient (σ). The first three terms have been discussed previously in Chapters 2 and 3. The reflection coefficient is a new term which takes into account the leakage of solute through the semipermeable membrane [5]. A perfect membrane would have a coefficient of unity. For membranes that offer no resistance to solute transport the coefficient would equal 0. Obviously, in terms of membrane performance and predictability, it is important to select a material whose reflection coefficient is close to 1. All these aforementioned terms combine to yield the following equation describing volume flow [6]:

$$\frac{dV}{dt} = \frac{A}{l} Lp[\sigma (\Pi_2 - \Pi_1) - (P_2 H - P_1 H)] \tag{17}$$

where PH is the hydrostatic pressure. From Eq. (17), one can see that volume flow into an osmotic delivery device can be used to deliver an equal volume of drug solution/suspension from the device.

PROTOTYPE SYSTEM DEVELOPMENT

The earliest application of osmotic pressure to drug delivery was by Rose and Nelson in 1955 [7]. The authors described two systems, one that delivered 0.02 ml/day for 100 days, and one that delivered 0.5 ml/day for 4 days, both for use in pharmacologic research. A schematic diagram of their prototype device is shown in Fig. 3. The device consisted of a drug solution in a rigid glass ampule (D) with a delivery orifice, an osmotic pressure unit made from an expandable latex bag (B) to contain an osmotic agent, and a rigid

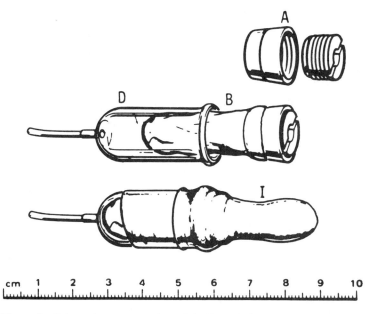

Figure 3 Schematic representation of the Rose-Nelson osmotic pump. A, membrane holder; B, latex bag; D, rigid glass ampule; and I, latex bag with water supply. (From Ref. 7.)

circular holder (A) which houses a semipermeable membrane. As illustrated, the osmotic pressure unit was inserted into the glass ampule and the system was completed by addition of a water source housed in a separate rubber bag (I). In 1971, Stolzenberg [8] received a U.S. patent for an osmotic system which was operationally similar to that of Rose and Nelson.

Drug delivery from both systems was dependent on expansion of the osmotic pressure unit, due to the influx of water, which resulted from the osmotic pressure difference between the unit and the environment. The delivery rate for such systems, given a constant concentration of drug (C_s), is defined by the following equation:

$$\frac{dm}{dt} = \frac{dV}{dt} Cs \tag{18}$$

To achieve a zero-order release rate, the differential osmotic and hydrostatic pressure terms in Eq. (17) must maintain a constant value. The internal hydrostatic pressure (P_2H) is a function of flexibility of the osmotic pressure unit as well as the rheology of the drug solution and the dimension of the delivery orifice. These factors will be discussed in greater detail in the following section, but for now we can assume that the hydrostatic pressure difference ($P_2H - P_1H$) across the semipermeable membrane, by design, approaches a constant value. One method for achieving a constant osmotic pressure difference is to maintain a saturated solution of the osmotically active agent in the osmotic pressure unit.

It was realized later that such osmotic pressure-controlled delivery systems could offer great potential in achieving zero-order delivery of drugs, which spurred development and eventual commercialization of the osmotic pump (OSMET), the mini-osmotic pump (ALZET), and the elementary osmotic pump (OROS) by Alza Corporation [9].

IMPLANTABLE OSMOTIC PUMPS

Description

The common element, from a design standpoint, for the two prototype systems discussed in the previous section was a flexible interface between the osmotic pressure and drug reservoir compartments. Operation of the devices was dependent on expansion of the osmotic pressure compartment due to water influx, with concomitant contraction of the drug reservoir compartment, resulting in the outflow of drug solution. Although both systems are useful for conducting laboratory research, they have limited practical utility because they are not amenable to mass production. However, the obvious advantages of osmotically controlled drug release spurred continued research by Alza Corporation in the early 1970s. Their efforts resulted in the development of osmotic pump systems.

A diagram of an osmotic pump, which consists of several discrete elements, is shown in Fig. 4. The outermost component is a rigid, semipermeable membrane fabricated from substituted cellulosic polymers. The innermost component is a drug reservoir compartment having a flexible wall membrane that is impermeable to water or the osmotic agent. Sandwiched between the two membranes is the dry osmotic energy source. A rigid, polymeric plug is used to form a leakproof seal between the drug reservoir and the semipermeable membrane. The flow modulator, which consists of a cap and a tube, is constructed from stainless steel. The osmotic pump is available unfilled. Filling the pump with drug solution is accomplished via a syringe with a specially designed filling tube.

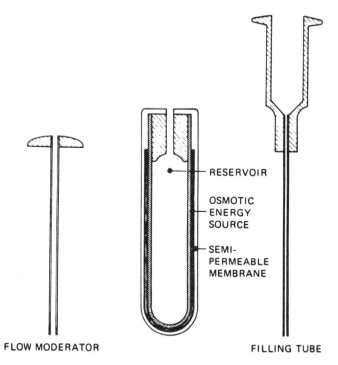

RESERVOIR

OSMOTIC
ENERGY
SOURCE

SEMI-
PERMEABLE
MEMBRANE

FLOW MODERATOR   FILLING TUBE

Figure 4 Miniature osmotic pump and its major components. (From Ref. 6 with permission of CRC Press, Inc.)

This basic design has proven versatile, as the dimensions can be adjusted to vary the fill volume of the drug reservoir compartment, while the steady-state release rate (zero-order) can be controlled by varying certain release-controlling factors [5]. The osmotic pump has been designed for oral, subcutaneous, or rectal drug delivery and is especially well suited for preliminary screening of new drugs in assessing their pharmacokinetic and pharmacodynamic properties (Table 1). Results from such studies are invaluable in the rational development of an optimized drug delivery system. Now, it would be beneficial to discuss, in greater detail, the key formulation factors that control the release of drug from the osmotic pump systems in the following section.

Release Controlling Factors

Based on the design description of the osmotic pump and recalling Eq. (17), which describes the volume flow, one can easily recognize the importance of the semipermeable membrane in controlling release of the drug. Hence, the membrane must meet several performance criteria. First, the material must possess sufficient wet strength ($\sim 10^5$ psi) and wet modulus ($\sim 10^5$ psi) so as to retain its dimensional integrity during the operational lifetime of the device [10]. Second, the polymer membrane must exhibit sufficient water permeability so as to attain water flux rates (dV/dt) in the desired range. The water vapor transmission rates, which can be used to estimate water flux rates for some polymer membranes, are given in Table 2 [11]. Third, the reflection coefficient (σ), or "leakiness" of the membrane

Table 1 Characteristics of Osmotic Pumps

Application	Route of administration	Duration of steady-state delivery (h)	Fill volume[a] (ml)	Steady-state delivery rate (μl/h)	Distinguishing terminology
Clinical research	Oral	12	0.1	15	Oral pump
	Oral	24	0.2	8	Oral pump
	Rectal/vaginal	30	2.0	60	Rectal pump
Animal research	Implantation	168	0.2	1	Mini-osmotic pump
	Implantation	336	0.2	0.5	Mini-osmotic pump
	Implantation	168	2.0	10	Osmotic pump
	Implantation	336	2.0	5	Osmotic pump

[a] Fill volume: 0.2 ml for mini-osmotic pump (2.5 cm length × 0.7 cm diameter); 2 ml for osmotic pump, 4.5 cm length × 1.3 cm diameter).
Source: From Ref. 10 by permission of Butterworth & Co.

Table 2 Water Vapor Transmission Rates of Some Polymer Membranes

Polymer membrane	Water vapor transmission rates (g/100 in.²/24 hr/mm thick)
Polyvinyl alcohol	100
Polyurethane	30–150
Methylcellulose	70
Cellulose acetate	40–75
Ethylcellulose	75
Cellulose acetate butyrate	50
Polyvinyl chloride (cast)	10–20
Polyvinyl chloride (extruded)	6–15
Polycarbonate	8
Polyvinyl fluoride	3
Ethylene vinyl acetate	1–3
Polyesters	2
Cellophane (polyethylene-coated)	>1.2
Polyvinylidene fluoride	1.0
Polyethylene	0.5–1.2
Ethylene/propylene copolymer	0.8
Polypropylene	0.7
Polyvinyl chloride (rigid)	0.7

Source: From Ref. 11 with permission.

to the osmotic agent, should approach the limiting value of 1; where the membrane is selectively permeable only to water (as the solvent). Unfortunately, polymer membranes that are more permeable to water are also, in general, more permeable to the osmotic agent [12]. Finally, the membrane should also be biocompatible. Keeping these criteria in mind, Alza selected cellulose esters to fabricate the semipermeable membrane. Once the membrane system is selected, further control over release rate is afforded by varying the area and the thickness of the membrane. Both factors are easily modified to optimize drug delivery from the device.

Another important term in Eq. (17) is ΔP, or the hydrostatic pressure difference between the osmotic pressure compartment and the external environment. The hydrostatic pressure term can be expressed as a summation of the internal pressure associated with volume flow (ΔP_d) and the internal pressure necessary to deform the collapsible drug reservoir (ΔP_c):

$$\Delta P = \Delta P_d + \Delta P_c \qquad (19)$$

Both terms on the right-hand side of Eq. (19) can be reduced to approximately zero by optimizing the size of the delivery orifice as well as by selecting a polymer with sufficiently low modulus. Moreover, the difference in magnitude between $\Delta\pi$ and ΔP can be tailored such that Eq. (17) reduces to

$$dV/dt = K\frac{A}{l}(\Delta\Pi) \qquad (20)$$

where K equals the product of Lp and σ.

The last release-controlling factor which should be considered is the osmotic pressure gradient between the osmotic pressure compartment and the external environment. For zero-order release, $\Delta\Pi$ in Eq. (20) must attain a constant value. The simplest and most predictable way to achieve a constant gradient of osmotic pressure is to maintain a saturated solution of osmotic agent in the compartment. Application of such a limiting condition can be expressed by modifying Eq. (16) to account for the concentration of solute at saturation, S:

$$\Pi_s = iSRT \qquad (21)$$

where i is a correction factor for the nonideal behavior of electrolyte solutions. The value of i varies as a function of solute concentration and must be determined experimentally. Substituting Eq. (21) into Eq. (20) gives the following expression:

$$\left(\frac{dV}{dt}\right)_s = K\frac{A}{l}(\Pi_s) \qquad (22)$$

The relationship in Eqs. (21) and (22) indicates that osmotic pressure, and hence dV/dt, can be increased by selecting an osmotic agent with higher solubility. However, Theeuwes and Yum [5] derived a functional relationship between volume of the osmotic pressure compartment (V_s) and volume of the drug reservoir (V_d):

$$\frac{V_s}{V_d} = \frac{S}{p_s - S} \qquad (23)$$

where p_s is density of the osmotic agent. Based on Eqs. (21) and (23), and in the instances where pump size is small, or the ratio of V_d to overall device volume is greater than 0.3, less soluble osmotic agents must be selected for use.

Performance Testing

Increased development and commercialization of controlled-release dosage forms has necessitated changes in performance testing, so as to provide in-house quality control and to furnish regulatory agencies with experimental evidence that the dosage form delivers drug in a controllable and reproducible manner. Moreover, in-vitro/in-vivo correlations in drug release profiles need to be established to assure that the ever-variable in-vivo environment does not induce dosage-form failure with subsequent dose dumping. In keeping with this philosophy, osmotic pumps have been subjected to a variety of performance tests, both in-vitro and in-vivo.

First, to verify that the drug delivery profile from osmotic pumps is independent of environmental pH, in-vitro performance tests have been developed to test osmotic pumps in isotonic saline, artificial gastric fluid USP (pH = 1.2), and artificial intestinal fluid USP (pH = 7.5) at 37°C [10]. This is an important performance test, because if the semipermeable membrane is truly selective, ions should not be able to diffuse into the osmotic pump and affect the release profile. These investigations were conducted in a specially designed rotating-tube apparatus. The apparatus consists of a series of 16 dissolution medium-containing tubes and a transfer device, which transfers the osmotic pump, after some finite time period determined by the expected release life, from one tube to another. Osmotic pumps were filled with water-soluble FD&C Blue #1 dye to allow quantitation of the amount released by visible spectroscopy. Results from the experiments conducted in isotonic saline (Fig. 5) are indicative of zero-order release over the claimed life of the osmotic pump. Results obtained in the simulated gastric fluid and simulated intestinal fluid were found to be statistically comparable to those in isotonic saline. This performance suggests that variation of pH does not affect release rate from the osmotic

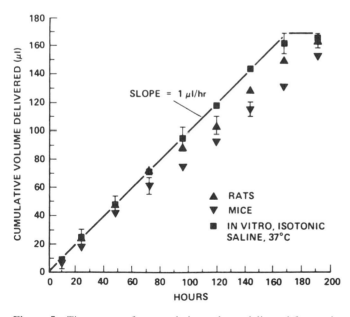

Figure 5 Time course for cumulative volume delivered from miniaturized osmotic pump under in-vitro and in-vivo conditions. (From Ref. 6 with permission of CRC Press, Inc.)

pump [5]. Results from the studies conducted in rats and mice indicated that the in-vivo release profiles are slightly lower than those in-vitro (Fig. 5).

In-vivo performance of the osmotic pump was also conducted in two animal models: dogs for 24-h oral osmotic pumps and rats for 1- and 2-week subcutaneously implantable mini-pumps. For the oral studies, individual pumps were administered to several dogs at 2, 4, 6, 12, 18, and 24 h. Exactly 2 h after the last dose, the animals were sacrificed and the pumps recovered from the gastrointestinal tract. The recovered samples were analyzed for the residual contents and the volume of solution released as a function of residence time. In the implantation studies, the mini-pumps were implanted subcutaneously in rats with a small catheter tube penetrating through the skin. Thus, external sample collection for quantitation of the volume released from implanted pumps was possible. Results from these studies, which are expressed as the ratio of in-vitro/in-vivo release rates versus normalized time, are shown in Fig. 6. The clustering of data points at 1, for the duration of release, is indicative of excellent in-vitro/in-vivo correlation [9].

Another important performance issue, especially for the implantable devices, is device biocompatibility. Specifically, the device must be noninflammatory, nontoxic, nonantigenic, noncarcinogenic, and nonthrombogenic [13]. To assess biocompatibility of the osmotic pump, both the pump and a piece of silastic rod, as the control, were implanted subcutaneously in rats. After 14 days, the rats were sacrificed, and the implant sites were photographed, excised, and examined microscopically to determine the differences in tissue histology. In general, the tissue reaction to the osmotic pump was identical to that of the control. Based on the results of the histological evaluation, the osmotic pump was deemed biocompatible [5].

Applications

Earlier in the chapter, the utility of osmotic pumps in screening new drugs and assessing their relevant pharmacokinetic and pharmacologic properties was introduced. The fol-

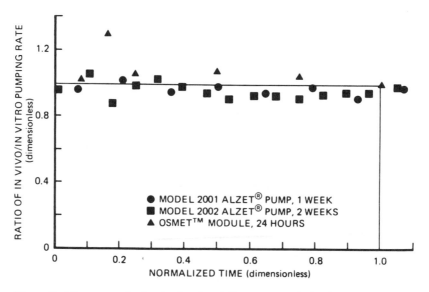

Figure 6 Time course for in-vivo/in-vitro delivery rates ratio for the 24-h oral pump (OSMET) and 1- and 2-week mini-osmotic pumps (ALZET). (From Ref. 9 with permission of CRC Press, Inc.)

lowing section describes several actual investigations that demonstrate the utility of osmotic pumps.

Nau and his co-workers used mini-osmotic pumps in mice to investigate the incidence of embryotoxicity of valproic acid, an antiepileptic agent suspected of causing embryotoxicity in pregnant women [14]. Since the biological half-life of valproic acid in mice is significantly different from that in humans (8–16 h in humans versus 0.8 h in mice), the conventional dosage regimen of two to three equally divided doses per day in humans would cause a substantial peak-to-valley fluctuation in mice. Thus, the authors used the zero-order delivery of valproic acid from implanted mini-osmotic pumps (1 week) to achieve plasma levels similar to those observed in humans with conventional therapy. Results from the study indicated that reduction in fetus weight and the incidence of fetus malformation is significantly lower in mice receiving constant-rate therapy, as opposed to mice on conventional therapy of valproic acid.

In another application, Sikic and his co-workers used mini-osmotic pumps to examine the effect of bleomycin dosage regimen on its pharmacological activity in mice with Lewis lung carcinoma [15]. In their experimental protocol, two intermittent injection frequencies, high and low, were compared with continuous administration from a mini-osmotic pump, having a similar dosage regimen ranging from 0 to 80 mg/kg of bleomycin, over a 7-day period. Results, shown in Fig. 7, indicate that continuous infusion of bleomycin is superior

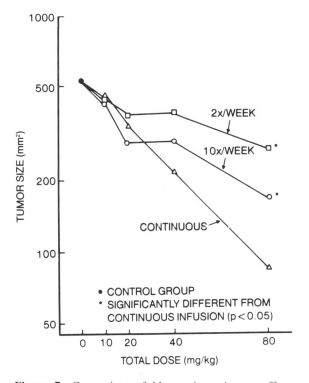

Figure 7 Comparison of bleomycin antitumor effect against Lewis lung carcinoma with three dosage regimens. These measurements, made on day 15 after implantation, are representative of differences that existed throughout the course of tumor growth. (From Ref. 9 with permission of CRC Press, Inc.)

to intermittent infusion in reducing tumor size, while in the control animals, which received mini-osmotic pumps delivering 0.9% saline, no reduction in tumor size was detected.

The readers are directed to other pharmacology studies in which implantable mini-osmotic pumps have been used to control the subcutaneous delivery of vasopressin or insulin in diabetic rats [16]; and in site-specific drug delivery to specific tissues such as the visual cortex of the brain [17], the uterine cavity [18], the cerebral ventricular system [19], and the fornix of the eye [20].

ORAL OSMOTIC PUMPS

Description

In the previous section, the utility of osmotic pumps for investigational research via implantation was described. However, mass production of these osmotic pumps is not economical because of their intricate nature. This fact led to the development of the elementary osmotic pump (OROS system) by Alza [21] for oral use.

Although the elementary osmotic pump operates under the same fundamental principles as the implantable osmotic pump, it is inherently less complex. The basic design is illustrated in Fig. 8. The unit consists of an osmotic core containing drug with or without an osmotic active salt, coated with a semipermeable membrane having a delivery orifice to permit the release of drug solution. In general, the system can assume any shape or size, as required by the dosage requirements. The semipermeable membrane exhibits sufficient strength to maintain a constant volume during pump operation. Delivery rate of drug is dependent on membrane permeability, the osmotic pressure of the core formulation, and the solubility of the drug in question. The delivery rate is independent of the release orifice size as long as the cross-sectional area (A_0) is within two critical limits [22]:

$$A_{min} \leq A_0 \leq A_{max} \tag{24}$$

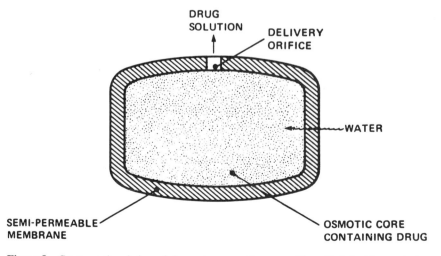

Figure 8   Cross-sectional view of elementary osmotic pump. (From Ref. 6 with permission of CRC Press, Inc.)

Operationally, the elementary osmotic pump delivers drug at a zero-order rate so long as saturated concentration of osmotic active agent is present inside the core. Pertinent release-controlling factors, along with the relevant equations, are discussed in the following section.

Release Controlling Factors

Substitution of Eq. (17) for dV/dt in Eq. (18) gives the following general equation for release rate from the elementary osmotic pump:

$$\frac{dm}{dt} = \frac{A}{l} Lp \, [\sigma \, (\Delta\Pi - \Delta P)]C_s \tag{25}$$

Similar to the implantable osmotic pump, a zero-order release rate is achieved by designing the system so that $\Delta\Pi >> \Delta P$, and that $\Pi_i >> 0$. Taking these conditions into account, and combining the reflection coefficient and membrane permeability coefficient, results in a simplified expression:

$$\left[\frac{dm}{dt}\right] = \frac{A}{l} K \Pi C_s \tag{26}$$

Therefore, the release rate defined in Eq. (26) is zero-order as long as the terms remain constant. Proper selection and optimization of the semipermeable membrane system maintain the first three terms on the right-hand side of Eq. (26) constant. For the case in which excess solid remains in the osmotic core, thereby maintaining saturation of drug and osmotic active agent in the system from $t = 0$ to $t = t_z$, and when the dissolution rate of the drug is greater than its delivery rate, Eq. (26) can be modified to:

$$\left[\frac{dm}{dt}\right]_z = \frac{A}{l} K \Pi_s S \tag{27}$$

where S is the solubility (or saturation concentration) of the drug and Π_s is the osmotic pressure of the saturated system.

Example 2. Calculate the zero-order delivery rate for drug X with S of 300 mg/ml, from an oral osmotic pump with $K\Pi_s$ of 1×10^{-3} cm²/h, A of 2.5 cm², and l of 0.03 cm.

Based on Eq. (27),

$$\left[\frac{dm}{dt}\right]_z = \frac{A}{l} K\Pi_s S$$

$$K\Pi_s = 0.001 \text{ cm}^2/\text{h} \qquad S = 300 \text{ mg/cm}^3$$

$$A = 2.5 \text{ cm}^2 \qquad l = 0.03 \text{ cm}$$

$$\left[\frac{dm}{dt}\right]_z = \frac{2.5 \text{ cm}^2}{0.03 \text{ cm}} \times 0.001 \text{ cm}^2/\text{h} \times 300 \text{ mg/cm}^3$$

$$= 25.0 \text{ mg/h}$$

At some finite time, when concentration of the osmotic agent drops below saturation, the osmotic pressure and hence the delivery rate will decline as a function of time. A rigorous mathematical derivation of the non-zero-order rate has been reported elsewhere [21]. The final result of that work is presented in the following equation:

$$\frac{dm}{dt} = \frac{(dm/dt)^2}{[1 + (1/SV)(dm/dt)_z \, (t - t_z)]^2} \tag{28}$$

Earlier, it was stated that delivery rate is independent of the orifice size as long as the boundary conditions in Eq. (24) are satisfied. The first boundary condition, A_{min}, is established to minimize hydrostatic pressure within the device. Again, this is a necessary step in achieving zero-order release. Obviously, the presence of any significant hydrostatic pressure would decrease the driving force for drug delivery. In addition, large hydrostatic pressure could deform the device, and produce unpredictable effects on the delivery rate. The A_{min} boundary is calculated using Poiseuille's Law [23].

$$A_{min} = 5\left(1\frac{dV}{dt}\frac{\eta}{\Delta P_{max}}\right)^{1/2} \tag{29}$$

where dV/dt is the volume flux, l is the length of the orifice, η is the viscosity of the dispersed solution, and ΔP_{max} is the maximum hydrostatic pressure across the device. Note that ΔP_{max} must be small enough so as not to deform the device, nor reduce the osmotic pressure driving force.

The second boundary condition, A_{max}, is established to minimize the diffusional contribution to the delivery rate. Theeuwes developed the following equation for A_{max} [21]:

$$A_{max} = \frac{1}{F}\left(\frac{dm}{dt}\right)_z\frac{1}{DS} \tag{30}$$

where F is the ratio of total delivery rate to diffusive rate, and D is the diffusion coefficient of drug in the dissolution medium. In general, when $F \geqq 40$, the diffusive contribution to delivery rate is negligible.

Example 3. Given the following information, calculate the boundary conditions, A_{max} and A_{min}, for an elementary osmotic pump.

$$\frac{dV}{dt} = 2.78 \times 10^{-5}\,cm^3/s \qquad 1 = 0.025\,cm$$

$$\eta = 1\,cP = 10^{-3}\,Ns/m^2 \qquad F = 40$$

$$\frac{dm}{dt} = 0.0069\,mg/s \qquad \Delta P = 1\,atm$$

$$S = 300\,gm/cm^3 \qquad D = 2 \times 10^{-5}\,cm^2/s$$

$$\Delta P_{max} = 1\,atm = 101,325\,N/m^2$$

$$A_{min} = 5\left[1\left(\frac{dV}{dt}\right)\left(\frac{\eta}{\Delta P_{max}}\right)\right]^{1/2}$$

$$= 5\left[(0.025\,cm)(2.78 \times 10^{-5}\,cm^3/s)\left(\frac{10^{-3}\,Ns/m^2}{101,325\,N/m^2}\right)\right]^{1/2}$$

$$= 1.309 \times 10^{-6}\,cm^2$$

$$A_{max} = \frac{1}{F}\left(\frac{dm}{dt}\right)_z\frac{1}{DS}$$

$$= \frac{0.025\,cm}{40}(0.0069\,mg/s)\left(\frac{1}{2 \times 10^{-5}\,cm^2/s \times 300\,mg/cm^3}\right)$$

$$= 0.7 \times 10^{-3}\,cm^2$$

Typically, there is a short lag time of 30 to 60 min as the system hydrates before zero-order delivery from the elementary osmotic pump is obtained. After that, about 60% of the dose is delivered at a zero-order rate. It should be noted that an overage on the order of 10% is required to deliver the target dose. This is due to the fact that as the osmotic pressure gradient approaches zero, drug flux becomes diffusion rate-limited.

Example 4. Using the following relationships from Theeuwes [21] to calculate:

   a.   The fraction of dose delivered in a zero-order manner
   b.   The length of time in which zero-order release occurs

$$\left(\frac{dm}{dt}\right)_z = 25 \text{ mg/h} \qquad \text{(from Example 2)}$$

$$m_z = 500 \text{ mg}$$

$$S = 300 \text{ mg/cm}^3$$

$$\rho = 1.5 \text{ g/cm}^3$$

   a.   The fraction of dose (F_z) delivered in a zero-order manner:

$$F_z = 1 - \left(\frac{S}{\rho}\right)$$

$$= 1 - \left(\frac{0.3 \text{ g/cm}^3}{1.5 \text{ g/cm}^3}\right)$$

$$= 0.80$$

   b.   The length of time (t_z) in which zero-order release occurs:

$$t_z = m_z(F_z)\left(\frac{1}{(dm/dt)_z}\right)$$

$$= 500 \text{ mg}(0.80)\left(\frac{1}{25 \text{ mg/h}}\right)$$

$$= 16 \text{ h}$$

When designing an elementary osmotic pump, several considerations must be addressed. First, the solubility of the drug and its osmotic pressure at saturation must be determined. For ionic drugs, this may require the use of an alternative salt form. An example of this was reported for oxprenolol [24]. Theeuwes and co-workers found the hydrochloride salt too soluble to maintain a saturated solution, and hence zero-order delivery, for the anticipated delivery life of the dosage form. Subsequently, the succinate salt form was identified as having the optimum solubility. Second, if a saturated solution of the drug does not possess sufficient osmotic pressure, an additional osmotic active agent must be added to the core formulation. Sodium bicarbonate was used as an osmotic driving agent in several oxprenolol formulations [24]. Finally, the permeability of the semipermeable membrane, as well as the dimension of the release orifice, must be optimized to obtain the desired zero-order release rate.

Example 5. Using the following information, calculate the permeability coefficient (K) for a semipermeable membrane required to give a delivery rate of 15 mg/h.

$$A = 2.0 \text{ cm}^2 \qquad\qquad \Pi_s = 60 \text{ atm}$$

$$S = 500 \text{ mg/cm}^3 \qquad\quad l = 0.02 \text{ cm}$$

Based on Eq. (7),

$$\left[\frac{dm}{dt}\right]_z = \frac{A}{l}(K)\Pi_s S$$

$$K = \frac{(15 \text{ mg/h})(0.02 \text{ cm})}{(60 \text{ atm})(500 \text{ mg/cm}^3)}$$

$$= 1.0 \times 10^{-5} \text{ cm}^2/\text{h atm}$$

There are several important features of osmotic pressure-activated drug delivery devices [21]: First, the attainable delivery rate is significantly greater than the rate that can be attained with a diffusion-based device with comparable size. This is especially important for cases where large doses of drug must be administered. Second, the delivery rate is independent of environmental hydrodynamic effects; i.e., there is no stagnant boundary layer effects. Third, as the semipermeable membrane is permeation-selective, the pH of the core is independent of the environmental pH. This suggests that oral drug delivery rate is independent of the variation in solution pH throughout the gastrointestinal tract. Performance testing of these system features is discussed in the following section.

Performance Testing

Based on the principles discussed above, the delivery of drugs from the elementary osmotic pump should be independent of the variation in pH and hydrodynamic conditions. In his pioneering work [21], Theeuwes demonstrated this with potassium chloride- and sodium phenobarbital-releasing elementary osmotic pumps. In Fig. 9, the delivery rate of potassium

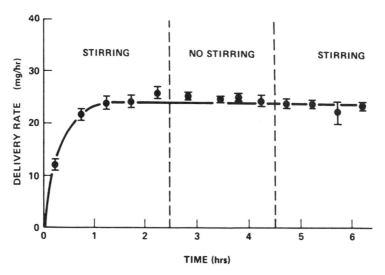

Figure 9 Effect of hydrodynamic conditions on the in-vitro delivery rate profile of potassium chloride from elementary osmotic pumps in water at 37°C. The vertical dashed lines indicate the time at which the systems were transferred from a stirred to a stagnant medium and back to a stirred medium. ♦ represents the range of experimental data obtained from five systems. (From Ref. 6 with permission of CRC Press, Inc.)

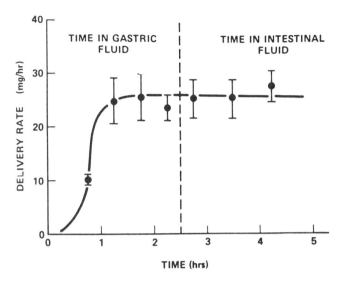

Figure 10 Effect of pH conditions on the in-vitro delivery-rate profile of sodium phenobarbital from elementary osmotic pump systems in gastric and intestinal fluid USP (without enzymes). The vertical dashed line indicates the time at which the systems were transferred from gastric to intestinal fluid. ⧫ represents the range of experimental data obtained from three systems. (From Ref. 6 with permission of CRC Press, Inc.)

chloride as a function of agitation or no agitation (i.e., infinite stagnant layer) is depicted. In Fig. 10, the delivery rate of sodium phenobarbital at low pH condition (simulated gastric fluid, pH 1.2) and moderate pH condition (simulated intestinal fluid, pH 7.5) is depicted. The results of both studies are in agreement with the earlier claims. Subsequent in-vivo studies were conducted in dogs by dosing coded units at specified intervals. At some time point, the animals were sacrificed, the dosage forms retrieved and analyzed for residual contents, and the in-vivo delivery rate calculated. Results, shown in Fig. 11, are indicative of excellent correlation in the in-vitro and in-vivo delivery rates. Finally, the utility of Eqs. (27) and (28) in estimating the release rate and release profile is depicted in Fig. 12.

Liu and co-workers [25] conducted in-vitro studies to compare the release of phenylpropanolamine hydrochloride (PPA) from the oral osmotic pump system and one marketed long-acting appetite-suppressant product. In this investigation, Acutrim tablets (Ciba-Geigy), an oral osmotic pump system, was compared with Dexatrim capsules (Thompson Medical), a spansule-type sustained-release product. Both products were purported to deliver PPA continuously over 16 h, thus giving once-a-day weight-control treatment. For both products, about one-third of the dose (or 24 mg) is designed to be rapidly released for immediate systemic delivery. This amount represents the loading dose that quickly raises blood drug concentration to the desired steady-state level. The remainder of the dose will be delivered in a controlled or sustained fashion to maintain blood drug concentrations in the therapeutic range. It was found that Acutrim tablets deliver PPA at a rate which is independent of the hydrodynamic conditions in a well-controlled, validated

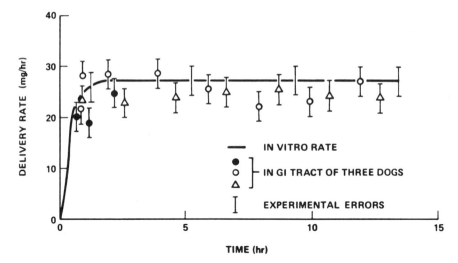

Figure 11 Correlation between in-vitro and in-vivo delivery-rate profiles of potassium chloride from oral osmotic pumps. Key: ———, average in-vitro rate from systems of the same batch; and △, ○, ●, average delivery rate of one system in the GI tract of dogs 1, 2, and 3, plotted as a function of the total time period each system resided in the dog. (From Ref. 9 with permission of CRC Press, Inc.)

dissolution apparatus. Additionally, no significant difference in drug delivery rate profiles was observed at pH 1.2 or pH 7.4. However, delivery of PPA was observed to vary as a function of the osmotic pressure in the dissolution medium (Fig. 13). A linear decrease in the delivery rate of PPA from the osmotic pump system was observed as increasing the osmotic pressure (Fig. 14). Therefore, as the osmotic pressure of the environment approaches that of the internal osmotic pressure, drug delivery ceases. Comparative drug release profiles of the two products, depicted in Fig. 15, indicate that Dexatrim gives essentially 100% PPA release after 7 h, while Acutrim controls the release of PPA for up to 18 h. In addition, Acutrim was demonstrated to deliver at a zero-order rate over several hours. The results led the authors to conclude that osmotic pressure-controlled drug delivery system provides better control over drug release than the sustained-release spansule system.

In a recent report by Ramadan and Tawashi [26], the effect of hydrodynamic conditions and orifice delivery size on drug release rate from an elementary osmotic pump system was investigated. In this study, neat potassium chloride tablets were prepared and coated in accordance with the patent literature. After curing the semipermeable membrane, various size orifices were mechanically created. Release characteristics were examined using the USP Basket method, at different rotation speeds, and a Turbula mixer, with equivalent volume of distilled water (200 ml) at 37°C. In this investigation, drug release was found to be dependent on the rotation speed of the particular apparatus, contrary to the theory discussed earlier. Moreover, release rate was considerably higher under turbulent conditions operating in the Turbula mixer. Orifice size was not a significant factor under laminar hydrodynamic conditions. The authors offered no explanation for the behavior of their osmotic system under conditions of turbulent flow.

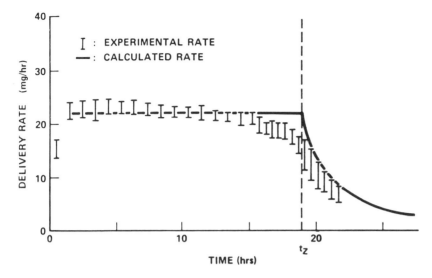

Figure 12 In-vitro delivery-rate profile of potassium chloride from elementary osmotic pumps in water at 37°C. I, range of experimental data obtained from five systems; ———, calculated delivery rate. (From Ref. 6 with permission of CRC Press, Inc.)

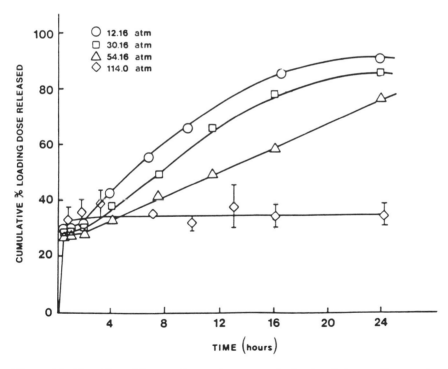

Figure 13 The effect of increased osmotic pressure in the dissolution medium on the release profile of phenylpropanolamine from Acutrim tablets at intestinal conditions. ○, simulated intestinal fluid with osmotic pressure at 12.16 atm, which was adjusted by adding NaCl to 30.16 atm (□), 54.16 atm (△), 114.0 atm (◇). (From Ref. 25.)

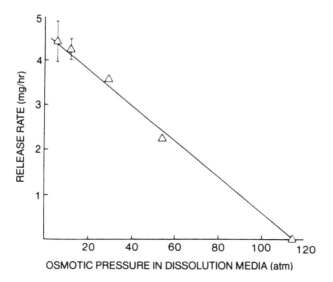

Figure 14 Linear relationship between the release rate of phenylpropanolamine from Acutrim tablets and the osmotic pressure
in the dissolution medium. (From Ref. 25.)

Bindschaedler and co-workers [27] reported their study on the elementary osmotic
systems with cellulose acetate coatings prepared from organic solutions or aqueous dispersions. Given the recent environmental regulations concerning the use and emission of
organic solvents, the application of aqueous dispersions is an attractive alternative. In
this study, potassium chloride tablets served as the osmotic core. The osmotic core was

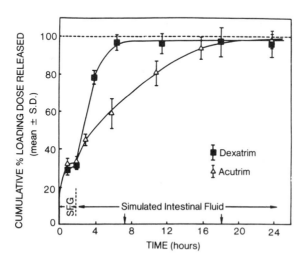

Figure 15 Comparative percent loading dose released versus time profile for the release of phenylpropanolamine from
Acutrim (△) and Dexatrim (■) in simulated gastric fluid
(SGF) for 2 h and then in simulated intestinal fluid for the
remaining 22 h. (From Ref. 25.)

coated with cellulose acetate (mean particle size 350 nm, polydispersity index 4), containing several plasticizers in a standard coating pan. Control tablets were coated with cellulose acetate and diethyl tartrate from di-chloroethane/ethyl acetate/methanol (60:25:15) in the same coating pan. Subsequently, a release orifice of 250 μm was created using a microdrill. Release experiments were conducted in distilled water (500 ml) at 37°C. Based on the results, the authors concluded that aqueous-based latex films exhibit a shorter lag time to constant release, with a higher release rate, when compared to organic-based coatings of the same film weight. One possible explanation for the results is the difference in swelling behavior between the two systems, which is likely due to the method of film formation.

Alternative Systems

The elementary osmotic pump benefits from its simple functional design, and it is well suited for the formulation and delivery of drugs with intermediate water solubility [28]. One finds, however, many examples of drugs with either poor or high water solubility. For drugs of this type, a new device, called the push-pull osmotic pump, was developed by Alza [29]. A cross-sectional view of the push-pull osmotic pump is shown in Fig. 16. In general, the system is constructed of a bilayer tablet core that contains drug in the upper compartment and an osmotic polymeric driving agent in the lower compartment [30]. The core system is coated with a semipermeable membrane in a manner similar to the elementary osmotic pump, with release coming from an orifice formed after the coating operation. In this device, the drug layer accounts for 60–80% of the tablet weight, while the osmotic polymer layer accounts for 20–40%. During operation, both the drug layer and the polymer layer imbibe water. In the case of a poorly soluble drug, an in-situ suspension is formed in the drug compartment, which is dispensed from the orifice by the expanding osmotic push layer. The basic mass delivery rate expression was presented in Eq. (18). However, in the push-pull osmotic system, the volume flow is the summation of the contributions from the polymeric compartment, Q, and from the drug compartment, F. Furthermore, drug concentration in the dispensed formulation is given by the following equation:

$$C_s = F_D C_0 \qquad (31)$$

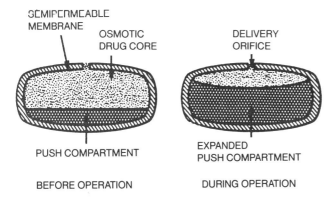

Figure 16   Cross-sectional view of the push-pull osmotic pump before and during operation. (From Ref. 30.)

where C_0 is the concentration of solids dispensed from the dosage form, and F_D is the fraction of the drug dose formulated in the drug compartment. Thus Eq. (31) implies that the formulated drug fraction in the drug compartment remains constant and is equal to the fraction of the drug dispensed from the orifice. The assumption here is that the ratio of drug to solids in the dispensed formulation and in the solid drug core are the same. Proper selection of excipients in the drug core formulation is necessary to ensure that there is no shift in the ratio during storage or operation of the push-pull osmotic pump.

Explicit equations for the osmotic flows Q and F are as follows:

$$Q = \frac{K}{l} A_p \Pi_p \tag{32}$$

$$F = \frac{K}{l} (A - A_p) \Pi_D \tag{33}$$

where the terms, K, h, A, l, and Π have been previously described; and the subscripts p and D denote the polymer and drug compartments, respectively. Substitution of Eqs. (31)–(33) into Eq. (26) results in the following expression:

$$\frac{dm}{dt} = (Q + F) F_D C_o \tag{34}$$

Equations (32), (33), and (34) are the fundamental equations for the operation of the push-pull osmotic pump. Typically, the push-pull system can deliver drug at a constant rate for 80% or more of its theoretical content; this is in contrast to 60–80% of the contents for the elementary osmotic pump. Push-pull technology has been applied to develop at least three formulations, with varying delivery rates and delivery duration, for nifedipine, a poorly soluble drug [30].

Recently, a controlled-porosity osmotic pump was described by Zentner and co-workers [31]. In this system, an osmotically active core was coated with a mixture of polymers with differing degrees of water solubility. In the presence of water, the soluble components in the coating dissolve, leaving a microporous film. Subsequently, water can diffuse into the core, setting up an osmotic gradient, which controls the release of drug. The rate of release was reported to be dependent on coating thickness, level of soluble components in the coating, solubility of drug in the tablet core, and the osmotic pressure difference between the core and the environment. As expected, release of potassium chloride from a prototype system was independent of medium agitation or pH. Zero-order release was observed over several hours.

Applications

To date, elementary or push-pull osmotic systems have been developed for several drugs and tested in humans. One commercial system that has enjoyed considerable success is Acutrim, which has already been discussed earlier in the chapter. In this section several examples will be illustrated.

Theeuwes and co-workers [24] developed several elementary osmotic pumps for metoprolol, which provided delivery rates of 14 and 19 mg/h with delivery times of 10 and 15 h. Drug release profiles for these prototypes are depicted in Fig. 17. Once again the familiar zero-order release pattern is evident for drug release up to 60–80% of the dose. In a subsequent study in humans, Godbillon and co-workers examined the in-vivo performance of metoprolol OROS systems with different durations of drug release [32].

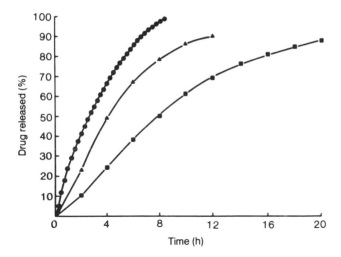

Figure 17 Comparison of the cumulative amount of metoprolol released in vitro, measured by the flow-through method and expressed as percent of the total content, from a slow-release tablet (●) and from 19/190 (▲) and 19/285 (■) metoprolol Oros systems. (From Ref. 24.)

In one study with six subjects, metoprolol OROS 19/190 or OROS 19/285 was dosed in a crossover fashion with resultant blood-level curves as shown in Fig. 18. Using the Nelson-Wagner method [33], the in-vivo delivery rate was determined. The in-vivo release was found to correlate well with the in-vitro release, but the lag times were found to differ by 1 h (Fig. 19). The authors concluded that the blood-level curves substantiate the claim of extended drug release from the devices, with subsequent absorption from the gastrointestinal tract.

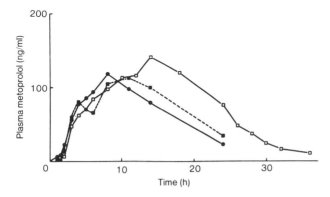

Figure 18 Mean plasma concentration profiles of metoprolol after administration of a 19/190 Oros delivery system on the first (●—●) and second (■—■) occasion, and of a 19/285 Oros system (□—□) on a separate occasion to the same six volunteers. (From Ref. 32.)

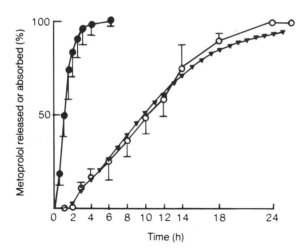

Figure 19 Comparison of the mean apparent in-vivo absorption (○—○) and in-vitro release (▼—▼) profiles for the 19/285 Oros system; the vertical lines represent 1 s.d. of the mean. The in-vitro profile is displaced by 1 h in this figure and the in-vivo profile for the conventional tablet formulation (●—●) is also shown for comparison. (From Ref. 32.)

In a recent paper, Swanson and co-workers detailed the development of three push-pull osmotic pumps for 24-h oral controlled delivery of nifedipine [30]. All the prototypes (1.7/30, 3.4/60, and 5.1/90) delivered drug for approximately 24 h. Zero-order release rates for the prototypes were 1.7, 3.4, and 5.1 mg/h, respectively. The respective total amount of drug released was 30, 60, and 90 mg. After successful in-vitro/in-vivo performance testing, human clinical trials were conducted. Chung and co-workers conducted in-vivo investigations to compare the osmotic pump system with conventional capsules by controlled delivery [34]. Nifedipine administered by controlled oral delivery was found to be well absorbed from the GI tract. In one instance, the relative bioavailability of the 3.4/60 pump was found equivalent to an equal dose of nifedipine in capsule formulation (3 × 20 mg/day). Moreover, the oral osmotic pump system was observed to achieve a sustained plasma level of nifedipine with minimal fluctuation, while conventional dosing showed substantial seesaw-shaped fluctuation (Fig. 20). In the second study, the 1.7/30-mg osmotic pump was compared to a 30-mg sustained-release tablet. Again, the resulting plasma-level curves, shown in Fig. 21, demonstrate that steady-state level is achieved by the zero-order release of nifedipine from the oral osmotic system, while the sustained-release Adalat tablet showed a non-zero-order burst delivery of nifedipine. In addition, the oral osmotic systems exhibited linear pharmacokinetics over the dosage range of 30 to 180 mg, regardless of the combination used, i.e., three 60-mg units or two 90-mg units. The authors concluded that the plasma levels of nifedipine obtained from oral osmotic pumps are predictable, based on the release kinetics, and are proportional to the amount delivered.

Vetrovec and co-workers evaluated the clinical efficacy of nifedipine-releasing oral osmotic pumps in the treatment of angina pectoris [35]. In this 14-week, multicenter,

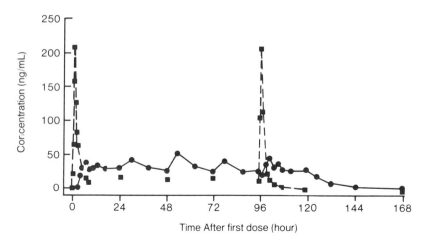

Figure 20 Mean plasma nifedipine concentration profiles in healthy volunteers after multiple 60-mg oral osmotic pump tablets, once daily, or two 10-mg conventional nifedipine capsules, once every 8 h (n = 23). Peaks represent nifedipine capsules only. ■—■, capsules; ●—●, oral osmotic pump system. (From Ref. 34.)

open-label, crossover trial, patients received 30–120 mg of nifedipine daily, using the appropriate combination of conventional dosage forms or oral osmotic systems. Results from the trial indicated that the nifedipine osmotic oral controlled-release system is more effective in reducing angina attacks than the conventional dosage form. This may result from the controlled delivery of nifedipine, which achieves a more consistent blood level within the therapeutically effective range for prolonged duration.

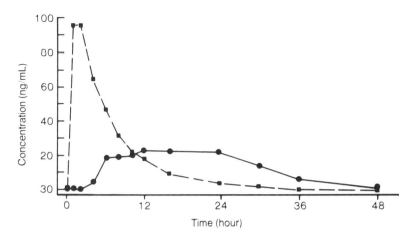

Figure 21 Comparative mean plasma nifedipine concentration profiles after single oral administration of 30-mg Adalat Retard tablet (■) versus 30-mg oral osmotic pump system (●) (n = 16). (From Ref. 34.)

SUMMARY

In this chapter, the fundamental principles underlying the application of osmotic pressure in the controlled delivery of systemically active drugs was presented. It should now be apparent that osmotic pressure-powered drug delivery has several advantages over diffusion-based drug delivery systems. First, osmotic pumps can be used as a useful experimental tool to determine important pharmacokinetic parameters of new drugs, which ultimately find use in the development of an optimized delivery system. Second, because osmotic systems deliver the drugs at zero-order release kinetics, they are, in many instances, superior to older sustained-release technologies, in that better control over their in-vivo performance is possible. Third, drug release from the osmotic systems is, to a large extent, independent of variation in environmental pH and hydrodynamic conditions. Fourth, it is possible to attain substantially higher release rates than with diffusion-based drug delivery systems. For these and other reasons, the future of osmotic technology in drug delivery is bright.

ACKNOWLEDGMENTS

The authors thank H. Martynowicz and K. Meltzer for their assistance in the preparation of this manuscript. The authors also thank J. Alcantara for graphical assistance and N. Jain, A. Serajuddin, and K. Morris for their constructive comments.

PROBLEMS

1. Assume that there is a drug which has an aqueous solubility of 1.2 g/ml and a diffusion coefficient of 5×10^{-6} cm^2/s in a solution with viscosity of 5 cP. You intend to deliver this drug at a zero-order delivery rate of 24.84 mg/h by using an elementary osmotic pump which has the following features:

   $$\frac{dV}{dt} = 0.1 \text{ ml/h} \qquad F = 40$$

   $$l = 250 \text{ } \mu\text{m} \qquad \Delta p = 5 \text{ atm}$$

   Determine the maximum and minimum boundary conditions for the cross-sectional area (A_0) of the delivery orifice for the pump to be developed.

2. Assume that there is a drug which has an aqueous solubility of 500 mg/ml and gives an osmotic pressure of 20 atm when it is totally dissolved. You want to deliver it at a zero-order rate profile using an oral osmotic pump which is coated with a semipermeable membrane with the following properties:

   $$A = 2.0 \text{ cm}^2 \qquad l = 400 \text{ } \mu\text{m}$$

   $$L_p = 10^{-3} \text{ cm/h} \qquad \sigma = 0.95$$

   Calculate the oral delivery rate for this drug.

3. For the same drug delivered by the oral osmotic pump with characteristics outlined in Problem 2, determine the period of time (t_z) for which the drug will be delivered at a zero-order delivery rate given a pump volume of 2 cm^3.

REFERENCES

1. F. Theeuwes and W. Bayne, *J. Pharm. Sci.*, *66*:1388 (1977).
2. Y. Chien, Rate-control drug delivery systems: Controlled release vs. sustained release, *Med. Prog. Tech.*, *15*:21 (1989).
3. Y. Chien, in *Novel Drug Delivery Systems*, Marcel Dekker, New York, chap. 9 (1982).
4. P. Hiemenz, in *Principles of Colloid and Surface Chemistry*, Marcel Dekker, New York, chap. 4 (1977).
5. F. Theeuwes and S. Yum, *Ann. Biomed. Eng.*, *4*:343 (1976).
6. F. Theeuwes, in *Controlled Release Technologies* (Λ. F. Kydonieus, Ed.), CRC Press, Boca Raton, Fla., chap. 10 (1980).
7. S. Rose and J. Nelson, *Aust. J. Exp. Biol. Med. Sci.*, *33*:415 (1955).
8. S. Stolzenberg, U.S. Patent 3,604,417 (September 14, 1971).
9. S. Yum and R. Wright, in *Controlled Drug Delivery*, Vol. II (S. D. Bruck, Ed.), CRC Press, Boca Raton, Fla., chap. 3 (1983).
10. B. Eckenhoff and S. Yum, *Biomaterials*, *2*:89 (1981).
11. J. Johnson, *Sustained Release Medications*, Noyes Data Corp., Park Ridge, N.J., p. 116 (1980).
12. H. Lonsdale, U. Merten, and R. Riley, *J. Appl. Pol. Sci.*, *9*:1341 (1965).
13. R. Langer and D. Wise, in *Medical Applications of Controlled Release*, CRC Press, Boca Raton, Fla.
14. H. Nau, R. Zierer, H. Spielmann, D. Neubert, and C. Gansau, *Life Sci.*, *29*:2803 (1981).
15. B. Sikic, J. Collins, E. Mimnaugh, and T. Gram, *Cancer Treat. Rep.*, *62*:2011 (1978).
16. S. Yum, S. Tillson, and F. Theeuwes, in *Proc. 5th Int. Cong. Endocrinology*, W. Germany, p. 366 (1976).
17. T. Kasamatsu, J. Pettigrew, and M. Ary, *J. Comp. Neurol.*, *185*:163 (1979).
18. B. Prat, R. Butcher, and E. Inskeep, *Animal Sci.*, *48*:1441 (1979).
19. E. Wei and H. Loh, *Science*, *193*:1262 (1976).
20. M. Falcon and B. Jones, *Trans. Ophth. Soc.*, *97*:330 (1977).
21. F. Theeuwes, *J. Pharm. Sci.*, *64*:1987 (1975).
22. A. Mehta, *Pharm. Manuf.*, *1*:23 (1986).
23. N. Lakshminarayanaiah, *Transport Phenomena in Membranes*, Academic Press, New York (1969).
24. F. Theeuwes, D. Swanson, G. Guittard, A. Ayer, and S. Khanna, *Br. J. Clin. Pharmacol.*, *19*:69S (1985).
25. F. Liu, M. Farber, and Y. Chien, *Drug Dev. Ind. Pharm.*, *10*:1639 (1984).
26. M. Ramadan and R. Tawashi, *Drug Dev. Ind. Pharm.*, *13*:235 (1987).
27. C. Bindschaedler, R. Gurny, and E. Doelker, *J. Control. Rel.*, *4*:203 (1986).
28. F. Theeuwes, *Pharm. Int.*, *12*:293 (1984).
29. R. Cortese and F. Theeuwes, U.S. Patent 4,327,725 (May 4, 1982).
30. D. Swanson, B. Barclay, P. Wong, and F. Theeuwes, *Am. J. Med.*, *83*(suppl. 6B):3 (1987).
31. G. Zentner, G. Rork, and K. Himmelstein, *J. Control. Rel.*, *1*:269 (1985).
32. J. Godbillon, A. Gerardin, J. Richard, D. Leroy, and J. Moppert, *Br. J. Clin. Pharmacol.*, *19*:213S (1985).
33. J. Wagner and E. Nelson, *J. Pharm. Sci.*, *53*:1392 (1964).
34. M. Chung, D. Reitberg, M. Gaffney, and W. Singleton, *Am. J. Med.*, *83*(suppl. 6B):10 (1987).
35. G. Vetrovec, V. Parker, S. Cole, P. Procacci, B. Tabatznik, and R. Terry, *Am. J. Med.*, *83*(suppl. 6B):24 (1987).

6

Oral Controlled-Release Delivery

Pardeep K. Gupta   *Philadelphia College of Pharmacy and Science, Philadelphia, Pennsylvania*

Joseph R. Robinson   *School of Pharmacy, University of Wisconsin, Madison, Wisconsin*

INTRODUCTION

Among all the routes of drug administration that have been explored for the development of controlled-release (CR) systems, the oral route has by far achieved the most attention and success. This is due, in part, to the ease of administration as well as to the fact that gastrointestinal physiology offers more flexibility in dosage-form design than most other routes. Development of an oral CR dosage form for a given drug involves optimization of the dosage-form characteristics within the inherent constraints of gastrointestinal (GI) physiology.

Although significant clinical advantages have been obtained for CR formulations, most such dosage forms are still designed on an empirical basis. An understanding of varied disciplines, such as GI physiology, pharmacokinetics, and formulation techniques, is essential in order to achieve a systematic approach to the design of oral CR products. The scientific framework required for development of a successful oral controlled drug delivery dosage form consists of an understanding of three aspects of the system, namely, (1) the physicochemical characteristics of the drug, (2) relevant GI anatomy and physiology, and (3) dosage-form characteristics. The anatomy and physiology includes insight into the basic physiology of the gut as well as the absorptive properties of the GI mucosa. Often one encounters additional factors, including the disease being treated, the patient, and the length of therapy. Given that it is usually not practical to alter the physicochemical characteristics of the drug, design of controlled-delivery systems generally optimizes dosage-form characteristics relative to the GI environment.

The objective of this chapter is to review oral CR systems, with a focus on dosage-form characteristics and GI physiology. Since an understanding of the basic concepts of CR systems is vital for future development, particular emphasis will be on the rationale and mechanism of such delivery systems.

Definitions

The term CR implies a system that provides continuous delivery of the drug for a pre-determined period with predictable and reproducible kinetics, and known mechanism of release. Also included in this term are systems that provide control over movement of the dosage form through the GI tract and/or deliver the drug to a specific area within the GI tract for either local or systemic effect. This chapter will deal only with dosage forms intended to be swallowed orally and will thus exclude buccal and rectal areas of delivery.

Advantages/Disadvantages of Oral CR Dosage Forms

The goal of oral CR products is to achieve better therapeutic success than with conventional dosage forms of the same drug. This goal is realized by improving the pharmacokinetic profile as well as patient convenience and compliance in therapy. Improvement is perhaps the major reason for so much attention being focused on drugs used in chronic therapy; e.g., diuretics, cardiovascular, and CNS agents. Some of the advantages of oral CR dosage forms are

1. Reduced dosing frequency
2. Better patient convenience and compliance
3. Reduced GI side effects and other toxic effects
4. Less fluctuating plasma drug levels
5. More uniform drug effect
6. Lesser total dose

The ideal system possesses all of the above advantages. In most cases, however, there is little direct evidence of a more uniform drug effect, and success has to be based on circulating plasma drug levels. Also, a lesser total dose is based on the assumption that the drug shows linear pharmacokinetics, which in many cases, as will be discussed below, may not be achieved.

On the other hand, oral CR formulations suffer from a number of potential disadvantages. These include:

1. Generally higher cost
2. Relatively poor in-vitro/in-vivo correlation
3. Sometimes unpredictable and often reduced bioavailability
4. Possible dose dumping
5. Reduced potential for dose change or withdrawal in the event of toxicity, allergy, or poisoning
6. Increased first-pass metabolism for certain drugs

Unpredictable and poor in-vitro/in-vivo correlations and bioavailability are often observed with such formulations, especially when the drug release rate is very low or drug absorption from the colon is involved. Dose dumping is a phenomenon where a large amount of the drug is released in a short period of time, resulting in undesired high plasma drug levels and potential toxicity.

Drug Candidate Criteria

A number of drug characteristics need to be considered in evaluating drug candidates for oral CR dosage forms. Some of these characteristics are discussed here.

Dose

Dose limitation is a major factor to consider in many routes, especially for transdermal and buccal patches. In oral systems, however, total drug dose is infrequently a limiting factor. A total dose of several grams may be administered orally as single or multiple units to obtain and maintain adequate drug levels. Nevertheless, for drugs with an elimination half-life of less than 2 h as well as those that are administered in large doses, a CR dosage form may need to carry a prohibitively large quantity of drug.

Biological Half-Life

In general, drugs with short half-lives (2-4 h) make good candidates for CR systems. For drugs with half-lives shorter than 2 h, a prohibitively large dose may be required to maintain the high release rate. Additional factors, such as the reduced rate of absorption from the distal small intestine and colon, may also reduce the rate of drug input to less than that required for adequate drug levels. On the other hand, drugs with elimination half-lives of over 8 h are commonly sufficiently sustained in the body after a conventional oral dose to make sustained release unnecessary.

Therapeutic Range

The range of plasma drug levels between the minimum effective and toxic levels is known as the therapeutic range. Oral CR formulations are valuable for maintaining plasma levels within a narrow therapeutic range. In fact, a valid rationale for formulating drugs with half-lives of over 8 h as CR formulations is to maintain plasma drug levels within a narrow range. By reducing the rate of drug release, it is possible to produce a flatter plasma-level curve and avoid toxic drug concentration in the body. Another means of expressing safe and effective plasma drug levels is the therapeutic index, which is discussed in detail later.

GI Absorption

Most CR formulations are dissolution-controlled, and drug release rate from the dosage form is the rate-limiting step. It is assumed that, once released, the drug is rapidly transferred from the gut lumen to blood. Therefore, efficient drug absorption from the GI tract is a prerequisite for a drug to be considered for use in an oral CR dosage form. In general, the absorption rate for most drugs decreases as the dosage form moves beyond the jejunum. As long as the absorption rate remains above that of the release rate, this change does not affect plasma levels. However, once past the ileocecal junction, a variety of factors generally reduce the drug absorption rate to below acceptable values. This creates a time limit of about 6–9 h during which the drug can be delivered in a predictable manner. For drugs that are absorbed passively, gut wall permeability shows a consistent pattern, even though the rate of drug absorption may decrease progressively. But for compounds that are absorbed via an active transport mechanism, absorption from the GI tract may not be consistent. For such drugs, and for many others, an acceptable rate of absorption may exist only from a limited portion of the small intestine, which may further limit their suitability for CR systems.

Aqueous Solubility

Absorption of poorly soluble drugs is often dissolution rate-limited. Such drugs do not require any further control over their dissolution rate and thus may not seem to be good candidates for oral CR formulations. However, the rate of dissolution of free drug particles decreases with time due to a reducing surface area. CR formulations of such drugs may be aimed at making their dissolution more uniform rather than reducing it. Drugs with good aqueous solubility make good candidates for CR dosage forms. Since the GI environment changes considerably in terms of pH, as well as viscosity, it is desirable that the dissolution rate be independent of such variables; indeed, with systems such as the elementary osmotic pump, dissolution may be rendered independent of pH and viscosity.

Stability to Wide pH Range, GI Enzymes, and Flora

Irrespective of the system employed, an orally administered drug must be exposed to the luminal contents of the gut before it is absorbed. Stability of the drug in the GI content is therefore important to ensure a complete and reproducible drug input into the body. Typically the drug must be stable in the pH range of 1 to 8. Unlike a conventional dosage form, a CR formulation is exposed to the entire range of GI pH, enzymes, and flora. If some degree of colonic absorption is expected, stability to the metabolizing effect of the colonic bacterial population is also required.

First-Pass Metabolism

Saturable hepatic metabolism may render a drug unsuitable for oral CR. This is because systemic availability for such drugs is highly reduced when the input rate is small. First-pass metabolism will be discussed in detail in the section on pharmacokinetic and pharmacodynamic considerations.

PERTINENT BIOLOGICAL PARAMETERS

Design of oral delivery systems, both conventional and CR, have to date been based largely on an empirical understanding of GI physiology. Insight into the biological aspects of oral delivery is more important for CR systems than it is for conventional dosage forms, because, in order to exert control over the rate of drug release, as well as movement of the dosage form through the GI tract, a number of factors such as motility, pH, ionic strength of luminal content, differential absorption, etc., come into play.

Listed below are some of the factors that influence delivery of drugs to the GI tract. These factors show considerable inter- and intrasubject variation, as well as variations due to disease state and circadium rhythm.

Some biological factors influencing the performance of oral CR products include:

1. GI motility and transit time
2. Blood flow
3. Environment of the GI tract
   (a) Luminal contents and pH
   (b) Mucus
   (c) Ileo-cecal junction
   (d) Gut flora
   (e) GI immunology

GI Anatomy

In order to set the stage for subsequent discussion of GI physiology, a brief overview of the functional anatomy of the human GI tract is presented.

Figure 1 shows a schematic representation of the GI tract, and Table 1 [1] lists some of the characteristics of the GI tract that are relevant to drug delivery.

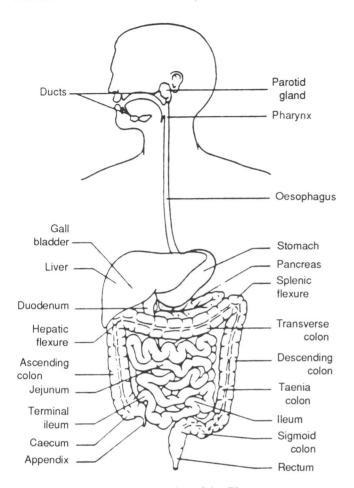

Figure 1   Schematic representation of the GI tract.

Under physiological conditions, gastric absorption of most drugs is insignificant. Factors contributing to limited absorption from the stomach include the limited surface area, the lack of villi on the mucosal surface, a relatively thick layer of mucus on the stomach lining, and the short residence time of most drugs in this organ.

Contents of the stomach pass through the antral area by an opening called the pylorus, into the proximal duodenal part of the small intestine. The gastroduodenal junction controls traffic between the stomach and duodenum, allowing unidirectional passage from the former to the latter, although a duodeno-gastric reflux has been observed in many species. Contents of the gall bladder (bile) and pancreas are emptied into the proximal duodenum, as are some duodenal secretions including bicarbonate. The other end of the small intestine, the terminal ileum, passes into the colon via a junction known as the ileo-cecal valve. Unlike the stomach, the small intestine has on its mucosal surface numerous villi, which impart an enormous surface area. There is a progressive decrease in surface area from the proximal to the distal small intestine and colon. As a result, most nutrients and drugs are absorbed predominantly from the proximal small intestine.

Table 1 Characteristics of the GI Tract

Section	Area (m²)	Liquids secretion (l/d)	Reaction (pH)	More important constitutents	Transit time of food (h)
Oral cavity	About 0.05	0.5-2	5.2-6.8	Amylase Ptyalin Mucins	Short
Esophagus		0	—	—	Very short
Stomach	0.1-0.2	2-4	1.2-3.5	Hydrochloric acid Pepsin Rennin Cathepsin Lipase Intrinsic factor	0.25-3
Duodenum	About 0.04	1-2	4.6-6.0	Amylase Glucohydrolase Galactohydrolase Lipase Trypsin Chymotrypsin Bile acids	1-2
Small intestine	4500[a]	0.2	4.7-6.5	Like in duodenum	1-10
Large intestine	0.5-1	About 0.2	7.5-8.0	Mucus Bacteriums	4-20

[a] Taking intestinal microvilli area into account; without them, about 100 m².

The primary function of the colon is to store indigestible food residues. The luminal content of the colon is much more viscous than that of the small intestine. The colonic mucosal surface lacks villi, thus reducing its exposed surface area. It also contains a variety of bacteria, which are normal residents of the GI tract.

Gastrointestinal Motility

An important consideration when contemplating use of CR dosage forms in the GI tract is the continuous motility of this organ. The pattern and force of the motility vary depending on whether the animal is in a fed or a fasted state [2].

Figure 2 [3] shows a representation of the typical motility patterns in the interdigestive (fasted) and digestive (fed) state.

It is now well documented that there are two modes of GI motility patterns in humans and animals that consume food on a discrete basis; the digestive (fed) mode and the interdigestive (fasted) mode [4]. The characteristic of fasting GI motility is a cyclic pattern which has been fully characterized in both dogs and humans. This cyclic pattern of motility, which originates in the foregut and propagates to the terminal ileum, can be divided into four distinctive phases: phase I, representing a quiescent period with no electrical activity and no contractions; phase II, the period of random spike activity or intermittent contractions; phase III, the period of regular spike bursts or regular contractions at the maximal frequency that migrate distally; and phase IV, the transition period between phase III and phase I.

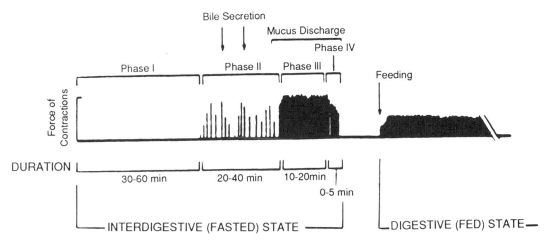

Figure 2   Pictorial representation of the typical motility patterns in the interdigestive (fasted) and digestive (fed) state. (From Ref. 3.)

The average length of one complete cycle, commonly known as the interdigestive migrating motor complex (MMC), ranges from 90 to 120 min in both humans and dogs [5]. Certain disease conditions, such as bacterial overgrowth, mental stress, and diurnal variations, or a combination of the above factors, can affect the length of the total cycle or its individual phases [6–8]. Phase III, also known as the housekeeper wave, serves to clear the digestive tract of all indigestible materials from the stomach and small intestine. Nondigestible solids when administered during phase I are emptied from the dog stomach only during phase III [9]. Shear forces involved during this phase can pose a problem for bioadhesive systems in the GI tract; consequently, any system that is designed to remain in the stomach during the fasted mode must adhere to the membrane strongly enough to withstand the force of the housekeeper wave.

A characteristic feature of cyclic motor activity is its association with the secretory gastrointestinal component. Gastric, pancreatic, and biliary secretory components of the MMC in the human duodenum indicates that the migratory motor and secretory activity constitutes two aspects of the same periodicity [10]. It may be concluded that under fasting conditions both motor and secretory activities of the stomach, gut, pancreas, and liver change periodically to provide both mechanical and chemical means of intestinal housekeeping.

Feeding results in interruption of the interdigestive motility cycle of the GI tract and in the appearance of a continuous pattern of spike potentials and contractions called postprandial motility. A minimum amount of gastric content appears to be necessary in order to change motility from an MMC to postprandial. It has been shown that oral administration of 150 ml of water during phase I changes the fasted motor activity to a fed-like pattern in dogs [11]. A normal meal changes the motility pattern to a fed state for up to 8 h, depending on caloric content of the food [12].

GI Transit

The single most limiting biological factor in the development of once-daily oral CR systems is the transit time of a dosage form through the GI tract. Of particular importance in this context is the residence time of a dosage form in certain parts of the GI tract, since drug absorption may not be possible through the entire lining of the gut. Like the motility pattern, transit patterns of both solids and liquids through the gut also vary depending on whether the person is in a fasted or a fed state. Accordingly, these two types of transit patterns will be discussed separately.

Transit Patterns in the Fasted State

Gastric emptying. *Liquids*: The process of distintegration and dissolution starts in the stomach. Transit of liquids already present in the stomach and administered with the dosage form can play an important role in this process. Gastric emptying of liquids in the fasted state is a function of the administered volume of liquid [11]. For small volumes, generally less than 100 ml, gastric emptying is controlled by the existing phasic activity. Liquids empty at the onset of phase II, and most of the fluid is gone before arrival of phase III. Volumes larger than 150 ml show a different transit pattern and empty with a characteristic discharge kinetics irrespective of phasic activity. These kinetics can be approximated by a plot of first-order or square root of volume remaining in the stomach versus time, the slope of the curve in both cases being a function of caloric content of the meal. Figure 3 shows the cumulative volume emptied as a function of time when different volumes of water are administered during phase I in the dog. This difference in transit behavior between large and small volumes is due to the fact that small volumes do not change the existing motility pattern in the stomach, while large volumes convert

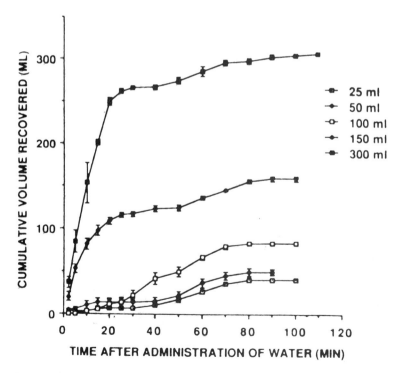

Figure 3 Commulative volume recovered as a function of time after administration of different volumes of water (25–300 ml) 15–20 min after cessation of high antral activity in fasted dogs. (From Ref. 11.)

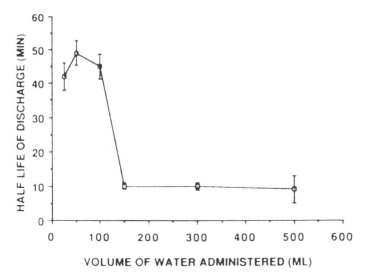

Figure 4   Time taken for the discharge of half of the administered volume of water given 15–20 min after cessation of high antral activity, plotted against the volume of test solution. (From Ref. 11.)

the fasted state to a fed-like state, which in turn creates the fed-state motility pattern. The time taken for discharge of 50% of a large volume of liquid is 8–12 min, as shown in Fig. 4.

Thus a dosage form given with a small volume of liquid can stay in contact with that liquid in the fasted state from 0 to 60 min, depending on the phase of activity at the time of administration. Dissolved drug in the media will then be emptied into the duodenum almost as a bolus. This emptying pattern of liquids in the fasted state is independent of the presence of any indigestible solids in the stomach [13].

Gastric emptying of indigestible solids: Indigestible solids, which include most solid dosage forms, empty from the stomach as a function of their size. Solids of particle size smaller than 1 mm can empty with the liquid, especially if the liquid viscosity is high. Solids of size 2 mm or more do not empty until arrival of phase III activity, at which time they empty as a bolus [13]. In the fasted dog, onset of the gastric emptying of solids is variable, depending on proximity of the time of ingestion to the next phase III activity. Thus, a solid dosage form can stay in a fasted stomach anywhere from 0 to 120 min. Also in fasted dogs, gastric emptying of solids is independent of size, density, and surface characteristics [14].

Thus, manipulation of density and shape of solids does not seem to be a viable approach, although some studies have claimed otherwise. The only possibility may be to increase the size of the dosage form to a degree that it cannot pass through the pylorus until degraded, or perhaps convert the stomach to a fed state, thus initiating the retropulsion phenomenon in the antral area which will keep the dosage form from emptying. Moreover, distribution of multiunit dosage forms administered during the fasted state is questionable in light of the observation that solids empty during phase III as a bolus.

Intestinal transit: During phase I of the fasted state, when contractions are minimal, there is little or no movement of content through the intestine. Flow of materials is progressively faster during phases II and III. Segregation of liquids and solids also occurs, so that fluids tend to move during phase II and solids during phase III.

Transit of solids through the small intestine is variable because motor activity may not be sufficiently strong to move the solids. This implies that, during the fasted state, there is relative motion between the dosage form and the luminal fluid content. Shear forces and constant fluid movement around the dosage form may explain the sometimes-observed higher in-vivo bioavailability compared with in-vitro release.

For multiunit dosage forms, once the particles have left the stomach, there is little, if any further spreading of particles in the intestine [15]. Since particles usually leave the stomach as a bolus during the fasted state, multiunit dosage forms may not serve their intended claim of dispersion. These findings are consistent in both humans and dogs. Once in the colon, however, particles do show a tendency to disperse, perhaps due to high viscosity.

Fed State Transit Behavior

Gastric emptying of liquids and solids.   Following ingestion of a meal the fundus expands to accommodate the meal without an appreciable increase in intragastric pressure. This phenomenon is known as receptive relaxation. Once in the stomach, food starts emptying almost immediately. Liquids empty faster, compared to solids, the rate of emptying being controlled by feedback mechanisms from the duodenum and ileum. Solids are handled differently by the stomach according to their particle size. In general, solids are not emptied in the fed state unless they have been ground to a particle size of 2 mm or less. Thus, there is a sieving mechanism in the fed stomach, and meal viscosity seems to influence this mechanism. Since grinding and mixing takes place in the antral area, dosage forms will tend to reside in this area due to their large size. Multiunit dosage forms, however will disperse and empty with food and thus achieve a considerable degree of distribution.

Another event that follows feeding is gastric secretions. Depending on the nature and volume of the ingested meal, gastric volume may actually remain constant during the first hour of emptying, the volume emptied being replaced by gastric secretions. Thus, in the fed state more fluid is available for dissolution. Total time for gastric emptying varies from about 2 to 6 h.

Intestinal transit during the fed state.   The contents of the small intestine move faster during the fed state compared with phase III transit during the fasted state. This helps move smaller particles more rapidly, but larger particles are unaffected by this flow and thus travel relatively slowly. Regardless of the nature of the digestible fluid and particles, the intestinal transit time for both liquids and solids is around 3–4 h, in both the fasted and fed state. The constancy of intestinal transit can be important in colonic drug targeting.

Studies of dosage forms such as tablets, capsules, and particles have shown transit patterns similar to those of nutrients. Thus most dosage forms administered in the fasted state empty in 0–90 min. In the fed state, nondisintegrating tablets and capsules stay in the stomach for 2–6 h and are discharged only at the onset of fasted activity. However, small particles and disintegrating dosage forms will empty with food. In all instances, the small bowel transit time is 3–4 h.

In summary, the total transit time of nutrients and dosage forms in humans from the stomach to the ileo-cecal junction is approximately 3–6 h in the fasted state, and 6–10 h in the fed state. This puts an approximately 10-h delivery limit on drugs that are absorbed only from the small intestine.

Environment of the GI Tract

Blood Flow

The GI tract is a well-perfused organ, receiving about 30% of the total cardiac output. Changes in blood flow can only affect the absorption of compounds with high intestinal permeability. Generally, lipid-soluble molecules and those small enough to penetrate the aqueous pores are absorbed rapidly and show blood flow-dependent absorption. Since the absorption rate of many drugs shows an intermediate dependence on blood flow, relatively large changes in blood flow are required to produce a significant change in the absorption rate [16]. Splanchnic blood flow increases considerably after a meal, reaching its peak after a heavy meal. In fact, any distention in the stomach causes an increase in intestinal blood flow to some degree, and this increase can last for up to 1 h. Therefore, dosage forms given with large volumes of water (200 ml or more) could facilitate drug absorption by inducing an increase in blood flow. Indeed, this may be a partial explanation of the higher bioavailability observed for some dosage forms when given with a large volume of water.

Luminal Contents and pH

The GI tract offers a pH range of 1–8 for drug delivery systems, and Table 1 lists the pH range for different parts of the GI tract. Table 1 also lists a variety of acids, enzymes, and special factors present in the gut. This varying pH and composition of the luminal content affects performance of orally administered dosage forms in several ways. The pH of the bulk solution not only affects the release rate and dissolution of drug, it also determines the ratio of charged and uncharged species for ionizable drugs, thus affecting the rate of absorption. The pH is different in the fasted and fed states, and certain disease states may also alter the pH.

The gastric pH in the resting state is generally between 1 and 3 in both dogs and humans [17]. Upon feeding or distending the stomach to over 150 ml, gastric secretion is stimulated. Ingested food, however, may have a considerable buffering effect on gastric content and help maintain the pH above 4 for up to 1 hour [18]. Eventually, gastric pH returns to its base level, i.e.,1–3. Thus, basic drugs have a better chance of dissolution in the stomach, provided the dosage form stays in the stomach for a sufficient time.

Intestinal pH varies between 4 and 7.5 depending on location. The duodenal pH ranges between 4 and 6, whereas most of the intestine has a pH near neutrality. Substantial bicarbonate and bile secretions during the fed state may push duodenal pH toward the basic side. The colonic pH is normally above 7 and can be as high as 8, and in certain cases, bacterial metabolism may alter the pH in the large bowel.

The pH of the mucus membrane has been shown to be fairly constant throughout the GI tract. The thick mucus layer is considered to be the pH barrier in the stomach, higher pH at the membrane being a result of the sodium-dependent transcellular bicarbonate transport system.

GI Mucus

Specialized goblet cells located throughout the GI tract continuously secrete mucus. Fresh mucus on the surface of the membrane is very thick. Away from the membrane and closer to the lumen, mucus is dilute and less viscous. Chemically, mucus is a glycoprotein network holding a variable amount of bound water.

The thickness of the mucus layer varies depending on the region of the GI tract. The primary function of mucus appears to be protection of the surface mucosal cells from acid and peptidases. In addition, it also serves as a lubricant for the passage of solids and as a barrier to antigens, bacteria, and viruses.

Mucus is considered to be an absorptive barrier in the GI tract, since it acts as a stagnant diffusion layer. For small-molecular-weight compounds, it merely adds to the stagnant diffusion layer through which compounds must diffuse before reaching the membrane. However, for large molecules such as peptides, it may offer some added resistance due to the expanded network of glycoproteins.

Many substances can interact with mucus and change its physicochemical characteristics. This interaction can also result in a change in absorption of these compounds. Tetracyclines have been shown to complex with mucus and have their transport delayed [3].

Ileo-Cecal Junction

The ileo-cecal junction serves primarily to ensure unidirectional flow of material from the small to the large bowel. Due to the large water-absorptive capacity of the colon, the colonic content is considerably more viscous than ileal chyme. This presents a problem for absorption of most drugs, since mixing and hence availability of drug to the absorptive membrane is not efficient.

Gut Flora

The intestinal flora may play an important role in the metabolism of certain foreign compounds. Many drugs are metabolized by enteric bacteria, including sulfasalazine and acetyl salicylic acid. The human colon has over 400 distinct species of bacteria and has up to 10^{10} bacteria per gram of content [19]. Among the reactions carried out by these bacteria are azoreduction and enzymatic cleavage, e.g., by glycosidases [20]. These bacterial degradation processes present an interesting concept of drug delivery, i.e., use of drug complexes that are degraded by bacteria. These will be discussed in greater detail later in this chapter.

Immunology of the GI Tract

The entire gastrointestinal tract is exposed to an immense and diverse range of potentially antigenic materials, primarily from food but also from a number of pathogenic and non-pathogenic microorganisms. In response to this antigenic challenge, the GI tract is populated by an abundance of immunological elements, both as individual lymphoid cells and as organized lymphoid tissue. When challenged by an antigen, these elements produce IgA antibodies, which by virtue of their secretory component are able to traverse the epithelial layer and appear at the epithelial surface and in the lumen of the GI tract [21]. IgA antibodies are primarily responsible for providing an immunological barrier against mucosal penetration of antigens encountered by the GI tract.

There are no studies available on antigen absorption from the stomach. In terms of nonspecific immune reactions, production of acid and pepsin is thought to denature most ingested antigens and bacteria and thus relieve the stomach of the necessity of having an active immune system. The intestinal tract, consisting of the small intestine, caecum, and large intestine, has a well-developed immune system that is both specific and nonspecific.

Antigen uptake from various parts of the intestine is important not only from an immunological point of view, but also in the context of drug delivery, because it may provide an opportunity to deliver large molecules, including peptides and proteins, provided they can be protected from degradation in the intestine and localized at the site of uptake. Two compartments through which antigen absorption might occur are the villous mucosa and Peyer's patches. Through the formation of pinocytotic vesicles at the base of the microvilli the villous mucosa is able to absorb molecules as large as horseradish peroxidase and ferritin [22]. This process has also been implicated in maternal immunoglobin uptake by the neonate. However, the amounts absorbed are generally too small to be of any therapeutic significance.

Gastrointestinal Absorption

Intestinal Permeability

Absorption in the GI tract takes place predominantly in the small intestine because of its large surface area and the high permeability of the lining membrane. The luminal surface of the mucosa is organized such that the surface area available for contact with intestinal contents is greatly amplified. Since the epithelium of the intestinal membrane consists of only a single layer of loosely packed cells, the permeability of this membrane is high. A variety of routes exist for drug and nutrient passage through this membrane. Drugs can pass directly through the cell membranes into the underlying blood vessels (transcellular) or permeate through the spaces between the cells (paracellular). Attempts have been made to correlate various physiochemical properties of molecules and the permeability in order to determine their rate and predominant route of passage through intestinal tissue. For relatively lipophilic drugs, which permeate largely via the transcellular route, the pH partition hypothesis and the three-aqueous-compartment model generally explain the absorption characteristics, unless some active transport mechanism exists in the system. The three-compartment model has additional refinements over the pH partition hypothesis. These include the treatment of the absorption process in dynamic terms rather than an equilibrium state, and better representation of the true physiological situation in that both the transcellular and paracellular pathways are included. Also, it does not involve a complex and ill-defined microclimate at the mucosal surface. Small hydrophilic compounds are apparently absorbed through aqueous pores or channels formed by protein components in the membrane. There can be segmental differences in absorption of different kinds of drugs in the intestine, but most absorption takes place in the first half of the small intestine.

For conventional dosage forms, the major concern is the bioavailability from each dose. As long as most of the administered dose is absorbed as completely as possible, within a given period of time, the outcome is acceptable. Thus, if absorption takes place only from a part of the GI tract, it may not have much significance for conventional dosage forms. However, the situation is not the same for CR formulations. Since these systems are designed to stay in the GI tract for longer periods of time and continuously release drug during their entire stay, efficient abortion from the entire GI tract is a prerequisite for optimal performance of such systems. In general, drugs with high membrane permeability are rapidly absorbed from the entire lining of the gut, although the rate of permeation may vary. Variation in drug absorption rate does not have significance as long as it is significantly greater than the drug-release rate from the dosage form. This is because, for most drugs, the rate at which the drug is presented at the site of absorption is the rate-limiting step in the absorption process. However, for drugs that show intermediate or low permeation through the GI lining, the situation may be different. There is potential for such drugs to have permeation rates less than their rate of release from the dosage form, thus making the rate of GI absorption limiting. If one considers the fact that most drugs are absorbed efficiently from the first half of the small intestine, there is the possibility that drug input into the body declines as the dosage form moves closer to the ileo-cecal junction.

Certain drugs show a differential absorption, being absorbed predominantly from a particular segment of the small intestine. This segment is also referred to as the "window of absorption." Such drugs are bioavailable only if released before or at the absortion window, because any drug released after the dosage form has cleared the absorption area is simply passed into the feces. This effect may be pronounced for drugs that are transported by an active transport mechanism, because active transport tends to show well-defined segmental differences in the GI tract.

Colonic drug absorption is typically believed to be poor and variable [23]. This is attributed to high viscosity of the colonic content and lack of microvilli on the mucosal surface. Unlike the small intestine, where a multitude of drug absorption mechanisms exist, colonic absorption appears to be primarily a simple diffusion process through the lipid membrane, with no evidence to date of any carrier mediation [24]. Bacterial degradation may also be a contributing factor to poor colonic absorption. This may be of particular importance to controlled drug delivery systems which are aimed at a once-daily dosing. Some degree of colonic absorption is necessary for oral delivery beyond 6–8 h. Recently, however, it has been shown that some drugs, including theophylline and metoprolol, are absorbed from the colon [23,25].

Effect of Food on Drug Absorption

The presence of food in the GI tract can often have a marked and sometimes variable effect on drug absorption [26,27]. Food can increase or decrease the rate or extent of absorption of a drug, or delay the onset of absorption. A change in the extent of absorption is usually due to the direct or indirect interaction of food with the formulation or drug, while a delay in absorption is usually a result of delayed gastric emptying. These effects

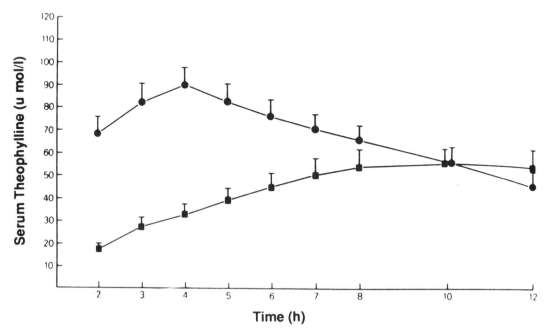

Figure 5 Serum theophylline concentrations after a single dose of a prolonged release product to fasted (●) and fed (■) children. (From Ref. 26.)

are so variable that the same drug may appear under different categories of food effects, depending on the nature of the food and the formulation.

In general, food prolongs the gastric residence time of nondigestible solids for up to 6 h. Thus, formulations designed to release their drug in the intestine only (enteric coated), when administered with food will show an increase in the lag time for absorption. However, for formulations designed to release drug independent of pH, gastric residence time does not affect drug release and subsequent absorption, unless the drug is unstable in an acid environment. For such formulations, food generally improves the bioavailability. This effect can be seen in Figure 5, which compares bioavailability of a sustained form of theophyllin in the fasted and fed states [26]. The relationship between variations in the gastric residence time and the absorption of procainamide from a wax-matrix sustained-release tablet in humans has been reported [5]. Unlike enteric-coated aspirin, gastric retention of a procainamide dosage form does not delay absorption of the drug, but shows a slight increase in AUC and C_{max}. Food can have a marked influence on the GI distribution of multiunit dosage forms, provided they are 2 mm or smaller.

PHARMACOKINETIC CONSIDERATIONS

Kinetic Parameters

From a release kinetics standpoint, there are three categories of oral delivery systems, conventional, first-order slow release, and zero-order release. Variations within these systems are possible in terms of free drug load for a burst effect or additional coatings to introduce a lag time before drug release begins. Figures 6a, 6b, and 6c show typical in-vitro/in-vivo mass balances of such systems. These curves are applicable assuming that drug release from the dosage form is the rate-limiting step. For a CR system, there is a small amount of drug in the form of solution in the gut. Most of the drug resides either in the dosage form or in the body. However, for conventional systems, there is a period of high drug content in the gut lumen as a solution, which is then rapidly absorbed to give a characteristic peak associated with administration of such dosage forms.

The goal of a CR formulation is to improve therapy by reducing the ratio of the maximum and minimum plasma drug concentration (C_{max}/C_{min}) while maintaining drug levels within the therapeutic window. In a conventional dosage form, a relatively large C_{max}/C_{min} is typically observed due to rapid absorption of drug into the body. This ratio is considered relative to another term known as the therapeutic index. For practical purposes, the therapeutic index of a drug can be defined as the ratio of the maximum drug concentration in blood that can be tolerated without appreciable side effects to the minimum drug concentration needed to produce and maintain a desirable pharmacological response. Therefore, the goal of any therapy is to give a drug with sufficient frequency and dose so that the ratio C_{max}/C_{min} in plasma at steady state is less than the therapeutic index and drug levels are always maintained at effective concentrations.

Two intrinsic properties of the drug determine the frequency with which a drug must be given as a conventional dosage form in order to keep the C_{max}/C_{min} ratio within the therapeutic index: the therapeutic index and the biological half-life. For a rapidly absorbed and distributed drug, the ratio of maximum to minimum concentration in plasma at steady state is given by [1]

$$\frac{C_{max}}{C_{min}} = e^{kT} \tag{1}$$

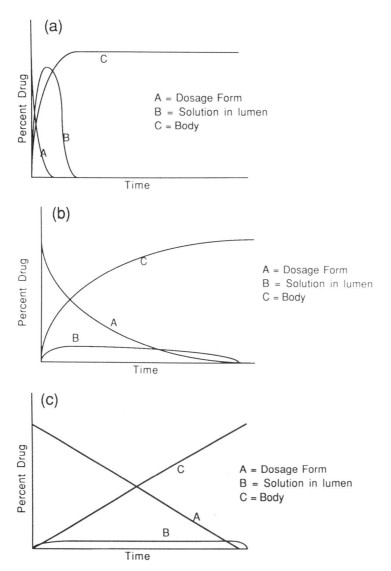

Figure 6 Hypothetical curves for drug fraction in dosage form, solution in gut, and body after oral administration of (a) a solid conventional-dosage form; (b) a first-order slow-release dosage form; (c) a zero-order-release dosage form.

where

k = first-order elimination rate constant

T = dosing interval

Example: A rapidly absorbed and distributed drug is administered twice a day and has an elimination half-life of 3 h. Calculate C_{max}/C_{min} at steady state.

Using Eq. (1):

$$\frac{C_{max}}{C_{min}} = e^{kT} = e^{(0.693/3)(12)} = 16.0$$

Since C_{max}/C_{min} should be less than the therapeutic index (TI), it follows that:

$$T < t_{1/2} \left(\frac{\ln TI}{\ln 2} \right) \tag{2}$$

For a drug with therapeutic index of 2, the dosing interval should not exceed more than one biological half-life of the drug. For drugs with short biological half-lives ($t_{1/2}$ = 2–5 h) and low therapeutic indices (T < 3), dosing has to be unreasonably frequent to maintain desired drug levels in the body. This situation is observed for a number of drugs, including theophylline and procainamide. Therefore, from a pharmacokinetic standpoint, there are two approaches to the design of formulations that give desirable therapeutic concentrations at a reasonable dosing frequency. The first approach is to select a drug that has a $t_{1/2}$ value long enough to be administered as infrequently as once or twice a day. Certain drugs, including warfarin, digoxin, and phenobarbital, fall into this category. But for a given medical condition, a drug or its analog with a suitable $t_{1/2}$ may not be available. In general, most drugs demonstrate relatively short half-lives and thus need to be formulated according to the second approach, which involves modification of the drug formulation in such a way that the drug input into the body is slowed. Such formulations are particularly suitable for drugs with short half-lives and low therapeutic indices, and are used in chronic therapy—e.g., antiarrhythmics.

In principle, in order to keep a constant plasma drug level, the drug input rate into the body should be zero-order. While many systems promise zero-order release based on in-vitro situations, in-vivo profiles are seldom the same due to a number of physiological constraints and variations as discussed earlier. In general, it is easier to design systems that release drug with first-order kinetics, but at a slow enough rate that the drug is delivered from a single dose over a period of up to 12 h. Drug release rate and plasma drug levels from zero-order or first-order release can be computed from equations derived with the following assumptions: (1) Drug absorption, metabolism, and elimination are first-order processes; (2) the rate-limiting step in drug input is the release rate of drug from the formulation; (3) the drug shows linear pharmacokinetics.

Plasma drug concentration, C, following a single dose of a first-order-release formulation can be calculated by [28]:

$$C = \left[\frac{Dk_r}{V(k_r - k_{el})} \right] [\exp(-k_{el}t) - \exp(-k_rt)] \tag{3}$$

where

    D = drug in sustained-release form

    V = volume of distribution

    k_r = rate constant for drug release ($k_r \ll k_a$)

    k_a = rate constant for drug absorption

    k_{el} = rate constant for drug elimination ($k_{el} > k_r$)

    t = time

In most situations, k_r is smaller than k_{el} and presents an example of a "flip-flop" model. From the above equation, it is obvious that the plasma drug levels are a function of k_{el} and k_r (assuming that $k_r \ll k_a$). As k_r decreases, the drug profile is lowered and prolonged for a given k_{el}.

C_{max} and C_{min} at steady state can be calculated as:

$$C_{max} = \frac{D[\exp(-k_{el}t_{max})]}{V[1 - \exp(-k_{el}T)]} \tag{4}$$

$$C_{min} = \left[\frac{k_r D}{V(k_r - k_{el})}\right]\left\{\frac{\exp(-k_{el}T)}{[1 - \exp(-k_{el}T)] - \exp(-k_r T)/[1 - \exp(-k_r T)]}\right\} \tag{5}$$

where

$$t_{max} = 2.3\log\left\{\frac{k_r[1 - \exp(-k_{el}T)]/k_{el}[1 - \exp(-k_r T)]}{k_r - k_{el}}\right\} \tag{6}$$

Using the above equations, one can calculate the dose and the dosing interval required in order to achieve a given dosage-form index. In general, drugs with short half-lives and low therapeutic indices must be given no less frequently than twice a day.

Following a single dose of a zero-order-release formulation, plasma drug concentration can be calculated as [28]:

$$C = \frac{K_0}{Vk_{el}[1 - \exp(-k_{el}t)]} \tag{7}$$

where K_0 = zero-order drug release rate constant.

Example: A zero-order-release device releases drug at a rate of 12 mg/h. Given that $k_{el} = 0.15$ and $V = 70$ liters, calculate the plasma drug concentration at 6 h after dosing and at plateau.

Using Eq. (7):

$$C_{6h} = \frac{12 \text{ mg/h}}{(70 \text{ liters})(0.15 h^{-1})(1 - e^{(0.15)(6)})} = 1.92 \text{ mg/liter}$$

$$C_{plateau} = \frac{12 \text{ mg/h}}{(70 \text{ liters})(0.15 \text{ h}^{-1})} = 3.42 \text{ mg/liter}$$

Example: A wax matrix tablet containing 250 mg of drug released 90% of its drug load as a zero-rate release over 12 h, the rest being eliminated in the feces. Calculate the plasma drug concentration at 3, 6, and 12 h after first dose. Given $V = 60$ liters, clearance (Cl) = 15 liters/h.

Using Eq. (7):

$$k_{el} = \frac{Cl}{V} = 0.25 \text{ h}^{-1}$$

$$K_0 = \frac{(0.9)(250)}{12} - 18.75 \text{ mg/h}$$

$$C_{3h} = \frac{18.75}{(60)(0.25)(1 - e^{(0.25)(3)})} = 0.33 \text{ mg/liter}$$

$$C_{6h} = 0.48 \text{ mg/liter}$$

$$C_{12h} = 0.59 \text{ mg/liter}$$

In both first-order- and zero-order-release systems, the time required to achieve desired drug levels in the body depends on the elimination-rate constant. The slower the elimination, the longer it takes to reach steady state.

Bioavailability

Factors affecting the bioavailability of a drug after its oral administration include incomplete absorption from the GI tract, presystemic clearance (gut metabolism and liver first-pass effect), and degradation of drug in the gut lumen. These factors may vary in their magnitude depending on whether a drug is given as a conventional dosage form or as a CR formulation. Incomplete drug release from a CR dosage form will constitute an additional factor contributing to the loss of drug prior to its absorption. Among these factors, first-pass liver metabolism is particularly susceptible to change when changing the drug input rate.

First-Pass Liver Metabolism

After absorption from the GI tract, the drug must first pass through the liver before it reaches systemic circulation. This is because blood drainage from the entire GI tract, with the exception of the buccal cavity and lower rectum, goes to the liver via the hepatic portal vein. Since the liver is the principal site of metabolism for a number of drugs, a fraction of the absorbed drug may be eliminated through metabolism by the liver before it reaches the general circulation. This fraction is a function of the susceptibility of the drug to liver microsomal enzymes for metabolism and is measured in terms of a parameter called extraction ratio. Because of this presystemic metabolism, which is also referred to as the "first-pass" effect, an oral dose of a drug may have incomplete bioavailability despite its complete absorption from the GI tract.

A number of drugs have been identified as having a significant first-pass effect, and many of these have been shown to obey Michaelis-Menten kinetics in the therapeutic dose range [29]. Factors that affect first-pass metabolism are (1) liver enzyme activity (2) blood flow (3) plasma protein binding, and (4) plasma drug concentration. All of these factors can play important roles, depending on the nature of the drug and its interaction with liver enzymes.

The major difference between conventional and CR oral dosage forms is the rate of drug input into the body. The amount of drug absorbed during any 24-h period is usually comparable. Therefore, if linear kinetics of drug metabolism are involved, one should expect no difference between the pharmacokinetic parameters of the two dosage forms. However, linear pharmacokinetics do not always apply in real situations. One such example is propranolol, which accumulates during repeated oral administration to a greater extent than predicted from its half-life and area under the curve after a single oral dose [30]. This type of nonlinearity is commonly referred to as "dose-dependent kinetic." Such nonlinearity may also arise from other saturable processes arising during the course of drug absorption and disposition [31]. In addition, certain disease conditions, such as renal insufficiency, can also lead to dose-dependent kinetics for certain compounds.

Dose-dependent kinetics can be an important factor in considering the design and evaluation of CR systems. This is because the rate and pattern of drug delivery with a conventional dosage form are considerably different from those with a CR dosage form. Most important among saturable processes from an oral delivery standpoint is the saturable first pass liver metabolism effect. Experimental observations indicating dose-dependent and saturable first-pass metabolism include: (1) increase in dose-normalized bioavailability with increase in dose and (2) decreased clearance at steady state compared to a single dose. A consequence of dose-dependent kinetics is that bioavailability will decrease with a decrease in the rate of absorption after oral administration of the same dose. If one considers that a decreased rate of drug absorption from the GI tract is the primary goal of most CR formulations, drugs showing saturable kinetics will need special attention, and indeed, they may be unsuitable for such formulations.

Michaelis-Menten enzyme kinetics can be employed to better understand saturable liver metabolism. The equation describing the rate of drug metabolism is

$$\text{rate of metabolism} = \frac{V_{max}C}{K_m + C} \tag{8}$$

where

V_{max} = maximum rate of metabolism

C = drug concentration in plasma

K_m = Michaelis-Menten constant measured as plasma drug concentration at metabolism rate of $V_{max}/2$

The K_m value is a measure of the approximate concentration above which saturability becomes evident.

For drugs like phenytoin, which show saturation of liver enzymes at relatively low concentrations (therapeutic concentration), increase in dose results in a disproportionate increase in bioavailability and circulating drug levels because both first-pass metabolism and systemic metabolism (clearance) are saturable. Propranolol and alprenolol show similar dose-dependent behavior [32,33].

Therefore, bioavailability from an oral dose is an important parameter to consider when contemplating a CR dosage form for oral use. Generally, drugs with medium to high extraction ratio and saturable first-pass metabolism make unsuitable candidates for CR. Alternatively, an appropriate change in the release rate may be incorporated into the dosage form to compensate for the increased loss due to first-pass effect. This approach may be possible for drugs with low to medium extraction ratios. Thus, dose-dependent nonlinearity can present a serious limitation for development of oral CR formulations.

Pharmacokinetic Analysis

An important consideration in oral CR formulations is the selection and use of appropriate models to assess in-vivo pharmacokinetic parameters. Most important in this regard is the measurement of the in-vivo release rate and its correlation with in-vitro dissolution profiles. Such information can help evaluate as well as refine oral delivery systems. One can use either a compartment model approach or a relatively recent "noncompartmental" or "model-independent" approach in such studies. In both approaches, the kinetic processes are assumed to be first-order, linear, and irreversible.

The compartment model methods assume that the drug concentration-time profile can be described by one of many pharmacokinetic models. The data are evaluated by using an equation consistent with the assumed model by using either the method of residuals or a nonlinear least-square regression analysis. Standard equations for one- or multicompartment models are used to estimate pharmacokinetic constants, including the absorption rate constant. The problem with model-based methods is that for drugs showing multicompartment kinetics, one cannot be sure about the relative nature of the absorption and distribution rate constants. Additional factors such as drug degradation and metabolism in the gut, gastric emptying, and GI motility can further complicate the analysis.

An alternative to curve-fitting the data is construction of percent absorbed-time plots, which do not require assumption of the order of the absorption rate. The Wagner-Nelson method [34] has been widely used for this purpose, but it gives best estimates for drugs showing one-compartment kinetics only. The Loo-Riegelman method [35] can be applied to linear multicompartment pharmacokinetic models. It requires blood-level data after both oral and intravenous administration.

However, in recent years, model-independent methods based on statistical moment theory have gained popularity for estimating absorption rate constants of orally administered drugs. Noncompartmental analysis assumes input, end elimination, as well as sampling from the central compartment. In addition, these methods assume that metabolism of the drug is exactly the same after oral and intravenous administration, which is not the case for certain drugs such as quinidine and propranolol.

Noncompartmental analysis based on statistical moment theory usually utilizes the area under a plot of drug concentration versus time as the basis for estimating the kinetic parameters. It can be applied to any compartmental model provided that the pharmacokinetics are linear. Its advantage is that it permits a wide range of analysis that is usually adequate to characterize the pharmacokinetics of a drug [36–38].

Mathematically, the moment method considers the time course of in-vivo drug concentration as a statistical distribution function F(t), whose n-th moment can be expressed as

$$t^n F(t) \, dt$$

In pharmacokinetics, F(t) is plasma drug concentration. The first three moments can be defined as follows:

$$AUC = \int c \, dt$$

$$MRT = \frac{\int tc \, dt}{\int c \, dt} = \frac{AUMC}{AUC}$$

$$VRT = \frac{\int t^2 c \, dt}{\int c \, dt}$$

where

> AUC = area under the curve
>
> MRT = mean residence time
>
> VRT = variance of the mean residence time of drug in body
>
> AUMC = area under the first moment curve

From these moments, one can calculate bioavailability and the absorption rate constant as follows:

$$\text{Bioavailability} = F = \frac{AUC_{oral}}{AUC_{i}v} \qquad \text{for equal dose}$$

$$MAT = MRT_{oral} - MRT_{iv}$$

where MAT is the mean absorption time for a first-order drug absorption:

$$MAT = \frac{1}{k_a}$$

given that k_a is the apparent first-order absorption rate constant.

For a zero-order absorption process,

$$MAT = \frac{T}{2}$$

where T is the total time during which the absorption takes place.

Other pharmacokinetic parameters, including clearance, half-life, apparent volume of distribution, and metabolism kinetics can be similarly calculated.

Another example of the statistical moment theory is the method of deconvolution, introduced by Rescigno and Segre [39]. The following scheme described a physical system that transforms input into output:

drug − (A)t → body − (B)t → plasma levels

For an orally administered drug, A(t) represents the rate for in-vivo drug release, and B(t) represents plasma drug concentration. Computing B(t) from A(t) is called convolution, and conversely, computing A(t) and B(t) is called deconvolution. This equation can be written as

B(t) = G(t − T) A(t) dT

The Laplace transform of the above equation is

B(s) = g(s) A(s)

where B(s), g(s), and A(s) are the Laplace transforms of B(t), G(s), and A(t), respectively. A number of numerical deconvolution methods have been reported [40]. A practical example of the application of this method is given in the next section.

Pharmacodynamics

While pharmacokinetic parameters provide useful information regarding the time course of the drug and its metabolites in the body, they may not be representative of the pharmacological response or therapeutic effectiveness of a dosage regimen. This is due, in part, to the fact that the plasma drug concentration is not necessarily at equilibrium with drug concentration at the receptor site. Also, owing to individual variations in drug-receptor interactions commonly observed with drugs other than antibiotics, plasma drug concentration may not be the best way to evaluate the success of therapy. Due to a combination of pharmacodynamic and pharmacokinetic variables, a single dose-response curve does not apply to a population. Factors that contribute to pharmacodynamic variability include intersubject variability in drug-protein binding in plasma, rate and pattern of metabolism for drugs forming active metabolites, drug concentration at the receptor site, affinity and/or activity of drug-receptor interaction, and balance between pharmacological response and toxicity.

IN-VITRO/IN-VIVO CONSIDERATIONS

In-Vitro Considerations

In a conventional oral delivery system, drug content is released within a short period of time and plasma drug levels peak at a given time, usually within a few hours after dosing.

Since the dosage form will encounter gastric content or possibly proximal duodenal content, in-vivo disintegration and dissolution conditions are relatively well defined. In such a case, an in-vitro dissolution profile is based on the fastest possible dissolution rate, and can have a direct correlation with in-vivo bioavailability. But this kind of arrangement is simply not possible with CR systems. For oral CR products, in-vitro testing is not aimed at how fast, but at how uniformly the drug is released. The uniformity of drug release is measured in terms of a predetermined rate of release. The deviations from release rate can be either too slow or too fast from the desired value. Optimal dissolution profile is determined by drug properties, which include the biological half-life, and therapeutic plasma levels of the drug. For drugs with longer half-lives, the initial release period should provide enough drug for a minimum effective plasma drug level. Subsequently, the release rate can drop to maintain drug levels. For drugs with relatively short half-lives, however, release rates may have to be more or less the same throughout due to rapid elimination of the drug from the body. A variety of in-vitro dissolution characteristics of CR dosage forms set them apart from conventional formulations. These include the following:

1. Dissolution is measured in terms of optimum drug release rate, not the fastest release rate.
2. The optimum release rate is usually an intermediate value and is related to the required biologic input function.
3. Dosage forms are designed for different release patterns, e.g., first-order versus zero-order, both with or without a rapid-release component.
4. Disintegration may or may not precede the process of dissolution.
5. In-vitro medium may not adequately mimic the pH, motility, and viscosity variations of the GI tract to which the dosage form will be exposed.

Despite a large variety of CR dosage forms, and variations in drug release rate, kinetic models that describe drug release are generally of two types, i.e., first-order and zero-order. In addition, both these models may have an initial period of rapid drug release conforming to first-order kinetics.

Testing Procedures

Current USP (XXI) guidelines for in-vitro dissolution tests for CR products are limited. CR products are referred to the USP as modified-release dosage forms and are further classified as sustained-release and delayed-release products [41]. Sustained-release systems are those that allow at least a twofold reduction in dosing frequency when compared to the same drug in conventional dosage form. A delayed-release dosage form is one that releases its drug at a time other than immediately after administration, e.g., enteric-coated tablets.

The dissolution test apparati for such formulations are basically the same as those or conventional dosage forms, i.e., the rotating basket method (apparatus 1) or the paddle method (apparatus 2). The following dissolution procedure and interpretation are quoted directly from USP XXI, chapter on drug release <724>, and apply to all modified-release dosage forms. A list of articles subject to extended-release definition has been published in *Pharmacopeial Forum*.

Acceptance Table 1

Stage	Number Tested	Criteria
S_1	6	No individual value lies outside the stated range and no individual value is less than the stated amount at the final test time.
S_2	6	The average value of the 12 units $(S_1 + S_2)$ lies within the stated range and none is more than 10% of labeled content outside of the stated range; and none is less than the stated amount at the final test time.
S_3	12	The average value of the 24 units $(S_1 + S_2 + S_3)$ lies within the stated range, and not more than 2 of the 24 units are more than 10% of labeled contents outside of the stated range; and not more than 2 of the 24 units are less than the stated amount at the final test time.

Extended-release Articles—General Drug Release Standard

Time—The test-time points, generally three, are expressed in terms of the labeled dosing interval, D, expressed in hours. Specimens are to be withdrawn within a tolerance of $\pm 2\%$ of the stated time.

Interpretation—Unless otherwise specified in the individual monograph, the requirements are met if the quantities of active ingredient dissolved from the units tested conform to *Acceptance Table 1*. Continue testing through the three stages unless the results conform at either S_1 or S_2. Limits are expressed in terms of the percentage of active ingredient dissolved. The limits embrace each value of Q_t, the amount dissolved at each specified fractional dosing interval.

Enteric-coated Articles—General Drug Release Standard

Use Method A or Method B and the apparatus specified in the individual monograph. Conduct the *Apparatus Suitability Test* as directed under *Dissolution* (711).
Method A.

Procedure (unless otherwise directed in the individual monograph)—

Acid phase—Place 750 mL of 0.1 N hydrochloric acid in the vessel, and assemble the apparatus. Allow the medium to equilibrate to a temperature of 37 $\pm$ 0.5°. Place 1 tablet or 1 capsule in the apparatus, cover the vessel, and operate the apparatus at the rate specified in the monograph for 2 hours (± 5 minutes).

After 2 hours of operation in 0.1 N hydrochloric acid, withdraw an aliquot of the fluid, and proceed immediately as directed under *Buffer phase*.

Perform an analysis of the aliquot using the *Procedure* specified in the test for *Drug release* in the individual monograph.

Unless otherwise specified in the individual monograph, the requirements of this portion of the test are met if the quantities, based on the percentage of the labeled content, of active ingredient dissolved from the units tested conform to *Acceptance Table 2*.

Acceptance Table 2

Stage	Number Tested	Criteria
A₁	6	No individual value exceeds 10% dissolved.
A₂	6	Average of the 12 units (A₁ + A₂) is not more than 10% dissolved, and no individual unit is greater than 25% dissolved.
A₃	12	Average of the 24 units (A₁ + A₂ + A₃) is not more than 10% dissolved, and no individual unit is greater than 25% dissolved.

In-Vivo Considerations

Despite recent improvements in and general guidelines for in-vitro evaluation of CR formulations, such tests are more useful in ensuring product uniformity than predicting in-vivo performance of a dosage form. In-vivo tests for CR formulations are based on drug concentrations in plasma/blood, or cumulative urinary excretion of a drug and/or its metabolities. These parameters can be treated in a number of ways to obtain different kinds of plots or kinetic constants, which can then be correlated with corresponding in-vitro parameters. However, in-vivo parameters are subject to a number of unpredictable physiological factors which can affect both drug release and absorption. Some of these factors have been outlined earlier in this chapter. Continuously changing conditions, as the delivery system moves through the GI tract, can exert a significant effect on performance of CR systems. A major objective, therefore, should be to design a system that compensates for the effect of the GI environment on the rate of release.

In-Vitro/In-Vivo Correlations

A valid in-vitro/in-vivo correlation is one which allows prediction of the in-vivo performance of a dosage form from its in-vitro dissolution profile. It is desirable to obtain a suitable mathematical equation describing a quantitative correlation of a particular in-vivo variable with an in-vitro variable. In-vivo variables are obtained from concentration-time plots of either plasma drug concentration or urine drug excretion. Parameters such as AUC, absorption rate constant, specific blood levels at a particular time, etc., may also be used for this purpose. Corresponding in-vitro parameters will typically be drug concentrations at specific times.

Most CR formulations are designed to release the drug at a rate slower than its rate of absorption, making the drug release rate limiting. Thus, plasma drug concentrations can be correlated with in-vitro drug release rate as long as this assumption is valid. Generally, model-independent methods, such as the moment method and convolution/deconvolution methods, have been shown to be suitable for this purpose.

The deconvolution method has been utilized successfully in estimating the in-vivo dissolution rate during development of a 24-h dosage form of a nonsteroidal anti-inflammatory agent [42]. Figure 7 shows in-vitro dissolution profiles of two experimental sustained-release tablets which were later tested in humans. The dissolution rates were determined in a spin filter apparatus. Figure 8 shows blood-level profiles of the two sustained-

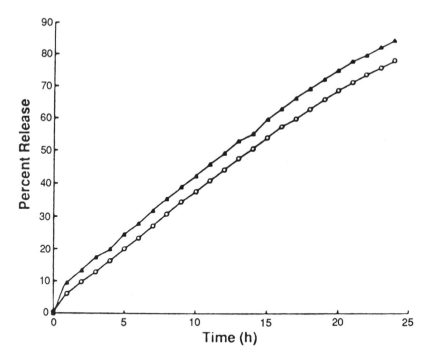

Figure 7 In-vitro profiles of two sustained-release tablets. (From Ref. 42.)

release tablets (A, B) and a fast-release 100-mg tablet given twice at 12-h time intervals. The in-vivo dissolution rate profiles were computed as shown in Fig. 9 using blood-level data and numerical deconvolution. It is obvious from the plots that both tablets show similar in-vivo release rates and complete bioavailability. Figure 10 shows a linear correlation that is observed between in-vitro and in-vivo dissolution rates.

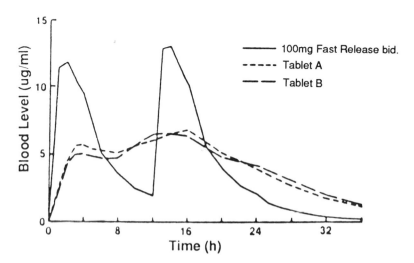

Figure 8 Serum levels of two 200-mg sustained-release (A, B) and 100-mg fast-release tablet. (From Ref. 42.)

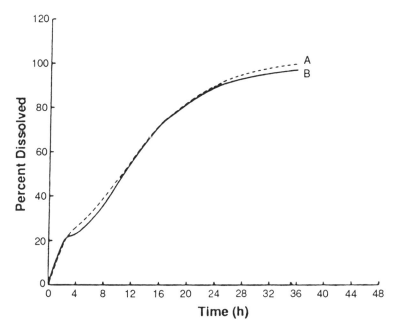

Figure 9   In-vivo dissolution profiles by numerical deconvolution. (From Ref. 42.)

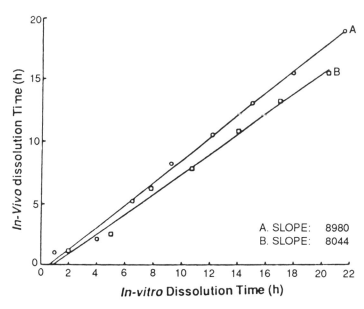

Figure 10   In-vitro/in-vivo correlation of a sustained-release formulation of a nonsteroidal anti-inflammatory drug. (From Ref. 42.)

BIOPHARMACEUTICAL CONSIDERATIONS

The success of a therapy depends on selection of the appropriate delivery system as much as it depends on selection of the drug itself. It is well recognized that a dosage form, whether conventional or CR, can have a significant effect on bioavailability, and indeed make a difference between success and failure of therapy.

For conventional oral dosage forms, a major concern is bioavailability of the drug. Selection of a dosage form is often based on how rapidly and completely the drug is available. In this regard, both from intuition as well as experimental observations, systemic availability of a drug is maximum from an aqueous solution and minimum from a coated tablet, with suspensions, capsules, and tablets showing intermediate bioavailabilities in that order. Deviations from this rule are sometimes observed. The picture, however, is different for CR formulations, where one rarely has a choice of solution or suspension dosage forms. The concern in CR drug delivery is not only bioavailability, but also uniformity of drug input into the body.

The rate and extent of drug absorption from CR dosage forms is determined by the rate of release from the dosage form. This is based on the assumption that absorption from the entire GI tract is efficient enough not to be rate-limiting, although this may not be true in many practical situations. In most situations, a common observation with respect to bioavailability from dosage forms is that in-vitro as well as in-vivo availability from CR formulations is less than from a conventional dosage form. Possible explanations for this observation are as follows:

1. Drug release is not complete from a CR formulation, especially for those designed to release drug for periods longer than 6 h at a low release rate.
2. There is a greater degree of preabsorption degradation and metabolism in the GI tract, particularly for saturable degradation processes or colonic delivery systems.
3. First-pass metabolism for CR formulations may be higher.
4. Drug release may be at a site of poor absorption, e.g., the colon.
5. Fewer dissolution media are available for CR dosage forms, especially in the terminal ileum.
6. There is differential absorption from the GI tract, i.e., drug absorption takes place in a limited area.
7. GI residence of the dosage form may be variable and unpredictable.

As with conventional dosage forms, considerable differences in performance among different CR products of the same drug are frequently observed. The variables responsible for the observed differences are more often in the case of CR dosage forms due to greater complexity involved in their design. Various kinds of devices may involve different mechanisms and kinetics of drug release, and they may be subject to different biological constraints in the GI tract. This will result in considerable variation in plasma drug profiles for similar doses. Theophylline has been formulated in a variety of sustained-release formulations that show significant variations in the rate as well as the extent of absorption from the GI tract. Indeed, some of these products show no reduction in release rate compared to that of a conventional dosage form of theophylline [43,44]. Certain classes of drug, including theophylline, show a considerable intersubject variation in pharmacokinetics. This fluctuation can result from variable absorption or different rates of metabolism. Thus, individualization of dose may be required for such compounds. Indeed, such a study has been done for TheoDur, a sustained-release formulation of theophylline

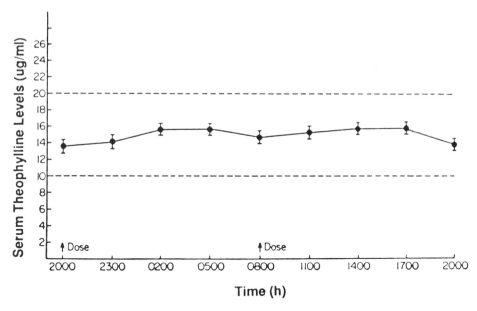

Figure 11 Mean 24-h serum theophylline levels for 20 asthmatic children on a mean TheoDur dosage of 10.0 mg/kg/dose every 12 h. (From Ref. 45.)

[45]. The daily dose needed to produce an average blood level of 15 μg/ml ranged from 6.1 to 16.3 mg/kg, showing an almost threefold difference in dose requirement. The blood levels resulting from such doses were remarkably stable, as shown in Fig. 11 [45]. A peak-to-trough ratio of less than 2.0 was observed in most cases.

An example of the formulation effect on blood levels of CR dosage forms is a report on metoprolol [46]. Two formulations containing the same amount of drug were compared, a conventional sustained-release tablet and an elementary osmotic pump with a release rate of 19 mg/h. As is evident from Fig. 12, the osmotic pump device shows relatively less variation in drug plasma levels [46]. Another interesting observation from this plot is that the plasma metoprolol levels from the osmotic pump show stability between 8 and 18 h, despite the fact that the pump was made to deliver the drug for only 10 h. This leads to the conclusion that either in-vivo drug release is slower than the in-vitro rate, or the drug-release rate from the pump may not be the rate-limiting step; instead, the drug continues to be absorbed long after its delivery has stopped.

No systematic study has been reported that evaluates the performance of different kinds of CR devices for oral use. An understanding of dosage-form interaction with the GI environment is necessary to explain the observed differences. In general, for solid CR dosage forms, one would expect flatter plasma drug levels from zero-order-release formulations, compared with first-order-release formulations. For multiunit formulations, appropriate distribution in the GI tract may be the key to obtaining consistent blood levels. Unless given with a large quantity of food, a dosage form designed to release drug for every 6 h will release some fraction of its drug load in the colon. Knowing that drug absorption from this part of the GI tract may be erratic and incomplete, the observed differences in bioavailability of such dosage forms should not be surprising.

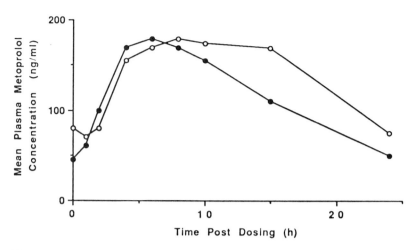

Figure 12   Mean steady-state plasma levels of metoprolol in healthy subjects after repetitive dosing of an osmotic pump (○) and a conventional CR product (●). (From Ref. 46.)

Drug devices that are designed to stay in a particular segment of the GI tract, e.g., bioadhesive systems and large size to delay gastric emptying, must take into account the stability of the drug in that environment. Degradation due to pH or enzymes may reduce bioavailability of such dosage forms. Also, single-unit forms, intended to stay at the pylorus or ileo-cecal junction, may release enough drug in their immediate vicinity to cause local toxicity or irritation. Design must also account for possible bacterial degradation, and variable and poor absorption from the colon and rectum.

STRATEGIES AND DESIGN OF ORAL CONTROLLED-RELEASE SYSTEMS

The design and fabrication of oral CR systems has been reviewed recently by a number of authors [47]. These reviews are extensive concerning the technology involved in the fabrication of such systems and the underlying mechanisms of release. Table 2 lists some of the technological approaches to the fabrication of oral CR systems [1]. The present section will focus on the basic principles involved in conception and development of new approaches to oral CR drug-delivery systems. Emphasis will be on the rationale of design of systems and their interaction with the GI environment.

Most oral CR systems are solids, although a few liquids, all of them suspensions, have recently been introduced. The following classification of such systems is chosen because it includes not only the conceptual approach of design, but some elements of physiology of the GI system as well.

1. Continuous-release systems
   a. Dissolution control
   b. Diffusion control
   c. Dissolution and diffusion control
   d. Ion-exchange resins
   e. Osmotically controlled devices
   f. Slow-dissolving salts or complexes
   g. pH-independent formulations

2. Delayed-transit and continuous-release systems
    a.   Density-based systems
    b.   Size-based systems
    c.   Bioadhesive-based systems
3. Delayed-release systems
    a.   Intestinal release
    b.   Colonic release

The design of oral CR dosage forms is aimed at presenting the drug to the absorptive membrane of the GI tract at a predetermined rate. The majority of such systems rely on dissolution, diffusion, or a combination of both mechanisms, to control drug release rate in the gastrointestinal lumen. Whatever the mechanism may be, as long as the drug release rate from the dosage form is significantly smaller than the rate of drug absorption, there is little drug in solution in the gut. The drug is absorbed by the GI mucosa as soon as it is released. On the other hand, when drug absorption is the rate-limiting step, there is a high concentration of drug in solution in the gut lumen. Plasma drug concentration in release rate-limited processes reflect drug release rate from the dosage form. Before a decision about a system, based on a particular mechanism, is made, drug properties such as solubility, dose requirements, stability, and absorption rate must be considered. These issues have already been discussed in previous sections. Desired in-vivo kinetics of drug release will also play a part in decision making.

Continuous-Release Systems

Dissolution Control

Continuous release for extended periods can be obtained by employing dissolution as the rate-limiting step in drug release. Certain drugs are slow-dissolving due to their intrinsic low aqueous solubility and thus act as natural sustained-release products. Digoxin and griseofulvin are examples of slow-dissolving drugs. A few others, such as aluminum aspirin and benzamphetamine pamoate, produce slow-dissolving forms when they come in contact with aqueous media [48].

For compounds with high aqueous solubility, one needs to reduce the solubility rate by some mechanism. Unless a chemical modification of the drug in question is involved, the approach to control the rate of dissolution of such compounds will be based on either or both of the following techniques:

1. Increase in the stagnant diffusion layer
2. Encapsulation or coating which erodes or slowly dissolves

Stagnatnt-layer control.   If the dissolution process is diffusion layer-controlled, i.e., the rate of diffusion through an unstirred water layer on the solid surface to the bulk of solution is rate-limiting, an increase in the stagnant diffusion layer works effectively. In such a system, flux J (mg/s) is given by

$$J = -D\left(\frac{dc}{dx}\right) \qquad (9)$$

where

D = diffusion coefficient (cm^2/s)

$\dfrac{dc}{dx}$ = concentration gradient from the solid surface to bulk solution (mg/ml/cm)

Table 2 Principles of Technological Possibilities for the Manufacture of Oral Extended-Release Dosage Forms

Method and type of factors used to achieve extended release	Examples of use excipients	Kind of drug release	Kinetics of drug release	Possibilities of release rate regulation	Examples of dosage forms
1. Binding					
(a) Chemical binding (slightly soluble salts or complexes)	Tannic acid, polygalacturonic acid, albumins, pectins	Slow dissolution of salts, esters hydrolysis, complex dissociation	First-order under, some conditions zero order	Selection of salt or complex forming substances	Tablets, capsules, liquid suspensions
(b) Physical-chemical binding	Ion-exchange resins	Ion exchange	First-order	Variation of binding strength depending on chemical stuctures of the resin or adsorbents and the drug	Tablets, capsules, liquid suspensions
	Absorbents	Desorption			
2. Coating					
(a) Insoluble membrane	Ethyl cellulose	Diffusion, partitioning	Zero- or first-order	Membrane porosity and/or thickness	Granules, pellets, microcapsules, film tablets
	Polymers	Diffusion	Zero-order	Membrane porosity and thickness	Oros osmotic pumps
(b) Soluble membrane					
(i) pH-Dependent solubility	Polymers of methacrylic acid and its esters, cellulose acetate phthalate, hydroxypropyl-methylcellulose phthalate	Drug dissolution after coating; disintegration, repeat release	First- or second-order	Substitution and/or polymerization degree, membrane thickness	Multilayer tablets, coated granules or pellets with varying disintegration time, ultiple-unit capsules, liquid suspensions

(ii) Enzyme-dependent solubility	Lipids, proteins	Drug dissolution after coating, disintegration	First-order	Changes in the chemical composition	Coated granules or pellets with varying disintegration time, tablets from mixed granules, multiple-unit capsules
(iii) Liquid membrane	Glycerides and surfactants	Diffusion and partitioning	Zero- or first-order	Oil-to-water phase ratio, droplet diameter in the dispersed phase	Multiple emulsion
3. Embedding					
(a) Hydrophilic carrier (gel-forming base)	Methylcellulose, gelactose mannate, alginic acid or sodium alginate, polyacrylic acid	Slow diffusion from viscous gel, very slow pH-dependent dissolution of the matrix	First- or second-order	Polymerization degree, drug-to-carrier ratio	Multilayer tablets with slow-release cores, capsules
(b) Hydrophobic carrier					
(i) Soluble carrier (digestible base)	glycerides, waxes, fatty alcohols, fatty acids	Release when the surface layer is continuously eroded in the gastrointestinal fluids	$Q = \sqrt{t}$ First- or second-order	Changes in the chemical composition influencing the lipase sensibility, melting point, self-emulsifying properties, drug-to-carrier ratio	Eroding tablets, multilayer tablets with slow-release cores, capsules, liquid suspensions
(ii) Insoluble carrier (non-digestible base)	Polyethylene, polyvinylchloride, polyvinylacetate, waxes, calcium sulphate	Immediate release from the surface, after that, continuous diffusion (leaching principle)	$Q = \sqrt{t}$	Tablet porosity, compression conditions, addition of soluble solids, drug-to-carrier ratio	Matrix tablets

The material flow rate through a unit area A from a dosage form can be defined as

$$J = \left(\frac{1}{A}\right)\frac{dm}{d_t} \tag{10}$$

The gradient dc/dx can be expressed in terms of diffusion-layer thickness, and the concentration gradient across this layer as

$$\frac{dc}{dx} = \frac{C_b - C_s}{h} \tag{11}$$

where

C_b = concentration in the bulk solution

C_s = concentration on the solid surface, which is usually the same as the saturated solution

h = diffusion-layer thickness

The above equation assumes that the concentration gradient across the diffusion layer is linear.

Thus, the rate of material flow will be

$$\frac{d_m}{d_t} = -\left(\frac{DA}{h}\right)(C_b - C_s) = kA(C_s - C_b) \tag{12}$$

where $k = D/h$ = intrinsic dissolution rate constant.

If A, D, h, and the concentration difference remain constant, the release rate will be constant. In practice, however, all of these parameters may change continuously, especially surface area.

For release rate from a diffusion layer-controlled system, the following general equation may be more useful:

$$\frac{M_t}{M} = 1 - \left(\frac{1 - K_0 t}{C_0 a}\right)n \tag{13}$$

where

M_t = amount released at time t (mg)

M = total amount released (mg)

a = half-thickness of dosage form (cm)

n = constant shape factor: n = 3 for a sphere, n = 2 for a cylinder, and n = 1 for a slab

Example: The intrinsic dissolution rate constant of a drug is 5×10^{-5} cm/s. Calculate the rate of dissolution in milligrams per hour from a tablet of surface area 2.5 cm^2 under sink conditions. The solubility of the drug is 50 mg/ml.

Using Eq. (12):

$$\frac{dm}{dt} = (5 \times 10^{-5}\text{ cm/s})(3600\text{ s/h})(2.5\text{ cm}^2)(50\text{ mg/cm}^3)$$

$$= 22.5\text{ mg/h}$$

Matrix dissolution control is the most commonly employed technique to achieve dissolution control. The rate of drug availability is controlled by the rate of penetration of the dissolution fluid into the matrix. This rate of penetration of the dissolution media can be controlled by the porosity of the tablet matrix, the presence of hydrophobic additives, and the wettability of the tablet. The porosity of the tablet can be altered by changing the compression force in a tablet. Size and shape of particles can also affect porosity of the dosage form.

Wax-impregnated tablets are examples of matrix dissolution systems. Wax-impregnated particles can be prepared either by aqueous dispersion or by a congealing process. The aqueous dispersion method simply involves spraying or placing the wax-drug mixture in water and collecting the resulting particles. Alternatively, one may use the spherical agglomeration technique, where drug particles are suspended in an aqueous media, stirred with wax at high temperature, and then cooled while stirring. Particle size can be controlled by the speed of stirring. In the congealing method, drug is mixed with wax material and either spray-congealed or congealed and screened.

A variety of wax matrix materials can be used for such formulations. Among these are hydrogenated castor oil and carnauba wax. Important factors affecting drug release are the physical properties and chemical composition of the wax used, and the composition of the dissolution media. Surfactants are typically added to improve the release rate. Sorbitan monostearate is used in a concentration range of 0.1 to 5% for this purpose.

A slow-release procainamide tablet, releasing drug through matrix dissolution, has been compared to conventional dosing [49]. Wax matrix tablets showed less plasma-level fluctuations of procainamide and could be administered every 8 h to keep drug concentration within therapeutic range. There is a good correlation between bioavailability and the in-vitro dissolution profile.

A major disadvantage of stagnant layer-controlled systems is that they fail to give a zero-order release; i.e., release rate progressively decreases with time. This is a result of an increased diffusional distance and decreased surface area at the penetrating solvent front. Geometry changes can help reduce this problem to some degree. Also, the drug release rate is influenced by the nature of the GI content, particularly by the viscosity of the dissolution media.

Encapsulation dissolution control. The basic approach in encapsulation is the coating of drug particles with a slowly dissolving material. Coated particles can be compressed directly into tablets or placed into gelatin capsules. Since the time required for dissolution of the surface coat is a function of coat thickness and its aqueous solubility, good control of the release rate can be achieved. One can obtain a repeat or continuous release of drug by using granules of varying coating thickness.

A wide range of drugs have been formulated as sustained-release coated granules and compressed into tablets. These include antispasmodic-sedative combinations, phenothiazines, and aspirin [48].

There are several ways to prepare drug-coated beads or granules. Usually inert beads are coated with drug, followed by coating with a slowly dissolving material. It is common practice to include some uncoated drug particles in the dosage form to give an initial priming dose.

An illustration of this approach is the formulation of dextroamphetamine sulfate. The release rate of dextroamphetamine sulfate could be effectively controlled by varying the wax coating thickness [50]. Also, by using a selected blend of different coating materials, a desired rate of release can be obtained.

Microencapsulation is another approach which is analogous to encapsulated dosage forms, except that it involves a much smaller size of particle. This process is normally used to convert liquid or semisolid materials into solid particles by coating them with another solid material. It appears that a portion of drug becomes embedded in the coating during this process, and this drug is provided in a sustained fashion as the coat dissolves.

Coacervation is one of the commonly employed techniques to microencapsulate materials. This process utilizes the interaction of two oppositely charged polyelectrolytes in water to form a polymer-rich coating solution called a coacervate. This coacervate encapsulates the liquid or solid to form an embryo capsule. Other techniques used for microencapsulation include interfacial polymerization, electrostatic method, precipitation, hot melt, salting out, and solvent evaporation.

The thickness of a microcapsule coat can be adjusted from less than 1 μm to 200 μm by changing the amount of coating material.

Microencapsulation has an additional advantage in that sustained drug release can be achieved with better GI tolerability. Microencapsulated aspirin and potassium chloride are examples of better GI tolerance. Core of the microcapsules can consist of pure drug, buffered drug mixtures, or wax-core formulations. A dual approach to dissolution control is possible by using a slowly dissolving coat on wax-core beads.

Diffusion Control

Diffusion-controlled systems fall into two basic categories:

1. Reservoir devices
2. Matrix devices

Reservoir devices.   In reservoir devices, a water-insoluble polymeric material encases a core of drug. Drug release through the system occurs by partition through the coating membrane. Drug penetrates the membrane and diffuses to the other side, and eventually into the dissolution media.

The rate of diffusion across the membrane is governed by Fick's law:

$$J = -D \frac{dc}{dx} \tag{14}$$

where

J = flux

D = diffusion coefficient

$\dfrac{dc}{dx}$ = change in concentration with distance x within the membrane

At steady state:

$$J = D \frac{c}{l} \tag{15}$$

This is an integrated equation where l is the diffusional path length, which in an ideal case is the membrane thickness. In terms of amount of drug released:

$$\frac{dm}{dt} = ADK \frac{c}{l} \tag{16}$$

where

A = area

K = partition coefficient of drug between solution and membrane

K is defined as the drug concentration ratio in the membrane and core. This is an important parameter controlling the rate of drug release. In such a system it is relatively easy to keep all parameters more or less constant so that a zero-order rate can be achieved. However, in an in-vivo situation, deviations are usually observed.

Example: Calculate the rate of flux in milligrams per minute from a diffusion-controlled device when A = 1.4 cm^2, D = 10^{-6} cm^2/s, c = 20 mg, l = 50 μm, and k = 5. Using Eq. (16):

$$\frac{dm}{dt} = (1.4 \text{ cm}^2)(10^{-6} \text{ cm}^2/\text{s})(60 \text{ s/min}) \frac{(20 \text{ mg/cm}^3)(5)}{0.005 \text{ cm}}$$

$$= 1.68 \text{ mg/min}.$$

Insoluble coatings can be applied to a drug core by a variety of techniques. Commonly used approaches are press coating and air-suspension coating. For smaller particles intended for tablets or capsules, microencapsulation techniques are generally used. Uncoated drug may be enclosed in the system to provide an initial rapid dose.

Figure 13 shows the release characteristics of a reservoir dosage form for salicylic acid [51]. Varying rate of release can be obtained depending on the membrane thickness. The coating material used in this case was hydroxypropyl cellulose.

Several parameters are crucial in maintaining a constant rate of drug release from the reservoir system. These include:

1. Polymer ratio in the coating
2. Film thickness
3. Hardness of microcapsules

Among these factors, film thickness is an easily manipulated parameter to obtain the desired release rate. Figure 14 shows the release rate of clofibrate as a function of the wall thickness of gelatin-sodium sulfate microcapsules [52].

Hydroxypropyl cellulose and polyvinyl acetate are commonly used polymers used as an insoluble permeable coat. A laminated diffusion controlled drug-hydroxypropyl cellulose matrix coated with hydroxypropyl cellulose and polyvinyl acetate has shown zero-order drug release [53]. The drug-containing core serves as a reservoir exerting some control over the duration of drug release, while the coat serves as a diffusion-controlled, rate-limiting membrane. Ratio of polymers in the coat determines the release rate of drug.

Drug particles can be coated in a fluidized bed with aqueous dispersions of polymers. These dispersions typically contain additives known as plasticizers to help the polymer stick to pellets uniformly. Dibutyl sebacate is a commonly used plasticizer in such formulations. Permeability additives may be needed to enhance the release rate of drugs.

Cellulose derivatives contain a few carboxyl groups. Therefore rate of diffusion through the membranes tend to be pH-dependent. This can make the drug release rate different in the stomach and intestines. However, the differences in permeabilities are generally small.

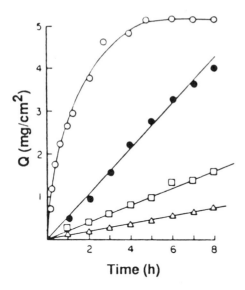

Figure 13   Drug release from films containing 20% salicylic acid in hydroxypropyl cellulose as the reservoir layer. ○, no membrane layer; (●), 0.164-mm hydroxypropyl cellulose-polyvinyl acetate (8:2) membrane; □, 0.204-mm hydroxypropyl cellulose-polyvinyl acetate (6:4) membrane; △, 0.164-mm hydroxypropyl cellulose-polyvinyl acetate (6:4) membrane. (Plotted with data obtained from Ref. 51.)

Some parameters controlling drug release from pellets of guaipnesin coated with an aqueous ethyl cellulose dispersion have been reported to include thermal post-treatment of the coating, plasticizer content, and the pH and ionic strength of the dissolution medium.

Different types of poly (vinyl alcohol) can be cross-linked to varying degrees to control the degree of swelling and hence the rate of drug release [54]. One such study reported cross-linking of three types of poly (vinyl alcohol) by glutaraldehyde to form water-swellable materials possessing a three-dimensional, molecular network [55]. With proxyphylline and theophylline as model drugs, release rates could be controlled by varying the degree of cross-linking of the polymers. Figure 15 shows the effect of type of alcohol on the release of proxyphylline from micromatrices at a fixed degree of cross-linking and drug load. Different rates of release of theophylline could be obtained by varying the cross-linking ratio of micromatrices produced from Elvanol 71-30 with a fixed drug load of 4% as shown in Figure 16.

Matrix devices.   The matrix approach employs a system where the drug is compressed with a slowly dissolving or insoluble polymer. The rate of drug availability is controlled by the rate of penetration of the dissolution medium through the matrix and to the surface of the unit. As the drug dissolves, the diffusional path length increases because the polymer matrix is insoluble. With proper design of the system, an initial loading dose can be provided from the drug particles on or near the surface of the tablet. Once pores have

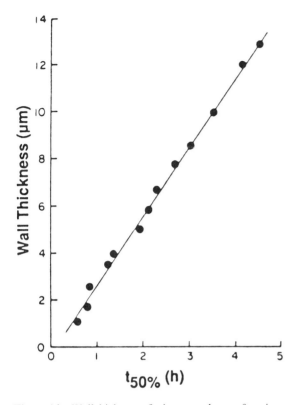

Figure 14 Wall thickness of microcapsules as a function of in-vitro $t_{50\%}$ release time of clofibrate. (From Ref. 52).

been created, drug release will slow down. Obviously, the rate of release will not be zero-order, as may be desired, because, as the diffusional length increases, the rate of dissolution falls. However, if one uses a slowly dissolving polymer matrix, where the matrix itself dissolves at a certain rate so as to keep the diffusional length more or less the same, it can result in a zero-order release.

In such a system, the rate of drug release is dependent on the rate of drug diffusion but not on the rate of solid dissolution. Higuchi's equation can be used to express the release rate from such systems:

$$Q = \left(\frac{DE}{T(2A - EC_s)C_s t}\right)^{\frac{1}{2}} \tag{17}$$

where

       Q = drug released in g per unit surface area

       D = diffusion coefficient of drug

       E = porosity of the matrix

       T = tortuosity of the matrix

       C_s = solubility of drug in release medium (g/ml)

       A = concentration of drug in the tablet (g/ml)

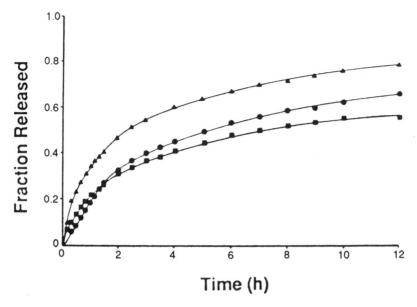

Figure 15 Effect of the type of poly(vinyl alcohol) on the release of proxyphylline from micromatrices cross-linked at a ratio, X, of 0.20 and loaded with 14% drug. Type of PVA: Elvanol 71–30 (▲); Mowiol 40–88 (■); Mowiol 66–100 (●). (From Ref. 55.)

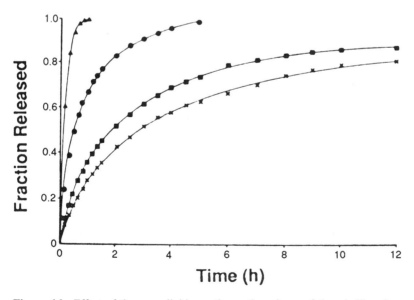

Figure 16 Effect of the cross-linking ratio on the release of theophylline from micromatrices produced from Elvanol 71–30 and loaded with 4% drug. Cross-linking ratio, X: 0.05 (▲); 0.10 (●); 0.15 (■); 0.20 (X). (From Ref. 55.)

The following assumptions are made in deriving this equation:

1. A pseudo-steady state is maintained during release.
2. $A >> C_s$; that is, excess solute is present.
3. $C = 0$ in solution at all times (perfect sink).
4. Drug particles are much smaller than those in the matrix.
5. The diffusion coefficient remains constant.
6. No interaction occurs between the drug and the matrix.

Since the goal is to keep all the parameters constant, for the purpose of treatment, Higuchi's equation can be reduced to:

$$Q = kt^{1/2} \tag{18}$$

Therefore, a plot of amount of drug released versus the square root of time should be linear if the rate of drug release is diffusion-controlled. The rate of release in a matrix system can be altered by varying any of the variables in the equation.

Three major types of matrix systems are fatty, plastic, and hydrophilic matrices. Fatty matrices consist of waxes and are generally prepared by dispersing the drug and excipients in molten wax, followed by congealing and coating. Nondigestible hydrophilic gums such as hydroxypropylmethylcelluose or sodium carboxymethylcellulose are mixed with drug and compressed to make hydrophilic matrix tablets. When such a tablet is exposed to an aqueous medium, rapid drug release occurs initially, but release slows as the gum swells. Such formulations generally have poor control over release, and their performance varies considerably with varying GI conditions.

Diffusion- and Dissolution-Controlled Systems

Some systems employ diffusion as well as dissolution control over the drug release rate. The dosage form consists of a drug core encased in a partially soluble membrane. When placed in the appropriate mileu, the soluble part of the membrane dissolves away, creating pores in the remaining coat. This allows for entry of aqueous media into the core and allows dissolution of the drug. An example of such a coating would be a polymer coating consisting of ethylcellulose and methylcellulose. The latter dissolves, leaving the ethylcellulose coat intact. The release profile from such a system can be described by the following equation:

$$\text{release rate} = AD \frac{(C_1 - C_2)}{1} \tag{19}$$

A = surface area

D = diffusion coefficient of drug

C_1 = concentration of drug in the core

C_2 = concentration of drug in the dissolution medium

1 = diffusion path length

Surface area in such a system can be easily controlled by varying the fraction of soluble material in the coating. Also, by incorporating more than one soluble material with different rates of solubility, one can increase the release rate after a certain period of time. This can be useful in oral systems designed to deliver for more than 12 h. Since colonic absorption may not be as efficient and complete as intestinal absorption, an increased release rate could compensate for reduced absorption to maintain a constant input of drug into the body.

Ion-Exchange Resins

Polymers containing groups of exchanging ions can be used to obtain extended delivery preparations of ionizable drugs. These polymers, also known as ion-exchange resins, contain ionizable groups. Thus, they may contain acidic-reacting groups such as phenolic, carboxylic, or sulfonic (cation ion-exchange resins), or basic groups, such as amino or quaternary ammonium groups (anion-exchange resins). These reacting groups of ion-exchange resins can be used to bind drugs, basic drugs to acidic cation ion exchangers, and acidic drugs to basic anion ion exchangers.

The following simplified equation describes the release of a basic primary amine drug from a cation exchanger when in contact with a dissolution medium containing an ionic compound XY:

$$(R—SO_3—H_3N^+—R) + (X^+Y^-) = (R—SO_3—H^+) + (H_3N^+—RY^-)$$
resin-drug complex   resin   active drug

Drug release, especially with strong acidic groups, is primarily a function of the ionic strength of the gastrointestinal fluid, with pH having little effect other than ionic. The extended release rate of drug is a result of slow diffusion of drug molecules through the resin particle structure. Release rate can be modified by alteration of the resin particle dimensions and chemical composition of the resin. The release rate can be further controlled by coating the drug-resin complex using one of the encapsulation processes described earlier. A mixture of the coated and uncoated complex can then be used to obtain a desired rate of release. A drug-resin complex of phenyl propanolamine administered every 12 h for 2 weeks has been shown to provide the same plasma concentrations as a solution of the drug administered every 5 h [56]. Prolongation of therapeutic effects have also been reported for noscaine, an antitussive, when the drug-resin complex was administered [57].

The preparation of drug/ion-exchange complex can be accomplished by either incubating the resin with the drug solution or passing drug solution through a column loaded with the appropriate resin. In both processes, enough time is allowed for the drug to displace the suitable ion from the resin. Resins are used in their salt form because they often swell as salts when placed in aqueous medium. This facilitates drug permeation into the resin.

Osmotically Controlled Devices

Osmotically controlled systems utilize osmotic pressure as the driving force to release drug at a constant rate. A cross-sectional view of an elementary osmotic pump is shown in Fig. 17 [47]. It consists of a drug core surrounded by a semipermeable membrane coating which has one orifice [58,59]. Water imbibed from the environment crosses the membrane at a controlled rate and causes the drug solution to exit through the delivery orifice. it delivers drug at a rate independent of gastrointestinal pH and motility. The delivery rate is controlled by osmotic properties of the core as well as membrane area, thickness, and permeability to water.

The mathematical relationship used to describe the drug release rate from an osmotic system can be written as

$$\left(\frac{dm}{dt}\right)_{t=T} = \frac{kA\pi S}{h} \tag{20}$$

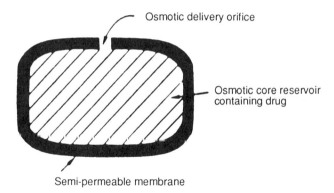

Osmotic delivery orifice

Osmotic core reservoir containing drug

Semi-permeable membrane

Figure 17 Schematic representation of an elementary osmotic pump. (From Ref. 47.)

where

  m = released amount of drug (mg/h)

  t = time from zero to T (h)

  T = time at which the entire drug core has gone into solution (h)

  k = osmotic permeability constant of the membrane (cm^3/h)

  A = area of the membrane (cm^2)

  h = membrane thickness (cm)

  π = total osmotic pressure

  S = drug solubility (mg/ml)

From the above equation, it is clear that the drug release rate will be zero-order between the starting time and the time when the entire drug core dissolves. After time T, drug release follows first-order kinetics. Generally, 80% of the drug is released at a constant rate and the rest as first-order release.

Example: An osmotic pump has a total surface area of 2.5 cm^2 and a membrane thickness of 200 μm. The solubility of the contained drug is 200 mg/cm^2. Calculate the zero-order delivery rate in milligrams per hour. The value of kπ is 5.0×10^{-2} cm^2/h.

Using Eq. (20):

$$\frac{dm}{dt} = \frac{(5.0 \times 10^{-2})(200)(2.5)}{0.02} = 12.5 \text{ mg/h}$$

Drug solubility is an important parameter in determining release rate. For compounds with low solubility, the osmotic pressure developed in the system may not be enough to ensure the desired drug release rate. In such situations, one can use highly soluble substances, generally salts, which serve to increase the osmotic gradient across the membrane and increase the release rate. Thus, the pump core may consist of pure drug, or drug and other additives, to achieve the desired release rate. Potassium chloride and mannitol are commonly employed osmogens to improve the release rate for poorly soluble drugs. On the other hand, certain drugs may be too soluble to provide a saturated solution for a long time. In such cases, the saturated solution is diluted too rapidly, causing a premature and

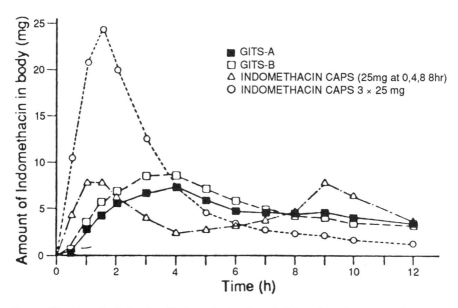

Figure 18 Mean body levels of indomethacin after administration of two osmotic pumps (GITS) each containing 75 mg with different release rates, three indomethacin capsules taken together, and three indomethacin capsules taken at 0, 4, and 8 h. (From Ref. 62.)

rapid fall in the pumping rate. For compounds with high solubility, the choice of a less soluble salt or ester can be a solution.

From a technology standpoint, osmotic devices are little more than coated tablets. A compressed core is coated with the water-insoluble but permeable polymer, and a small hole is drilled through this coating on one side of the tablet [60].

One such system has been developed for indomethacin [61,62]. As shown in Fig. 18, plasma-level excesses encountered with capsules were avoided by using osmotic devices. Bioavailability of indomethacin from osmotic pumps was 85% relative to capsules.

Typical membrane materials for osmotic devices are derivatives of polysaccharides, which include cellulose esters and cellulose ethers. Examples are various cellulose acylates, cellulose acetoacetate, etc. In addition to the polymeric material for the wall, the coating solution usually contains a stabilizing agent that imparts physical and chemical integrity to the wall, a flux enhancer to achieve the desired rate of fluid permeation, a plasticizer that gives flexibility to the wall, and a dispersant to blend the materials well. The final coating membrane must be rigid and capable of maintaining the structural integrity of the drug delivery system during the course of drug release.

Several modifications of the elementary osmotic pump pressure-controlled drug delivery system have been developed. One such system consists of two compartments separated by a movable partition. The osmotically active part imbibes water, swells, and moves the partition to expel the contents of the other compartment. Another modification is a pump without an orifice so that the osmotic pressure simply ruptures the device to release the contents as a bolus. By controlling membrane permeability, such a device could be used to target certain areas of the GI tract.

Osmotic devices can also be designed as multiunit dosage forms. Such formulations will consist of relatively small particles of drug core, coated with a water-permeable membrane, and dispensed in a capsule. A delivery orifice can be made by either laser drilling as in an osmotic tablet, or by using a channeling agent in the coat which dissolves in the dissolution media to create tiny holes. Such devices have the potential of being less irritating, especially when strongly GI-irritating additives or drugs such as potassium chloride are used. Although osmotic devices are essentially independent of the GI environment, they may be unpredictable in a high-viscosity region such as the colon.

Slow-Dissolving Salts or Complexes

A salt or complex of drug that is only slightly soluble in gastrointestinal fluids can provide an extended release of drug without further control over its release rate. For such a function amine drugs can form slightly soluble salts with tannic acid [63]. The process of complex formation is usually a simple acid-base reaction, as in the case of amines and tannic acid. Solutions of both compounds in suitable solvents are mixed together and the resulting complex precipitated by the addition of another solvent or salt.

Drug-polymer complexes can be employed to provide extended release of drug. The complex releases the drug molecule either due to its degradation, as in the case of certain dye complexes with dextran, or simply due to equilibrium because drug is not bound covalently to the polymer.

pH-Independent Formulations

Since pH in the GI tract varies considerably and continuously as the formulation moves through it, pH-independent formulations are particularly attractive for oral use. These formulations are prepared by blending an acidic or basic drug with one or more buffering agents; e.g., primary secondary, or tertiary salts of citric acid, granulated and coated with appropriate materials. These materials are permeable to GI fluids so that dissolution can occur through the dosage form, but the dosage form cannot disperse and lose buffer. When gastrointestinal fluid passes through the membrane, buffering agents adjust the pH to an appropriate, predetermined, constant pH at which the drug dissolves and permeates out at a constant rate regardless of the external pH.

The buffer ingredients chosen must be compatible with the drug and other excipients and should also be physiologically acceptable. The amount of buffer must be sufficient to ensure a buffer effect throughout the drug-release period. The proportion of buffer material and drug depends on the relative permeability of both through the coating membrane.

Delayed-Transit and Continuous-Release Systems

As discussed previously, the length of in-vivo delivery by oral CR products is severely limited due to a short GI-transit time of solids and liquids. In addition, GI transit time tends to show considerable inter and intra-subject variation. This can also make drug delivery both variable and unpredictable. As a result, most oral dosage form are limited to a 12-hour period.

Several efforts have been aimed at prolonging residence time of the delivery devices in the GI tract. Given the nature of GI motility, the only viable approach appears to be to delay gastric emptying; because once a dosage form is emptied from the stomach, little can be done to retard its movement through the intestine. Indeed, most approaches have been aimed at delaying gastric emptying, although success in this regard has been limited to date. Such devices would use any of the mechanisms discussed so far to control the rate of the drug delivery except that they will be modified to stay in the GI tract for longer periods of time. Since the stomach is the most likely target, it is obvious that the drug in question must be stable to gastric contents. Some of these approaches to prolong GI residence time are discussed in this section.

Density-Based Systems

Results of studies using variable-density dosage forms or pellets have been conflicting. The basic approach will be either high- or low-density pellets.

High-density approach. In the high-density approach, the density of the pellets must exceed that of normal gastric contents, i.e., approximately 1.2 g/cm^3, and therefore should be more than 1.4 g/cm^3. Claims have been made about longer gastric as well as intestinal residence times of such pellets when density is increased from 1 to 1.6 [64]. From studies done in this laboratory, however, such delays in gastric emptying or intestinal transit were not observed for pellets of densities of up to 2. In fact, glass beads with a density of 4 were emptied from the stomach in the same manner as pellets with a density of 1.

Low-density approach. The low-density approach forms the basis of formulations known as buoyant tablets or capsules. The approach is based on the assumption that a formulation with a density less than that of gastric contents will float on the surface of the gastric content and thus escape gastric emptying. It sounds reasonable in principle but neglects the basic physiology of gastric emptying. Gastric fluid empties fast, usually in a matter of minutes, and one would have to continuously drink prohibitively large volumes of water in order to keep enough volume in the stomach to prolong retention. Also, gastric motility would make it impossible for any device to stay afloat, regardless of its density.

Certain low-density materials such as polystyrene may be used for such systems. A modification of low-density materials may be a drug reservoir containing entrapped air to make it lighter than water, as shown in Fig. 19 [47]. From our basic knowledge of the process of gastric emptying, it seems unlikely that a density-based system will be viable.

Size-Based Systems

Studies have consistently shown that the size of a dosage form administered in the fasted state has little effect on its transit time through the GI tract. In the fed state, however, dosage forms of size greater than 2 mm show a longer transit time, the difference being due entirely to delayed gastric emptying. In order to achieve delay in gastric emptying

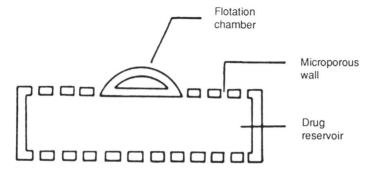

Figure 19 Schematic representation of a drug delivery system with flotation chamber. (From Ref. 47.)

long enough to allow once-a-day dosing, the dosage form has to be 2.5 cm or larger to prevent it from passing through the pylorus. Degradation of the device, after a certain period of time, will enable it to pass through. Such dosage forms may not be practical to swallow, unless they are made to swell or somehow inflate in the stomach.

Bioadhesive Systems

Bioadhesives are materials that can bind to a biological membrane and are capable of being retained on that membrane for an extended period of time. This binding, which usually takes place due to interfacial forces between two surfaces, can be added directly to the membrane surface (cell layers), or to a coating on the membrane surface, such as the mucin layer. The bioadhesive material itself may be biological or nonbiological in nature and source, although in a drug-delivery context, it is usually a nonbiological macromolecular or hydrocolloid material. The term "mucoadhesive" is commonly used for materials that bind to the mucin layer on a biological membrane, but throughout this section, the general term bioadhesive will be used. Besides acting as platforms for sustained-release dosage forms, bioadhesive polymers can themselves exert some control over the rate and amount of drug release, and thus contribute to the therapeutic design of such systems.

A bioadhesive delivery system that adheres to the stomach will be able to provide a continuous dose of drug into the intestine for extended periods of time. However, there are a number of problems, as listed below, associated with development of a suitable adhesive for the stomach.

1. Gastric motility will be a dislocating force for the adhesive. This motility is particularly strong during phase III of the fasted state. During the fed state also, the stomach is in a state of continuous motility, with substantial retropulsive forces acting, particularly in the antral area. In addition, the presence of food may make it difficult for such polymers to attach to the gastric mucosa. Only bioadhesives that bind strongly enough to withstand these shear forces will be practical.
2. Most adhesive polymers studied thus far actually attach to the mucin layer on the mucosal membrane. In the stomach, the mucin turnover rate is substantial, in both the fed and fasted states. Thus, the adhesive will attach to mucus, and be detached along with the mucus when it is released from the membrane. Further attachment of the polymer may not be possible because all the active binding sites on the adhesive will be covered with mucin.
3. The pH of the stomach, which normally ranges between 1.5 and 3.0, may not be suitable for bioadhesion. This is not the case for the polyacid polymers such as cross-linked polyacrylic acid, where the predominant mechanism of bioadhesion is through hydrogen bond formation.
4. Unlike areas such as the buccal cavity, the GI tract is not directly accessible to place an adhesive system on the mucosa. In the absence of a mechanical force to achieve the initial attachment, such systems may have trouble attaching to the membrane.

However, all of these problems can be overcome, either by designing suitable polymers, or by incorporating certain ingredients in the dosage form which will modify conditions in the immediate vicinity of the dosage form to maximize bioadhesion. One approach would be to develop adhesive polymers that attach to the epithelial membrane, instead of mucin. Incorporation of a mucolytic agent in the formulation may create a local mucosal-free surface and attach to it, although it will raise the question of causing physical insult to the membrane or making it more susceptible to attack by acid and enzymes in the stomach.

Similar problems can be anticipated in the intestine, but the pH may be more helpful in this region. The key to success of a bioadhesive polymer in these areas seems to lie in an understanding of the adhesive phenomenon at a molecular level, to a degree that suitable adhesives can be designed to attach to specific areas in the GI tract. This area of research is ongoing and needs to be pursued vigorously.

Delayed-Release Systems

Delayed-release systems for oral controlled delivery are the ones aimed at delivering drug to a particular area of the Gi tract, instead of delivering the drug continuously immediately after ingestion. This site-specific delivery can be aimed at systemic absorption, as in case of enteric-coated tablets, or for local effects. Certain disease conditions of the colon and rectum could be treated by delivering drugs specifically in the desired area. These systems can provide one or more of the following advantages over other oral CR systems:

1. Bypass areas of potential drug degradation, e.g., the stomach for acid-labile drugs, the stomach and jejunum for peptidase-labile drugs
2. Achieve local effects in the lower GI tract without much systemic absorption or side effects
3. Reduce GI discomfort in the upper area
4. Deliver drugs to a specific absorption site to achieve a high concentration at the absorptive membrane, e.g., delivery to Peyer's patches or colon bacteria

Delayed-release devices can be divided into two categories, intestinal-release and colonic-release devices.

Intestinal Release

Enteric-coated tablets are examples of the intestinal-release approach. This approach is usually used for acid-labile drugs. In the case of aspirin, prevention of gastric irritation is the aim. However, enteric-coated formulations tend to be unpredictable in their bio-availability. Enteric-coated erythromycin tablets are well known for their unpredictable and variable bioavailability. In addition to protection from stomach acid, a drug can also be protected from most of the intestinal enzymes if it is released in the terminal ileum. In this area, additional routes of absorption could be utilized to deliver certain drugs such as macromolecules via Peyer's patches, and very hydrophobic drugs via the lymphatic route.

Peyer's patches, which are organized mucosal lymphoid tissues of the gut, play an important role in regulating the immune response to orally presented antigens. They are generally larger and more numerous in the distal than in the proximal small intestine and are usually present on the antimesenteric circumference of the intestinal wall. Their size and number vary from species to species, with as many as 100 in humans [65]. Peyer's patches consist of a collection of lymphoid follicles that occupy the full thickness of the small intestinal mucosa.

It is well documented that Peyer's patches are able to internalize particulate matter, bacteria, and marker proteins [22]. Both soluble and colloidal substances enter Peyer's patches by vesicular transport through specialized epithelial cells. Some of the cells covering Peyer's patches have microfolds and have been called microfold or "M" cells. These cells have been demonstrated to be involved in antigen uptake, and serve as an explanation for uptake of high-molecular-weight soluble and colloidal proteins.

Due to their capability of absorbing large molecules, Peyer's patches present a potential site for delivery of macromolecules. The exact nature of the surface characteristics of these patches is not yet fully understood, but if adhesive systems are to be devised to attach to or around them, a successful delivery system for large molecules may emerge.

Lymphatic absorption presents a viable route of absorption for compounds with certain characteristics, chiefly the hydrophobicity or partition coefficient of the drug. Factors controlling lymphatic uptake of drugs are not well understood yet, but the partition coefficient of the compounds seem to play a dominant role. It appears that a compound has to have a very high partition coefficient in order to be taken up predominantly by the lymphatic route. Most absorption into the lymphatic system is via two mechanisms, chylomicrons and large-molecule uptake by pinocytosis. Chylomicrons are small spherical particles made up almost entirely of dietary fat and cholesterol and are specifically internalized into the lymphatic vessels. Thus, any compound that can be incorporated into chylomicrons will also be taken up by the lymphatic system. Since lymphatic drainage does not go through the liver, drugs absorbed by this route will not be subject to first-pass liver metabolism.

Colonic Release

Despite a small absorptive surface area, the potential for delivery through the colonic mucosa still exists because the desired rate of absorption from CR formulations is generally not very high. Another factor in drug absorption through the colon is the physical nature of the luminal content. Once past the ileo-cecal junction, gut content thickens quickly and considerably due to increased absorption of water. This puts an additional constraint on the dosage form from a drug-release standpoint. Although no systematic study has been reported to date to evaluate drug diffusion through the viscous colonic contents, there is little doubt that both the rate and the extent of drug release are compromised.

There are basically two approaches toward delivering drugs through the colon: (1) use of bioerodible polymers to protect drug during its passage through the upper GI tract, and (2) use of prodrugs that are activated by bacterial degradation or metabolism. The use of bioerodible polymers to control the release of drugs is based on a pH gradient which exists in the GI tract or on high levels of enzymatic activity in the lower GI tract.

Copolymers of methacrylic acid and methyl methacrylic acid, and cellulose acetate phthalate are examples of pH-sensitive bioerodible polymers. They have been used to coat 5-aminosalicylic acid, an anti-inflammatory agent, for its selective delivery to the colon to treat inflammatory bowel disease [66].

Copolymers of styrene and hydroxyethylmethacrylate, cross-linked with divinyl-azobenzene, can be designed to be susceptible to cleavage by the azo-reductase activity of the colonic microflora. Through the use of such polymers, attempts have been made to deliver insulin and other peptides, including vasopressin, via colonic absorption [67].

An example of a colon-specific prodrug is sulfasalazine, an azo compound degraded by the azo-reductase activity in the colon. One of its degradation products, 5-aminosalicylic acid, is thought to be the ingredient active against local inflammation. Glycoside-linked drugs are another example of prodrugs designed for activation in the colon by glycosidase activity.

SUMMARY

The oral route of drug delivery continues to attract the most attention with respect to development of CR systems. Research during the last three decades has established the scientific framework leading to development of a number of oral CR systems. Most of these are polymer-based systems in which the drug release rate is controlled by a membrane or matrix of a polymeric material. However, relatively few of these devices have proved to be useful therapeutic systems, due mainly to biological constraints of the gastrointestinal tract. The next challenge, therefore, is to increase our understanding of the physiology of this route in order to optimize drug delivery. This task will involve the incorporation of tissue, cellular, and molecular elements of GI physiology into the design of oral CR systems. Of particular significance among these are transit time studies, a detailed picture of the GI permeability to drug molecules, and GI motility. This will enable the pharmaceutical scientist to either optimize the dosage form to the GI environment or find ways to perturb GI physiology in a noninvasive way to deliver drugs. Specific regions of this organ should also be explored for delivery of peptides and proteins. Control over the movement of a dosage form and achieving some degree of colonic absorption are necessary in order to design systems to deliver drugs for more than 12 h with a single dose.

PROBLEMS

1. The therapeutic dose of a rapidly and completely absorbed drug is 450 mg/day, given 8 h apart as an osmotic pump device which delivers 75% of its load as a zero-order release, and the rest is retained in the dosage form. Half-life of the drug is 3 h, and rate of clearance is 50 liters/h. Calculate the amount in each dose, the release rate from the pump, and plasma drug concentration at 8 h after the first dose (μg/ml).

2. A drug shows a DI index of 3.2 from a controlled-release dosage form, and has a therapeutic index of 5. Can you evaluate the therapeutic value of the dosage form?

3. List three potential advantages of multiunit dosage forms over single-unit dosage forms.

4. An orally administered drug as a conventional tablet form shows 60% bioavailability when given as 250 mg t.i.d., and 80% bioavailability when given as 375 mg b.i.d., indicating a saturable presystemic drug elimination process. Suggest some noninvasive or invasive animal experiments to study the contribution of saturable, enzymatic degradation in the GI tract, gut wall metabolism, and first-pass liver metabolism. Note that more than one process can be involved.

5. The pharmacokinetic parameters of a drug, as determined from i.v. data, are as follows:

$$\text{Desired SS blood level} = 10 \text{ mg/liter.}$$
$$V = 15 \text{ liters.}$$
$$k_{el} = 0.2 \text{ h}^{-1}.$$
$$F = 1.$$

(a) If you were to device a zero-order-release formulation for oral administration, what rate of drug release would be required in order to maintain the desired blood levels in the body?

(b)   How long will it take to reach 90% of plateau levels with the above release rate?

6.   A controlled-release drug device releases the drug by a zero-order-release rate of 12 mg/h. It contains 288 mg of drug load.

(a)   Complete the following table. The pharmacokinetic parameters are as follows:

$$ka = 12 \, h^{-1}.$$
$$V = 1000 \, liters.$$
$$k_{el} = 0.12 \, h^{-1}.$$

Time (h)	Amt. in body (mg)	Amt. in dosage form (mg)	Amt. in gut lumen (mg)
0.25			
0.5			
1			
2			
4			
6			
8			
12			
16			
20			
24			

(b)   Obtain a drug amount-versus-time plot for body (curve A), dosage form (curve B), and gut lumen (curve C).

(c)   In an in-vivo study, curve B was observed to shift to the right 6 h after dosing, with a corresponding downward shift in curve A. Curve C showed no change. Give some possible explanations for this shift.

(d)   After another 6 h (12 h after dosing), curve B shifts further to the right, curve C shows a slight rise, and curve A falls further down compared to the in-vitro results. Explain the events that could be responsible for this observation.

(e)   Give two approaches that could be used to keep the in-vivo curves closer to those of in-vitro curves.

(f)   Assuming that the dosage form is administered every 24 h and the in-vitro and in-vivo curves overlap, what will be the drug amount in the body at steady state?

7.   A typical osmotic device comes in the shape of a tablet, with a tiny hole drilled through the semipermeable membrane on one side of the tablet. The usual range of the size of this orifice is 100–250 μm. Explain why such a range is chosen. What might happen if the size is smaller than 100 μm or larger than 250 μm?

8.   The following information is available about two new antidepressant drugs:

	Drug A	Drug B
MW	260	580
pK_a (base)	9.5	8.4
Aqueous solubility (mg/ml)	10	2.0
Oral bioavailability (% from soln.)	55	75
Absorption rate constant (k_a)	10	5
Therapeutic index	4	8
Apparent volume of distribution (liters)	80	2200
Clearance (liters/h)	50	170
Minimum effective concentration (ng/ml)	170	60

(a) Evaluate, on a comparative basis, the first six parameters above in order to assess the suitability of both drug candidates for formulation as an oral controlled-release system. Based on your evaluation, indicate the better candidate.

(b) Based on the overall evaluation, which drug candidate would you choose and why?

(c) Complete the following table:

	Drug A	Drug B
Target blood level (μg/liter)		
Zero-order release rate desired (mg/h)		
Loading dose (mg)		

(d) Both of the above drugs were formulated as slow first-order-release formulations as follow:

	Drug A	Drug B
Dose/unit	250 mg	250 mg
Release rate constant (k_{rel})	0.3 h^{-1}	0.2 h^{-1}

Assume the same F as from solution. No loading dose is given. Calculate the dosage form index at steady state for both drugs for the following dosage-form regimens:

(i) One unit given every 6 h

(ii) Two units given every 12 h

(iii) Four units given every 24 h

(e) Which dosage regiment of those above would you choose for Drugs A and B? Explain your choice.

ANSWERS

1. (a) Amount needed $= 450/3 = 150$ mg, but since only 75% of drug load is delivered, the amount required in dosage form is $150 \times 100/75 = 200$ mg.
   (b) Release rate $= 200/8 = 25$ mg/h.
   (c) Using Eq. (7), $c = 0.019$ μg/ml.

2. In order to assess the therapeutic success of a dosage regimen, one needs to know the desired blood levels in addition to the TI and DI. The fact that DI $<$ TI for the above formulation does not necessarily mean that the plasma drug levels fall within the effective range. It is possible that C_{max} and C_{min} at steady state overlap with the toxic levels or with minimum effective levels of drug. In both these cases, therapy will fail. Therefore, from a pharmacokinetics standpoint, the goal of a therapy is to maintain the drug levels within therapeutic window as well as to keep DI $<$ TI.

3. (a) Better gastrointestinal tolerance
   (b) Less chances of dose dumping
   (c) Better colonic absorption

4. Perform an in-situ intestinal perfusion experiment, using well-washed-out intestine to eliminate gut lumen enzymes. Collect blood samples from the portal vein in one experiment, and from the hepatic vein in the other.
   (a) 100% bioavailability from the hepatic vein will indicate that the gut lumen enzymes are solely responsible for degradation.
   (b) The difference between the portal and hepatic venous drug levels will indicate the contribution of liver metabolism.
   (c) Less than 100% availability from portal vein drug levels will indicate a contribution from gut wall metabolism.

5. (a) $c = k_0/Vk_{el}$; $k_0 = 30$ mg/h
   (b) 90% of 10 mg/liter $= 9$ mg/liter
   $c = k_0(1 - e^{-k_{el}t})/Vk_{el}$; $t = 1.1$ h

6. (a) The equations are as follows:
   Amount in body $= (k_0/k_{el})(1 - e^{-k_{el}t})$
   Amount in dosage form $= 288 - k_0t$
   Amount in gut lumen $= (k_{el}/k_a)(1 - e^{-k_a t})$

Time (h)	Amt. in body (mg)	Amt. in dosage form (mg)	Amt. in gut lumen (mg)
0.25	2.95	285	0.0095
0.5	5.82	282	0.0099
1	11.3	276	0.01
2	21.3	264	0.01
4	38.1	240	0.01
6	51.3	216	0.01
8	61.7	192	0.01
12	76.3	144	0.01
16	85.3	120	0.01
20	90.9	48	0.01
24	94.4	0	0.01

(c) Since the shift in curve B is to the right, the drug release rate is falling (k_0 < 12 mg/h). This is further evident from the fact that curve A shifts downward, as a result of reduced drug input. However, no shift in curve B means that k_a holds up steady. Possible reasons for this observation could be increased gut lumen viscosity or lower GI motility. The dosage form at this time is probably in the terminal ileum.

(d) The rise in curve C means that k_a has lowered, in addition to a further drop in k_0. The dosage form at this time is definitely in the colon, and highly viscous contents contribute to slower dissolution as well as diffusion of drug.

(e) (i) Use a multiunit dosage form, with a portion of particles releasing drug only after 6 or 12 h to compensate for reduced k_0.

   (ii) Administer a nonabsorbable, nondegradable, and highly hydrophillic polymer, e.g., polymethacrylic acid, to maintain adequate fluid levels around the dosage form and help sustain drug dissolution and diffusion.

   (f) Using Eq. 7, c = 100 mg.

7. Drug diffusion takes place with osmotic pressure as well. For this range, the rate of drug release due to simple diffusion is negligible compared to the total release rate, but does not restrict the solution movement. Size smaller than 100 μm may make the orifice the rate-limiting step for drug release. Size larger than 250 μm may allow an unacceptably high rate of drug diffusion. Since drug diffusion is affected by a variety of GI variables including viscosity and the hydrodynamics of GI contents, this may result in an unpredictable release rate.

8. (a) (i) MW: Drugs with MW less than 600 usually do not pose problems with respect to absorption from the GI tract. Drug A, with a MW of 260, does not seem to be a problem at all, but B, with a MW of 580, approaches 600. Yet it shows a slightly higher bioavailability than A. Since a complete absorption profile is not given, drug B could pose some absorption problems in the terminal ileum and colon. Consequently, although both drugs are below the 600 limit, based on MW alone, drug A will be a better choice.

   (ii) Pk_a: Both drugs are basic, with Pk_a higher than the pH of GI tract at all times. Thus they will be mostly in an un-ionized form through most of the GI tract. In this respect, both drugs are equally suitable for controlled-release formulations.

   (iii) Aqueous solubility: An aqueous solubility of 2 mg/ml seems to pose no problem for controlled-release dosage forms because the desired rate of release is usually small. Also, k_a for both drugs is fairly high, so not much drug will stay in the GI tract during the entire process of drug release. Thus, solubility does not seem to be an important factor for these drugs.

   (iv) Bioavailability: Less than 100% bioavailability from solution usually indicates drug loss due to degradation or metabolism. Liver first-pass metabolism could be a contributing factor. Since the data are only from solution dosage form and no effect of dose on bioavailability has been evaluated, a saturable presystemic clearance process is a possibility, especially for drug A. Apart from that, drug B will be a better choice simply due to its higher F.

(v) k_a: Since most controlled-release formulations are designed to have the drug release rate as the rate-limiting step, the value of k_a should be much higher than that of k_0 (the drug release rate). Values of 5 and 10 h^{-1} seem high enough for this purpose. Nevertheless, k_a has been determined from a solution dosage form, and, thus, reflects absorption kinetics only from the upper small intestine. Considering that k_a generally falls from the proximal to the distal part of the small intestine, drug A, with a value of 10 h^{-1}, will be a better choice.

(vi) TI: TI alone is of limited value in determining suitability for controlled release. From the given values of V_{app} and Cl, one can get an estimate of k_{el}, which seems much higher for drug A (50/180 liters/h) compared with drug B (170/2200 liters/h). Thus, DI for B will be much smaller compared to that of A. Given that the TI for B is also higher, controlled release may not offer much advantage over the conventional dosage form. [This will be more evident in part (d) of this problem.] Therefore, once again A will be a better choice.

(b) Based on the above parameters, drug A makes a better candidate for formulation as a controlled-release product in all respects except oral bioavailability. Unless a saturable metabolic clearance is documented for A, it should be chosen.

(c)

	Drug A	Drug B
Target blood level (MEC × TI/2)	340 μg/liter	240 μg/liter
Zero-order release rate (C·Cl/F)	31 mg/h	30.6 mg/h
Loading dose (C·V/F)	111 mg	704 mg

(d) Use Eq. (6) for t_{max}, Eq. (4) for C_{max}, and Eq. (5) for C_{min}.
For drug A:

One unit every 6 h: $t_{max} = 2.1$ h
                         $C_{max} = 0.52$ mg/liter
                         $C_{min} = 0.45$ mg/liter
                         $$DI = \frac{0.52}{0.45} = 1.2$$

Two units every 12 h: $t_{max} = 3.0$ h
                         $C_{max} = 0.67$ mg/liter
                         $C_{min} = 0.21$ mg/liter
                         $$DI = \frac{0.67}{0.21} = 3.2$$

Four units every 24 h: $t_{max} = 3.5$ h
                         $C_{max} = 1.16$ mg/liter
                         $C_{min} = 0.01$ mg/liter
                         $$\frac{1.16}{0.01} = 116$$

For drug B:

One unit every 6 h:

$$t_{max} = 2.7 \text{ h}$$
$$C_{max} = 0.18 \text{ mg/liter}$$
$$C_{min} = 0.18 \text{ mg/liter}$$
$$DI = \frac{0.18}{0.18} = 1.0$$

Two units every 12 h:

$$t_{max} = 4.4 \text{h}$$
$$C_{max} = 0.20 \text{ mg/liter}$$
$$C_{min} = 0.15 \text{ mg/liter}$$
$$DI = \frac{0.20}{0.15} = 1.3$$

Four units every 24 h:

$$t_{max} = 6.4 \text{ h}$$
$$C_{max} = 0.24 \text{ mg/liter}$$
$$C_{miin} = 0.09 \text{ mg/liter}$$
$$DI = \frac{0.24}{0.09} = 2.6$$

(e)   The choice should be the most infrequent dosing where DI < TI and plasma levels stay between MEC and toxic levels.

Drug A: The choice is regimen (ii), i.e., two units every 12 h, because TI = 0.68/0.17 = 4 and DI = 0.67/0.21 = 3.2.

Drug B: The choice is regimen (iii), i.e., four units every 24 h, because TI = 0.48/0.06 = 8 and DI = 0.24/0.09 = 2.6.

REFERENCES

1.   Leszek Krowczynski, *Extended-Release Dosage Forms*, CRC Press, Boca Raton, Fla., chap. 6, p. 96 (1987).

2.   F. Azpiroz and J. R. Malagelada, Pressure activity patterns in the canine proximal stomach: Response to distension, *Am. J. Physiol.*, *247*:G265 (1984).

3.   A. Rubinstein, V. H. K. Li, P. Gruber, and J. R. Robinson, Gastrointestinal-physiological variables affecting the performance of oral sustained release dosage forms, in *Oral Sustained Release Formulations: Design and Evaluation*, A. Yacobi and E. Halperin-Walega, Eds.), Pergamon Press, New York (1987).

4.   E. M. M. Quigly, S. F. Phillips, and J. Dent, Distinctive patterns of interdigestive motility at the canine Ilio-colonic junction, *Gastroenterology*, *87*:836 (1984).

5.   Z. Itoh and T. Sekiguchi, Interdigestive motor activity in health and desease, *Scand. J. Gastroenterol.*, *18*(Suppl. 82):497 (1982).

6.   J. E. Kellow, T. J. Borody, S. F. Phillips, R. L. Tucker, and A. C. Hadda, Human interdigestive motility: Variations in patterns from esophagus to colon, *Gastroenterology 91*:386 (1986).

7.   S. McRae, K. Younger, D. G. Thompson, and D. L. Wingate, Sustained mental stress alters human jejunal motor activity, *Gut*, *23*:404 (1982).

8.   C. T. Sekiguchi, T. Nishioka, Mi. Kogure, M. Kusano, and S. Kobayashi, Interdigestive gastroduodenal phasic contractions and intraluminal pH in gastric and duodenal ulcer patients, in *Gastrointestinal Function: Regulation and Disturbances*, Vol. 2 (Y. Kasuya, M. Tshuchiya, F. Hagao, and Y. Matsuo Eds.), Excerpta Medica, Amsterdam, p. 93 (1984).

9.   H. J. Ehrlein, A new technique for simultaneous radiography and recording of gastrointestinal motility in unanesthetized dogs, *Lab. Animal Sci.*, *30*:879 (1980).

10. S. J. Konturek, P. J. Thor, J. Bilski, W. Bielanski, and J. Laskiewicz, Relationships between duodenal motility and pancreatic secretions in fasted and fed dogs, *Am. J. Physiol.*, *250*:G570 (1986).

11. P. K. Gupta and J. R. Robinson, Gastric emptying of liquids in the fasted dog. *Int. J. Pharm.*, *43*:45 (1988).

12. I. DeWever, C. EecKhout, G. Vantrappen, and J. Hellemans, Disruptive effect of test meals on interdigestive motor complex in dogs, *Am. J. Physiol.*, *235*:E661 (1978).

13. R. A. Hinder and K. A. Kelly, Canine gastric emptying of solids and liquids, *Am. J. Physiol.*, *233*:E335 (1977).

14. P. Gruber, A. Rubinstein, V. H. K. Li, P. Bass, and J. R. Robinson, Gastric emptying of non-digestible solids in the fasted dog. *J. Pharm. Sci.*, *76*:117 (1986).

15. G. A. Digenis, Gamma scintigraphy in development of CR oral delivery systems, in *Proc. 13th Int. Symp. Bioactive Materials* (I. Caudry and C. Thies, Eds.), Control Release Society, p. 115 (1986).

16. D. Winne, Influence of blood flow on intestinal absorption of xenobiotics, *Pharmacology*, *21*:1 (1980).

17. C. A. Youngberg, R. R. Beradi, W. F. Howatt, M. L. Hyneck, G. L. Amidon, H. J. Meyer, and J. B. Dressman, Comparison of gastrointestinal pH in cystic fibrosis and healthy subjects, *Dkg. Dis. Sci.*, *32*:472 (1987).

18. J-R. Malagelada, G. R. Longstreth, W. H. J. Summerskill, and V. L. M. Go, Measurement of gastric function during digestion of ordinary solid meals in man, *Gastroenterology*, *70*:203 (1976).

19. R. R. Scheline, Toxicological implications of drug metabolism by intestinal bacteria, *Eur. Soc. Study Drug Tox., Proc.*, *13*:25 (1972).

20. R. R. Scheline, Drug metabolism by intestinal microorganisms, *J. Pharm. Sci.*, *57*:2021 (1968).

21. J. L. Gowans and E. J. Knight, The route of recirculation of lymphocytes in the rat, *Proc. Roy. Soc. B.*, *159*:257 (1964).

22. D. E. Bockman and M. D. Cooper, Pinocytosis by epithelium associated with lymphoid follicles in the bursa of Fabricus, appendix and Peyer's patches. An electron microscope study, *Am. J. Anat.*, *136*:455 (1973).

23. P. Gruber, M. A. Longer, and J. R. Robinson, Some biological issues in oral controlled drug delivery, *Adv. Drug Del. Rev.*, *1*:1 (1987).

24. D. W. Powell, Intestinal water and electrolyte transport, in *Physiology of Gastrointestinal Tract*, 2nd ed., Vol. 2 (L. R. Johnson, Ed.), Raven Press, New York, p. 1267 (1986).

25. M. Tomita, S. Masaharu, M. Hayashi, and S. Awazu, Enhancement of colonic drug absorption by the paracellular permeation route, *Pharm. Res.*, *5*:341 (1986).

26. S. Pedersen, Delay in the absorption rate of theophylling from a sustained release theophylline preparation caused by food, *Br. J. Clin. Pharmacol.*, *12*:904 (1981).

27. P. G. Welling, Influence of food and diet on gastrointestinal drug absorption: A review: *J. Pharmacokinet. Biopharm.*, *5*:291 (1977).

28. P. G. Welling and M. R. Dobrinska, Dosing considerations and bioavailability assessment of controlled drug delivery systems, in *Controlled Drug Delivery*; *Fundamentals and Applications*, 2nd ed., (J. R. Robinson and V. H. L. Lee, Eds.), Marcel Dekker, New York, p. 253 (1987).

29. J. G. Wagner, Effect of first-pass Michaelis-Menten metabolism on performance of CR formulations, in *Oral Sustained Release Formulations*; *Design and Evaluation*, (A. Yacobi and E. Halperin-Walega, Eds.), Pergamon Press, New York (1987).

30. P. A. Routeledge and D. G. Shand, Clinical pharmacokinetics of propranolol, *Clin. Pharmacokinet.*, *4*:73 (1979).

31. T. M. Tozer and G. M. Rubin, Saturable kinetics and bioavailability determination, in: *Pharmacokinetics: Regulatory, Industrial, Academic Perspectives* (P. G. Welling and F. L. S. Tse, Eds.), Marcel Dekker, New York (1988).

32. B. Silber, N. Holford, and S. Riegelman, Dose dependent elimination of propranolol and its major metabolites in humans, *J. Pharm. Sci.*, *72*:725 (1983).

33. B. Ablad, M. Ervik, J. Hallgren, G. Johnsson, and L. Solvell, Pharmacological effects and serum levels of orally administered alprenolol in man, *Eur. J. Clin. Pharmacol.*, *5*:44 (1972).

34. J. G. Wagner and E. Nelson, Kinetic analysis of blood levels and urinary excretion in the absorptive phase after single doses of drug, *J. Pharm. 53*:1392 (1964).

35. J. C. K. Loo and S. Riegelman, New method for calculating the intrinsic absorption rate of drugs, *J. Pharm. Sci.*, *57*:918 (1968).

36. K. Yamaoka, T. Nakagawa, and T. Uno, Statistical moments in pharmacokinetics, *J. Pharmacokin. Biopharm. 6*:547 (1978).

37. D. J. Cutter, Theory of the mean absorption time, an adjunct to conventional bioavailability studies, *J. Pharm. Pharmacol.*, *30*:476 (1978).

38. S. Riegelman and P. Collier, The application of statistical moment theory to the evaluation of in-vivo dissolution time and absorption time, *J. Pharmacokin. Biopharm.*, *8*:509 (1980).

39. A. Rescigno and G. Segre, *Drug and Tracer Kinetics*. Blaisdell, Waltham, Mass., p. 109 (1966).

40. J. G. Wagner, Pharmacokinetics: Past developments, present issues, future challenges, in *Pharmacokinetics: Regulatory, Industrial, Academic Perspectives* (P. G. Welling and F. L. S. Tse, Eds.), Marcel Dekker, New York (1988).

41. *The United States Pharmacopeia* (USP XXI); *The National Formulary* (NF XVI), 1985.

42. A. C. Shah, Design of oral sustained release drug delivery systems; In-vitro/in-vivo considerations, in *Oral Sustained Release Formulations; Design and Evaluation* (Yacobi and E. Halperin-Walega, Eds.), Pergamon Press, New York (1987).

43. M. Weinberger, L. Hendles, and L. Bighley, The relation of product formulation to absorption of oral theophylline, *N. Engl. J. Med.*, *299*:852 (1978).

44. R. A. Upton, J-F. Thiercelin, R. W. Guentert, L. Sanson, J. R. Powell, and P. E. Coates. Evaluation of the absorption from some commercial sustained release theophylline products, *J. Pharmacokin. Biopharm.*, *8*:131 (1980).

45. H. W. Kelly and S. Murphy, Efficacy of a 12 hour sustained release preparation in maintaining therapeutic serum theophylline levels in asthmatic children, *Pediatrics*, *66*:97 (1980).

46. M. J. Kendall, D. B. Jack, K. L. Woods, S. J. Laugher, C. P. Quarterman, and V. A. John, Comparison of the pharmacokinetic and pharmacodynamic profiles of single and multiple doses of a commerical slow release metoprolol formulation with a new Oros delivery system, *Br. J. Clin. Pharmacol.*, *13*:393 (1982).

47. Ho-Wah Hui, J. R. Robinson, and V. H. L. Lee, Design and fabrication of oral CR drug delivery systems, in *Controlled Drug Delivery; Fundamentals and Applications*, 2nd ed. (J. R. Robinson and V. H. L. Lee, Eds.), Marcel Dekker, New York, p. 373 (1987).

48. Leszek Krowczynski, *Extended-Release Dosage Forms*, CRC Press, Boca Raton, Fla., chap. 6, p. 106 (1987).

49. C. Graffner, G. Johnson, and J. Shogren, Pharmacokinetics of procainamide intravenously and orally as conventional and slow-release tablets, *Clin. Pharmacol. Ther.*, *17*:414 (1975).

50. J. A. Bakan and J. L. Anderson, Microencapsulation, in *The Theory and Practice of Industrial Pharmacy*, 2nd ed. (L. Lachman, H. A. Lieberman, and J. L. Kaing, Eds.), Lea & Febiger, Philadelphia, chap. 13, part III (1967).

51. S. Borodkin and F. E. Tucker, Drug release from hydroxypropyl cellulose-polyvinyl acetate films, *J. Pharm. Sci.*, *63*:1359 (1974).

52. P. L. Madan, Clofibrate Microcapsules II: Effect of wall thickness on release characteristics, *J. Pharm. Sci.*, *70*:430 (1981).

53. S. T. Borodkin and F. E. Tucker, Linear drug release from laminated hydroxypropyl cellulose-polyvinyl acetate films, *J. Pharm. Sci.*, *64*:1289 (1975).

54. B. H. Lippold, B. K. Sutter, and B. C. Lippold, Parameters controlling drug release from pellets coated with aqueous ethyl cellulose dispersion, *Int. J. Pharm.*, *54*:15 (1989).

55. B. Gander, R. Gurny, E. Doelker, and N. A. Peppas, Effect of polymeric network structure on drug release from cross linked poly (vinyl alcohol) micromatrices, *Pharm. Res.*, 6(7):578 (1989).

56. Y. Raghunathan, L. Amsel, O. Hinvack, and W. Bryant, Sustained release drug delivery systems I: Coated ion exchange resin system for phenylpropanolamine and other drugs, *J. Pharm. Sci.*, 70:379 (1981).

57. O. Wulff, Prolonged antitussive action of a resin bound noscaine preparation, *J. Pharm. Sci.*, 54:1058 (1965).

58. F. Theeuwes and T. Higuchi, U.S. Patent 3,916,899, Nov. 4, 1975.

59. F. Theeuwes, Evolution and design of rate controlled osmotic forms, *Curr. Med. Res. Opi.*, 8(Suppl.):220 (1983).

60. F. Theeuwes, Oral dosage form design: Status and goals of oral osmaotic systems technology, *Pharm. Int.*, 5:293 (1984).

61. F. Theeuwes, D. Swanson, P. Wong, P. Bonsen, V. Place, K. Heimlich, and K. C. Kwan, Elementary osmotic pump for indomethacin, *J. Pharm. Sci.*, 72:253 (1983).

62. J. D. Rogers, R. B. Lee, P. R. Souder, R. K. Ferguson, R. O. Davies, F. Theeuwes, and K. C. Kwan, Pharmacokinetic evaluation of osmotically controlled indomethacin delivery systems in man, *Int. J. Pharm.*, 16:191 (1983).

63. C. J. Cavallito and R. Jewell, Modification of rates of gastrointestinal absorption of drugs, I. Aines, *J. Am. Pharm. Assoc. Sci. Ed.*, 47:165 (1958).

64. H. Bechgaard and K. Ladefoged, Distribution of pellets in the gastrointestinal tract. The influence on transit time exerted by the density or diameter of pellets, *J. Pharm. Sci.*, 30:690 (1978).

65. R. A. Good and J. Finstad, The phylogenetic development of the immune responses and the germinal center system, in *Germinal Centers in Immune Response* (H. Cottier, N. Odartchenko, R. Schindler, and C. C. Congdon, Eds.), Springer-Verlag, Berlin (1967).

66. M. J. Dew, R. E. J. Ryder, N. Evans, B. K. Evans, and J. Rhodes, Colonic release of 5-aminosalicylic acid from an oral preparation in active ulcerative colitis, *Br. J. Clin. Pharmacol.*, 16:185 (1983).

67. M. Saffaran, G. S. Kumar, C. Savariar, J. C. Burnham, F. Williams, and D. C. Neckers, A new approach to the oral delivery of insulin and other peptide drugs, *Science*, 223:1081 (1986).

7

Parenteral Drug Delivery: Injectables

Thomas R. Tice *Southern Research Institute, Birmingham, Alabama*

S. Esmail Tabibi *Micro Vesicular Systems, Inc., Nashua, New Hampshire*

INTRODUCTION

Parenteral controlled-release dosage forms have been and will continue to be useful for the treatment of disease. But, it is important to realize that parenteral controlled-release dosage forms will not provide all of the answers for drug delivery. In fact, no single controlled-release technology will provide all of the answers for the effective treatment of disease because of the diversity of drug properties, dosing levels, and treatment durations, as well as other factors such as patient acceptability and cost. The challenge, therefore, is to select the best controlled-release technology for each drug and associated disease treatment.

When a parenteral controlled-release dosage form appears to be the best approach for a specific drug, the task of the pharmaceutical scientist is to sort out and balance the many factors associated with designing, developing, and manufacturing such dosage forms. This task is a complex one that requires a strategy and involves a diversity of disciplines ranging from medicine and chemistry to regulatory affairs and marketing.

The purpose of this chapter is to highlight some of the key factors in designing and developing parenteral controlled-release dosage forms. It begins by addressing some of the reasons why one would select a parenteral controlled-release dosage form over other controlled-release dosage forms.

In doing so, the advantages and disadvantages of parenteral controlled-release dosage forms are brought forth. Also, this chapter gives examples of some of the key approaches that have been taken to afford safe and efficacious parenteral controlled-release dosage forms. These examples range from older, but still useful, technologies involving emulsions, drug solutions, and drug suspensions to new and more sophisticated technologies such as biocompatible, biodegradable, injectable microsphere formulations.

SELECTION OF PARENTERAL CONTROLLED-RELEASE DOSAGE FORMS

Conventional Dosing

When developing a *conventional* dosage form for a new drug, one usually prefers to use the oral route of administration. An oral dosage form is easily administered by the patient, which gives complete control to the patient to take the medication. Moreover, the manufacturing requirements for an oral dosage form are less rigorous than other dosage forms, in particular, less rigorous than the manufacture of parenteral dosage forms.

A drug, however, cannot be given orally if it degrades substantially in the gastrointestinal tract, if it undergoes extensive first-pass metabolism, or if it is poorly absorbed. Very important drugs that cannot be taken orally are the "biotech drugs," peptides and proteins that are synthetically prepared or made using recombinant DNA technology. When taken orally, peptides and proteins are quickly broken down by proteolytic enzymes such as pepsin, trypsin, and chymotrypsin. Some peptides and proteins are degraded or inactivated due to the low-pH environment of the stomach. Moreover, the high molecular weight of peptides and proteins prevents them from being effectively transported across the epithelial barrier of the gastrointestinal tract.

If a drug cannot be taken orally due to poor bioavailability, then the parenteral route of administration can be a viable alternative approach, as long as the dose is of a reasonable size and the drug is distributed appropriately by the circulatory system with subsequent absorption by the target tissue. A parenteral formulation is commonly given as a continuous intravenous infusion or as an intramuscular or subcutaneous injection. Because of the risks involved, a health care professional usually administers parenteral dosage forms. As a result, the patient has less control in taking the medication in comparison to oral dosing and the patient must make the effort to visit a physician's office or stay in a hospital to receive treatment. Moreover, a large responsibility rests with manufacturers of parenteral dosage forms to ensure that their parenteral products are of extreme high purity. That is, parenteral dosage forms must be sterile, and free of pyrogens and foreign particulates. High-purity product is obtained by requiring high standards and precise specifications on raw materials and by rigorously controlling manufacturing procedures to produce and package parenteral dosage forms. A more detailed overview of the development and manufacture of conventional parenteral dosage forms can be found [1].

Controlled-Release Dosing

Oral *controlled-release* dosage forms, more correctly termed modified-release dosage forms, have been available for many years. These formulations, depending on the drug employed, provide efficacy for up to 12–24 hr per administration. Even in a best-case situation, where a drug is absorbed throughout the entire gastrointestinal tract, the main limitation to developing a useful oral modified-release dosage form is the short transit time of the dosage form through the gastrointestinal tract (about 12 h). And, if the drug is only absorbed within a specific region of the gastrointestinal tract, the duration of action can be less than 12 h. This relatively short duration of action is the primary limitation of modified-release oral dosage forms. Looking at the advantages, modified-release oral formulations share the advantages of conventional oral formulations with respect to patient administration and manufacturing requirements.

Why use a parenteral controlled-release dosage form over an oral modified-release dosage form? One reason is patient compliance. In practice, poor patient compliance is one of the causes, if not the key cause, why patients do not benefit fully from drug treatment. Part of this compliance problem occurs because an adult patient, an elderly patient, or a child's parent, school nurse, or day-care worker must not only remember when to interrupt the day's normal activities but also must remember to administer sometimes several different drugs that are given at different times. So, even if an oral dosage form has good bioavailability, a long-acting parenteral dosage form that is safe and efficacious for a long period of time (days or months) can be beneficial because it ensures that the patient is receiving medication.

Why use a parenteral controlled-release dosage form over a conventional parenteral dosage form? For one reason, injections are not favored by patients, especially if multiple injections are involved each day for several days. If chronic treatment is required, a patient must make the effort to go to a physician's office or the hospital to receive treatment, which in practice is difficult to do on a long-term, routine basis. Furthermore, if a drug is given by intravenous infusion, the cost of hospital care, safety issues, as well as the time required for such treatment are obvious disadvantages to parenteral dosage forms.

Therefore, with respect to improved patient compliance, safety, and less cost, controlled-release parenteral dosage forms that are efficacious for days or months after a single administration can be better than conventional parenteral dosage forms. It should also be mentioned that as with other controlled-release dosage forms, parenteral controlled-release dosage forms can be beneficial by decreasing the dose, minimizing side effects, and improving efficacy. On the other hand, the main disadvantage to parenteral controlled-release dosage forms is that, once administered, they cannot be easily removed. Thus, a problem could arise for the patient if a drug caused an undesirable reaction or if a drug was no longer needed; for instance, because treatment was no longer required, the drug was not effective, or the drug was not tolerated with other medication being taken by the patient.

Another way to improve the efficacy and reduce the undesirable side effects of a parenteral formulation is to place a parenteral controlled-release formulation at the desired site of action. This approach, local drug delivery, eliminates the need for the circulatory system to carry the drug to the target tissue. In turn, the amount of drug required is reduced and toxicological effects are minimized.

Lastly, another advantage of parenteral controlled-release formulations is that they can deliver drugs that will not penetrate the skin. That is, they can be used as long-acting systems to deliver skin-impermeable, low-molecular-weight drugs as well as high-molecular-weight drugs such as peptides and proteins. In this regard, parenteral controlled-release dosage forms have a distinct advantage over transdermal controlled-release dosage forms.

Design and Development of Parenteral Controlled-Release Forms

Duration of Action

There are many parameters to consider in order to properly design and develop a parenteral controlled-release dosage form. The "duration of action" is one of these parameters. To achieve the maximum duration of action one must consider the potency of the drug, the physical/chemical limitations of the controlled-release technology employed, and marketing data (patient acceptability, physician requirements, profitability).

The amount of drug required, which relates to drug potency and the efficiency of the delivery system, as well as the amount of rate-controlling excipient required, will determine the maximum duration of release one can formulate into the dosage form. That is, if the required amount of drug plus rate-controlling excipient is too large in terms of total mass, then the desired duration of action must be shortened. In other words, the more potent a drug and the less amount of rate-controlling excipient required, the smaller the total mass needed per day and the longer the duration of release one has the potential to provide.

For instance, for an intramuscular or subcutaneous injection of a particulate formulation in humans, one typically prefers not to exceed a total injection volume of 1.5 mL. This volume includes the particulate controlled-release dosage form and the liquid injection vehicle. So, with the limitation of a 1.5-mL injection volume and assuming a maximum solids content of 20–30%, the total mass of particulate one would be able to administer to a patient is about 300–500 mg. A 300-mg mass is preferred. Now continuing with the example, if the 300-mg particulate mass consisted of 10% drug and 90% rate-controlling excipient, and if the formulation was 100% efficient to where it perfectly released drug every day at the desired constant rate, then a 30-day formulation would be possible for a treatment requiring 1 mg of drug per day.

Usually, the potency of the drug and the physical/chemical limitations of the controlled-release technology limits the desired duration of action. But, other parameters can play a role as well. For example, the duration of action may be selected so as to correlate to a patient's visits to the physician's office. For instance, for a patient making once-a-month office visits for a particular disease treatment, a 1-month formulation would be preferred. Or for a dental application, a 6-month formulation would be preferred so as to correspond to established routine 6-month checkups. Of course, 1-week, 2-week, 1-month or 3-month formulations would be easier to implement, rather than odd number of weeks and months, 1.3 week or 1.7 month.

Sometimes it is possible to achieve a very long duration of action, for example, 6 or 12 months. However, the time to develop such a formulation as well as the time to perform quality-control tests following the manufacture of each batch can make a 6- or 12-month formulation impractical to commercialize. As a result, 1- to 3-month formulations or shorter-term formulations are usually preferred.

Materials

As with conventional parenteral formulations, the number of regulatory-approved materials that can be used in parenteral controlled-release dosage forms are less than those available for other dosage forms. Of course, new materials can be developed for parenteral controlled-release applications, but these new materials must be developed, tested in the clinic, and approved by regulatory agencies. This process can be very costly and time consuming. Therefore, if an approved material will function for the desired dosage form, it is best to consider this material first. In addition to being biocompatible and pyrogen free, the materials must biodegrade into nontoxic compounds within an appropriate time, preferably close to the duration of action of the dosage form. A useful reference for the selection of materials for parenteral controlled-release dosage forms is the *Handbook of Pharmaceutical Excipients* [2].

Manufacturing

As with any parenteral dosage form, the manufacturing processes used to make parenteral controlled-release formulations must be pharmaceutically acceptable and reproducible processes that afford sterile product of high purity without compromising the integrity of the drug [1]. Sterile product can be obtained by using aseptic manufacturing or terminal sterilization such as gamma irradiation. If the manufacturing process requires processing aids such as solvents or pharmaceutically unacceptable surfactants, these processing aids must not be present in the final dosage form. Either the process does not introduce a processing aid into a formulation or after fabrication, a processing aid must be removed. If small amounts still remain, the processing aid cannot be toxic and should be at levels acceptable to regulatory agencies.

Even if all of the above-described criteria are met, and after considering all of the ramifications of marketing, a parenteral controlled-release product will not be made available to the public unless it is cost effective. One can now better understand why it is a challenge to develop and commercialize long-acting, parenteral controlled-release dosage forms. This task is a complex one which requires a strategy and involves a diversity of disciplines ranging from medicine and chemistry to regulatory affairs and marketing.

POLYMERIC MICROSPHERE DRUG DELIVERY SYSTEMS

Injectable, biodegradable microsphere products are examples of the more recent technology developed for parenteral controlled-release dosage forms. Microsphere products are free-flowing powders consisting of spherical particles less than 250 μm in diameter—ideally less that 125 μm (see Fig. 1). Particles of this size can be administered easily by suspending them in a suitable aqueous vehicle and injecting them using a conventional syringe with an 18- or 20-gauge needle.

In reviewing the literature, one finds that there are several names given to these particles. Among these names include: microcapsules, microspheres, microparticles, nanocapsules, nanospheres and nanoparticles, as well as various trade names.

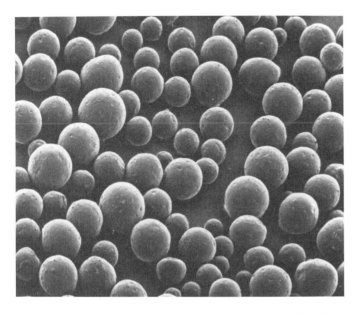

Figure 1   Scanning electron micrograph showing injectable, biodegradable microspheres.

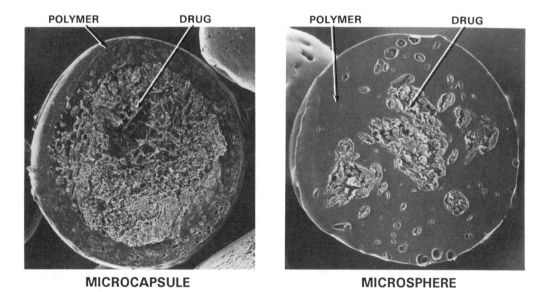

MICROCAPSULE   **MICROSPHERE**

Figure 2 Cross-sections showing the internal structure of a microcapsule (left) and a microsphere (right).

A *microcapsule* has its drug centrally located within the particle, where it is encased within a unique polymeric membrane (Fig. 2). Whereas, a *microsphere* has its drug dispersed throughout the particle; that is, the internal structure is a matrix of drug and polymeric excipient (Fig. 2). Theoretically, high-quality microcapsules will release their drug at a constant rate (zero-order release), whereas microspheres typically give a first-order release of drug. With proper formulation design and process manipulation, however, one skilled in the art of microencapsulation can fabricate microspheres to release drug at practical and efficacious rates.

Design and Development of Parenteral Controlled-Release Microspheres

One of the key factors in designing an injectable microsphere delivery system is to choose the appropriate polymeric excipient. Polymers chosen as excipients for parenterally administered microcapsules must meet several requirements, including suitable tissue compatibility, biodegradation kinetics, drug compatibility, drug permeability, mechanical properties, and ease of processing.

One class of polymers that is popular for this purpose includes the thermoplastic polyesters of poly(lactide) and copolymers of lactide and glycolide referred to as poly(lactide-co-glycolide).

$$H\left[O-\underset{CH_3}{CH}-\overset{O}{\underset{\|}{C}}-O-\underset{CH_3}{CH}-\overset{O}{\underset{\|}{C}}\right]_x\left[O-CH_2-\overset{O}{\underset{\|}{C}}-O-CH_2-\overset{O}{\underset{\|}{C}}\right]_y OH$$

Chemical structure of poly(lactide-co-glycolide)

First, these polymers are biocompatible. For example, Visscher et al. [3–5] have reported the biocompatibility of injectable microspheres made with poly(DL-lactide) and poly(DL-lactide-co-glycolide). Second, these polymers are biodegradable. They biodegrade by undergoing random, nonenzymatic, hydrolytic scissioning of their ester linkages to form lactic acid and glycolic acid, which are normal metabolic compounds.

One of the earliest applications of these polymers was for synthetic resorbable sutures. They have also been used for surgical clips and other surgical implants. Because these polymers have proved to be biocompatible and to have extensive toxicological documentation, their approvals by the regulatory authorities for use as microsphere excipients are usually less costly and more straightforward than approvals of new polymers for excipients. For this reason, lactide homopolymers and lactide/glycolide copolymers are often selected as excipients for parenteral controlled-release dosage forms.

To determine the biodegradation kinetics of microspheres prepared with lactide/glycolide excipients, studies were performed in rats [3,6]. Rats were injected intramuscularly in the leg with norethisterone or lypressin microspheres prepared with radiolabelled poly(DL-lactide-co-glycolide). The label was incorporated into the excipient by using ^{14}C-DL-lactide monomer during its polymerization. For each excipient tested, all soft tissue from the entire leg that had been injected with microspheres was removed from the bone. The soft tissues were digested, homogenized, and brought to near dryness. The homogenates were oxidized and the ^{14}C-carbon dioxide generated was sorbed and quantified. The results of the biodegradation studies are shown in Fig. 3. The biodegradation curves are expressed as the percent of the radioactivity remaining at the injection site at various times following treatment. Evaluation of these curves reveals that increasing the amount of glycolide in the polymer excipient increases the rate of biodegradation. (Note: 50:50 refers to the mole ratio of lactide-to-glycolide.) With these data, one has an ap-

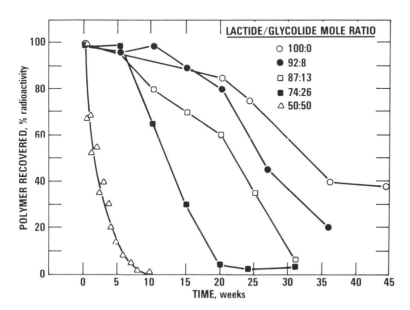

Figure 3 In vivo resorption of radiolabeled poly(DL-lactide-co-glycolide) microspheres designed to release (△) lypressin [3] and (○●□■) norethisterone [6].

proximation from which to determine the appropriate lactide/glycolide polymer to use to achieve the desired duration of action from a microsphere formulation.

In addition to poly(lactide-co-glycolide), many other biocompatible, biodegradable polymers have also been examined as excipients for parenteral controlled-release microspheres. Among these other polymers are albumin, polyanhydrides, and other polyesters such as polycaprolactone and poly(β-hydroxybutyrate). Albumin microspheres have been examined for the delivery of anticancer agents. In particular, the efficacy of albumin microspheres has been studied following their intra-arterial administration to target and control the release of drugs to the liver and lungs.

To emphasize their small size, microspheres (microparticles) and microcapsules smaller than 1 μm are usually referred to as nanospheres (nanoparticles, pseudo-latices) and nanocapsules. For the most part, the excipients used to make nanoparticles have been based on cyanoacrylate polymers which are polymerized as the nanoparticles are made. Preform polymers such as polylactide have also been used as well to make nanoparticles.

After selecting the excipient, the next step is to choose an appropriate microencapsulation process. There are many techniques available to make microcapsules and microspheres. The technique one chooses depends on many factors, for instance, how the microcapsules/microspheres are to function, the desired microcapsule/microsphere size, the physical and chemical properties of the drug and polymeric excipient, compatibility of the process conditions with the drug and polymeric excipient, and the ability to afford a product with the highest ratio of drug-to-polymer to minimize the amount of mass administered to the patient without compromising release kinetics.

Although there are many processes available to make microcapsules/microspheres, there are only a few processes that afford product suitable for parenteral use. To date, the commercialized parenteral microspheres have been made by one of three processes. These processes are spray-drying, solvent-evaporation and phase-separation techniques.

The processes available to make parenteral microcapsules/microspheres are limited because of the rigorous manufacturing requirements placed on parenteral products. For instance, there are a limited number of excipients acceptable for parenteral use and there are a limited number of pharmaceutically acceptable solvents and other processing aids. Also the microcapsules/microspheres must be made small enough to be injected with a conventional syringe and needle. And like any other pharmaceutical processes, the microencapsulation process must afford product that is pharmaceutically acceptable with respect to residual solvent and processing aids, batch-to-batch reproducibility, scale-up, encapsulation efficiency, and yields. Not all microencapsulation processes can provide product under these circumstances.

Other pharmaceutical criteria for the product include method and validation of sterilization, composition of the injection vehicle, vial- or syringe-loading method, and product stability. In addition to being a pharmaceutically acceptable process, the microencapsulation process should be cost-effective with respect to product isolation/drying and disposal or recycling of process media and solvents (environmental impact). Plant safety for workers is another factor to consider as well.

Examples of Parenteral Controlled-Release Microspheres

Peptides and Proteins

Peptides and proteins promise to have a profound effect on human health and on the future of the pharmaceutical industry. Because oral administration of peptides and proteins is not practical, the parenteral route is the most effective means of administration. With parenteral administration, however, one or more injections are required daily because of the characteristically short half-lives of these protein/peptides, and the initial delivery of peptide/protein to the target tissue results in high concentrations followed by a continuous decline of protein/peptide in the serum. These drugs, therefore, make themselves amenable for parenteral, controlled-release microsphere formulations, where a single administration will maintain constant, efficacious levels of these potent drugs for desired durations of action.

One of the first peptides that was successfully microencapsulated for the purpose of controlled release was luteinizing hormone-releasing hormone (LHRH), a highly water-soluble, 1200-dalton peptide. This microsphere formulation was designed to release the LHRH agonist, Nafarelin, for 30 days. The polymeric excipient used was 50:50 poly(DL-lactide-coglycolide). The microspheres are made by a phase-separation technique. As shown in Fig. 4, a single administration of Nafarelin microspheres suppressed LH (lu-

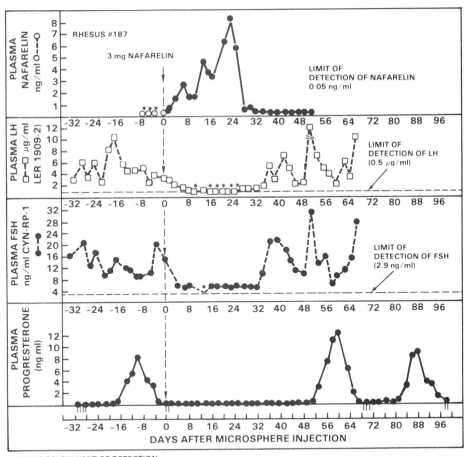

*VALUE BELOW LIMIT OF DETECTION

Figure 4 Nafarelin, luteinizing hormone (LH), follicle-stimulating hormone (FSH) and progesterone plasma levels in a female rhesus monkey after a single administration of Nafarelin microspheres. (Adapted from Ref. 9.)

teinizing hormone) and FSH (follicle-stimulating hormone) with a subsequent suppression of progesterone in females. Although the Nafarelin, plasma levels are not perfectly constant, the suppression of progesterone is perfectly acceptable and, in fact, quite optimal. It is thought that the LHRH is released from these microspheres initially by diffusing through water-filled pores followed by release due to the biodegradation of the copolymer excipient [7–11].

The first injectable, peptide microsphere formulation to be marketed delivered another LHRH agonist, [D-Trp⁶]-LHRH. This product was put on the market in 1986. It is called Décapeptyl and is manufactured and marketed by Ipsen Biotech of France. The polymeric excipient used is 50:50 poly(DL-lactide-co-glycolide). With a single injection, a total of 3.75 mg of [D-Trp⁶]-LHRH is delivered over a 1-month period. It is administered for the treatment of prostate cancer, a testosterone-dependent cancer. The microspheres are made by a phase-separation microencapsulation technique. The product is terminally sterilized with gamma radiation. (Note: A similar LHRH microsphere product, Leupron Depot, is produced aseptically by Takeda Chemical Industries, LTD., Japan. These microspheres are made by a solvent-evaporation microencapsulation process.)

It has also been shown that LHRH analogs can be formulated into microspheres that release LHRH for periods of up to 3 or 6 months (Fig. 5). This result is achieved by mixing microspheres. Each microsphere component within this mixture is made with an excipient having a different ratio of lactide-to-glycolide [12,13].

As more peptides and proteins are discovered and developed, work will continue to develop peptide and protein microsphere formulations with poly(DL-lactide-co-glycolide) and other biocompatible, biodegradable polymeric excipients. The proteins remain more difficult to formulate than peptides. This difficulty arises because proteins are higher in molecular weight and lower in solubility than peptides, which makes release kinetics difficult to manipulate. Moreover, proteins have conformational structures that must be preserved during microsphere manufacturing. And this conformation structure must be preserved when the protein drug is sequestered in the microspheres while in the body before being released into the bloodstream.

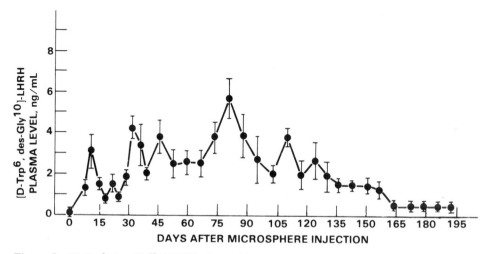

Figure 5 [D-Trp⁶, des-Gly¹⁰]-LHRH ethylamide plasma levels following a single administration to rats of a 6-month mixed-microsphere formulation.

Another concern is the potential of microsphere delivery systems to increase the immunogenicity of peptide or protein drugs. For instance, if peptide/protein microspheres are less than 10 μm, they will be easily taken up by macrophages followed by a production of antibodies against the peptide or protein. For vaccines, this adjuvant phenomenon is advantageous because one can use a microsphere delivery system to increase the immunogenicity of a poor antigen [14,15]. For peptide or protein drug delivery, however, producing antibodies against peptide or protein drugs is a disadvantage, because the antibodies will bind to the peptide or protein drug and prevent the drug from reaching target cells. (Note: For more information about parenteral, controlled delivery of peptides and proteins see reference [16]).

Bromocriptine

The other pioneer injectable microsphere formulation that appeared on the market in 1986 was Sandoz's Parlodel. This is a poly(L-lactide)-based formulation that is designed to release bromocriptine for one month. These microspheres are made by spray-drying.

Bromocriptine, a potent dopamine receptor agonist is used to treat neurological disorders such as Parkinson's disease and endocrinological indications such as prolactin-related dysfunctions and pituitary adenomas. By reducing serum prolactin (PRL) levels, it effectively prevents or suppresses lactation in postpartum women and in patients with hyperprolactinaemia of any etiology. Bromocriptine is also effective in reducing the size of PRL-secreting pituitary adenomas (prolactinomas). During oral bromocriptine treatment, side-effects including hypotension, nausea, vomiting, dizziness, and headaches are often observed making patient maintenance and compliance to treatment difficult.

Results of clinical studies indicate that Parlodel microsphere formulation effectively inhibits or reduces PRL secretion. Therapeutic plasma levels of bromocriptine were obtained within 24 hr of administration of a single intramuscular injection of 50 mg of Parlodel. These levels were maintained, with a corresponding reduction in plasma levels of PRL for up to 35 days. Parlodel has also been shown to be effective in reducing the size of pituitary adenomas. In one study, treatment with Parlodel required 30% less drug than needed with oral treatment for effective inhibition of PRL secretion. And, in most studies, no significant side effects or local reactions have been reported with the Parlodel microsphere product [17–24].

Other Parenteral Microsphere Formulations

Steroids, particularly the synthetic steroids, are potent enough to be useful in parenteral microsphere dosage forms. Norethisterone, for instance, has been formulated into poly(DL-lactide-co-glycolide) excipients to afford 3- and 6-month-releasing microspheres for contraception. Other steroids have been encapsulated as well; they include progesterone, testosterone, testosterone propionate, estradiol, norgestimate, and levonorgestrel. The higher-dose requirement of the natural steroids progesterone and testosterone make these steroids more difficult to formulate to achieve reasonable durations of action.

Other pharmaceuticals that have been incorporated into parenteral microcapsules and microspheres include anticancer drugs, narcotic antagonists, antibiotics, anesthetics, and vaccines.

Example of a Peptide Microsphere Preparation

As an illustration, the following section describes one way to microencapsulate a peptide to afford a parenteral, 1-month microsphere formulation. In this illustration, the peptide is an LHRH analog and the excipient is 50:50 poly (DL-lactide-co-glycolide) (DL-PLG).

Dissolve 4 g of 50:50 DL-PLG in 196 g of methylene chloride. This solution is then placed in a 300-mL resin kettle equipped with a true-bore stirrer having a 2.5-inch Teflon turbine impeller driven by a precision motor. In a 1-dram glass vial dissolve 0.06 g of LHRH analog in 1.3 g of deionized water. This LHRH analog solution is then added to the resin kettle. During this addition, the dilute DL-PLG solution is stirred at 3200 rpm in order to form a water-in-oil (w/o) emulsion. With continued stirring at this rate, 80 mL of silicone oil is added at the rate of 4.0 mL/min by means of a peristaltic pump. The silicone oil causes the DL-PLG to phase separate, and deposit onto the surface of the water-LHRH microdroplets. As the process proceeds, the coated water-LHRH droplets coalesced such that several water-LHRH analog microdroplets are within each embryonic microsphere. The microspheres are then hardened by pouring the contents of the resin kettle into a beaker containing 2000 mL of heptane. This mixture is stirred at 1000 rpm for 30 min with a stainless-steel impeller. Next, the heptane-methylene chloride-silicone oil solution is removed by filtering the solution. The isolated microspheres are then washed repeatedly with heptane to ensure complete removal of the silicone oil. The microspheres can then be dried at room temperature. The resultant microsphere product will consist of particles ranging in size from 10–40 μm. And, the product will have a LHRH analog content of about 2 wt %.

DISPERSED DRUG DELIVERY SYSTEMS

Dispersed drug delivery systems are from the earlier technologies that were used to extend the duration of action of drugs. In general, a dispersion is a heterogeneous mixture consisting of two or more immiscible phases. The "continuous phase" of a dispersion is the medium that *is not* subdivided. The "discontinuous phase" of a dispersion is the medium that *is* subdivided; it consists of either small particles (solids) or droplets (liquids). These particles (or droplets) have well stabilized boundaries which keep them separate from one another (Fig. 6). A variety of dispersed systems exists. Most of them are pharmaceutically important.

When administered parenterally, dispersed drug delivery systems can be given intramuscularly, subcutaneously, or intravenously. When given intravenously, uptake of these systems by the reticuloendothelial system may be a concern or may be advantageously used to target drugs to the reticuloendothelial system.

Emulsions

An emulsion is a mixture of two immiscible liquids, one finely divided into the other. Generally an emulsion drug delivery system consists of the following ingredients [26]:

Discontinuous phase.   A solution of an active ingredient in a suitable solvent and/or a liquid active material.

Continuous phase.   A solvent and/or solution of nonactive ingredients.

Dispersant.   A material used as an aid to disperse the discontinuous phase within the continuous phase; it acts in most cases as a physical stabilizer for the total system.

When water is one of the liquids used in an emulsion composition, there may exist two types of simple emulsions, namely, water-dispersed-in-oil emulsions (w/o) and oil-

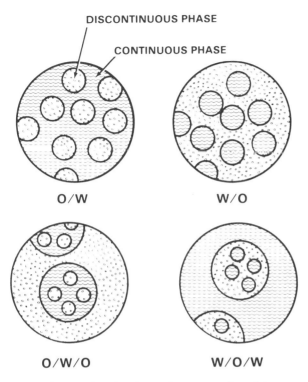

DISCONTINUOUS PHASE

CONTINUOUS PHASE

O/W   W/O

O/W/O   W/O/W

Figure 6 Drawing showing various types of emulsions.

dispersed-in-water emulsions (o/w) (Fig. 6). The oils that can be used for parenteral emulsions are limited and include soybean oil, safflower oil, peanut oil, cottonseed oil, corn oil, and sesame oil. Also multiple emulsions such as w/o/w and o/w/o may be prepared. The droplets of the discontinuous phase in such emulsions are stabilized by the presence of surface active agent(s). These surface active agents, due to their chemical and physical nature, will reduce the interfacial tension between the two phases by positioning themselves at the interface between the discontinuous and continuous phases.

Oil-in-Water Emulsions

Obviously, parenteral emulsions to be administered intravenously, should be the oil-in-water type in order to satisfy the miscibility of the water-based continuous phase with serum. These types of emulsion formulations (o/w) have been used, in general, to deliver poorly water-soluble drugs and nutrients [27,27].

No parenteral controlled-release o/w emulsions have been commercially produced because the partitioning of the oil-soluble drug away from the oil discontinuous phase into the serum is rapid. This rapid partitioning is due to the extremely low oil-to-water ratio which occurs after intravenous injection and is due to the well agitated state in the body which extracts almost all of the drug out of the oil phase. For example, Scieszka et al. [29] used an extremely lipophilic prodrug of flurbiprofen in an o/w emulsion. The presence of a very large interfacial area in the emulsion system caused raid partitioning of the flurbiprofen to take place.

However, if partitioning of a drug is favored toward the oil droplets and the emulsion is given intravenously, the reticuloendothelial system will likely remove the oil droplets from the circulation. If this uptake phenomenon occurs prior to the partitioning of the drug into the serum, then an o/w emulsion can be used as a passive drug-targeting system to the reticuloendothelial system.

Water-in-Oil-in-Water Emulsions

Because the blood, lymph, cerebrospinal fluid, synovial fluid, and urine are all basically aqueous media, sustained-release or controlled-release claims from heterogeneous systems can be made, if the rate of partitioning from oil into aqueous media or dissolution of solid into aqueous media is slow and controllable. In order to reduce diffusion of the drug from an emulsion formulation, Davis et al. developed a multiple emulsion of w/o/w type for intramuscular injection using iodohippuric acid as a model drug [30]. They incorporated the model drug in the internal water phase (w_1) and emulsified this phase within an oil phase comprised of arachis oil and a mixed surfactant system. A stable w_1/o emulsion was obtained. The w_1/o emulsion was then emulsified in an outer aqueous phase (w_2) to obtain $w_1/o/w_2$ type emulsion. The system was stabilized by means of aluminum stearate and cetostearyl alcohol. They have shown that both w_1/o and $w_1/o/w_2$ type emulsions can be utilized as sustained-release delivery systems for intramuscular administration of water-soluble drugs.

Fukushima et al. have also studied this type of emulsion system to evaluate the parameters effecting release rate of antineoplastic agents [31]. They studied glucose, cytarabine, 5-fluorouracil, and peplomycin as model drugs. The release studies revealed that the decrease in size of oil droplets caused an increase in the release rate of cytarabine. On the other hand, neither the volume of the inner aqueous phase, nor the concentration of cytarabine had any effect in its rate of release from w/o/w emulsions. Comparative studies showed that the rate of release of 5-fluorouracil was very fast. Glucose and cytarabine formulations showed a sustained rate of release of drug, while peplomycin was hardly released from these multiple emulsions. They concluded that difference in permeability of various drugs through the oil phase is the reason for the above-mentioned results. A release-rate study, therefore, must be performed for each drug within a given formulation.

Water-in-Oil Emulsions

Water-in-oil type emulsions have been investigated intramucularly as sustained-release delivery systems [32]. But because most pharmaceutical formulators favor aqueous-continuous-phase systems, there is a limited number of nonaqueous emulsions described in the literature [25].

Oil-in-Water-in-Oil Emulsions

Intaarterial use of emulsions for sustained-release purposes have also been explored. For example, Fukushima et al. [33] and Konno et al. [34] have used the hepatic artery in the treatment of hepatic cancer and have found that Lipiodol, an oily lymphographic agent, has remained selectively at a hepatic cancer site after being injected as an o/w/o type emulsion into the hepatic artery [35].

Suspensions

A suspension is a dispersed heterogeneous system in which a solid ingredient is dispersed in a nonsolubilizing liquid. Some interesting reviews have been published in this field

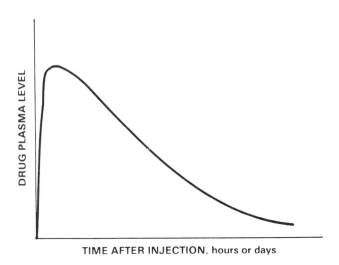

Figure 7   Typical pattern of drug plasma levels following administration of a dispersed drug delivery formulation.

[16–40]. A variety of combinations exists which can help to adjust the rate of release for a particular drug. For example, a water-soluble drug may be dispersed in an oil to reduce its diffusion rate from the formulation into the body and, as a result, obtain an optimum sustained-release effect. While a water-insoluble active ingredient may be suspended in water in order to obtain a sustained-release activity by controlling the rate of dissolution, the release rate of drugs from such systems can be further decreased by incorporating viscosity enhancers in the continuous phase. Figure 7 shows the kind of release profile one would typically obtain using a controlled-release suspension formulation.

Ericsson et al. [41], using liquid crystalline phases, have achieved a sustained-release of the peptide somatostatin after administering the system intramuscularly. They concluded that liquid-crystalline phases may offer a controlled-release parenteral delivery system for active ingredients having a very fast biological degradation rate (short half-life).

Particle size and suspending agents affect the rheological properties of suspensions. Rheologically, an injectable suspension presents a formidable task to achieve. Although an injectable suspension can be formulated to have desirable properties, rheological changes may occur during filling, shipping, injecting, and storage. Therefore, rheological characteristics of injectable suspensions should be carefully and continuously monitored to prevent any possible future difficulties. These properties may be characterized by injectability or syringeability. Syringeability relates to such characteristics as foaming, ease of withdrawal from the container by the syringe, and dose-measurement accuracy. Injectability relates to the properties of the suspension during injection, such as the required force of injection, qualities of aspiration, lack of clogging, and evenness of flow [42].

Preparation of Dispersed Drug Delivery Systems

Generally, a dispersed drug delivery system may be prepared by adding the ingredients in a particular order and mixing them with a given intensity. But, it must be mentioned that parenteral products must be sterile and pyrogen free. The sterility is the subject which needs further exploration. Sterile heterogeneous products can be processed by one of the following ways:

Combining sterile ingredients aseptically
Terminally sterilizing the product either by filtration, radiation, or autoclaving
Using sterile solutions and in situ crystallization

Any of the above methods can be selected after careful consideration of the various parameters affecting the physical and chemical integrity of the delivery system, drug, and method of preparation.

Stability Studies for Dispersed Drug Delivery Systems

Stability is an important subject to be considered when developing heterogeneous products, such as dispersed drug delivery systems. There are various types of stability questions to address [37]; among them are:

Chemical stability of active ingredients
Chemical stability of excipients
Physical stability of the total system
Microbiological stability

The testing of *chemical stability* of active ingredients and excipients is similar to that used for other dosage forms and should quantify not only the active ingedient(s) and excipients but also any degradation product(s).

The testing of *physical stability* required will undoubtedly depend on the particular dosage form [37,43]. In this regard, one source to consult is the United States' Food and Drug Administration Draft Guidelines for Stability Studies for Human Drugs and Biologics. Visual examinations to observe creaming, coalescence, and phase separation of an emulsion is similar to observation for sedimentation and caking of suspensions. Resuspendability on the other hand is a characteristic test for suspension physical stability. Size and size-distribution determinations will give an early warning for eventual phase separation in emulsions and crystal growth in suspensions. Change in pH may indicate hydrolytic degradation. Rheological evaluation, especially in suspensions, will indicate the stability with respect to syringeability and injectability.

The testing of *microbiological stability* include biological assay, pyrogen testing, sterility testing, and effectiveness of preservatives (if present).

LIPOSOMES

Although Bangham [44] was the first to make liposomes and stated that ". . . phase structure appears to be that of a layer latice giving rise to spherulites . . . consisting of many concentric bimolecular layers of lipid each separated by an aqueous compartment," Weissmann [45] coined the term "Liposome" to define "Phospholipid Spherules." The creation of the word liposome probably comes from the terminology of subcellular particles like glycosome, lysosome, and ribosome. The term liposome meaning "lipid body" may generally be defined as any structure with enclosed volume that is composed of lipid bilayers. A variety of lipid molecules have been used to create liposomal structures. And as a result, different names have been devised to differentiate between them. In Table 1, some of the chemicals used to prepare these lipid structures along with the designated name/term are presented. Generally, liposomes consist of bilayer forming lipids, cho-

Table 1 Partial List of Chemicals Used in the Preparation of Various Lipid Sructures and Their Designated Names

Lipid structure used	Designated name	Ref.
Phospholipids[a]	Liposome	Many
Oleic acid	Ufasomes	53
Ether derivatives of ammonium		54
amphiphiles (two headed)	Niosomes	55
Cationic surfactants (double tail)	—	56
Cationic with amino acid residue	—	57
Zwiterionic (two chain)	—	57
Sucrose fatty acids (two tail)	—	58
Cationic and anionic (single tail)	—	59
Polyoxyethylene alkyl ethers	Novasome	60
Sorbitan alkyl ester	Novasome	60
Polyoxyethylene alkyl esters	Novasome	61
Diethanolamides	Novasome	61
Long-chain acyl hexasomides	Novasome	61
Long-chain acyl amino acid amides	Novasome	61
Long-chain amides	Novasome	61
Polyoxyethylene(20) sorbitan monooleate	Novasome	61
Polyoxyethylene(20) sorbitan trioleate	Novasome	61
Polyoxyethylene glyceryl monostearate	Novasome[b]	61
Sphingolipids and ceramides	Sphingosome	62

[a] See text for details. Source of phospholipids may also vary, and includes egg, soybean, and synthetic sources with a different degree of purity depending on use.
[b] Various ingredients in this category have been used.

lesterol, and a charge-generating molecule. But, it should be emphasized that only the bilayer-forming lipid is the essential part of the formulation and the other two components are added to the formula to impart certain characteristics to the vesicles. For example, cholesterol is added to stabilize the bilayer structure and to reduce its permeability. Figure 8 shows an electron micrograph of freeze fractured liposomes.

Phospholipids were among the first chemicals used to form bilayer structures in order to mimic the cell membrane [44–46]. Soon the potential of these structures as a drug-delivery system was realized. Liposomes as drug-delivery systems will be discussed briefly in the following section. The chemical structure of a phospholipid molecule (phosphatidyl choline) is shown.

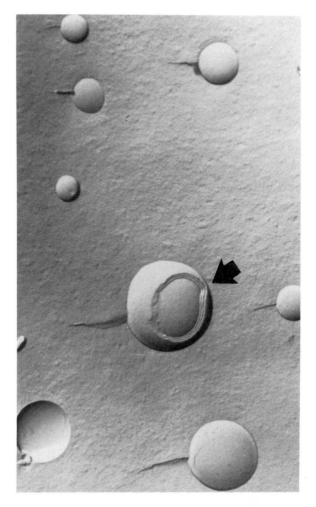

Figure 8   Scanning electron micrograph of freeze-fractured
liposomes. The arrow points out the multilamellarity of one
particular liposome.

It is important to note that many thousands of scientific papers containing the word
''liposome'' have been published. Therefore, the purpose of the following section is to
concentrate on the basic understanding of liposomes and to discuss the use of liposomes
as controlled-release or sustained-release drug-delivery systems. Interested readers should
refer to scientific reviews on this subject [47–51]. These reviews indicate that liposomes
display therapeutic advantage in many therapeutic areas when combined with antifungal,
anticancer, antiviral, and antimicrobial compounds. As with dispersed drug-delivery systems, uptake of liposomes by the reticuloendothelial system limits their therapeutic applications.

Liposomes have even been considered for the topical delivery of drugs to enhance
the penetration of the active ingredients deep into the stratum corneum [47–51]. While
claims of enhancement of penetration have been advanced, it is not clear that the difficulties
in manufacturing liposomes outweigh the simplicity of emulsions.

There are various terms and methods used in the literature to classify the different morphological categories of liposomes. The following is the classification which combines most of the definitions:

MLV. Multilamellar vesicles where multiple layers of "onion-like" lamellae structure surrounds a relatively small internal volume, and as mentioned above, produced and defined by Bangham [44].

OLV. Oligolamellar vesicles where the large central aqueous compartment is surrounded by two to ten bilayer structures, also called Paucilamellar vesicles (PLV) [52].

ULV. Unilamellar vesicles where there is only a single bilayer structure surrounding the internal aqueous compartment. This particular category has several subcategories based on the sizes of the vesicles.:

SUV. Small unilamellar vesicles with the size range of 20–40 nanometer, having very little use for drug delivery.

MUV. Medium unilamellar vesicles with a size range of 40–80 nanometer.

LUV. Large unilamellar vesicles with a large internal aqueous compartment and in the size range of 100–1000 nanometer.

GUV. Giant unilamellar vesicles where the size is larger than 1000 nanometer.

MVV. Multi-vesicular vesicles where a large vesicle contains smaller and usually unilamellar vesicles.

Although the method of preparation of MLVs have been described by Bangham [44], different methods of preparing liposomes have been reviewed by many investigators in the field. Many methods have been developed to prepare multilamellar and unilamellar liposomes and are partially listed in Table 2. Due to the existence of various types of liposomes, the characterization of these lipid vesicles become extremely important if one desires reproducible results. Four major characterization methods are briefly described here.

Size and Size Distribution

Most phospholipids adapt a bilayer structure spontaneously upon dispersing in water which is not necessarily true in case of other lipids. The resultant liposomes, irrespective of the

Table 2 Partial List of Preparation Methods and Entrapped Active Ingredients

Type	Method of preparation	Entrapped agent	Ref
MLV	Solvent evaporation, sonication	Oxytetracycline	51,63
MLV	Sonication, dehydration-rehydration	Melphalin, vincristine	64
MLV	Sonication, freeze-thaw	Asparaginase	63
MLV	Hydration by injection	—	52
MLV	Mechanical processing	Carboxyfluorescein	65
MLV	Freeze-thaw technique	Inulin	65
LUV	Solvent evaporation	Glucose, DNA	50,51
LUV	Detergent removal	Cytochrome C	50
LUV	Reverse-phase evaporation	Insulin	66
LUV	Extrusion	Methotrexate	67
SUV	Sonication	Cytosine arabinoside	67
SUV	Detergent removal	Carboxyfluorescein	50
SUV	French press	Carboxyfluorescein	50

lipid used in their preparation, may be of a large and/or small size depending upon a variety of factors including lipid composition, lipid structure, and processing technique and may differ with respect to size distribution. It is, therefore, important to determine the size and size distribution of liposome preparations and to determine their stability after storing for a certain time period. The determination of size and stability is important because the size and size distribution of a batch of liposomes may change during storage, thus affecting their function. The laser, light-scattering technique is the most common method used for size analyses of liposome formulations.

Lamellarity

The various types of liposomes have different degrees of lamellarity. Therefore, it seems reasonable to characterize liposomes by their lamellarity. Different experimental procedures have been employed to determine the lamellarity such as labeling and binding studies. Labeling with ^{31}P and then employing nuclear magnetic resonance (NMR) to determine the phosphorus signal intensity in phospholipid liposomes is one of the best and most precise methods of characterizing the lamellarity of liposomes. It is obvious that this method of labeling will not be suitable for nonphospholipid vesicles and the presence of ^{14}C in the head group may be required. Another method, but not as precise a method of characterizing lamellarity, is scanning electron microscopy.

Entrapped Volume

The entrapped volume is generally defined as the amount of entrapped volume per mole of lipid. This parameter can vary from about 0.5–30 liter per mole of lipid (generally expressed as μL/μmol). It should also be noted that the entrapped volume is much smaller for MLV liposomes. Entrapped volume is generally determined by entrapping an impermeable radiolabeled molecule such as inulin within the liposome, and removing the external radioactivity by dialysis, gel filtration, or centrifugation, and then determining the residual radioactivity. This method assumes that there is no binding of the inulin to the liposome.

Solute Distribution

Solute may partition between the lipid and aqueous phase which is not specific for liposomes, but varying degrees of distribution of the solute among the various bilayers within the MLV system is a subject for consideration. This varying degree of solute distribution is probably due to the hydration sequence of dry lipid film coming in contact with aqueous buffer solution. Also, various materials with different degrees of diffusivity permeate in different rates through the sequence of hydrating lipid bilayers in the MLV. The solute distribution may be determined by NMR.

Although various factors affect the above-mentioned liposome characteristics, the pharmacokinetics and biodistribution of entrapped active ingredient mainly depends on the route of administration. Due to their particulate nature, liposomes, upon intravenous injection, will be taken up by the reticuloendothelial system, rendering little sustained-release activity. On the other hand, if liposomes are injected intramuscularly or subcutaneously, they will act as sustained-release delivery systems. For example, Fig. 9 illustrates the performance of a liposomal, sustained-release delivery system given intramuscularly [68]. It has also been reported that one may dissolve a water-insoluble protein in a lipid based solubilizer and then entrap the solubilized protein within a liposomal structure to provide controlled release. This formulation was then injected either intramuscularly or subcutaneously to obtain a sustained-release action for the period of about one month [69]. Also reported is a sustained-release liposomal ophthalmic delivery system that will

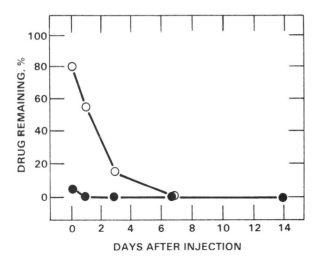

Figure 9 ^{125}I-human growth hormone remaining at an intramuscular injection site following administration of (●) free drug and (○) drug-loaded liposomes. (Adapted from Ref. 68.)

deliver an antimicrobial agent to the infected eye for three days to treat infectious bovine keratoconjunctivitis [69]. The major advantages of liposomal drug delivery systems are their ability to either deliver drugs *to* a target organ or reduce the delivery of drugs *from* a certain organ [70]

SUMMARY

Parenteral controlled-release dosage forms vary from the older, but still useful, technologies involving emulsions, drug solutions, and drug suspensions to new and more sophisticated technologies, such as liposomes and biodegradable microspheres. There is no doubt that parenteral controlled-release dosage forms will continue have a profound effect on the pharmaceutical industry especially because of their potential to efficaciously deliver the biotech drugs of the future.

PROBLEMS

1. What are the advantages and disadvantages of conventional/controlled-release dosage forms given orally and parenterally?
2. What factors should be considered to determine if a drug can be formulated into a parenteral controlled-release dosage form?
3. What is the difference between a microcapsule and microsphere?
4. Why are injectable microspheres an ideal formulation approach to deliver peptides and proteins?
5. Describe the characteristics of dispersed drug-delivery systems.
6. How do emulsions and suspensions sustain the release of drugs?
7. Describe the characteristics of liposomes.
8. What are the advantages and disadvantages of the reticuloendothelial system interacting with nanosphere, emulsion, suspension, and liposome drug-delivery systems that are administered intravenously?

REFERENCES

1. K. E. Avis, L. Lachman, and H. A. Lieberman, Eds., *Pharmaceutical Dosage Forms: Parenteral Medications*, Vol. 1, Marcel Dekker, New York (1984).

2. *Handbook of Pharmaceutical Excipients*, American Pharmaceutical Association, Washington, D.C. and The Pharmaceutical Society of Great Britain, London, England (1986).

3. G. E. Visscher, R. L. Robison, H. V. Maulding, J. W. Fong, J. E. Pearson, and G. I. Argentieri, Biodegradation of and tissue reaction to 50:50 poly(DL-lactide-co-glycolide) microcapsules, *J. Biomed. Mater. Res.*, *19*:349–365 (1985).

4. G. E. Visscher, R. L. Robison, H. V. Maulding, J. W. Fong, J. E. Pearson, and G. I. Argentieri, Note: biodegradation of and tissue reaction to microcapsules, *J. Biomed. Mater. Res.*, *20*:667–676 (1986).

5. G. E. Visscher, R. L. Robison, and G. I. Argentieri, Tissue response to biodegradable injectable microcapsules, *J. Biomat. Appl.*, *2*:118–131 (1987).

6. L. R. Beck and T. R. Tice, *Long-Acting Steroid Contraception*, Raven Press, New York, pp. 175–199 (1983).

7. L. M. Sanders, G. I. McRae, K. M. Vitale, B. H. Vickery, and J. S. Kent, An injectable biodegradable controlled release delivery system for nafarelin acetate, *Int. Congr. Ser.-Excerpta Med.*, *656*:53–58 (1984).

8. L. M. Sanders, J. S. Kent, G. I. McRae, B. H. Vickery, T. R. Tice, and D. H. Lewis, Controlled release of a luteinizing hormone-releasing hormone analogue from poly(d,l-lactide-co-glycolide) microspheres, *J. Pharm. Sci.*, *73*(9):1294–1297 (1984).

9. L. M. Sanders, G. I. McRae, K. M. Vitale, and B. A. Kell, Controlled delivery of an LHRH analogue from biodegradable injectable microspheres, *J. Control. Rel.*, *2(1–4)*:187–195 (1986).

10. L. M. Sanders, B. A. Kell, G. I. McRae, and G. W. Whitehead, Prolonged controlled-release of nafarelin, a luteinizing hormone-releasing hormone analogue, from biodegradable polymeric implants: influence of composition and molecular weight of polymer, *J. Pharm. Sci.*, *75*(4):356–360 (1986).

11. L. M. Sanders, R. Burns, K. Vitale, and P. Hoffman, Clinical performance of nafarelin controlled release injectable: influence of formulation parameters on release kinetics and duration of efficacy, *Proc. Intl. Symp. Control. Rel. Bioact. Mater.*, *15*:62–63 (1988).

12. T. R. Tice, D. W. Mason, J. H. Eldridge, and R. M. Gilley, *Novel Drug Delivery and Its Therapeutic Application*, Wiley, Chichester, England, pp. 223–235 (1989).

13. D. Lacoste, F. Labrie, D. Dubé, A. Bélanger, T. R. Tice, R. M. Gilley, and K. L. Pledger, Reversible inhibition of testicular androgen secretion by 3-, 5-, and 6-month controlled-release microsphere formulations of the LHRH agonist [D-Trp^6, des-Gly-NH$_2^{10}$] LHRH ethylamide in the dog, *J. Steroid Biochem.*, *33*(5):1007–1011 (1989).

14. J. K. Staas, J. H. Eldridge, J. D. Morgan, O. B. Finch, T. R. Tice, and R. M. Gilley, Microsphere vaccines: enhanced immune response through adjuvant effect and multiple-pulse capability, *Proc. Intl. Symp. Control. Rel. Bioact. Mater.* Paper to be presented at the 18th International Symposium on Controlled Release of Bioactive Materials; July 8–11 (1991).

15. E. Fernandez-Repollet and A. Schwartz, *Microspheres: Medical and Biological Applications*, CRC Press, Boca Raton, Florida, pp 139–163 (1988).

16. C. G. Pitt, The controlled parenteral delivery of polypeptides and proteins, *Int. J. Pharm.*, *59*:173–196 (1989).

17. T. Kissel, W. Bruckner, Z. Brich, J. Lancranjan, J. Rosenthaler, F. Nimmerfall, W. Prikoszovich, and P. Vit, Microspheres as depot-injections, an industrial perspective, *Proc. Intl. Symp. Control. Rel. Bioact. Mater.*, *15*:260–261 (1988).

18. F. Peters, E. del Pozo, A. Conti, and M. Breckwoldt, Inhibition of lactation by a long-acting Bromocriptine,'' *Obstet. and Gynec.*, *67*:82–85 (1982).

19. A. Grosman, J. A. H. Wass, and M. Besser, The rapid diagnosis of sensitivity or resistance to dopamine agonists with depot bromocriptine, *Acta Endocrinol.*, *116*:275–281 (1987).

20. L. Svanberg, I. Lancranjan, T. Arvidsson, and B. Andersch, Single dose bromocriptine microcapsules in postpartum lactation inhibition, *Acta Obstet. Gynecol. Scan.*, *66*:61–62 (1987).

21. A. M. Landolt, E. del Pozo, and J. Hayek, Injectable bromocriptine to treat acute, oestrogen-induced swelling of invasive prolactinoma, *Lancet*, *2*:111 (1984).

22. M. Montini, G. Pagani, D. Gianola, M. D. Pagani, M. Salmoiraghi, L. Ferrari, and I. Lancranjan, Long lasting suppression of prolactin secretion and raid shrinkage of prolactinomas after a long-acting injectable form of bromocriptine, *J. Clin. Endocrinol. Metab.*, *63*:266–268 (1986).

23. M. D. Bronstein, C. S. Cibele, and R. Marino, Jr., Short-term management of macroprolactinomas with a new injectable form of bromocriptine, *Surg. Neurol.*, *28*:31–37 (1987).

24. E. J. Van Cutsem and S. W. J. Lamberts, Long-acting injectable bromocriptine treatment for macroprolactinomas, *Neth. J. Med.*, *32*:112–117 (1988).

25. L. H. Block, in *Pharmaceutical Dosage Forms: Disperse Systems*, vol. 2 (K. E. Avis, L. Lachman, and H. A. Lieberman, Eds.), Marcel Dekker, New York, pp. 335–378 (1989).

26. S. E. Tabibi, in *Specialized Drug Delivery Systems* (P. Tyle, ed.), Marcel Dekker, New York, pp. 317–331 (1990).

27. P. K. Hansrani, S. S. Davis, and M. J. Groves, The preparation and properties of sterile intravenous emulsions, *J. Parent. Sci. Technol.*, *37*:145 (1983).

28. L. C. Haynes and M. J. Cho, Mechanism of nile red transfer from o/w (oil-in-waer) emulsions as carriers for passive drug targeting to peritoneal macrophages in vitro, *Int. J. Pharm.*, *45*:169 (1988).

29. J. F. Scieszka, T. J. Vidmar, L. C. Haynes, and M. J. Cho, *Int. J. Pharm.*, *45*:165 (1988).

30. S. S. Davis, L. Illum, and I. M. Walker, *Int. J. Pharm.*, *38*:133 (1987).

31. S. Fukushima, M. Nishida, and M. Nakano, Preparation of and drug release from w/o/w type double emulsions containing anticancer agents using an oily lymphographic agent as an oil phase, *Chem. Pharm. Bull.*, *35*:3375 (1987).

32. J. J. Windheuser, M. L. Best, and J. H. Perrin, Evaluation of sustained parenteral emulsions, *Bull. Parenter. Drug Assoc.*, *24*:286 (1970).

33. S. Fukushima, M. Nishida, T. Shibuta, K. Juni, M. Nakano, N. Uchara, T. Ohkuma, Y. Yamishita, and M. Takahashi, Preparation and evaluation of w/o type and w/o/w type emulsions containing and anticancer drug, *J. Pharmacobio-Dyn.*, *9*:s–18 (1986).

34. T. Konno, H. Maeda, K. Iwai, S. Tashiro, S. Maki, T. Morinaga, M. Mochinaga, T. Hiraoka, and I. Yokoyama, Effect of arterial administration of high-molecular-weight anticancer agent, *Eur. J. Cancer Clin. Oncol*, *19*:1053 (1983).

35. K. Iwai, H. Maeda, and T. Konno, Use of oily contrast medium for selective drug treatment to tumor: enhanced therapeutic effect and x-ray image, *Cancer Res.*, *44*:2115 (1984).

36. S-H. S. Leung, J. R. Robinson, and V. H. L. Lee, in *Controlled Drug Delivery, Fundamentals and Applications*, 2d ed. (J. R. Robinson and V. H. L. Lee, Eds.), Marcel Dekker, New York, pp. 433–480 (1987).

37. J. R. Boyett and C. W. Davis, in *Pharmaceutical Dosage Forms; Disperse Systems*, Vol. 2 (K. E. Avis, L. Lachman, and H. A. Lieberman, Eds.), Marcel Dekker, New York, pp. 379–416 (1989).

38. Y. W. Chien, *Novel Drug Delivery Systems, Fundamentals, Development Concepts, Biomedical Assessments*, Marcel Dekker, New York, pp. 219–310 (1982).

39. M. J. Akers, A. L. Fites, and R. L. Robison, Formulation design and development of parenteral suspensions, *J. Parent. Sci. Technol.*, *41*:88 (1988).

40. S. S. Davis, C. Washington, P. West, L. Illum, G. Liversidge, L. Sternson, and R. Kirish, in *Biological Approaches to the Controlled Delivery of Drugs*, The New York Academy of Science, New York, pp. 75–88 (1987).

41. B. Ericsson, S. Leander, and M. Ohlin, *Proc. Intl. Symp. Control. Rel. Bioact. Mater.*, *15*:382 (1988).

42. P. P. DeLuca and J. C. Boylan, in *Pharmaceutical Dosage Forms: Parenteral Medications*, Vol. 1 (K. E. Avis, L. Lachman, and H. A. Lieberman, Eds.), Marcel Dekker, New York, pp. 139–201 (1984).

43. G. Zografi, "Physical Stability Assessment of Emulsions and Related Disperse Systems: a Critical Review," *J. Soc. Cosmet. Chem.*, *33*:345 (1982).

44. A. D. Bangham, M. M. Standish, and J. C. Watkins, Diffusion of univalent ions across the lamellae of swollen phospholipids, *J. Mol. Biol.*, *13*:238 (1965).

45. G. Sessa and G. Weissmann, Phospholipid spherules (liposomes) as a model for biological membrane, *J. Lipid. Res.*, *9*:310 (1968).

46. D. F. H. Wallach, P. A. Maurice, B. A. Steele, and D. M. Surgenor, Studies on the relationship between the colloidal state and the clot-promoting activity of pure phosphatydylethanolamines, *J. Biol. Chem.*, *234*:2829 (1959).

47. G. Gregoriadis, Drug entrapment in liposomes, *FEBS Lett.*, *36*:292 (1973).

48. A. D. Bangham, Introduction, in *Liposomes from Physical Structure to Therapeutic Applications*, Elsevier, New York (1981).

49. M. J. Ostro, *Liposomes*, Marcel Dekker, New York (1983).

50. G. Gregoriadis, Ed., *Liposome Technology*, Vols 1–3, CRC Press, Boca Raton, Florida (1984).

51. M. J. Ostro, *Liposomes, From Biophysics to Therapeutics*, Marcel Dekker, New York (1987).

52. D. F. H. Wallach, Paucilamellar vesicles. U.S. Patent 4,911,928 (1990).

53. J. M. Gebicki and M. Hicks, Ufasomes are stable particles surrounded by unsaturated fatty acid membranes, *Nature*, *243*:232 (1973).

54. R. M. Handjani-Vila, A. Ribier, and G. Vanlenberghe, Les niosomes, *Les Liposomes*, *Technique et Documentation Lavoisier*, 297–312 (1985).

55. T. Kunitake and Y. Okahata, A totally synthetic bilayer membrane, *J. Am. Chem. Soc.*, *99*:3860 (1977).

56. K. Kano, A. Romero, B. Djermouni, H. J. Ache, and J. H. Fendler, Characterization of surfactant vesicles as membrane mimetic agents. 2. temperature-dependent changes of the turbidity, viscosity, fluorescence polarization of 2-methylanthracene, and positron annihilation in sonicated dioctadecyldimethylammonium chloride, *J. Am. Chem. Soc.*, *101*:4030 (1979).

57. Y. Murakami, A. Nakano, and H. Ikeda, Preparation of stable-single-compartment vesicles with cationic and zwitterionic amphiphiles involving amino acid residues, *J. Org. Chem.*, *47*:2137 (1982).

58. Y. Ishigami and H, Machida, Vesicles from sucrose fatty acid esters, *J. Am. Oil Chem. Soc.*, *66*:599 (1989).

59. E. W. Kaler, K. Murthy, B. E. Rodriguez, and J. A. Zasadziniski, *Science*, *245*:1371 (1989).

60. D. F. H. Wallach, and J. Philippot, Method of manufacturing unilamellar lipid vesicles. U.S. Patent 4,853,228 (1989).

61. D. F. H. Wallach, Lipid Vesicles Formed of Surfactants and Steroids, U.S. Patent 4,917,951 (1990).

62. D. Brunke, Specific properties of sphingosomes, *SOFW*, *116*:53 (1990).

63. T. Ohsawa, H. Miura, and K. Harada, Evaluation of a new liposome preparation technique, the freeze-thawing method, using L-asparginase as a model drug, *Pharm. Bull.*, *33*:2916 (1985).

64. C. J. Kirby and G. Gregoriadis, Dehydration-rehydration vesicles: a simple method for high yield drug entrapment in liposomes, *Biotechnology*, *2*:979 (1984).

65. L. D. Mayer, M. J. Hope, R. P. Cullis, and A. S. Janoff, Solute distribution and trapping efficiencies observed in freeze-thawed multilamellar vesicles, *Biochim. Biophys. Acta*, *817*:193 (1985).

66. F. Szoka and D. Papahadjopoulos, Procedure for preparation of liposomes with large internal

aqueous space and high capture by reverse-phase evaporation, *Proc. Natl. Acad. Sci. U.S.A.*, 75:4194 (1978).

67. M. J. Hope, M. B. Bally, G. Webb, and P. R. Cullis, Production of large unilamellar vesicles by a rapid extrusion procedure, characterization of size distribution, trapped volume and ability to maintain a membrane potential, *Biochim. Biophys. Acta*, *812*:55 (1985).

68. A. L. Weiner, S. S. Carpenter-Green, E. C. Soehngen, R. P. Lenk and M. C. Popescu, Liposome-collagen gel matrix: a novel sustained drug delivery system, *J. Pharm. Sci.*, *44*:922 (1985).

69. Anon., The Liposome Company, Brochure, May, 1984.

70. M. Riaz, N. Weiner, and F. Martin, *Liposomes in Pharmaceutical Dosage Forms: Disperse Systems*, Vol. 2 (M. M. Reiger and G. S. Banker, eds.), Marcel Dekker, New York, 1989, pp. 567–603.

8

Transdermal Delivery

Thomas J. Franz   *University of Arkansas for Medical Sciences, Little Rock, Arkansas*

Kakuji Tojo   *Kyushu Institute of Technology, Iizuka, Japan*

Kishore R. Shah and Agis Kydonieus   *ConvaTec, A Bristol-Myers Squibb Company, Princeton, New Jersey*

INTRODUCTION

Transdermal delivery may be defined as the delivery of drugs through intact skin to reach the systemic circulation in sufficient quantity to administer a therapeutic dose.

During the last decade, transdermal delivery has received increasing attention in the face of growing awareness that drugs administered by conventional means are frequently excessively toxic and sometimes ineffective. Thus, conventionally administered drugs in the form of pills, capsules, injectables, and ointments are introduced into the body as pulses that usually produce large fluctuations of drug concentrations in the bloodstream and tissues and, consequently, unfavorable patterns of efficacy and toxicity.

As shown in Fig. 1, transdermal delivery affords an improved approach to the administration of drugs by maintaining a therapeutic but constant concentration of drug in the blood for a desired period of time, usually between 1 and 7 days [1]. Theoretically, the two most important advantages of transdermal delivery are (1) reduction of side effects due to optimization of the blood concentration-time profile; and (2) extended duration of activity, which allows greater patient compliance owing to elimination of multiple dosing schedules. Transdermal delivery may also increase the therapeutic value of many drugs by obviating specific problems associated with the drug, e.g., gastrointestinal irritation, low absorption, decomposition due to hepatic ''first-pass'' effect, formation of metabolites that cause side effects, and short half-life necessitating frequent dosing.

The choice of drugs to be delivered transdermally is a most difficult one, and careful consideration should be given to each application before large expenditures are committed to clinical testing. Table 1 shows the important criteria that should be considered in the drug selection process. Currently, only a few drugs can be delivered transdermally, owing to three basic limitations: inadequate permeability through the skin, inadequate tolerability

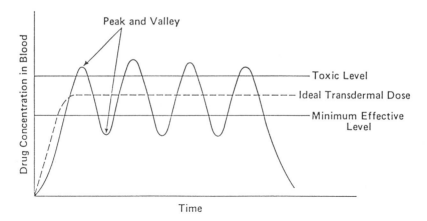

Figure 1 Hypothetical blood level pattern from a conventional multiple dosing schedule, and the idealized pattern from a transdermal controlled-release system. (From Ref. 1.)

of the drug by the skin, and clinical need. The stratum corneum layer of the skin is an exceptionally effective barrier to most chemicals, including drugs. Thus, very few drugs can permeate this horny layer in amounts sufficient to deliver a therapeutic dose. Fortunately, skin permeation models and adequate experimental data exist that allow one to make a reasonable prediction of a drug's permeability through skin. Figure 2 shows one such predictive model, which is based on the experimental data of 25 drugs [2]. Chemical and nonchemical enhancers used to increase permeation exacerbate the skin's tolerability of the drug and enhancers. We now appreciate that the skin performs a variety of tasks that maintain a homeostatic balance between the "inside" of the body and the "outside" environment. Tasks significant in transdermal delivery include regulation of heat and water loss (hydration, occlusion), resistance to mechanical stress (ultrasonic enhancement), and protection of the host from toxic materials (drugs, enhancers). Protection of host from drugs and enhancers comes in the form of skin irritation and sensitization. Skin irritation

Table 1 Important Criteria in the Drug Selection Process

1. Adequate skin permeability:
    Drugs with low molecular weight
    Drugs with low melting point
    Drugs with moderate oil and water solubility
    Potent drugs
2. Adequate skin acceptability:
    Nonirritating drugs
    Nonsensitizing drugs
    Nonmetabolizing drugs
3. Adequate clinical need:
    Need to prolong administration
    Need to increase patient compliance
    Need to reduce side effects on nontarget tissues

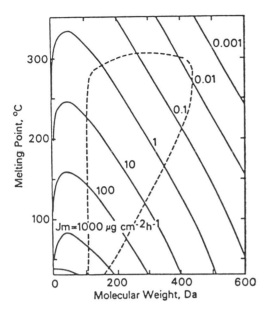

Figure 2   Constant flux contours for percutaneous absorption of drugs from saturated vehicles. (From Ref. 2.)

(or contact irritant dermatitis) results from the direct toxic injury to cell membranes, cytoplasm, or nuclei. To repair the damage, the arachidonic acid cascade is initiated through which the lipoxygenase and cycloxygenase pathways produce very potent inflammatory chemicals, such as leukotrienes, prostaglandins, and thromboxanes. Skin sensitization, or contact allergic dermatitis, involves mainly a host immunologic activity which is mediated by the different types of skin cells, e.g., keratinocytes, Langerhans, macrophages, and mast cells, which can synthesize a variety of cytokines and other bioactive materials that are responsible for the local skin immune reaction and systemic immune responses. It is expected that over 80% of all drugs, including complete drug families such as antihistamines, beta blockers, ACE inhibitors, antiasthmatics, and nonsteroidal anti-inflammatories, will have significant irritation or sensitization problems.

Clinical need is another important area of consideration and possible limitation. Most drugs are adequately administered by oral medication, so it is imperative that an important clinical need be addressed for any proposed transdermal delivery system. Of the commercial products available, bypassing the hepatic ''first pass'' (inactive metabolites) was the clinical need for the nitroglycerine and ISDN systems; extending the duration of activity (t.i.d. versus one patch every 7 days) was needed for clonidine; minimizing toxic side effects (central nervous system, tachycardia, drowsiness, dry mouth) and greater duration of activity (four to six doses daily versus one patch every 3 days) was needed for scopolamine; and bypassing the hepatic ''first pass'' (toxic proteins), minimizing side effects (hypertension, hyperlipidemia, hypercoagulability), and prolonging duration of activity (once per day versus $3\frac{1}{2}$ days per patch) was needed for estradiol.

In addition to skin, drug, and drug-skin interactions, drug pharmacokinetics and design of the transdermal delivery device are also very important. In transdermal drug delivery, pharmacokinetics are important because target tissues are seldom directly accessible, and drugs must be transported from the portal of entry on the body through a variety of biological interfaces to reach the desired receptor site. During this transport, the drug can undergo severe biochemical degradation and, thereby, produce a delivery pattern at the receptor site that differs markedly from the pattern of drug release into the system.

The process of molecular diffusion through polymers and synthetic membranes has been used as an effective and reliable means of attaining transdermal controlled release of drugs and pharmacologically active agents. Although many types of devices have been developed, including monolithic, membrane-controlled liquid reservoir, and membrane-controlled polymeric laminates, all commercial systems are based on molecular diffusion as presented in Chapter 3.

Based on the parameters most important to transdermal delivery, the rest of this chapter is divided into three sections: (1) skin as a barrier membrane and an an immunologic organ; (2) the pharmacokinetics of skin permeation, and (3) transdermal devices.

SKIN AS BARRIER MEMBRANE AND AS IMMUNOLOGIC ORGAN

Historical Perspective

Like all epithelial systems of the body, the prime function of skin is to serve as a barrier, to keep water and other vital substances in and foreign material out. However, in contrast to the epithelia that cover internal structures (e.g., oral and nasal mucosae, lungs, and gastrointestinal tract), skin is the *only* epithelial system that functions in a *dry* environment. To survive and protect its own integrity from the ravages of desiccation and to fulfill its functional obligations to the body, skin developed a specialized structure of *unique physical-chemical composition*, the stratum corneum (horny layer).

The stratum corneum, which is the dead outermost layer of skin, is a multilayered structure consisting of flat anucleate cells totally devoid of normal intracellular structures (e.g., microsomes and mitochondria). Though metabolically inactive, enzymatic activity essential to its function persists. That the "barrier" properties of skin do indeed reside in the stratum corneum was not clearly demonstrated until the work of Winsor and Burch [3]. Through a series of simple in-vitro experiments, they were able to show that destruction of the stratum corneum by sandpapering led to a large increase in the permeability of skin to water. Others have subsequently duplicated the finding using a more sensitive technique, cellophane tape-stripping, in which the stratum corneum can be removed layer by layer. Destruction of the stratum corneum not only results in increased permeability of the skin to water, but to most other substances as well.

Although the first studies to demonstrate the barrier properties of stratum corneum all made use of destructive techniques, a more definitive experiment was conducted using a pure preparation of intact stratum corneum. Scheuplein measured the diffusion of water in vitro through the isolated horny layer and found that the permeability coefficient obtained from this pure tissue was the same as that of full-thickness skin [4]. Thus, it is clear that the resistance to water movement resides *entirely* in the stratum corneum, and it is this layer that is the *rate-limiting barrier* of the skin.

Ensuing work during the next 25 years has served to better define the nature of the stratum corneum barrier and has shown that the stratum corneum is also the rate-limiting barrier to the diffusion of virtually all molecular species. During this time, our concept of the barrier has changed dramatically. Whereas during the early part of this century the skin was thought impenetrable, skin is now known to be a barrier of variable permeability, and, in fact, can serve as a route for the systemic delivery of therapeutic substances as well as the accidental delivery of toxic materials.

Most recently, the unique structural composition underlying barrier function has been elucidated and its biochemical nature defined. It has been shown that *lipid bilayers*, analogous to those of cell membranes, fill the extracellular space of the stratum corneum [5]. Since the stratum corneum is a multilayered structure, 15 to 25 cells thick over most of the body, all permeating substances must traverse this space at some point (ignoring for the moment the role of shunts in the process of percutaneous absorption). There is no pure transcellular pathway available through the barrier.

Thus, the multiple lipid bilayers that encase the cells of the stratum corneum create a continuous hydrophobic environment that is most effective for limiting loss of water and other physiologic (hydrophilic) substances from the body. Not surprisingly, this structural element is not found anywhere else in the body, since free and rapid water movement is requisite to normal cellular function in all other organs.

Anatomy and Biochemistry

Skin Structure

Overview. The skin is the largest organ of the body, 15,000 to 20,000 cm^2 in area in most adults, varying in thickness from approximately 1.5 to 4 mm, and weighing approximately 2 kg. It consists of two parts: (1) the cellular outermost layer, epidermis; and (2) the relatively acellular connective tissue matrix, dermis (Fig. 3). Lying between these two layers is a submicroscopic structure, the basal lamina or basement membrane zone, which is derived from both the epidermis and dermis and serves as the anchoring structure by which the two layers of skin are held together. The blood supply to the skin resides exclusively in the dermis, and the nutritional requirements of the epidermis are met by diffusion.

The epidermis is a continually renewing, stratified squamous epithelium covering the entire outer surface of the body and varying in thickness regionally from 0.04 to 1.5 mm. Regional variation in thickness is largely the result of variation in stratum corneum thickness, since the horny layer is greatly thickened over the palms and soles to subserve the additional mechanical needs of those two sites. Over most of the body, the epidermis is approximately 0.06 to 0.1 mm thick.

The epidermis is composed of two parts: the living cells of the Malpighian layer, which in turn can be subdivided into several strata, and the dead cells of the horny layer (stratum corneum). The prime function of the viable cells of the epidermis is to move progressively through a process of differentiation, eventually to die (terminal differentiation), and through this mechanism generate the barrier layer. In the skin, the process of differentiation has been equated to keratinization, though many other complex biochemical and structural changes unrelated to the synthesis and deposition of keratin also occur. The major identifiable changes that take place as cells undergo differentiation are (1) loss of mitotic activity; (2) synthesis of new organelles (lamellar and keratohyaline granules) and subsequent loss of all cell organelles; (3) total remodeling of cell architecture as it increases in size, flattens, and loses almost all of its water; (4) modification of cell membrane and cell surface antigens and receptors; and (5) synthesis of new lipids, as well as structural and enzymatic proteins.

Keratin. All eukaryote cells possess a complex intracellular cytoskeleton comprised of three major types of structural proteins: actin-containing microfilaments about 7 nm in

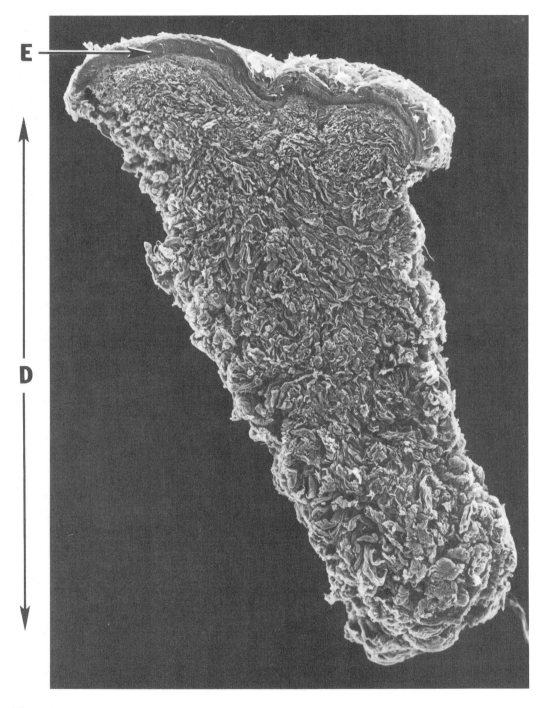

Figure 3   Low-magnification (×80) scanning electron micrograph of skin showing relative size of dermis (D) and epidermis (E). Fibrous nature of dermis is evident. (Courtesy of Dr. Karen Holbrook.)

diameter, tubulin-containing microtubules about 20 to 25 nm in diameter, and a class of filaments 7 to 10 nm in diameter (intermediate filaments) into which the keratins fall. The keratins represent a family of alpha-helical, water-insoluble proteins, ranging in size from 40 to 70 kDa, that make up a major part of the cytoskeleton of all epithelial cells. Although more than 20 different human keratins have been identified, the particular types expressed depend on the tissue and stage of differentiation, as well as on the health of the tissue. Analysis of the keratins isolated from various epithelial tissues has revealed that only 2 to 10 keratins are expressed in any single tissue. One unique characteristic of these proteins is their propensity to aggregate. Even in vitro, under the appropriate conditions, they will form intermediate filaments that are morphologically similar to those found in the epidermis.

Keratins can be divided into two subfamilies: acidic keratins having an isoelectric point < pH 5.5, and basic keratins having an isoelectric point > pH 6.5. Individual keratin filaments are known to be composed of at least one of each subfamily, since it has been shown through in-vitro reconstitution studies that a single type of keratin will not form an intermediate filament [6]. All keratins are rich in serine, glutamic acid, and glycine but have a low cystine content.

Epidermal keratinocytes are unusually rich in keratin. In the case of both human and bovine epidermis, up to two-thirds of the total dry weight is made up of keratin proteins. Four major keratins are found in normal epidermis: acidic 50 kDa and 56.5 kDa, and basic 58 kDa and 65 to 67 kDa. These are co-expressed as pairs; the 50/58-kDa keratins (acidic/basic) are synthesized in the basal layer, and the 56.5/65 to 67-kDa (acidic/basic) proteins are synthesized in suprabasal cells. Thus, it can be seen that the biochemical composition of the keratin filaments varies with the state of differentiation of the cells [7].

Epidermis. The epidermis is composed of four cell types: the "native" or ectodermally derived keratinocyte, which constitutes approximately 80% of the epidermis; plus the three other cell types, which arise in other germ layers and migrate into the epidermis during fetal development. These cells are (1) melanocytes, the source of melanin pigment, which gives the skin its color and protects us from the damaging effects of ultraviolet radiation; (2) Langerhans cells, which are the outermost arm of the immunologic system and serve in host defense; and (3) Merkel cells, which are thought to function as mechanoreceptors for the sensation of touch.

The Langerhans cell is part of the immune surveillance system of the body, and, as such, plays a key role in the recognition of foreign invaders. Topically applied drugs and chemicals are sometimes perceived by the Langerhans cell as "foreign," and for this reason its position in the skin is pivotal to the development of contact allergic dermatitis, a not uncommon adverse reaction that can occur following the use of topical drugs and transdermal delivery systems. This problem will be discussed later in the section on inflammatory processes of skin.

The major cell of the epidermis is the keratinocyte, so named because of the family of fibrous proteins (keratins) contained within. It is the terminal differentiation of this cell that leads to the production of the nonviable, fully keratinized stratum corneum. In addition to the presence of keratin filaments, keratinocytes can be characterized by the presence of desmosomes, highly developed, discrete structures that occur at irregular intervals along the cell membrane and serve as points of attachment between adjacent cells (akin to spotwelds). Each desmosome is composed of two electron-dense plaques (attachment plaques) located in adjoining keratinocytes immediately adjacent to the cell membrane, the intervening cell membranes, and an intercellular cement substance. The desmosome appears

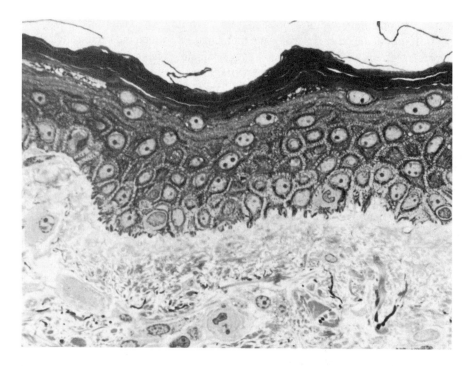

Figure 4 Photomicrograph ($\times 250$) illustrating cellular nature of epidermis and a portion of the upper dermis. The amorphous dark appearing layer at the top is the stratum corneum. (Courtesy of Dr. Karen Holbrook.)

as a series of alternating electron-dense and electron-lucent zones under electron microscopy. Keratin filaments, loosely organized into bundles and often referred to as tonofilaments, insert into the attachment plaques and radiate from there to the perinuclear region of the cell.

Keratinocytes are organized into layers (strata) within the epidermis, each named either because of location or the presence of a unique structural or functional characteristic (Fig. 4). These are, from inside to outside, the stratum germinativum (also referred to as the stratum basale or basal layer), stratum spinosum or spinous layer, stratum granulosum or granular layer, and stratum corneum.

The mitotically active keratinocytes reside in the stratum germinativum, giving rise to daughter cells which progress upward through the other layers of the epidermis and are eventually shed as nonviable cells of the stratum corneum (corneocytes). The different strata of epidermis, all of which are morphologically distinct and represent different stages of the differentiation process, are in fact a functional continuum.

Stratum germinativum: The keratinocytes of the basal layer, which are cuboidal to low columnar in shape with their long axis perpendicular to the basal lamina, have an undulating basal surface that conforms to the underlying connective tissue layer (Fig. 5). They are firmly attached to the basal lamina by hemidesmosomes and to adjacent keratinocytes by desmosomes. Interspersed along the basal layer are melanocytes, whose dendritic processes intertwine with adjacent keratinocytes and through which melanin pigment is transferred.

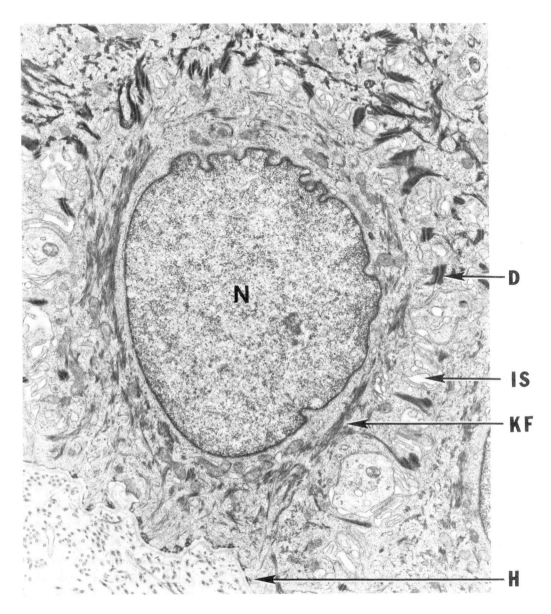

Figure 5   Electron micrograph ($\times$ 14,700) of basal cell with large nucleus (N) and bundles of keratin filaments (KF). Attachments to adjacent cells by desmosomes (D) and, to the basal lamina, by hemidesmosomes (H) are visible. Convoluted intercellular space (IS) appears empty. (Courtesy of Dr. Karen Holbrook.)

The cells of the germinativum contain a large nucleus, prominent nucleolus, and all the usual intracellular organelles (mitochondria, lysosomes, Golgi apparatus, rough endoplasmic reticulum, and ribosomes). In addition, the pigment-containing melanosomes transferred from the melanocyte can be found. Basal cells also contain keratin filaments, organized into bundles, which are distributed around the nucleus and attach to both desmosomes and hemidesmosomes. The keratins present in the basal layer, like those of simple epithelia elsewhere in the body, are of low molecular weight.

Stratum spinosum: The cells of the spinous layer are so named because of their spine-like shape when viewed by light microscopy. Routine histologic preparation leads to dehydration of the tissue, causing the cells to shrink and pull away from each other except where they are firmly attached by desmosomes.

Though the basal cells are not unlike those of other epithelial tissues, as the keratinocytes differentiate and move upward in the spinous layer, major morphologic changes occur. The cells begin to flatten and assume a polyhedral shape, more so in the upper than in the lower layers. All of the usual cellular organelles are present, and the keratin filaments become more prominent. In addition, higher-molecular-weight keratins begin to appear.

New cellular organelles, the *lamellar granules*, are found in the cells of the upper spinous layer. These 0.1- to 0.3-μm-diameter organelles are distinguished by the presence of a series of alternating lamellae. Histochemical studies show that they contain acid hydrolases, neutral sugars, and lipids [5]. It is the extrusion of these lipids into the intercellular spaces of the granular layer that initiates formation of the barrier. In both the spinous and basal layers, the intercellular spaces are freely permeable, as judged by the movement of large mass tracers such as lanthanum and horseradish peroxide [8,9]. However, their outward diffusion through the skin is blocked upon reaching the granular layer.

Stratum granulosum: Major changes in cellular architecture leading to the final stages of differentiation occur in the granular layer, named for the basophilic granules (keratohyaline) that are so prominently seen under both the light and electron microscope (Fig. 6). The cells in this layer continue to flatten and become much wider in diameter than the cells of lower layers, and new proteins appear.

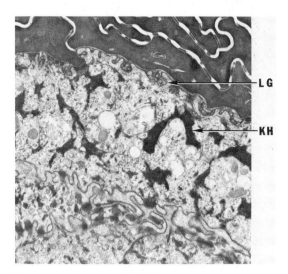

Figure 6   Electron micrograph ($\times$ 21,000) of granular-layer and lower-stratum corneum showing lamellar granules (LG) and masses of amorphous keratohyaline (KH). (Courtesy of Dr. Karen Holbrook.)

The major new protein of this layer, contained in the keratohyaline granules, is a high-molecular-weight, histidine-rich precursor substance, pro-fillaggrin. It will be converted to fillaggrin, deposited in the interstices of the keratin filaments, and promote their aggregation. It is thought to be the matrix in which the keratin filaments are enmeshed in the stratum corneum. Another protein, involucrin, appears in the granular layer. This is a cystine-rich protein that will become a major component of the thickened cell envelope of the stratum corneum cells.

The lamellar granules become more numerous in the granular layer, and it is here that they migrate to the cell membrane and release their contents into the intercellular space. As will be discussed later, this serves to seal the intercellular space as a pathway for diffusion and imparts the unique characteristics to the skin that are not found in other epithelial barriers.

The transition from granular layer to horny layer must occur abruptly, as intermediate cell types are seldom seen. The nucleus and all cell organelles are broken down, a thick band of protein is deposited on the inner surface of the plasma membrane to form the cell envelope, and the entire cell is filled with keratin filaments and associated matrix proteins.

Stratum corneum: The stratum corneum is the end product of epidermal differentiation and consists of 15 to 25 cell layers over most of the body surface, though it is much thicker over the palms and soles. Each cell (corneocyte) is approximately 0.5 μm in thickness and 30 to 40 μm in width, the largest cell in the epidermis. Its contents are devoid of all the usual cellular organelles and consist largely of protein; i.e., 80% of the cells are high-molecular-weight keratins ($>60,000$). In addition, it has a very low water content, though it is capable of taking up approximately three times its weight in water. The intercellular space is filled with lipids organized as bilayers, and these lipids are of unusual composition, as will be discussed subsequently. Approximately 14% of the stratum corneum, by weight, are lipids.

The corneocytes are joined together by modified desmosomes and overlap with each other at their edges to form a mechanically strong layer (Fig. 7). All the mechanical strength of the epidermis derives from the stratum corneum, and it can be prepared in pure form as an intact "membrane" suitable for permeability studies [10].

A particularly intriguing feature of stratum corneum architecture is its orderly arrangement into columns or stacks [11], a feature not seen in routine histologic preparation of the skin in which defatting solvents have been used. However, when frozen sections of unfixed skin are swollen with dilute acid or alkali, the cells of the horny layer are neatly aligned (Fig. 8). This is most evident in regions where the epidermis is thin (and in animal skin) and the turnover rate low. It is not seen in areas such as the palms and soles, where the stratum corneum is unusually thick and the turnover rate high. This orderly arrangement of corneocytes may have some relationship to barrier function, as the diffusion coefficient of water in palmar and plantar skin is much greater than elsewhere.

Turnover time.   The turnover time of the epidermis is often stated to be about 28 days, with the turnover time of both the Malpighian layer and stratum corneum each being approximately 14 days. However, several factors have confounded interpretation of the original data, and it is now clear that the turnover time of the epidermis must be much longer than 28 days. The problems arise not in the measurement of stratum corneum turnover, but in the measurement of turnover of the viable layers.

Tritiated thymidine has been used to label all basal cells undergoing DNA synthesis and subsequently to measure transit time of the cells through the Malpighian layer. With

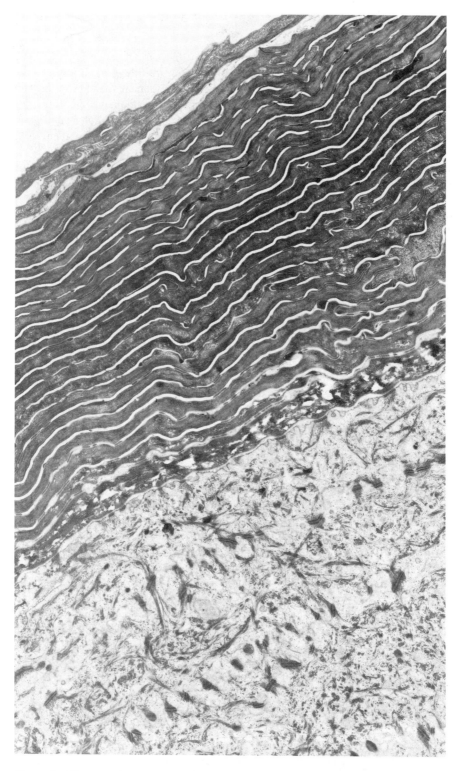

Figure 7 Electron micrograph (×9,600) through full-thickness of stratum corneum and part of the upper layers of the living epidermis. The multilayered nature of this structure and relative thinness of individual cells is apparent. The absence of structure in the intercellular space is the result of lipid solvents used to prepare the specimen. (Courtesy of Dr. Karen Holbrook.)

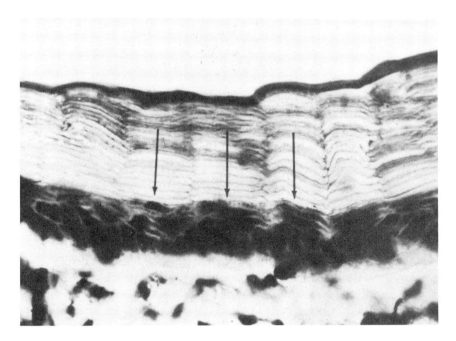

Figure 8 Light micrograph of alkali-swollen mouse-ear skin showing corneocytes arranged in vertical columns. (From Ref. 11.)

this technique, 10 to 14 days elapsed before the first labeled cells reached the stratum corneum [12]. It was noted, however, that not all labeled cells moved synchronously. Some labeled cells were found in the spinous layer up to 42 days post-injection. Thus, a 10- to 14-day transit represents the minimum rather than the average time for movement of the cells from basal layer to horny layer.

Another problem associated with the determination of turnover time of the viable epidermis is that not all basal cells normally undergo cell division. There exists a population of noncycling or slowly cycling stem cells that are held in reserve and respond only to such stimuli as wounding or other insults that upset the normal homeostasis of the skin. Experiments in which tritiated thymidine was continuously infused into human subjects suggest that approximately 50% of the basal cells are noncycling [13]. When these and other factors are taken into consideration, it is estimated that the true turnover time of the viable Malpighian layer is 26 to 42 days.

Estimation of the turnover time of the horny layer has been somewhat simpler because cells appear to move through this layer synchronously. An average renewal time of 14 days has been determined using two separate techniques. Impregnation of the stratum corneum with the fluorescent dye tetrachlorsalicylanilide revealed that it took approximately 10 to 15 days for the surface of the skin to lose its fluorescence in regions such as the abdomen, back, and forearm [14]. Labeling of the proteins of the viable epidermis with ^{14}C-glycine revealed that it took 13 to 14 days for the first labeled cells, i.e., those coming from the granular layer, to reach the skin surface [15]. It should be noted, however, that there is a significant regional variation in stratum corneum renewal (Table 2).

Dermis. The dermis is largely an integrated fibro-elastic, acellular structure consisting

Table 2 Regional Differences in Stratum
Corneum Turnover Time

Average turnover region	Time	(Days)
Back of hand	20.8	± 2.3
Shin	19.6	± 2.5
Back	15.3	± 5.6
Forearm	13.3	± 1.5
Abdomen	9.6	± 2.9
Scalp	9.6	± 1.0
Forehead	6.3	± 1.4

Data from Ref. 14.

of interwoven fibrous, filamentous, and amorphous connective tissue. The dermis is the largest component of skin, and it is from this layer that the skin derives its mechanical strength. The dermis is arbitrarily divided into two parts: the upper, papillary, and the lower, reticular. The papillary dermis is the thinner of the two and is distinguished from the reticular dermis by having fiber bundles of much smaller diameter. Its interface with the epidermis is irregular and thrown into folds. The most elevated portions of the papillary dermis are referred to as dermal papillae, and each contains a capillary loop arising from the underlying arteriole plexus.

Collagen is the major fibrous protein of the dermis, making up more than 70% of the dry weight of this layer. Bundles of collagen fibers are woven into a network within the dermis, which accounts for the great tensile strength of skin. Interwoven among the collagen fabric is a network of elastic fibers which give skin its resilience, i.e., the ability to restore normal structure following deformation by external forces. It makes up only 1 to 2% of the dry weight of the dermis. Other nonfibrous components of the dermis are the glycosaminoglycans (the amorphous ground substance) and the finely filamentous glycoproteins. Together they account for the water-binding properties of the dermis.

Though the dermis is largely acellular, three cell types are regular inhabitants of this layer: fibroblasts, mast cells, and macrophages. Fibroblasts are responsible for the synthesis of the dermal fabric (collagen, elastin, ground substance), whereas mast cells and macrophages participate in inflammatory reactions.

From the standpoint of percutaneous absorption and transdermal delivery, perhaps the most important element of the dermis is the vasculature it contains, because absorbed materials are distributed throughout the body by this route. The blood supply for the skin comes from cutaneous branches of musculoskeletal arteries, which ascend from the underlying musculature, penetrate the subcutaneous fat and enter the dermis. In the deeper regions of the dermis, branches spread horizontally to form a deep vascular plexus. In addition, the parent vessel ascends to the papillary dermis and divides into smaller arterioles which form a superficial plexus running parallel to the surface of the skin. Arising from the superficial plexus are smaller arterioles that give rise to capillary loops running at right angles to the surface of the skin and traversing the dermal papillae (Fig. 9). These latter vessels appear as arcades in three dimensions and serve as the ''sink'' for percutaneous absorption. They are the component of the vasculature closest to the base of the epidermis, coming within 10 to 20 μm of the basement membrane.

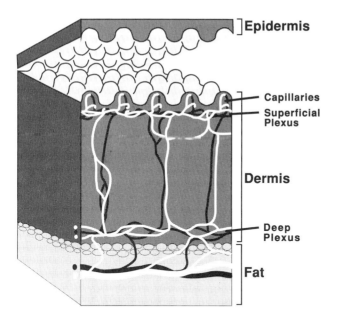

Figure 9 Illustration of the cutaneous vasculature showing both a superficial and deep plexus. Capillary loops arise from the superficial plexus and serve as the nutritional source for the living epidermis and the sink for percutaneous absorption.

Role of Lipids in Barrier Function

One major element underlying the impermeability of skin is the hydrophobic nature of the stratum corneum. Major changes in the lipid composition of the epidermis occur as cells move from the basal layer to the horny layer. There is a shift from polar lipids to neutral lipids and almost complete loss of phospholipids. With respect to barrier function, the most important changes occur in the stratum granulosum and lead to the deposition of lipids in the intercellular space. These lipids derive from a specialized 0.1- to 0.3-μm-diameter ovoid structure called a lamellar granule (Odland body, keratinosome, membrane-coating granule), whose first appearance is noted in the upper stratum spinosum. The number of lamellar granules becomes more numerous in the stratum granulosum, where it has been estimated that they make up 15% of the cell's cytoplasm. Along with the increased synthesis of keratin and filaggrin, this helps to explain the hypertrophy of keratinocytes as they move from spinous to granular layer.

In the granular layer, lamellar granules move to the periphery of the cell, fuse with the cell membrane, and discharge their contents into the intercellular space. As this material moves up into the stratum corneum, modifications that can be visualized by electron microscopy occur. In the intercellular space of the granular layer, the discharged material appears as short disks, just as it did in the lamellar granule. However, in the stratum corneum this material appears as broad, multilaminate sheets, which are postulated to form by edge-to-edge fusion of the disks (Fig. 10).

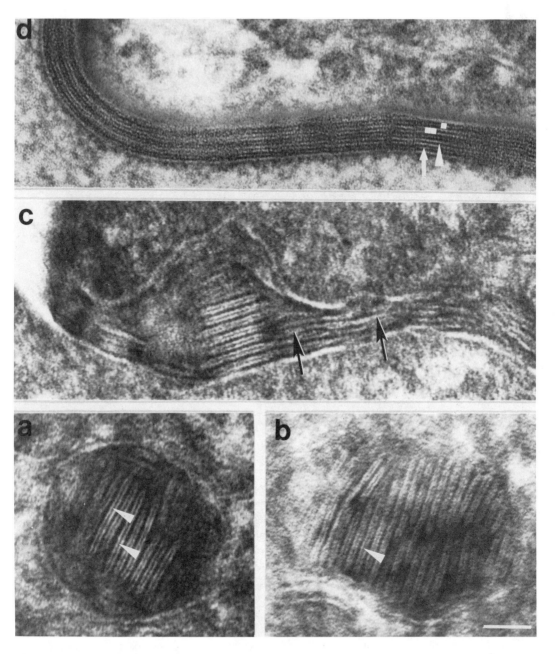

Figure 10 Electron micrograph of lamellar granules from neonatal mouse skin showing their changing structure. In a and b they appear as "stacked discs" intracellularly in the granular layer, with alternating electron-dense electron-lucent bands. A minor dense band (arrowhead) splits the lucent band centrally. In c, following extrusion to the intercellular space, fusion of discs (between arrows) begins to occur. In d, in the outer stratum corneum, only broad lamellar sheets are seen and the appearance of the dense and lucent bands are changed. Bar = 50 nm. (From Madison et al. (1987), *J. Invest. Dermatol. 88*:714–718.)

Synchronous with this change in morphology are major biochemical changes in lipid composition, which can best be characterized as movement from a "polar" lipid profile to a very nonpolar lipid profile. The lipid domain of the stratum corneum is unique when compared to cell membranes or that found in other epithelial structures. It consists mostly of neutral lipids (70–80%), and ceramides (15–20%), with small amounts of cholesterol sulfate (2–5%). Phospholipids are virtually absent!

Lipid profile in human skin. Knowledge of the lipid content of human epidermis and changes that occur with differentiation have been derived from the work of Lampe et al. [16]. A complete analysis of the lipids obtained from abdominal skin was made at three levels within the epidermis, the combined stratum basale/stratum spinosum (SB/SS), the stratum granulosum (SG), and the stratum corneum (SC). In addition, specimens of only the outermost portion of the stratum corneum were obtained from patients following orthopedic cast removal or from subjects exfoliating from sunburn. Using techniques of separation that resulted in homogeneous layers of greater than 99% purity, total lipids were extracted from each layer, fractionated, and quantitated using thin-layer chromatography. Subsequently, the fatty acid content of some fractions was determined using gas-liquid chromatography. The results are summarized in Tables 3 and 4.

The phospholipids isolated from epidermis included phosphatidylethanolamine, phosphatidylcholine, phosphatidylserine, sphingomyelin, and lysolecithin. With differentia-

Table 3 Variations in Lipid Composition During Human Epidermal Differentiation and Cornification[a]

Fraction	Strata basale/spinosum (n = 5)	Stratum granulosum (n = 7)	Stratum corneum Whole (n = 4)	Stratum corneum Outer (n = 8)
Polar lipids*	44.5 ± 3.4	25.3 ± 2.6	4.9 ± 1.6	2.3 ± 0.5
Cholesterol sulfate**	2.4 ± 0.5	5.5 ± 1.3	1.5 ± 0.2	3.4 ± 0.5
Neutral lipids	51.0 ± 4.5	56.5 ± 2.8	77.7 ± 5.6	68.4 ± 2.1
Free sterols	11.2 ± 1.7	11.5 ± 1.1	14.0 ± 1.1	18.8 ± 2.1
Free fatty acids***	7.0 ± 2.1	9.2 ± 1.5	19.3 ± 3.7	15.6 ± 3.0
Triglycerides	12.4 ± 2.9	24.7 ± 4.0	25.2 ± 4.6	11.2 ± 1.5
Sterol/wax esters[b]	5.3 ± 1.3	4.7 ± 0.7	5.4 ± 0.9	12.4 ± 1.9
Squalene	4.9 ± 1.1	4.6 ± 1.0	4.8 ± 2.0	5.6 ± 2.1
n-Alkanes	3.9 ± 0.3	3.8 ± 0.8	6.1 ± 2.6	5.4 ± 0.8
Sphingolipids****	7.3 ± 1.0	11.7 ± 2.7	18.1 ± 2.8	26.6 ± 2.3
Glucosylceramides I	2.0 ± 0.3	4.0 ± 0.3	trace	trace
Glucosylceramides II	1.5 ± 0.3	1.8 ± 0.2	trace	trace
Ceramides I	1.7 ± 0.1	5.1 ± 0.4	13.8 ± 0.4	19.4 ± 0.5
Ceramides II	2.1 ± 0.3	3.7 ± 0.1	4.3 ± 0.4	7.2 ± 0.5
Total	99.1	101.1	99.3	100.7

[a] Weight percent ± SEM.
[b] Sterol/wax esters present in approximately equal quantities, as determined by acid hydrolysis. Significant differences: *, SS/SB versus SG (P < 0.01); SG versus WSC or SS/SB (P < 0.001). **, SG versus WSC (P < 0.02); WSC versus OSC (P < 0.01). ***, SG versus WSC (P < 0.05). ****, SS/SB versus SG (P < 0.05); WSC versus OSC (P < 0.05).
Source: From Ref. 16, with permission.

Table 4 Fluctuations in Straight-Chain Fatty Acid Composition of Major Epidermal Fractions During Differentiation[a]

| Fraction | Neutral lipids | | | Sphingolipids | | Phospholipids | | |
| | Layer | | | Layer[b] | | Layer | | |
	SS/SB	SG	SC	SG	SC	SS/SB	SG	SC[c]
12:0	0.03	0.3	0.1					
14:0	1.9	3.5	2.7	0.7				
16:0	24.1	25.3	24.1	13.1	11.9	25.8	9.4	2.8
16:1	6.7	7.4	7.3	1.8	1.5			
18:0	10.7	16.7	24.7	11.4	8.3	14.1	20.6	14.9
18:1	36.8	31.1	24.1	32.3	28.2	42.1	31.0	20.0
18:2	14.5	14.3	14.7	18.8	15.9	12.3	26.5	20.1
20:0	0.5	0.3	0.3	1.2	2.0		2.1	
20:1				0.4	tr			
20:2	0.5	0.3	0.12					7.5
20:3								
20:4				1.8	tr		3.6	
22:0	0.9	0.4	0.4	2.5	4.7			
22:1								
24:0	3.8	0.7	1.6	6.8	1.6			
26:0				9.3				
Total	100.0	100.0	100.0	99.9	100.0	99.9	100.0	100.0

[a] Mole percent of identified fatty acid methyl ester. Because of methods used, branched and hydroxy acids are not included [14].
[b] Too little material in SS/SB fraction for GLC.
[c] Very little material present for GLC; comparisons may not be meaningful.
Source: From Ref. 16, with permission.

tion, each strata of the epidermis showed progressively less phospholipid, going from 45% in the SB/SS to 25% in the SG to less than 5% in the SC.

The neutral lipids consisted of six fractions: free sterols, free fatty acids, triglycerides, sterol and wax esters (not separable with the solvent systems used), squalene, and *n*-alkanes. Neutral lipids were found to increase with differentiation, going from 51% in the SB/SS to 57% in the SG to 78% in the SC. The increase was largely accounted for by increases in the free fatty acid and triglyceride fractions.

Sphingolipids, separable into at least 10 bands with the solvent system used, were identified and found to consist of a mixture of ceramides and glycosphingolipids. For convenience of presentation, they were combined into four groups. The relative quantities of sphingolipids increased with differentiation, from 7% in the SB/SS to 12% in the SG to 18% in the SC. Furthermore, though both the glycosphingolipid and ceramide fractions increased in going from the basal/spinous layers to the stratum granulosum, glycosphingolipids virtually disappeared in the stratum corneum, with the entire increase in sphingolipid content being ceramides.

Cholesterol sulfate, a highly amphipathic compound, was found at all levels of the epidermis. Its relative quantity increased in going from the SB/SS to the SG but, like the highly polar phospholipid fraction, its relative quantity decreased in going from the SG to the SC, from 5.5 to 1.5%.

In general, the fatty acid composition of all epidermal layers ranged from C12 to C24, with the C16 and C18 fractions predominating. Among the neutral lipids, shorter-chain fatty acids (C12:0 and C14:0) were most abundant, and no major shifts in composition were noted with differentiation. Among the sphingolipids, a greater proportion of longer-chain fatty acids (C22 to C24) and fewer short-chain fatty acids (C12 to C16) were noted in comparison with neutral lipids. Among the phospholipids, the fatty acid composition was largely C16 to C18, although, in the small amount remaining in the SC, 7.5% was found to be C24:0.

In summary, this review of human epidermal lipids demonstrates that *major changes in lipid composition occur during the process of differentiation*. The generation of a fully formed stratum corneum is accompanied by an enrichment of neutral lipids and sphingolipids and a virtual elimination of the most polar lipids, i.e., phospholipids and cholesterol sulfate. Whereas the neutral lipids and sphingolipids make up only 58% of the SB/SS layers, they account for nearly 96% of the SC layer (Fig. 11). It is logical to assume that these changes occur largely to subserve barrier function.

Regional variation in lipids.   In a second study of human epidermal lipids, Lampe et al. [17] examined regional variations in lipid content and composition. Human skin was

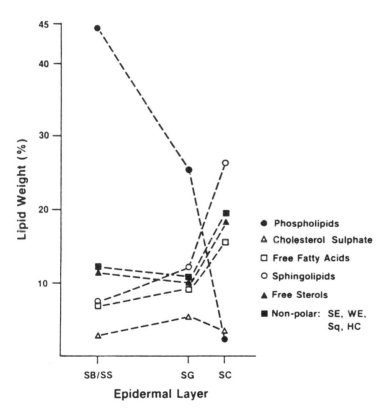

Figure 11   The composition of lipid species within the epidermis change as cells differentiate and move from the stratum basale/stratum spinosum (SB/SS), to the stratum granulosum (SG), and the stratum corneum (SC). (From Ref. 16.)

Table 5 Regional Variations in Percent Lipid Weight and in Major Lipid Species

	Site			
	Abdomen (n = 4)	Leg (n = 4)	Face (n = 3)	Plantar (n = 3)
Lipid weight, %	6.5 ± 0.5	4.3 ± 0.8	7.2 ± 0.4	2.0 ± 0.6
Major species				
Polar lipids	4.9 ± 1.6	5.2 ± 1.1	3.3 ± 0.3	3.2 ± 0.89
Cholesterol sulfate[a]	1.5 ± 0.2	6.0 ± 0.9	2.7 ± 0.3	3.4 ± 1.2
Neutral lipids	77.7 ± 5.6	65.7 ± 1.8	66.4 ± 1.4	60.4 ± 0.9
Sphingolipids[b]	18.2 ± 2.8	25.9 ± 1.3	26.5 ± 0.9	34.8 ± 2.1

[a] Significant differences: abdomen versus leg, $P < 0.01$; leg versus face, $P < 0.02$; abdomen versus face, $P < 0.02$; face versus plantar, $P < 0.01$; abdomen versus plantar, $P < 0.01$; plantar versus leg, $P < 0.02$. Cholesterol sulfate: leg > plantar > face > abdomen.
[b] Significant differences: abdomen versus leg, $P < 0.05$; abdomen versus face, $P < 0.05$. Sphingolipids: plantar > face > leg > abdomen.
Source: From Ref. 17, with permission.

obtained from four sites known to have different permeability properties (abdomen, face, leg, and sole) and the stratum corneum lipids were extracted and analyzed. The data are summarized in Table 5.

Total lipid content varied among the four sites, with face skin, the most permeable site, having the greatest lipid content, and plantar skin, the least permeable site, having the lowest lipid content. Neutral lipids comprised the largest percentage of SC lipids at all four sites, varying from 60 to 78%. Significant differences between sites were found in the free fatty acid, free sterol, and triglyceride fractions, with the latter showing the greatest differences. Sphingolipids were the second most common lipid class, found at all four sites (18–35%), with plantar SC containing significantly more and abdomen containing significantly less than the other three sites.

In trying to correlate these observations with the known regional variations in percutaneous absorption, an inverse relationship between the total amount of neutral lipid present and skin permeability is noted. Neutral lipids are highest in the abdomen, where water permeability is lowest, and lowest in plantar skin, where water permeability is the highest [18]. Also noted is the fact that sphingolipids show a direct correlation with skin permeability, being lowest in abdominal skin and highest in plantar skin.

Inflammatory and Immunologic Processes of Skin

The skin is subject to a variety of insults, both endogenous and exogenous, and, as a result, it is a frequent site of inflammatory reactions. The topical application of drugs and chemicals to the skin is also a common cause of inflammatory reactions, which may be divided into two types: those that have an underlying immunologic or allergic basis (contact allergic dermatitis) and those without an immunologic basis (contact irritant dermatitis).

Cutaneous inflammation is mediated by the interactions of a complex array of both cells and soluble factors, with the final expression of these events being the four classical signs of inflammation: heat, swelling (edema), redness (erythema), and pain. The visible appearance of cutaneous inflammation can vary depending on the stage of the reaction.

The early or acute phase of the reaction is characterized by erythema, edema, and burning or itching (pruritus). This may progress to vesiculation (blister), oozing, crusting, and scaling, depending on the severity of the reaction. If contact with the inciting agent continues, the acute weeping stage will be supplanted by a chronic dermatitis characterized by thickened (lichenified), dry, or scaly skin, continued pruritus, and frequently excoriations. Even after the reaction subsides, the site will appear darker (hyperpigmented) than the surrounding skin for some time owing to cellular damage with subsequent loss of melanin pigment into the surrounding tissue. Occasionally, severe dermatitic reactions lead to destruction of melanocytes and result in hypopigmentation or permanent depigmentation, a particularly distressing problem for darkly pigmented races.

Allergic Contact Dermatitis

Allergic contact dermatitis is a form of delayed or cell-mediated hypersensitivity. The immune system of the body is divided into two parts. One part, humoral immunity, serves as a defense against bacterial invasion and is derived from the production of circulating antibodies (immunoglobulins) by B-lymphocytes. The second part of the immune system, cell-mediated immunity, serves primarily as a defense against intracellular organisms (viruses, fungi, yeast, mycobacteria, and other intracellular parasites) and is derived from the function of T-lymphocytes. Allergic contact dermatitis is associated with cell-mediated immunity and, since the "invading" chemical is not a pathogenic microorganism, it can be viewed as a misdirected form of immunity.

Allergic contact dermatitis is a specific acquired hypersensitivity of the delayed type. Through repeated exposures to a particular chemical, a person becomes sensitized (allergic) to it, i.e., develops an immunologic mechanism to recognize and react to that specific chemical allergen (antigen). Not all chemicals are capable of stimulating an allergic reaction, nor is it usual for any one chemical to sensitize all individuals. Most contact allergens sensitize only a small percentage of those exposed. For example, the antihypertensive drug, clonidine, evokes allergic reactions in only one of five patients who use the transdermal patch (1990 *Physicians' Desk Reference*). However, some chemicals are known to be strong sensitizers. The antigens of poison ivy and poison oak will sensitize more than 70% of individuals who come into contact with them. In the past, the topical use of many drugs has had to be discontinued because of the occurrence of allergic contact dermatitis. The sulfonamides, penicillins, and antihistamines are examples of drug classes that commonly induce sensitization following repeated application to the skin [19].

The first step in the development of allergic contact dermatitis is the binding of the antigen to the cell membrane of an antigen-presenting cell. T-lymphocytes are incapable of interacting directly with antigens even though they possess the appropriate surface receptor for that antigen. In the skin, the prime antigen-presenting cell is the Langerhans cell, which, with its dendritic processes, forms a continuous network in the epidermis. About 3 to 4% of epidermal cells are Langerhans cells. As drugs and chemicals diffuse through the skin, some molecules interact with the cell membrane of the Langerhans cell, where a poorly understood change referred to as "processing" occurs. The net effect is that the antigen is converted to a form suitable for "presentation" to T-lymphocytes, presumably as they move through the skin as part of their normal traffic pattern.

Subsequent to binding of antigen to the T-cell surface, these cells enter the circulation and move to regional lymph nodes where they undergo clonal proliferation and produce many daughter cells genetically programmed to respond to the specific antigen that activated their parent. When sufficient numbers of daughter cells have been formed, they reenter the circulation and return to the skin and other peripheral tissues. This phase of the sensitization process (induction) normally takes 1 to 3 weeks.

When the sensitized daughter cells again encounter the specific "processed" antigen on the surface of a Langerhans cell or other antigen-presenting cell (macrophage), they become activated. The lymphocytes enlarge (blast transformation) and release a number of substances called lymphokines into the surrounding tissue. These substances act as mediators of the inflammatory process and induce migration and activation of other inflammatory cells. Following recontact with the antigen, a cell-mediated response usually takes 12 to 48 h to develop—thus the origin of the term "delayed" hypersensitivity.

An immediate form of hypersensitivity can occur following topical application of a wide range of agents. Referred to as the contact urticaria syndrome, it describes a wheal-and-flare response occurring within 30 to 60 min of application and is consistent with histamine release from mast cells. Contact urticaria can result from both immunologic (mediated by immunoglobulin-E) and nonimmunologic causes. For example, the penetration enhancer dimethyl sulfoxide directly stimulates histamine release through a non-immunologic mechanism, as do certain substances released by plants (nettles) and animals (caterpillars, jellyfish) [20].

Irritant Contact Dermatitis

Contact dermatitis may be produced on an allergic or nonallergic basis. The most common form of contact dermatitis occurs on a nonallergic basis and is referred to as irritant contact dermatitis. Its pathogenesis is poorly understood, but it can be thought of as resulting from cell injury or death. The irritant action of caustic or corrosive materials, such as strong acids or alkalis, is easily understood, since they produce cell death and elicit a reaction after a single application. The mechanism by which mild irritants such as surfactants act, and the need for repeated application in many instances, is not clearly understood. Many chemicals can act as irritants, provided the concentration and duration of exposure are sufficient. Penetration enhancers such as dimethyl sulfoxide, decylmethyl sulfoxide, ethanol, propylene glycol, azone, and other surfactant materials are all examples of compounds that will produce irritant reactions in a dose-dependent manner following repeated use.

Cells and Mediators of Inflammation

Localized vascular dilation and increased vascular permeability are an essential part of the early inflammatory reaction. These changes serve to increase the delivery of white blood cells and plasma proteins to the involved tissue. Derived from these cells and proteins are the mediators that, in concert, drive and modulate the inflammatory reaction. Some of these are listed in Table 6.

Many of the inflammatory mediators (the leukotrienes and prostaglandins) have a common precursor, arachidonic acid. They are collectively referred to as eicosanoids, since they all derive from 20-carbon polyunsaturated fatty acids. Some injurious event (physical, chemical, or immunologic) leads to activation of the enzyme phospholipase A_2, which liberates arachidonic acid from its esterified form in the 2-acyl position of cell membrane phospholipids. Once released, arachidonic acid is converted by microsomal cyclooxygenase to the prostaglandins, or by cytoplasmic lipoxygenase enzymes (5- and 12-lipoxygenase) to a series of hydroxy fatty acids including the leukotrienes (Fig. 12).

Both the cyclooxygenase and lipoxygenase pathways exist in human skin, and the products of these pathways have been clearly shown to have inflammatory activity. In-

Table 6   Mediators of Inflammation

Mediators that enhance vascular permeability
   Derived from plasma proteins:
      Bradykinin
      Complement components (C3a, C5a)
   Derived from cells:
      Histamine
      Serotonin
      PAF
      Leukotrienes (LTC4, LTD4, LTE4)
      Prostaglandins (PGD_2, PGE_2, PGI_2)
      Cationic polypeptides
      Neutral peptide
Mediators that enhance cell movement
   Derived from plasma proteins:
      Kallikrein
      Complement components (C5a, $\overline{C567}$)
      Fibrinogen degradation products
   Derived from cells:
      Leukotriene B_4
      12-HETE
      HMW-NCF
      ECF-A
      Histamine
      PAF

---

[a] PAF, platelet-activating factor; 12-HETE, 12-hydroxyeicosatetraenoic acid; HMW-NCF, high-molecular-weight-neutrophil chemotactic factor; ECF-A, eosinophil chemotactic factor of anaphylaxis.

tradermal injection of nanogram amounts of prostaglandins D_1, D_2, E_1, and E_2 produce erythema. The prostaglandins also seem able to amplify the increased vascular permeability of other mediators such as histamine and bradykinin, and to initiate the sensation of pain or itch. The leukotrienes C_4 and D_4 induce wheal-and-flare reactions when injected intradermally in picomolar amounts, and B_4 and 12-HETE are chemotactic for neutrophils and serve to recruit these cells from the vascular compartment. Increased levels of both LTB_4 and 12-HETE have been found in the skin of patients with allergic contact dermatitis [21].

The most widely used drugs in the treatment of contact dermatitis are the corticosteroids. It is interesting to note that one of their prime mechanisms of action is to inhibit the activity of the enzyme phospholipase A_2, which sits at the top of the arachidonic acid cascade and, therefore, controls the rate of generation of both cyclooxygenase and lipoxygenase products. Nonsteroidal anti-inflammatory agents (NSAIDs) have also been tried in various dermatitic conditions and have been found to be relatively inactive. The common link that all NSAIDs share is that they are cyclooxygenase inhibitors. Thus, products of the lipoxygenase pathway may be of greater significance to cutaneous inflammation than products of the cyclooxygenase pathway.

Almost all cells in the skin participate in the inflammatory reaction as either source,

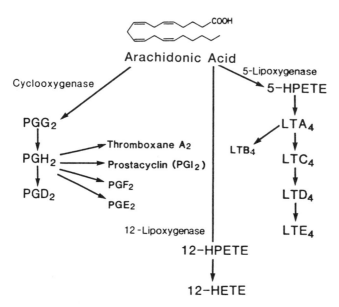

Figure 12 Pathways of arachidonic acid metabolism. P stands for prostaglandin and L for leukotriene. HPETE is hydroperoxyeicosatetraenoic acid and HETE is hydroxyeicosatetraenoic acid.

responder, or modulator of mediator release. The extent, control, and eventual resolution of the reaction is the resultant of the complex interactions between all of the participating cells, and it is clear that each of the cells and mediators serves more than one function. In Table 7 the major functions of cutaneous cell types are listed.

Permeation Enhancement and Reduction of Inflammatory and Immunologic Reactions

Permeation Enhancement

As discussed earlier, the permeation of drugs through the stratum corneum is the rate-controlling barrier to transdermal permeation for both lipophilic and hydrophilic drugs. Depending on the properties of the drug, both the lipids and the proteins in the stratum corneum could offer resistance to permeation. The protein and lipid regions form the polar and nonpolar pathways, respectively. Permeation through the polar pathway takes place by partitioning into the protein regions swollen by water. Permeation through the nonpolar pathway is more important for practical applications, since most drugs are lipophilic. Transport through this pathway takes place by dissolution and diffusion of the drug through the complex structure of the lipids of the stratum corneum.

Recently, Barry [22] proposed a theory of penetration enhancement activity which suggests that both the polar pathway and the nonpolar pathway reside within the intercellular lipids. He proposes that the dominant pathway for polar molecules resides in the aqueous region of the intercellular lipids, with the nonpolar pathway residing within the lipid chains of the hydrophobic region.

Looking at the permeation equations derived in Chapters 2 and 3 and assuming that the skin is a homogeneous structure, the factors that would affect skin permeability are

Table 7 Immunologic Functions of Cell Types Found in Skin

Cell type	Location	Cell surface determinant			Immunologic functions
		Class II	FcR	C3bR	
Keratinocyte	Epidermis	−/+[a]	−	−	Unknown
Melanocyte	Epidermis	−	−	−	Unknown
Merkel cell	Epidermis	−	−	−	Unknown
Langerhans cell	Epidermis	+	+	+	Antigen presenting cell (APC)
Indeterminant cell	Epidermis/dermis	+	?	?	Langerhans cell precursor?
Veiled cell	Dermis/afferent lymph	+	+	+	APC
Macrophage	Dermis	+	+	+	Phagocytosis, APC?
Mast cell	Dermis	−	−	−	Histamine release
Granulocytes	Dermis	−	+	+	Phagocytosis
Endothelial cells	Blood vessels	−/+	−	−	APC
Fibroblasts	Dermis	−/+	−	−	APC

[a] Although keratinocytes in normal skin do not express Class II determinants, they can be induced to express Class II molecules. FcR, receptors for the Fe portion of immunoglobulins; C3bR, receptors for the C3b component of complement. (From Lynch et al. (1987) J. Control Rel. 6:39–50.)

the concentration of drug dissolved in the uppermost layer of the stratum corneum (C_{sc}) and the diffusion coefficient for the drug through the stratum corneum.

There are several ways to maximize C_{sc}. First, the transdermal device should be designed in such a way that the concentration of the drug in the matrix of the device is above saturation. Second, the lipophilicity (partition coefficient) of the drug can be increased by preparing prodrugs (e.g., from salt to base) or preparing prodrugs with lower melting points. Ho et al. [23] prepared esters, amides, and ethers of a series of drugs and determined the lipophilicity change. Dunn [24] investigated the influence of the prodrug structure modification on partition coefficient. Sloan et al. [25] prepared prodrugs of 6-thiopurines, which are very insoluble due to their high melting points, and increased the permeability by preparing pivaloyloxymethyl derivatives. Valia et al. [26] increased the permeability of β-estradiol by the preparation of lower-melting, 17-monoester and 3,17-diester prodrugs. Higuchi and Yu [27] have presented a review of prodrugs for use in transdermal delivery. Finally, Michaels et al. [28] showed that transdermal permeability increases linearly (on a log-log scale) with increase in partition coefficient. Third, a compound with a high affinity for the drug, as well as for the lipids and/or proteins of the stratum corneum, can be incorporated into the stratum corneum. This type of compound could be considered a chemical enhancer.

The diffusion coefficient is a constant specific to a drug-membrane combination. If one assumes that the stratum corneum is not altered by the permeation of a series of drugs through it, the diffusion coefficient will change as a function of the properties of the drug only.

The molecular weight of the drug has been found to be the most important factor affecting the diffusion coefficient. As can be seen from the equation derived in accordance to the free-volume theory of Bueche for rigid molecules, presented in the section of this chapter on models of skin permeability, the diffusion coefficient decreases logarithmically with increase in molecular weight. Michaels et al. [28] have shown experimentally that the negative logarithm of the diffusion coefficient increases linearly with increase in drug molecular weight. Considering that the molecular weight of a specific drug cannot be substantially reduced by any means including prodrug preparation, the only way to increase the diffusion coefficient is to alter the stratum corneum structure. This can be accomplished by the use of chemical as well as physical enhancers.

Chemical Enhancement

A chemical enhancer can be defined as a compound that, by its presence in the stratum corneum, alters the skin as a barrier to the flux of the desired drug. This definition allows for the increase in flux due to the alteration of stratum corneum structure or due to the increase in drug concentration in the skin. In addition to increasing the permeability of the desired drug, an ideal enhancer would have many other attributes [29], e.g., it would be systemically nontoxic, nonirritating, or sensitizing, it would not elicit pharmacological responses, and its action would be reversible and its effect prompt, with a predictable duration.

The search for the ideal enhancer continues. Many solvents, detergents, oils, and other compounds have been tested [30–37]. The patent literature has been expanding at the rate of over 50 patents per year [38–42]. There are 15 patents just for the enhancement of β-estradiol. Since our understanding of the mode of action of enhancers is still in its infancy, these patents are limited in scope, i.e., a specific chemical for a specific drug.

Enhancers affecting the polar pathway.   We view stratum corneum as being in the form of lipid-depleted keratinocyte "bricks" surrounded by a lipid-enriched "mortar." The transcellular route or polar pathway takes place through the hydrated proteins of the keratinocytes. It is therefore believed that penetration enhancers affect this pathway by hydration and swelling of the proteins and by inducing conformational changes in these proteins. Water, being a polar molecule, permeates through this pathway and has the special property of increasing its own permeation [43]. Several investigators have also shown that water enhances the permeability of a great number of lipophilic and hydrophilic drugs [44–49]. Knutson et al. [50], however, concluded that hydration- and temperature-enhanced permeability of lipophilic molecules is associated with transitions involving the hydrocarbon chain of the lipids, with increased hydration fluidizing the lipid hydrophilic domain.

Surfactants act to dissolve lipophilic substances such as grease and oil in aqueous solutions. They are also known to disrupt the hydrophilic pockets in the tertiary structure of proteins, thereby disrupting the natural conformation and even denaturing the protein. Such substances would be expected to affect the polar pathway. They have indeed been shown to enhance the permeation of water, glucose, sodium salicylate, nicotinic acid, and several other hydrophilic drugs. Their preferential effect on the polar pathway was studied by Cooper [51], who showed that decylmethyl sulfoxide substantially increased the permeation of ionized salicylic acid, but marginally that of the un-ionized form. Surfactants are composed of a hydrophilic head group and a lipophilic chain. It has been shown that a 12-carbon lipophilic chain imparts the best enhancement, probably because it is lipophilic enough to dissolve in the lipids of the stratum corneum yet hydrophilic enough to alter the protein structure. Cooper and Berner [52] have shown that the more hydrophilic the head group, the higher the interaction with proteins and the higher the enhancing capability. The enhancing capability of surfactants increases with the following head groups: alcohols, ethoxylates, sulfoxides, zwitterionic, and ionic.

Enhancers affecting the nonpolar pathway.   At physiologic temperatures the lipids in the intercellular space of the stratum corneum are mainly in the gel state, with some lipid in a liquid-crystalline state. The nature of the lipid pathway, as well as the mode of action for permeation enhancement through this pathway, are not well understood. Unsaturated long-chain fatty acids are known to increase the permeation of lipophilic drugs through this pathway. Golden et al. [53] suggested that the *cis*-9- and *cis*-11-octadecenoic acids permeate into the lipid bilayers, thus reducing the lipid order and increasing fluidity. Bodde et al. [29] suggest that this effect is probably due to the kink in the *cis*-alkenyl chain, which might result in increased fluidity due to perturbations in the lipid packing of the bilayers. The same conclusion was reached by Ghenem et al. [54] in their studies in the enhancement of the permeation of hydrocortisone by ethanol. Knutson et al. [55] concluded that enhanced permeation of lipophilic solutes by unsaturated fatty acid and short-chain alcohols could be through gross lipid fluidization; however, alcohol-induced lipid polar head perturbations resulting in possible limited regions of interdigitated lipids are suggested as plausible mechanisms.

Solvents such as 2-pyrrolidone, N-methylformamide, and DMSO are capable not only of disrupting the lipid pathway but also of promoting permeation through the polar pathway. At least for DMSO, it is believed that, being a polar compound, it interacts with the polar heads of the lipid bilayers and thus interrupts the lipid order.

Binary enhancement systems have been shown to increase the permeability of both lipophilic and hydrophilic drugs. These binary systems may affect both polar and lipid pathways, or they may open a heterogeneous multilaminate pathway as suggested by Cooper and Berner [52]. The best binary systems are composed of an unsaturated fatty acid such as oleic acid and a diol such as propylene glycol. The effect of the unsaturated fatty acids on the lipid pathway has been discussed above. Propylene glycol, although it does not act on the lipids, is capable of solvating keratin and competing for the hydrogen-binding sites of the proteins. Other diols are also effective, but their effectiveness diminishes as the number of carbons in the chain increases.

Finally, it should be remembered that chemical enhancers will in general cause skin reactions, and possibly only those similar to the compounds already in the stratum corneum would prove to be completely acceptable. Table 8 shows a number of enhancers together with their mechanism of action and skin reactions [29].

Physical Enhancing Methods

Several nonchemical methods have been proposed for enhancing the permeability of drugs through skin. Iontophoresis and phonophoresis, the two methods most extensively studied, will be discussed here. A third approach recently introduced and worth mentioning involves controlled removal of the stratum corneum by pulsed laser light [56,57]. When the proper wavelength, pulse-length energy, and pulse rate are applied, the stratum corneum is removed without significantly damaging the underlying epidermis. When perfected, this method may be useful for the delivery of large-molecular-weight proteins in a hospital environment.

Iontophoresis.

Background—Iontophoretic drug delivery: The fundamentals, development, and biomedical applications of iontophoretic delivery of drugs have been recently reviewed by Banga and Chien [58].

An early clincal use was for the iontophoretic delivery of pilocarpine for the deletion of sweat in the diagnosis of cystic fibrosis of the pancreas [59], which is still the method of choice [60]. In another clinical application, the anesthetics lidocaine and epinephrine

Table 8 Some Penetration Enhancers Described in the Literature

Name	Mechanism of action	Skin reaction	Amount
Alkyl sulfoxides			
Dimethyl sulfoxide	Concentration gradients of DMSO cause a marked swelling of the stratum corneum	Erythema, burning sensation	DMSO 60%
Hexylmethyl sulfoxide	DMSO and hexylmethyl sulfoxide probably displace water from polar protein side chains	—	—
Decylmethyl sulfoxide	Decylmethyl sulfoxide probably has hydrophobic interactions with protein	—	—
Fatty acids			
cis- and trans-Octadecanoic acids	Increase in lipid fluidity, especially for the monounsaturated cis isomers, e.g., cis-vaccenic acid	No	—
Caprylic acid	—	—	—
Mono-, di-, triesters of glycerin and fatty acids, e.g., myglyol	—	No	—
Sufactants			
Ionic surfactants, e.g., sodium laurylsulfate, benzalconium chloride	Damage to skin integrity, interaction of micelles with keratin formation, swelling of stratum corneum	Strong irritation	0–1.5%
Nonionic surfactants	Probably formation of micelles; interactions with keratin, fluidization of lipids	Mild irritation, increased cell metabolism	—
Ion-pair formers			
Sodium laurylsulfate	Fat-soluble, ion-pair formation with drug	Strong irritation, sensitization	0–1.5%
Tetrabutyl ammonium bromide			
Benzalconium chloride	—	Strong irritation, sensitization	0.5%

Amphoteric surfactants			
Ethomeen S12		—	—
Alcohols			
Methanol	Lipid dissolution		0–100%
Ethanol			0–100%
Alkanols	Optimum around n = 6, lipid solubility		0–100%
Ether derivatives of PEG 400			
Glycerol	Humectant, effective in combination with water	50% allergic reaction, 25% no reaction, sensitization, mild reactions, strong irritation under occlusion	0–5%
Diols	Humectant, keratolytic properties; diols PG, EG		
	Butane diol shows strong enhancement		
Miscellaneous			
1-Dodecylazacyclopheptan 2-one-lauracapram, azone	Increase in lipid fluidity, possibly formation of micelles, synergism with propylene glycol and other diols	No irritation	1–5%
		Under occlusion possible irritation	
1-Butyl azacyclopentan-2-one	—	—	—
N-methyl-2-pyrrolidone	Formulation of drug resevoir in stratum corneum	Erythema	0–100%
2-Pyrrolidone	—	—	—
Dimethyl formamide	—	Less reaction than DMSO, no sensitization, no burning	—
Dimethyl acetamide	—	No burning	—
Dimethyl lauramide	—		—
N.N-Diethyl-m-toluamide	—		5% (10–100%)
Cyclohexanone derivatives	—		—
Urea	Keratolytic and protein denaturizing properties, moisturization	Erythema	2–10%
1,1,1 Trichlorethane	Skin delipidization (treatment of skin prior to drug application)	Strong irritation	100%
Salicylic acid	Keratolysis	Strong irritation	10%

Source: From Ref. 29, with permission.

were delivered to the ear canal [61]. In dermatological applications, iontophoresis has been used for the treatment of hyperhydrosis [62]. Another major application considered possible is the transdermal delivery of insulin to diabetics by iontophoresis. This technology is currently under development using animal models [63,64].

The advantage of iontophoretic drug delivery is the possibility of delivering ionic drugs, which are only poorly absorbed through the skin by the mechanism of passive diffusion alone or which require the use of chemical enhancers that are irritating and/or sensitizing to the immune system. According to Banga and Chien, "iontophoresis is by definition a process or technique which involves the transport of ionic (charged) molecules into the tissue by the passage of a direct electric current through an electrolyte solution containing the ionic molecules to be delivered using an appropriate electrode polarity." Iontophoresis may also be associated with an increased transport of water (iontohydro-kinesis), which may indirectly enhance the transport of nonelectrolytes [65].

Theoretical consideration in iontophoretic drug delivery: The Nerst-Planck equation is used to describe the membrane transport of ions,

$$J = -D\frac{dC}{dx} + \frac{DZeEC}{kT}$$

where J is the flux of ions across the membrane and D is the diffusion coefficient. C is the concentration of ions with valence Z and electric charge e. E is the electric field, k is Boltzmann's constant, and T is the absolute temperature.

For transport of a nonelectrolyte, $Ze = 0$, and the equation reduces to Fick's first law of Diffusion,

$$J = -D\frac{dC}{dx}$$

For an ion with a uniform concentration throughout the system ($dC/dx = 0$), the Nerst-Planck equation becomes the equation for electrophoresis,

$$J = \frac{DZeEC}{kT}$$

The Nerst-Planck equation assumes that the only driving force on the chemical species is the negative gradient of chemical potential for that species alone. It neglects coupling of chemical potential with another species, j, in the flux of the i-th species. For example, the gradient of chemical potential for water activates not only the flow of water (elec-troosmosis), but a flow of solute dissolved in water, i.e., solvent drag. Pikal recently described a theoretical model for the effect of electroosmotic flow on the enhanced flux in transdermal iontophoresis; it proposes counterion flow as the probable mechanism for enhanced bulk fluid flow [66].

The movement of an electrically charged ion in an electric field occurs under the influence of a gradient in electrical potential and may be opposed by a variable amount of resistance. These relationships are given by Ohm's law,

$$V = IR$$

where V is the voltage, I is the electric current, and R is the resistance. Ohm's law holds only for electric current flowing along a single path. For current flow through the skin, Ohm's law holds only if the applied voltage does not exceed 2 V. In direct-current mode, the direction of current flow is unidirectional, while in alternating currents it changes periodically. However, DC currents may be interrupted in a periodic fashion to produce a DC pulse or periodic current. In addition, the waveform, current intensity, duty cycle (on-off ratio), and frequency may be altered.

Electrical properties of skin: The stratum corneum or topmost layer of skin is a good insulator, i.e., a good barrier to electrical conductivity. The water content of skin is approximately 20%. The electrical conductivity of living tissue is proportional to its water content. The stratum corneum has a high electric resistance, which is an important element for the development of a high skin impedance, a correlate of skin capacitance. Biological tissues, such as skin, have a capacitance because they are able to store electrons and thereby act as capacitors.

When an electric current contains both capacitive and resistant elements, it is called reactive. Reactive currents are said to present impedance rather than resistance. Human skin has been reported to show a high impedance to alternating currents of low frequency, which decreases greatly if the skin is abraded. Direct currents also induce polarizing currents, which can mask the true electrical resistance of tissues and oppose effective current across tissues, thereby reducing the efficiency of iontophoretic delivery. Polarizing currents, however, can be employed through use of a pulse current with cycle periods of microseconds.

The route and mechanism of iontophoretic transdermal delivery (TDD) of drugs is influenced by many factors. If the drug bears a charge, an ionogenic compound, TDD is complicated by the presence of ionized and un-ionized species in solution, each permeating the skin at different rates. For ionic drugs, hair follicles and sweat glands act as diffusion shunts; i.e., they follow the path of least resistance. Pore transport properties and tissue alteration have been examined in excised human skin during iontophoresis [67]. A pore density of 2 to $5/cm^2$ was visualized by cathodic iontophoretic transport of fluorescein from the epidermis into the dermis. Microelectrode measurements showed that the pores were the locus of charge transport. After application of current of 0.16 mA/cm^2 for 1 h, the decrease in resistance was 0.2 Hz. The decrease in skin resistance was not due just to tissue hydration effects, indicating that the passage of electric currents through excised human skin at clinically acceptable current densities may irreversibly damage the tissue.

Another phenomenon is the development of potential-dependent pore formation in the stratum corneum that contributes to the action of a "flipflop" gating mechanism, whereby the potential across the stratum corneum induces a flipflop parallel alignment of polypeptide helices. Pores are opened as a result of repulsion between neighboring dipoles and water, and ions are induced to flow into the pores to neutralize the dipoles. The pH of the drug solution can affect TDD by iontophoresis. It is known that the pH of skin is between 3 and 4. Thus, pores will have a positive charge when exposed to a medium with a pH below 3, while exposure to a medium above 4 will give the pores a negative charge.

The possibility of programmed delivery from an iontophoretic device increases the interest in drug delivery requiring nonuniform rates, e.g., small peptide hormones. One difficulty indicated above is lowered efficiency because much of the electric current is carried by nondrug ions, e.g., the substantial concentrations of relatively small and mobile solvent ions such as sodium and chloride ions in skin. A recent paper examined the transport of sodium and chloride ion transport in frozen excised human skin using DC iontophoresis [68]. They found that the current-voltage relationship in this tissue was time-dependent, highly nonlinear, and asymmetric with respect to the sign of the applied potential. Of interest was the finding that the skin resistance decreased with increased current or voltage for current densities less than 15 $\mu A/cm^2$ and exposure time of 10 to 20 min, with the decrease being reversible.

Experimental applications—facilitated transdermal delivery of small drug molecules: There is extensive literature on the iontophoretic delivery of small drug molecules. The list of agents discussed below is not meant to be exhaustive and reflects only a small portion of the uses reported. Some recent applications include the iontophoretic TDD of copper ions in male animals for reproductive contraception [69], silver ions in the treatment of chronic osteomyelitis [70], zinc ions to treat ischemic skin ulcers [71], and iodine to reduce scarring [72]. Among active agents, the vasodilator histamine was used as a counterirritant and for the relief of painful joint and muscle spasms. In addition, anti-inflammatory steroids have been delivered both systemically and locally in the treatment of Peyrone's disease [73]. The efficacy of steroid TDD has recently been questioned by experiments that failed to demonstrate the penetration of corticosteroids through the skin by iontophoresis [74]. The beneficial effects seen in clinical trials were ascribed to anti-inflammatory and analgesic effects of the applied direct current. Negative effects of iontophoresis have also been seen in the treatment of hyperhidrosis [75,76], including plugging of sweat glands, and possible abnormal patterns of keratinization and hyper-keratoses. Nevertheless, iontophoresis continues to be the method of choice for diagnosis of cystic fibrosis [77].

Other uses include the iontophoretic delivery of hyaluronidase in the treatment of lymphedema [78], the iontophoretic delivery of sodium salicylate in treatment of plantar warts [79], adenosine arabinoside against herpetic lesions [80], and antibiotic administration in burns [81].

Experimental applications—transdermal delivery of peptides and proteins: There is now a growing list of therapeutic peptides and protein hormones that have been successfully delivered by transdermal iontophoresis, including the hormone insulin [82]. Transdermal iontophoretic delivery of peptides is preferred to the oral route owing to extensive pro-teolytic degradation and/or poor absorption in the gastrointestinal tract. In general, trans-dermal delivery of peptides/proteins is very restricted due to their relatively large size (molecular weight), and if they are charged or very polar, skin permeation is not sufficient enough to achieve therapeutic levels. In theory, small peptides and growth factors could be satisfactorily delivered, as nanomolar amounts are active, or they may be targeted only to the skin itself. Burnette and Marrero [83] have recently reported results comparing iontophoretic and passive transport of thyrotropin-releasing hormone (TRH), a weakly basic tripeptide, L-pyroglutamyl-L-histidyl-L-proline amide, across excised nude mouse skin. They report that iontophoretic fluxes across skin were greater for both uncharged and charged TRH than for passive diffusion alone. The steady-state fluxes of both species of TRH were directly proportional to the applied current density, and the uncharged TRH had the higher flux.

There was indirect evidence that the increased transport occurred through pores, and that positive ions are preferentially passed through skin. However, these results, as they relate to human skin, must be interpreted with caution, as the thickness of the stratum corneum of humans is far greater and more complex than that of the mice. And, as noted above, Burnette and co-workers have recently shown that clinically acceptable current densities can nevertheless lead to tissue damage in human skin. The large current densities required for efficient transport of charged peptides coupled with preferential pore passage create unwanted side effects, including an uncomfortable tingling effect in the patient.

Phonophoresis. Phonophoresis is the application of ultrasonic energy to increase the permeation of drugs through the skin. The mechanical energy delivered by ultrasonic treatment is obtained by passing an alternating current through a piezoelectric crystal, causing it to vibrate. The use of ultrasound to enhance permeability through polymer membranes and control the delivery from polymer implants has been reported [84–87]. Furthermore, it has been determined that the increased permeability could be attributed to the ultrasonic perturbation and not to temperature increase, disruption of the boundary layer, or any irreversible changes in the polymer structure. Julian and Zentner demonstrated that the ultrasonically enhanced diffusion was a result of the decrease in the activation energy barriers within the membrane [88].

As early as 1962, it was known that ultrasound could drive hydrocortisone through the skin of pigs [89]. Enhanced permeation for fluocinolone acetonide, lidocaine, prilocaine, mannitol, insulin, physostigmine, and ibuprofen has also been demonstrated [90–93]. Kost et al. [94] explored the effect of phonophoresis on the stratum corneum structure and composition using lipophilic and hydrophilic drugs. They concluded that phonophoresis alters and affects the polar as well as the nonpolar pathway.

Phonophoresis with topical anti-inflammatories or local anesthetics is currently utilized by physical therapists as part of their treatment plans [89]. Clinically, the dose intensity is below 2 W/cm^2 and the duration of treatment between 5 and 20 min. The drugs can be delivered on continuous or pulsed mode, with the latter being gentler to the skin. The ultrasound frequencies used are usually less than 3 MHz. Recently, however, it has been shown that by using frequencies of 1, 7.5, and 15 MHz, permeation can be progressively enhanced over the control [95]. Additional research is required before a practical system for transdermal delivery can be evolved. A balance will have to be reached between intensity, frequency, pulse period, and duration, plus the ability of the patient to tolerate the treatment.

Reduction of Inflammatory and Immunologic Reactions

As discussed earlier, most drugs cannot be delivered transdermally because they do not permeate through the skin in amounts sufficient to administer a therapeutic dose or because they irritate or sensitize the skin. Furthermore, when chemical or physical enhancers are used to increase permeation, in most cases, they also increase skin irritation and sensitization. It is therefore very surprising that until recently not much work has been done on this problem.

Obermeyer et al. circumvented the irritation caused by chlorpheniramine delivered from cellulose triacetate membranes to humans [96]. The irritation, as well as the percutaneous absorption, were substantially reduced by reacting chlorpheniramine with undecylenic acid to form a fatty acid salt. The absorption of the salt was enhanced by dissolving it in a nonpolar nonvolatile solvent such as isopropyl oleate.

Patel and Ebert found that glycerine reduces the primary irritation of moderately irritating drug/enhancer compositions [97]. Enhancers claimed consist of solvents and envelope-disordering compounds and belong to the binary systems discussed earlier.

Govil and Kohlman patented a nicotine patch containing an antipruritic to counteract pruritus observed with the transdermal administration of nicotine [98]. Antipruritics disclosed include bisabolol, oil of chamomile, chamazulene, allantoin, D-panthenol, glycyrrhetenic acid, corticosteroids, and antihistamines.

Corticosteroids have also been shown to suppress cutaneous delayed hypersensitivity [99,100]. Hydrocortisone and triamcinolone acetonide 21-oic acid methyl ester have been shown to counteract the sensitization potential of a variety of sensitizers, such as clonidine, scopolamine, tetracaine, chlorpheniramine maleate, naloxone, naltrexone, nalbuphine, levorphanol, hydromorphone, buprenorphine, and 1-chloro-2,4-dinitrobenzene.

Recently, Cormier was granted a patent for the elimination of irritation or sensitization caused by a drug that is metabolized by an enzyme in the skin to form an irritating or sensitizing metabolite [101]. The method involves the co-administration of a metabolic modulator capable of inhibiting the enzyme from metabolizing the drug, such metabolic modulators being selected from a group consisting of tranylcypromine and phenyl alcohols. Propranolol and tetracaine are shown to be drugs that should be co-administered with the metabolic modulators.

As indicated above, some progress has been achieved during the last 2 years in this most important area of transdermal delivery, with much remaining to be accomplished. These accomplishments will come through systematic research and study of the inflammatory processes and agents responsible for the initiation of the arachidonic cascade as well as through the understanding of the cellular network and signaling cascade as it pertains to the development of irritation and sensitization.

Models of Skin Permeability

Introduction

The skin is a heterogeneous membrane consisting of the stratum corneum, the viable epidermis, and the dermis [102]. The multilayer structure of the skin efficiently protects the body against the entry of foreign toxic compounds. The stratum corneum, the outermost layer of the skin, is the major physical barrier for passage of most drugs, although the viable skin may act as a metabolic barrier for some drugs. It is essential, for developing transdermal drug delivery, to minimize the diffusion resistance of the stratum corneum.

Since the stratum corneum itself has a multilayered heterogeneous structure which consists of the flattened, keratinized, protein-rich cells and the intercellular lipid layers [103], percutaneous absorption is influenced by the chemical composition and geometric arrangements of the stratum corneum, such as the lipid layer content, the number of cell layers stacked, and the thickness of the stratum corneum, as well as by the physicochemical properties of the penetrant.

As discussed earlier, there are two major pathways for drugs traveling through the stratum corneum: the intercellular route and the intra- (or para-)cellular route [104]. In addition, the shunt pathways such as hair follicles and sweat glands may have an appreciable influence, particularly during the transient period before establishing steady-state permeation. Under steady state, however, the effect of shunt pathways has been reported to be insignificant [104].

A variety of mathematical models for describing the permeability across the stratum corneum have been reported in the literature. These models are broadly classified into two categories: (1) homogeneous models and (2) heterogeneous models. Several models are briefly reviewed in this section.

Models of Permeation across Stratum Corneum

Homogeneous diffusion model.   The homogeneous diffusion model considers the stratum corneum as a homogeneous diffusion barrier. The concentration profile at steady state is shown in Fig. 13. If the concentration of the drug on the surface of the device (at x = 0) is constant (C_d), which condition is usually the case for skin-controlled drug

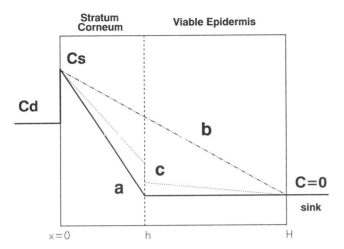

Figure 13 Steady-state concentration profile in the skin based on the homogeneous skin model. (a) Single-layer stratum corneum model, (b) single-layer whole-skin model, (c) bilayer whole-skin model.

delivery or in-vitro permeation experiments for the infinite dose system, the steady-state rate of permeation under the sink condition is given by

$$\left(\frac{dQ}{dt}\right)_{ss} = \frac{D_s K C_d}{h} \tag{1}$$

where K is the partition coefficient of the drug between the stratum corneum and the donor solution, and D_s is the effective diffusion coefficient across the stratum corneum. When the diffusion coefficient D_s, the stratum corneum thickness h, and the partition coefficient K are known or can be determined from the separate experiments, the steady-state rate of skin permeation is calculated using Eq. (1).

In recent years, several approaches have been proposed for handling the physicochemical data on skin permeation [105,106] in order to predict the skin permeation rate of specific drugs from Eq. (1). Since the diffusivity, partition coefficient, and thickness are difficult to separate for fragile membranes such as the skin, the permeation data are commonly expressed by the effective permeability coefficient, defined as follows [107,108]:

$$P = \frac{D_s K}{h} = \left(\frac{dQ}{dt}\right)_{ss} \frac{1}{C_d} \tag{2}$$

The diffusion coefficient across the stratum corneum can be calculated from the time lag t_d determined in the in-vitro experiment:

$$D_s = \frac{h^2}{6 t_d} \tag{3}$$

If the viable skin is also considered, the diffusion coefficient can be evaluated using a bilayer model:

$$D_s = \frac{1}{1 - \eta} \frac{h}{C_2} \left(\frac{dQ}{dt}\right)_w \tag{4}$$

where

$$C_2 = \frac{(1 - 3\tau + 2\eta\tau)}{(1 + 2\eta)(1 - \eta)} \frac{6t_{dw}}{h} \left(\frac{dQ}{dt}\right)_w \tag{5}$$

$$\eta = \frac{(dQ/dt)_w}{(dQ/dt)_v} \tag{6}$$

$$\tau = \frac{t_{dv}}{t_{dw}} \tag{7}$$

and the subscripts w and v stand for the intact skin and the viable skin, respectively. The steady-state flux and time lag for viable skin can be determined from the in-vitro permeation experiment using stripped skin [105].

Example: The in-vitro permeation experiment of desoxycorticosterone through hairless mouse skin was carried out, and the following results were obtained: Steady-state rate of permeation [μg/cm^2-h] = 4.73 for the intact skin and 12.1 for the stripped skin; time lag [h] = 3.9 for the intact skin and 1.2 for the stripped skin. The thicknesses of the stratum corneum and of the intact skin are 0.0010 cm and 0.0380 cm, respectively. Calculate the diffusion coefficient across the stratum corneum based on (a) the single-layer stratum corneum model, (b) the single-layer whole-skin model, and (c) the bilayer whole-skin model (see Fig. 13). [*Ans.*: (a) 1.2×10^{-11} cm^2/s; (b) 1.7×10^{-8} cm^2/s; (c) 6.6×10^{-11} cm^2/s.]

Two-phase diffusion model. Albery and Hadgraft [109,110] proposed a two-phase model as shown in Fig. 14 for describing drug diffusion through the stratum corneum. The permeation of the drug was analyzed based on Fick's second law of diffusion by assuming constant concentration on the surface of the skin and the sink condition in the receiver. The model considers two independent pathways: (1) diffusion through the keratinized cells (transcellular route) and (2) diffusion through the intercellular route.

At steady state, the rate of permeation through each route was given by

$$\left(\frac{dQ}{dt}\right) = \frac{C_d}{(2/\alpha k_1) + (h/\alpha\beta K_m D_s)} \tag{8}$$

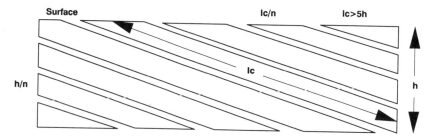

Figure 14 Heterogeneous stratum corneum model; two-phase model [109]. The stratum corneum consists of n layers of dead cells with interstitial channels of length ℓ_c.

where α is the area fraction, equal to unity for the transcellular route and 7×10^{-3} for the intercellular route; D_s is the diffusion coefficient in the stratum corneum; K_m is the partition coefficient of the drug between mineral oil and water; β is an empirical parameter; and k_1 is the interfacial transfer rate constant (10^{-5} m/s). By comparing the model solution with the experimental data of Michaels et al. [118], the authors suggested $\beta = 5$ for the intercellular route, and $\beta = 1$ for the transcellular route.

By assuming parallel pathways, the total flux can be calculated by

$$\left(\frac{dQ}{dt}\right)_T = \left(\frac{1}{(dQ/dt)_1} + \frac{1}{(dQ/dt)_2}\right)^{-1} \tag{9}$$

where the subscripts 1 and 2 stand for the transcellular route and the intercellular route, respectively.

Berner and Cooper [111] also proposed a two-parallel-pathways model for skin permeation. This model combines a polar or aqueous pathway and a nonpolar or lipophilic pathway. The drug diffusion through each pathway occurs independently, and no interaction is assumed between the two routes of penetration.

The steady-state flux through the polar pathway is the same as that given in Eq. (1):

$$J_p = \left(\frac{dQ}{dt}\right)_p = \frac{D_p C_w}{h}$$

The partition coefficient should be near unity; i.e., compounds that permeate by this pathway should penetrate independent of partition coefficient. C_w is the solubility in water, and D_p is the effective diffusion coefficient in the polar pathway. D_p is a function of the size and shape of the molecule and, according to the free-volume theory of Bueche for rigid molecules [112], may be written as

$$D_p = D_p^0 e^{-B_p M}$$

where M is the molecular weight and D_p^0 and B_p are constants.

To increase the flux through this polar pathway one must increase the solubility in water. Yalkowsky has derived an equation for water solubility as a function of partition coefficient and melting point [113]. This equation is limited to molecules that have a partition coefficient greater than 1, and for polar molecules, one must know the solubility in water.

The Bueche equation indicates that the diffusion coefficient will increase exponentially as the size of the molecule decreases.

The lipophilic pathway is the more important one, since most drugs and cosmetics are lipophilic and migrate via this route. As before, the steady-state flux may be written as

$$J_L = \frac{D_L C_L}{h}$$

where C_L is the concentration of drug in the lipids at the outer interface of the stratum corneum and D_L is the lipid diffusion coefficient in the nonpolar pathway. To increase the flux, one must increase C_L, however, since concentration is not usually measured in skin lipids, a suitable lipophilic solvent is used. This solvent, for example, octanol, is also used to determine the partition coefficient K. D_L for rigid molecules may again be written as [112]

$$D_L = D_L^0 e^{-B_L M}$$

where D_L^0 and B_L are constants.

Example: Derive the equation for the flux of a drug through the stratum corneum, knowing that the flux is the sum of the fluxes from the continuous polar and nonpolar regions. The area fraction of the polar region is A_p, and that of the nonpolar region is A_L.

We can begin with

$$J = J_p + J_L \tag{1a}$$

Combining with the above equations,

$$J = A_p \frac{D_p C_w}{h} + A_L \frac{D_L C_L}{h} \tag{2a}$$

Knowing the partition coefficient $K = C_w/C_L$, we can write

$$J = (A_p D_p + A_L D_L K) \frac{C_w}{h} \tag{3a}$$

The water solubility for lipophilic drugs may be predicted using the theory of Yalkowsky [113] stated earlier; the equation is

$$C_w = \frac{M \exp[-(0.023 T_f + 1.59)](5.5 \times 10^{-2})}{K\{1 - \exp[-(0.023 T_f + 1.59)]/K\}} \tag{4a}$$

where T_f is the melting point in degrees Celsius. For $K < 1$, the water solubility should be measured.

From the above equations, it is possible to predict skin permeation rate knowing the parameters M, K, and C_w or T_f.

Example: Using the equations derived for the flux of drug through the stratum corneum, prepare a plot of flux versus molecular weight for a series of lipophilic drugs having melting points of 100, 150, 200, and 250°. The partition coefficient may be taken as 5. The approximate values of the constants D_p^0 and D_L^0 are 3.8×10^{-5} and 1.7×10^{-5} cm²/h, respectively [114]. The thickness of the stratum corneum is approximately 25 μm and the constants $B_p = B_L = 0.016$ [10]. The constants A_p and A_L are taken as 0.1 and 0.9, respectively [111,114]. Substituting the Yalkowsky and Bueche equations into Eq. (3a) gives an equation in which the flux is found to be a function of molecular weight only after the parameters given above are introduced into the equation. The solved equation is plotted in Fig. 15.

Three-parallel-pathways model. Berner and Cooper [111] proposed a three-phase model of the stratum corneum as shown in Fig. 16. This model considers three pathways: (1) a nonpolar (lipid) pathway, (2) a polar (aqueous) pathway, and (3) an alternating nonpolar-polar pathway. The drug diffusion through each pathway occurs independently, and therefore no interaction or partitioning is assumed between two adjacent routes of penetration. The simplified equation for describing the skin permeability through the three-phase model has been reported elsewhere [111].

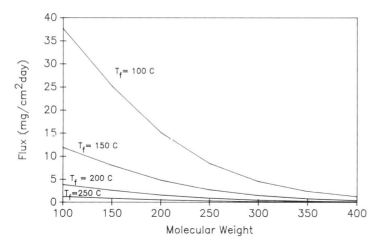

Figure 15 Flux versus molecular weight and melting point.

Single-compartment model. A single-compartment model has been widely used to investigate skin pharmacokinetics for predicting the plasma concentration following transdermal drug delivery. In a multicompartment model for skin pharmacokinetics, Guy and Hadgraft [115] assumed the stratum corneum to be a single compartment in which the drug molecules distribute homogeneously. In addition to the stratum corneum, they considered the blood compartment, the skin surface compartment, and the urine compartment (Fig. 17). By solving a set of first-order differential equations together with the model parameters, the authors successfully described the plasma concentration profile of clonidine, nitroglycerine, and scopolamine following transdermal delivery. This model would be useful both for screening transdermal drugs and for predicting an optimum delivery system. However, it is essential to develop a reliable method for determining the model parameters. Although the clinical data were well described after evaluating the intercompartmental rate constant between the stratum corneum and the viable epidermis based on

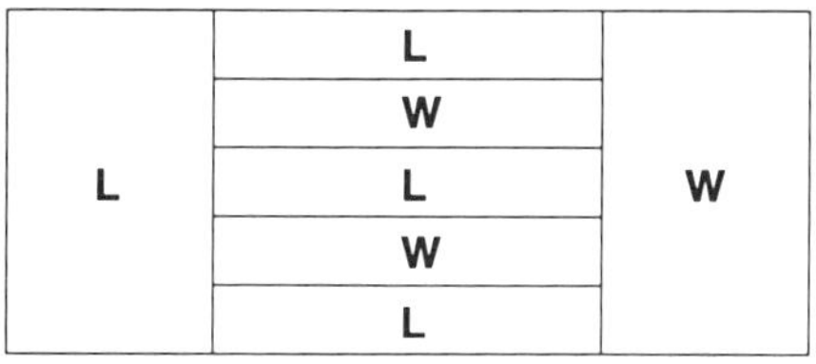

Figure 16 Three-parallel-pathways model for stratum corneum [111]: L, continuous lipid pathway; W, continuous aqueous pathway; L-W, lipid-aqueous multilaminate pathway.

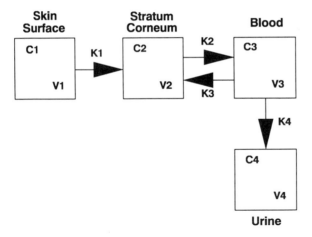

Figure 17   Single-compartment model for stratum corneum [115]: C_i, concentration; k_i, transfer rate constant between compartments; V_i, volume of compartment.

the Stokes-Einstein equation, it is not likely that the diffusion coefficient across the stratum corneum is inversely proportional to the molecular weight of the penetrant [116].

Multicompartment model.   Higuchi et al. proposed a multicompartment model of the stratum corneum as shown in Fig. 18 [117]. In this model, the stratum corneum is divided into many compartments (elements) and arranged in a series. The authors used this model to elucidate the effect of a skin enhancer (ethanol), which is asymmetrically distributed in the stratum corneum. In principle, the multicompartment model becomes identical to the diffusion model when the number of elements is infinite. It is frequently reported that the degree of enhancement in skin permeability differs between animal models and human. To understand the effect of different enhancers on skin permeability, the concentration of the enhancer in the stratum corneum must be taken into account since the drug solubility as well as the diffusion coefficient may be significantly influenced by the level of enhancer

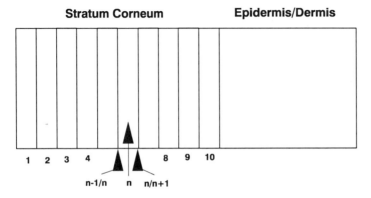

Figure 18   Multicompartment model for stratum corneum [117].

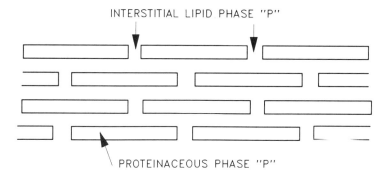

Figure 19 Ordered brick model for stratum corneum [118].

concentration in the stratum corneum. The multicompartment stratum corneum model is a simple but useful model for gaining a better understanding of the effect of enhancers on skin permeability. The effect of the enhancer concentration in the stratum corneum on the rate of skin permeation may be analyzed more rigorously by solving the diffusion equations for both the drug and the enhancer simultaneously. This approach may require complicated numerical methods analysis, however. If the enhancer is incorporated in the delivery system or applied on the skin surface as a pretreatment, it is probable that the skin structure varies during the medication and hence the solubility and the diffusion coefficient across the stratum corneum are no longer constant but rather space- or time-dependent as a result of the concentration gradient of the enhancer in the skin. At this stage, however, little is known with respect to the dynamic effect of enhancer in the stratum corneum. The multicompartment model can be employed to elucidate this phenomenon.

Brick model.   Michaels et al. [118] proposed a brick model as shown in Fig. 19 to describe the steady-state permeation rate of drugs across the stratum corneum. This model simplified the complex geometric arrangements of the stratum corneum to an ordered brick structure in which flattened protein cells (hydrophilic phase) are orderly stacked with a cement of lipid phase (continuous). The model considers two pathways, (1) lipid-protein cell in series and (2) lipid phase only, as the major routes of permeation. A theoretical equation was then developed for the steady-state flux. The results indicated that the permeability of stratum corneum to any specific penetrant should be determined by only two physicochemical parameters: the quantity $K_p D_p$, which is the specific permeability of the protein phase; and the quantity $\gamma (= \sigma D_L/D_p)$, which is the product of the lipid-protein partition coefficient (σ) and the ratio of the diffusion coefficient in the two phases.

For human stratum corneum with lipid content of 15% by volume, the steady-state flux was described by the following equation:

$$\left(\frac{dQ}{dt}\right)_{ss} = \frac{k_p D_p}{h}\left(\frac{1.16}{0.16/\gamma + 1} + 0.0017\gamma\right)C_d \tag{10}$$

where C_d is the drug concentration in the aqueous donor solution. By assuming that $k_p = 0.75$ and $D_p = 2 \times 10^{-7}$ cm^2/s for hydrated skin, they obtained

$$\left(\frac{dQ}{dt}\right)_{ss} = \frac{1.5 \times 10^{-7}}{h}\left(\frac{1.16}{0.16/\gamma + 1} + 0.0017\gamma\right)C_d \tag{11}$$

where the dimensions of the flux $(dQ/dt)_{ss}$ are $\mu g/cm^2/s$. When $D_l/D_p = 10^{-3} \sim 10^{-2}$, P is the mineral oil/water partition coefficient, and $h = 40 \ \mu m$ for hydrated stratum corneum, the model described the skin permeability of various drugs well [118].

By extending the ordered brick model, a random brick model was recently developed [119]. The stratum corneum was assumed to consist of protein-rich thin plates (cells) separated from one another by a thin-layer intercellular lipid, as shown in Fig. 20. The side length of individual cells may vary; however, the total average surface area of cells was assumed to be constant. The thickness of the cell element $(n\delta)$ and lipid layer (2δ) is assumed to be constant. In this model, the major routes of drug transport across the stratum corneum are divided into three pathways: (1) through the flattened protein-rich cells sandwiched with thin lipid-rich layers; (2) through the lipid intercellular route; and (3) through the cell/intercellular in series. The model considers the effect of the geometric arrangements and chemical composition of the stratum corneum on the skin permeability of drugs.

The steady-state flux was given by

$$\left(\frac{dQ}{dt}\right)_{ss} = \frac{C_a D_e}{h} \tag{12}$$

where

$$D_e = R\left(\frac{1 - \epsilon}{D_p} + \frac{\gamma\epsilon}{D_l}\right)^{-1} \tag{13}$$

$$R = \frac{2\epsilon(1 - \epsilon)(2\eta + 4)}{\eta + (\eta + 4)/\gamma} + \epsilon^2\gamma + \frac{(1 - \epsilon)^2(2\eta + 4)}{2\eta + 4/\gamma} \tag{14}$$

$$C_a = C_l(1 - \epsilon) + C_p\epsilon \tag{15}$$

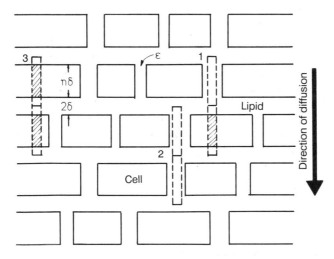

Figure 20 Random brick model for stratum corneum [119]. The number of total cell layers is usually about 20; ϵ is the average fraction of diffusion area of the lipid phase on the skin surface; f_v is the total fraction of lipid phase; $n = (2f_v - 2)/(\epsilon - f_v)$.

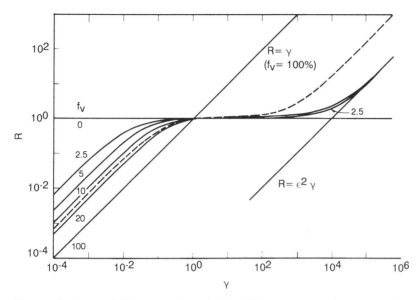

Figure 21 A plot of R versus γ from Eq. (14). The numbers on the lines are the percent of f_v; $\epsilon = 1\%$. (-----) Ref. 118.

where D_p and D_l are the diffusion coefficients across the protein layer and the lipid layer, respectively, k is the lipid-protein partition coefficient, ϵ is the average surface area fraction of the lipid phase, and γ has already been defined as $\sigma D_t/D_p$.

Equation (12) indicates that the steady-state flux across the heterogeneous stratum corneum can be approximated to a simple diffusion if the effective diffusion coefficient and the average surface concentration are defined by Eqs. (13) and (15), respectively.

Figure 21 shows the ratio R for a usual skin structure ($\epsilon \approx 0.01$) against γ as a parameter of the volume fraction of the lipid layer f_v. The contribution of the intercellular route can be neglected for $\gamma < 1$. If γ exceeds 10^4, however, the overall rate of drug transport across the stratum corneum is controlled by the intercellular route. In this figure, the result calculated from the ordered brick model ($\epsilon = 0.8\%$; $f_v = 15\%$) [Eq. (14)] is also plotted as a dashed line for comparison. The random brick model was used to elucidate the enhancement mechanism of skin permeability by various enhancers, as well as the effect of the geometric structure of the stratum corneum.

Recently, Osborne [114] reported that the mathematical models published occasionally predicted steady-state fluxes of skin permeation that were unacceptably high or low when compared with the experimental data. He therefore recommended use of more than one of the models for predicting skin permeability. The solubility and the diffusion coefficient of the drug in the stratum corneum may be significantly influenced, in a complicated manner, by the device design and materials incorporated, and for this reason it is better to determine the physicochemical properties such as the diffusion coefficient and the partition coefficient by separate experiments.

Conclusion

Since the major resistance to skin permeation resides in the stratum corneum, a variety of mathematical models of the stratum corneum have been proposed for predicting skin permeability of drugs. In this section, some of these models were reviewed briefly. Each model has advantages and weaknesses with respect to simplicity and rigorousness. In general, models with more parameters provide better fits with the experimental data. However, better fitting may not assure superiority over simpler models, unless the model parameters can be determined by a scientifically sound approach, either experimentally or theoretically. Before solving a skin model aimed at predicting the skin permeability of a drug in question, we must have reliable model parameters. Further work is needed to establish the equations for correlating model parameters with simple physicochemical properties and to gain a better understanding of the microscopic structure of the stratum corneum.

PHARMACOKINETICS OF SKIN PERMEATION

Introduction

A drug administered transdermally partitions on the surface of the skin, with some penetrating the skin and finally being transported away by the circulation system. Once in the blood, the drug molecules are distributed throughout the body or eliminated following the body pharmacokinetics. Since the skin generally serves as a major barrier membrane either physically or enzymatically against the entry of xenobiotics including drugs, the drug concentration in the body or plasma following transdermal delivery responds slowly compared to that following oral or intravenous administration. A long time lag in transdermal delivery may be more significant for drugs that are bound and/or metabolized in the skin tissue. Basically, the pharmacokinetics of skin permeation of drugs can be described by combining the drug release kinetics from the delivery system, the permeation kinetics through the skin, and the elimination/distribution kinetics in the body.

Release of a drug from the delivery system and permeation across the skin have been widely investigated in both in-vitro and in-vivo experiments, while the body elimination/distribution of drugs can obviously be studied under in-vivo conditions.

Before developing a transdermal delivery system for a drug, the effective or targeted plasma level must be clearly defined. Both steady-state and transient profiles of the plasma concentration following transdermal delivery are influenced primarily by the following two factors: (1) skin permeation/binding/bioconversion and (2) body elimination/distribution.

The purpose of this section is to discuss the pharmacokinetics of transdermal systemic delivery of drugs to gain a quantitative understanding of the effects of the pharmacokinetic factors on the time course of the drug concentration profile in the plasma. Understanding skin pharmocokinetics is essential for developing an optimum design for a delivery system with respect to the content of drug or loading dose, as well as the duration of application.

Events Taking Place in the Skin

Although many physical and biochemical processes take place in the skin [120], the major events influencing the plasma profile are diffusion, partitioning, binding, and metabolism in the skin. Each process influences, more or less, not only a steady-state or peak concentration of the drug in the plasma, but also the time course of the concentration profile. Generally speaking, drug binding in the skin delays the appearance of the drug in the plasma, while metabolic reactions may decrease the time lag [121]. Skin metabolism of the parent drug may also significantly reduce therapeutic effectiveness. For a prodrug, however, the enzymatic reaction is evidently essential [122].

Models of Skin Permeation

The skin is not a homogeneous membrane but a heterogeneous, multilaminated structure [123], which consists of the stratum corneum, viable epidermis, and dermis [124]. The stratum corneum is believed to provide the major physical barrier capacity for most drugs, particularly hydrophilic compounds. For highly lipophilic compounds, on the other hand, the hydrophilic viable tissue may provide a major barrier capacity.

For a drug that is metabolized negligibly in the skin, the permeation data can be analyzed on the basis of a single-layer model in which only the stratum corneum is considered. When metabolic reaction (bioconversion) is appreciable in the viable skin, we must consider the skin as a multilayer membrane with at least two layers, i.e., the stratum corneum and the viable skin. However, the stratum corneum itself is not a simple homogeneous layer [125], as discussed in the section on models of skin permeability. The heterogeneous structure of the stratum corneum can nevertheless be approximated as a homogeneous membrane with respect to skin permeability, if the effective diffusion coefficient as well as an average solubility are properly defined [126] after taking into account the microscopic structure of the stratum corneum. The effect of the geometric arrangements of the stratum corneum on the permeation of drugs has been discussed in the section on models of skin permeability. Despite the variety of skin permeation models reported in the literature, the discussion in this section will be limited to diffusion models because the drug molecule travels through the skin basically in accordance with the diffusion process. Various models, however, including multicompartmental skin tissue models and single-compartment models other than the diffusion model, have already been discussed in the section on models of skin permeation.

Single-Layer Skin Model

For a single-layer skin model, the concentration of the drug in the stratum corneum is described by

$$\frac{\partial C}{\partial t} = \frac{\partial}{\partial x}\left(D\frac{\partial C}{\partial x}\right) \tag{16}$$

where the diffusion coefficient D is in general a function of space and time, because the skin structure may change during medication. Such is probably the case in in-vitro permeation experiments using various solutions and enhancers. The enhancer, after partitioning on the skin surface, may develop a significant concentration gradient across the stratum corneum. The different levels of enhancer concentrations may then affect the microstructure of the stratum corneum as well as the solubility of the drug in a manner that depends on the concentration level.

If the skin remains intact during the entire period of medication and a constant diffusion coefficient and constant concentration on the skin surface are assumed, Eq. (16) can be solved with the following boundary and initial conditions [127]:

$$C = C_s(\text{constant}) \qquad \text{at } x = 0 \text{ (skin surface)} \tag{17}$$

$$C = 0 \qquad \text{at } x = h \text{ (sink)} \tag{18}$$

$$C = 0 \qquad \text{at } t = 0 \text{ and } 0 \le x \le h \tag{19}$$

where h is the thickness of the stratum corneum and C_s is the drug concentration on the surface of the skin.

The cumulative amount of drug permeated is then given by

$$Q(t) = \int_0^t \left(-D \frac{dC}{dx} \Big|_{x=h} \right) dt$$

$$= \frac{DC_s t}{h} + \frac{2hC_s}{\pi^2} \sum_{n=1}^{\infty} \frac{\cos n\pi}{n^2} \left[1 - \exp \left(-\frac{n^2\pi^2 Dt}{h^2} \right) \right] \tag{20}$$

At steady state ($t \to \infty$),

$$Q(t) = \frac{DC_s t}{h} + \frac{2hC_s}{\pi^2} \sum_{n=1}^{\infty} \left(\frac{\cos n\pi}{n^2} \right)$$

$$= \frac{DC_s}{h} \left(t - \frac{h^2}{6D} \right) \tag{21}$$

It has been widely observed that the in-vitro permeation profile across the stratum corneum can be described by Eq. (20) or the simplified Eq. (21), as shown in Fig. 22 [128].

The diffusion coefficient and the steady-state flux are evaluated from the time intercept and the slope of the permeation profile (Fig. 23), respectively:

$$D = \frac{h^2}{6t_d} \quad \text{or} \quad t_d = \frac{h^2}{6D} \tag{22}$$

$$\left(\frac{dQ}{dt} \right)_{ss} = \frac{DC_s}{h} \tag{23}$$

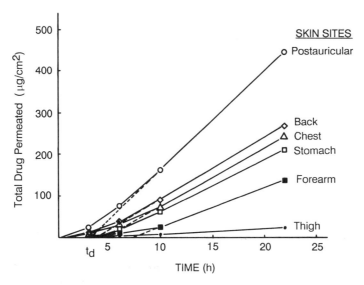

Figure 22 A typical profile (time course) of cumulative amount of drug permeated across the skin under in-vitro infinite-dose conditions. Permeation of scopolamine free base through human skin from various body areas. It is interesting to see that the time lag is relatively constant (3–5 h), in spite of the variety of skin sites. (Modified from Ref. 128, Fig. 9.)

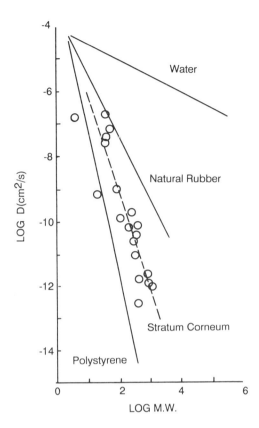

Figure 23 Relationship between the diffusion coefficient in the stratum corneum and the molecular weight of penetrants.

where the thickness of stratum corneum h is 10 to 40 μm as in normal human skin. If the thickness of the whole skin is used to evaluate the diffusion coefficient, the value of the diffusion coefficient may be significantly overestimated. Scheuplein and other researchers have indicated that the diffusion coefficient across the stratum corneum is usually 10^{-10} to 10^{-11} cm²/s for many drugs with molecular weights of 200 to 400 Da [129–131], while the diffusion coefficient across the viable skin is approximately 1000 times greater than that across the stratum corneum [130]. The diffusion coefficient of drugs across the stratum corneum is, in general, affected by the molecular weight of the penetrant, as shown in Fig. 23 [132]. Although the deviation is significant, the diffusion coefficient across the stratum corneum was found to be inversely proportional to the molecular weight of the penetrant, with the exponent being 3 to 4. A similar relationship was also obtained by Anderson [133].

Effect of Binding

It is well established that the time lag across the whole skin is approximately 3 to 5 h for many drugs that show little binding in the skin [130,134,135]. It has been found, however, that some compounds permeate the skin after a markedly long lag time, such as 30 to 40 h or longer [136,137]. This effect has been attributed to significant binding of the drug in the stratum corneum.

If a dual sorption model [138] based on the Langmuir isotherm is applied, the concentration of the free (unbound) drug in the stratum corneum is described by the following equation [139]:

$$\left[1 + \frac{K}{(1 + bC)^2} \right] \frac{\partial C}{\partial t} = D \frac{\partial^2 C}{\partial x^2} \tag{24}$$

where K and b are the binding constants based on the Langmuir isotherm.

Equation (24) indicates that the flux under steady-state conditions is not affected by the binding process if the Langmuir isotherm is assumed. The transient profile, however, may be significantly affected by the presence of skin binding.

The skin binding of several steroids was found to follow Eq. (24) [121]. The parameters K and b can be determined from equilibrium absorption experiments [140] or by mathematical simulation [121].

In the presence of binding, the apparent diffusion coefficient evaluated from the time lag [Eq. (22)] may be underestimated appreciably [141]. The apparent diffusion coefficient was found to be concentration-dependent and to approach the intrinsic diffusion coefficient when the donor (solution) concentration or skin surface concentration increased [142].

Chandrasekeran et al. [142] had previously demonstrated the applicability of the dual sorption model in the transport of scopolamine through human skin. The binding permeation of steroids, verapamil, and some amino acids across mouse skin has been successfully described by the dual sorption model [121,143,144]. Typical in-vitro profiles are shown in Fig. 24.

Since a variety of drug binding mechanisms may occur in the skin, the dual sorption model based on the Langmuir isotherm does not always apply to skin binding. It was observed that vitamin E binds strongly in the hairless mouse skin and the bound molecules release with great difficulty from the bound site of the skin under normal permeation-desorption experiments [145]. Under such conditions, the transient profile of skin permeation of vitamin E was hardly influenced by the binding [145].

Effect of Enzyme Reaction

The activities of various skin enzymes have been investigated, and some have been compared to hepatic activities. Montagna, in histologic investigations, reported on the localization and activity of skin enzymes in the viable epidermis [146]. It is widely believed that the major site of skin metabolism is in the viable skin, and the enzymatic activity in the stratum corneum is very minimal, although some exceptions are reported in the literature [120].

The enzymatic activity in human viable epidermis can be broadly classified into four categories [147]: (1) uniform distribution throughout the viable skin; (2) upper-layer distribution in the viable skin (mainly in the viable epidermis); (3) lower-layer distribution (mainly in the stratum basale); and (4) no distribution. The esterase, responsible for the bioconversion of many ester prodrugs, was found to follow upper-layer (type 2) distribution [147].

When the stratum corneum is removed or the barrier capacity of this layer is eliminated, the concentration of drug in the viable skin in the presence of bioconversion is expressed by the following partial differential equation:

$$\frac{\partial C}{\partial t} = D \left(\frac{\partial^2 C}{\partial x^2} \right) - R(C) \tag{25}$$

where D is the diffusion coefficient in the viable skin and R(C), the skin metabolism term, is usually described by the following Michaelis-Menten kinetics:

$$R(C) = \frac{V_{max}C}{K_m + C} \tag{26}$$

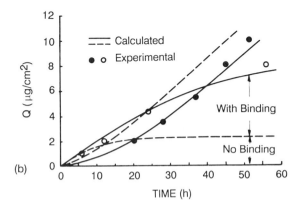

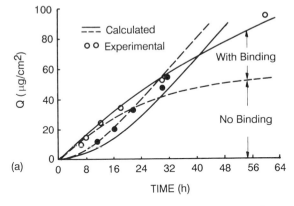

Figure 24 Stratum corneum reservoir capacity of steroids. Experiment: After establishing a steady-state permeation, the donor solution was completely removed at 48 h, the skin surface was quickly washed by buffer solution, and thereafter the desorption experiment was carried out. Solid lines and dashed lines are calculated by taking into account binding and no binding, respectively. (a) Progesterone, $D_{sc} = 7.8 \times 10^{-11}$ cm^2/s, $D_{vs} = 4.1 \times 10^{-8}$ cm^2/s, $P_c = 55$, $K = 8$, $b = 0.5$, $h = 0.0010$ cm, $H = 0.0380$ cm. (b) Hydrocortisone, $D_{sc} = 3.9 \times 10^{-11}$ cm^2/s, $D_{vs} = 4.3 \times 10^{-8}$ cm^2/s, $P_c = 0.105$, $K = 8$, $b = 0.5$, $l = 0.0010$ cm, $H = 0.0380$ cm.

where V_{max} is the maximum velocity and K_m is the Michaelis-Menten constant. If the drug concentration is small compared to the enzyme concentration, the bioconversion term Eq. (26) can be approximated to the first-order irreversible reaction and thus Eq. (25) becomes

$$\frac{\partial C}{\partial t} = D\left(\frac{\partial^2 C}{\partial x^2}\right) - kC \tag{27}$$

The solution of Eq. (27) under the boundary and initial conditions [$C = C_0$ at $x = 0$ (skin surface), $C = 0$ at $x = h$ and $t = 0$] was given by Leypoldt and Gough [148]. The cumulative amount of drug observed after skin permeation is given by

$$Q = \int_0^t \left(\frac{dQ}{dt}\right) dt$$
$$= \frac{DC_0}{h}\frac{\phi}{\sin h(\phi)}t - 2C_0 h \sum_{n=1}^{\infty}\frac{n^2\pi^2(-1)^{n+1}}{(n^2\pi^2 + \phi^2)^2}\exp\left[-\frac{(n^2\pi^2 + \phi^2)\,Dt}{h^2}\right] \tag{28}$$

The dimensionless time lag θ_L is then derived as

$$\theta_L = \frac{Dt_d}{h^2} = \frac{1}{2}\left[\frac{\cot h(\phi)}{\phi} - \frac{1}{\phi^2}\right] \tag{29}$$

The steady-state permeation (appearance) rate of the parent drug becomes

$$\left(\frac{dQ}{dt}\right)_{ss} = \left[\frac{\phi}{\sin h(\phi)}\right]\frac{DC_0}{h} = \eta\frac{DC_0}{h} \tag{30}$$

where DC_0/h is the steady-state rate of permeation without metabolic reaction and η is defined as a reduction factor due to the metabolism.

The above analysis indicates that the steady-state flux of the parent drug following skin bioconversion can be characterized by a dimensionless group defined as $\phi(= h\sqrt{k/D})$ and that the time lag decreases as a result of an irreversible first-order reaction in the skin. A typical profile of η and θ_L is shown as a function of ϕ in Fig. 25.

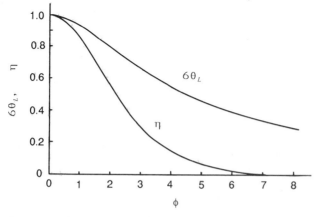

Figure 25 The variation of the dimensionless time lag (θ_L) and the reduction factor (η) as a function of dimensionless modulus (ϕ) according to Eqs. (29) and (30), respectively.

Bilayer Membrane Model of Intact Skin

Like the metabolic reactions that take place in the viable skin, binding of drugs does occur in the stratum corneum in transdermal drug delivery. Accordingly, a homogeneous membrane model, as described above, can be modified to a bilayer membrane model to achieve a better quantitative understanding of drug permeation through the skin.

Combining the skin binding in the stratum corneum based on the Langmuir isotherm and the enzymatic reaction in the viable skin, the concentration of free (unbound) drugs in the skin can best be described by the following equation [149]:

$$\left[1 + \frac{K}{(1 + bC)^2}\right] \frac{\partial C}{\partial t} = \frac{\partial}{\partial x}\left(D\frac{\partial C}{\partial x}\right) - \frac{V_{max}C}{k + C} \tag{31}$$

where the Michaelis-Menten kinetics are assumed for skin metabolism. The diffusion coefficient D is space-dependent:

$$D = \begin{cases} D_{sc} \ll D_{vs} & 0 \le x \le h \\ D_{vs} & h < 0 \le H \end{cases} \tag{32}$$

On the boundary between the stratum corneum and the viable skin (x = h), the following mass balance and partitioning equations are satisfied:

$$D\frac{\partial C}{\partial x}\bigg|_{x=h-} = D\frac{\partial C}{\partial x}\bigg|_{x=h+} \tag{33}$$

$$C_{x=h-} = P_c C_{x=h+} \tag{34}$$

The binding parameters K and b, diffusion coefficient D, and the reaction mechanism and its rate constant k can be determined by either in-vivo or in-vitro experiments [130,131,145,150]. Equation (31) can then be solved numerically under appropriate initial and boundary conditions to evaluate the permeation rate, which is generally time-dependent, following various modes of transdermal delivery.

Higuchi and his co-workers developed a skin bioconversion and diffusion model to describe the time profile of the permeation of a prodrug in hairless mouse skin [151,152]. Their model provided a good explanation of the experimental data on the bioconversion and permeation of estradiol esters through hairless mouse skin. Because they assumed equal concentration on the stratum corneum-viable skin boundary, their model may be limited to the use of drugs with partition coefficients between the stratum corneum and viable skin of approximately 1. More general situations using numerical methods have recently been reported [149,153,154]. Higuchi developed a general model for skin bioconversion under steady-state conditions [155]. He assumed a three-layer skin, composed of stratum corneum, viable epidermis, and dermis, and also took into account partitioning of the drug between the stratum corneum and the viable epidermis. Higuchi further considered the different modes of enzyme distribution or space-dependent enzyme activity. As indicated earlier, it is believed that the esterase responsible for esterification is widely distributed in the viable epidermis of the skin. Higuchi's findings suggest that enzyme distribution must be properly accounted for to better understand skin bioconversion and permeation.

The decrease in activity of enzymes in the skin that may occur during long-term transdermal medication is particularly important in in-vitro experiments using excised skin. Thus, the bioconversion rate of estradiol esters in hairless mouse skin was found to decrease gradually during an in-vitro permeation experiment, with no bioconversion being observed beyond 12 h [156]. This decay of enzyme activity due to skin aging may not be significant in in-vivo or clinical situations.

Animals Versus Humans

The fundamental studies on transdermal delivery of drugs are frequently carried out using animal models under either in-vivo or in-vitro conditions. In mammals, physical processes such as diffusion, partitioning, and physical binding often vary in a predictable manner. The plasma concentration profiles of nitroglycerine and clonidine in humans following transdermal delivery were successfully predicted using physical properties such as diffusion coefficients and partition coefficients which were determined from in-vitro experiments using excised hairless mouse skin [149,153]. However, chemical processes, such as metabolic reactions, may show large and unpredictable variations among species. Depending on the amount of background information available on the interaction of the physical and chemical processes, the difference in pharmacokinetics of any drug in question between one species and another may be predictable.

Body Elimination and Distribution

After permeating the skin, drugs are effectively passed into the systemic circulation, and are distributed and eliminated following body pharmacokinetics. Concentration profiles in the skin indicate that drug concentration decreases very rapidly in the viable epidermis, much less rapidly in the dermis [120,150]. A comparison of the concentration gradient near the boundary between the viable epidermis and the dermis [120] shows that the flux in the viable epidermis is much greater than that in the dermis, which implies efficient movement of drug molecules into the microcirculation, which is located near the boundary between the viable epidermis and the dermis. It has been observed that a significant aggregation of lipophilic drug molecules may accumulate in the subcutaneous fatty tissue adhering to the innermost layers of the dermis [120]. The effects of this accumulation on the plasma profile are not known at this stage of research in skin pharmacokinetics. However, it is possible that these compounds, once accumulated in the fatty tissue, may behave as a reservoir for further release after removal of the delivery source.

The movement of drugs into the systemic circulation in transdermal delivery is a much slower process than that for gastrointestinal or intravenous absorption. The time lag in skin permeation for most drugs is of the order of hours, while that in the body circulation is a matter of minutes. The objective here is to achieve a constant concentration in the plasma by properly designing the transdermal delivery system. Multicompartment models are frequently utilized to describe the drug concentration-time profile in the plasma following transdermal delivery [157]. The pharmacokinetic parameters for many drugs, such as the number of compartments, the volume of distribution, and the elimination rate constant, are available in the literature [158].

By combining the skin permeation model discussed earlier and the body elimination/distribution model, the plasma or tissue concentration-time profile can be described for any mode of transdermal administration (Fig. 26).

Considering the body as a single compartment in which a drug is introduced transdermally leads to the following differential equation:

$$V\frac{dC}{dt} = J(t)S_a - kVC \tag{35}$$

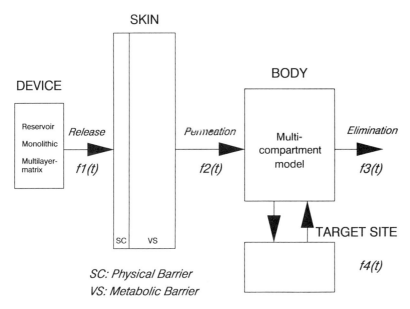

Figure 26   Drug movement from the delivery system device to the target site in transdermal systemic delivery. An optimum transdermal delivery system can be designed by solving the skin pharmacokinetic model for the time-dependent target concentration $f_4(t)$.

where V is the apparent volume of distribution, J(t) is the flux of skin permeation, which is time-dependent in general, S_a is the effective surface area for transdermal delivery, and k is the elimination rate constant.

If the transdermal delivery system provides a constant flux of the drug, that is, $J(t) = J_s = $ constant, the plasma concentration under steady state becomes constant:

$$C = \frac{J_s S_a}{kV} = \frac{J_s S_a}{C_L} \tag{36}$$

where C_L is defined as the plasma clearance. Since the flux J_s is usually controlled by the stratum corneum, and the permeation rate reaches a steady state after a lag time of several hours, a constant plasma concentration for an extended period of time can be achieved.

Pharmacokinetic parameters, such as the volume of distribution and the elimination rate constant and clearance, are generally assumed to be constant during the entire period of application of transdermal delivery systems. This assumption, however, may not always hold. It was found for cardiovascular drugs that blood flow changes in response to the pharmacologic action of the drug in a manner that depends on the level of plasma concentration [159]. The change in the blood flow alters the volume of distribution or plasma clearance during long-term medication. At this stage of research, however, little is known about the dynamic characteristics of skin pharmacokinetics. Further study is certainly needed to elucidate this aspect of time-dependent pharmacokinetics.

If the flux of skin permeation is assumed to be constant, Eq. (35) can be solved as follows:

$$C = \frac{J_s S_a}{kV} [1 - \exp(-kt)] \tag{37}$$

Equation (37) indicates that a steady-state plasma concentration for a drug with a small elimination rate constant is reached after a long time lag, and therefore the transient profile of the plasma concentration may be controlled by the pharmacokinetics in body distribution/ elimination. On the other hand, if the elimination rate constant of the drug is very large, such as for nitroglycerine, the transient profile of the plasma concentration following transdermal delivery may be controlled by the skin permeation kinetics. For such drugs, not only steady-state concentration but transient profile as well can be controlled by the design parameters of the transdermal drug delivery system [153]. Since the plasma profile for many drugs is controlled mainly by the skin permeation and/or the body elimination/ distribution kinetics, it is important to understand the quantitative effect of the physicochemical properties controlling skin permeation as well as the pharmacokinetic parameters affecting the body elimination/distribution in order to optimize the design of transdermal delivery systems.

Figure 27 shows a skin pharmacokinetic model based on the multicompartment body distribution/elimination model together with the bilayer skin diffusion/binding/bioconversion model [132,153]. This model was successfully employed for predicting the plasma concentration-time profile of various drugs following transdermal delivery [132,153,160]. A typical example for nitroglycerine is shown in Fig. 28.

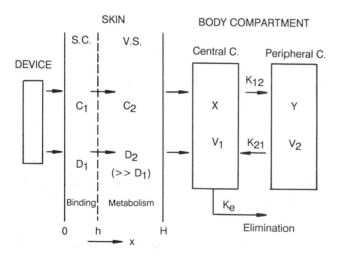

Figure 27 Bilayer skin diffusion/multicompartment body elimination/distribution model for pharmacokinetics of transdermal delivery. A two-compartment open model is shown as an example. S.C., stratum corneum; V.S., viable epidermis; X, drug concentration in the central compartment (plasma); Y, drug concentration in the peripheral compartment (tissue); V, volume of distribution; k_{ij}, intercompartmental rate constants; k_e, elimination rate constant in plasma.

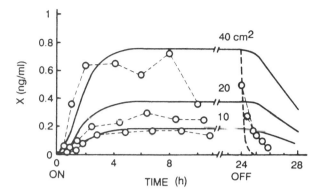

Figure 28 Comparison of time course of plasma concentration of nitroglycerin after transdermal drug delivery: ○, experimental (clinical data for Transderm-Nitro [173]; ——, calculated; $D_{sc} = 1.9 \times 10^{-10}$ cm^2/s, $D_{vs} = 2.1 \times 10^{-7}$ cm^2/s, $P_c = 15$, h $= 0.0020$ cm, H $= 0.0200$ cm, $C_s = 5.2 \times 10^4$ µg/ml, no binding, no metabolism; $V_1 = 2.8 \times 10^4$ ml, $V_2 = 1.4 \times 10^5$ ml, $k_e = 0.0082$/s, $k_{12} = 0.0053$/s, $k_{21} = 0.0011$/s. The numbers on the curves are the surface area of the device. ON, system applied; OFF, system removed.

Stratum Corneum Reservoir Capacity

A number of researchers have reported observations on the reservoir capacity for drugs or toxins in human stratum corneum since Vickers clearly demonstrated this phenomenon in 1963 [161]. Both lipophilic and hydrophilic drugs may be trapped in the matrix structure of the skin and diffuse out very slowly due to strong binding and/or markedly great diffusion resistance in the stratum corneum. This reservoir capacity affects significantly the tailing of the plasma concentration profile after removal of the transdermal drug delivery system [149]. Not only bound drugs but also unbound drugs give rise to the reservoir phenomenon because of the extremely slow diffusion rate in the stratum corneum. It is, however, the bound molecules that cause the major buildup of the reservoir, which may then release the drugs for weeks or even longer [162]

As was shown in Fig. 24, the Langmuir isotherm was successfully used to describe the stratum corneum reservoir capacity for various steroids. However, it was also found that the drug molecules of certain compounds bind so strongly to skin tissue that their desorption is negligible in normal permeation/desorption experiments [145].

Dupius et al. reported a simple stripping method to predict the stratum corneum reservoir function in vitro [163]. They found that, for many drugs, the amount of drug present in the stratum corneum 30 min after transdermal administration is linearly related to the total amount of drug that permeated in the 4 days after administration. It was shown theoretically that drug molecules released from the transdermal delivery system require 30 min or longer to traverse the entire thickness of the human stratum corneum [150]; the drug concentration near the boundary between the stratum corneum and viable skin remains zero or at a very low level during the first 1 h after the onset of transdermal administration (Fig. 29). In order to evaluate the skin permeability and to predict the

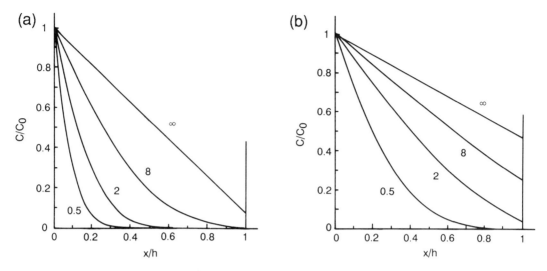

Figure 29 Transient profiles of drug concentration in the stratum corneum calculated from bilayer skin diffusion model [150]. Numbers alongside lines are values of time after drug application on the surface of the stratum corneum, h = 0.0020 cm, H = 0.0200 cm. (a) $D_{sc} = 10^{-11}$ cm^2/s; (b) $D_{sc} = 10^{-10}$ cm^2/s.

effect of permeation on the plasma concentration profile, the steady-state rate of permeation should be investigated. It is therefore important to determine all the physicochemical parameters that control the steady-state rate of permeation, including the diffusion coefficient and the partition coefficient.

Iontophoretic Transdermal Delivery

For many years, iontophoresis has been applied to enhance the permeation of charged molecules across various biological membranes. As mentioned earlier, this technique has been increasingly investigated as a means of noninvasive transdermal administration of systemically effective drugs [164,165]. The literature indicates that iontophoretic transdermal delivery is a promising approach for achieving therapeutic plasma levels, even for large molecules such as polypeptides and proteins, which are usually not lipid-soluble and penetrate the stratum corneum with great difficulty [166]. Meyer et al. concluded in their luteinizing hormone (LH) iontophoretic delivery studies that the absorption of LH occurs by the process of electroosmosis, in which the convective movement of water in an electric current leads to the secondary flow of solutes [167]. Burnette and Ongpipat-tanakul reported that the convective transport term is a second-order effect and plays a major role only when both the direct electrostatic repulsion term and the passive term are relatively small [168]. From analyzing in-vitro data of iontophoretic delivery of arginine vasopressin, Tojo demonstrated that the significance of the convective term can be estimated from the value of the Peclet number, defined as the (linear velocity of convective flow) × (stratum corneum thickness)/(stratum corneum diffusion coefficient) [160]. Since the diffusion coefficient for large molecules such as polypeptides is usually very small, convective flow can make a major contribution to the enhancement of skin permeability under iontophoretic transdermal delivery. Wearley et al. [144] and Lelawongs et al. [169] measured convective flow under iontophoresis in their in-vitro experiments. They found that convective flow was increased significantly by the application of an electric field.

Under the influence of an electric field, as described earlier, the flux of drug molecules (ionized or charged) can be described by [170]

$$j = C_i u - D_i \text{ grad } C_i + \frac{D_i z_i e F C_i}{RT} \tag{38}$$

where C is the concentration, D is the diffusion coefficient, e is the electric field, F is the Faraday constant, R is the gas constant, T is the absolute temperature, u is the velocity of the convective flow caused by the electric field, and z is the charge number of the ionized drug molecule.

For one-dimensional skin permeation with concurrent occurrence of binding in the stratum corneum, Eq. (31) can be modified for iontophoretic delivery as follows [171]:

$$\left[1 + \frac{K}{(1 + bC_1)^2} \right] \frac{\partial C_1}{\partial t} = D_1 \frac{\partial^2 C_1}{\partial x^2} + \frac{zFD_1 E}{RTh} \frac{\partial C_1}{\partial x} - u_1 \frac{\partial C_1}{\partial x} \tag{39}$$

Equation (39) was solved numerically under various initial and boundary conditions [171]. The model has been used to describe the in-vitro experimental data of the iontophoretic transdermal delivery of vasopressin, verapamil, and several amino acids. It was found that the in-vitro data, not only for steady-state flux but also for transient or pulsed profiles, were explained well by Eq. (39), indicating that convective flow under the influence of an electric field cannot be overlooked. The cumulative amounts of vasopressin permeated under in-vitro iontophoretic delivery are shown in Fig. 30 for various modes of application

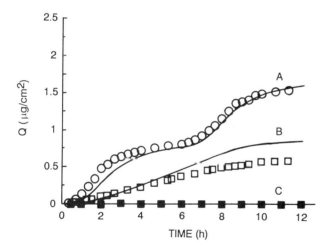

Figure 30 Time course of the cumulative amount of vasopressin permeated across hairless rat skin under two different modes of iontophoretic delivery [172]. Current density, 0.16 mA/cm²; donor solution concentration, 0.050 mg/ml. A: $t_a = 2$ h, $t_b = 6$ h during the period of 12 h. B: $t_a = 10$ m, $t_b = 40$ m ($0 \leq t \leq 8$ h) and $t_a = 0$ ($t > 8$ h). C: Passive transport ($E = 0$ and u $= 0$). Circles, open squares, and filled squares are experimental. Lines are calculated. $D_{sc} = 1.1 \times 10^{-11}$ cm²/s, $D_{vs} = 1.2 \times 10^{-7}$ cm²/s, $P_c = 0.40$, $C_s = 0.077$ mg/ml, h $= 0.0010$ cm, $H = 0.0400$ cm, $S_a = 0.64$ cm², Pe ($= uh/D_{sc}$) = 147.

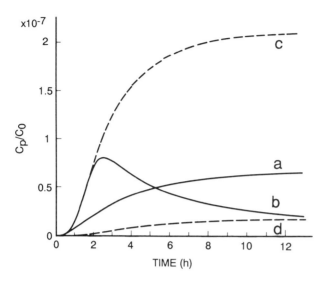

Figure 31 Simulation of the effect of application mode of electric field on plasma concentration profile following iontophoretic transdermal delivery [171]. $S_a = 20 \, cm^2$, $k_e = 0.0016/s$, $k_{12} = 0.0020/s$, $k_{21} = 0.0010/s$, $V_1 = 3 \times 10^4 \, ml$, $V_2 = 5 \times 10^4 \, ml$. (a) Periodic application ($t_a = 6 \, m$, $t_b = 60 \, m$), (b) continuous application ($t_a = 72 \, m$), (c) continuous application ($t_a = 24 \, h$), (d) passive transport with no electric field ($E = 0$).

of the electric field. The solid lines are calculated from Eq. (39) together with the physicochemical parameters determined by the in-vitro passive permeation experiments [172]. By combining the iontophoretic skin permeation model of Eq. (39) with the multicompartment body distribution/elimination model, it was found that the application mode of the electric field plays an important role in controlling the time course of the plasma concentration (Fig. 31). This finding may suggest that iontophoretic transdermal delivery is particularly effective for achieving not only enhanced permeability but a pulsed profile as well.

Future Study

The drug concentration in the plasma or at the target site following transdermal systemic delivery should be determined based on well-defined models for skin permeability and body elimination/distribution.

During the last two decades, much emphasis was placed on the steady-state flux after transdermal delivery of drugs, since a constant plasma level was believed to be the optimum concentration profile for medication. Recently, however, due to advances in pharmacodynamics, more general time-dependent concentration profiles in the plasma have become an important objective so as to maximize the therapeutic effect.

To optimize transdermal drug delivery systems, more rigorous experimental and theoretical investigations are needed to elucidate complex skin permeation mechanisms, including metabolism, binding, and pulsed delivery, together with physical or chemical enhancement.

TRANSDERMAL DEVICES

Devices for transdermal drug delivery are generally fabricated as multilayered polymeric laminate structures in which a drug reservoir or a drug/polymer matrix is sandwiched between two polymeric layers. The outer backing layer, comprising an impermeable polymer or a foil, is designed to prevent loss of drug through the backing surface. The other polymeric layer may function as an adhesive or a rate-controlling membrane (Fig. 32). Based on the mechanism by which the drug is released, the device can be classified into one of the following two categories.

Monolithic (or matrix) system: The drug is dissolved or dispersed in the polymer phase. The drug release from the drug/polymer matrix controls the overall rate of its release from the device (Fig. 32a).

Reservoir (or membrane) system: Diffusional resistance across a polymeric membrane controls the overall drug release rate (Fig. 32b).

The selection of either a monolithic or a reservoir system for transdermal drug delivery depends on which is a major contributing factor controlling the rate of drug transport and its delivery to the systemic circulation. The total resistance to drug transport from the device across the skin can be considered to be the sum of the diffusional resistances through the device (R_1) and that through the stratum corneum (R_2). It may be assumed that the epidermal resistance is negligible compared with that of the stratum corneum. In general, analogous to the current flow in an electrical circuit having resistances connected in series, the overall rate of drug transport will be inversely proportional to the sum of the two resistances ($R_1 + R_2$). If $R_2 \gg R_1$, the overall rate of drug transport will be governed by the rate of drug permeation through the stratum corneum. And if $R_1 \gg R_2$, the device will control the overall rate.

When the desired rate of drug transport is considerably less than that through the stratum corneum, a device control of drug delivery is required to attain therapeutic steady-state concentrations of drug in the blood plasma and prevent overdosing. In such cases, a device having a rate-controlling membrane is needed. On the other hand, if drug permeation through the stratum corneum is the rate-controlling step, a monolithic or matrix type of delivery system may suffice. The therapeutic index of the drug and the variations in its skin permeability for any given population are also important criteria in selecting the type of transdermal device. If the therapeutic index is narrow in relation to the range of systemic drug concentration attained due to the variability in its skin permeation characteristics, a monolithic system would not be suitable.

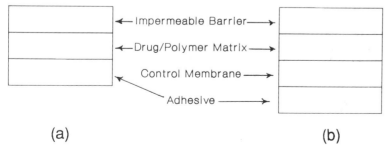

Figure 32 Schematic diagrams of laminated transdermal devices: (a) matrix system; (b) membrane system.

Reservoir Devices

The Transderm Nitro (Ciba Geigy) system for transdermal delivery of nitroglycerine for prophylaxis of angina pectoris is an excellent example of a typical constant-activity-source reservoir device. In this system, the drug reservoir, a dispersion of nitroglycerine-lactose triturate in silicone fluid, is encapsulated between an impermeable plastic laminate and a drug-permeable ethylene vinyl acetate copolymer membrane (Fig. 33). A drug-compatible silicone adhesive layer laminated to the membrane provides the means of attaching the device to the skin. Thus, in the reservoir, nitroglycerine is contained in an evenly dispersed separate phase, resulting in maintenance of drug saturation in the continuous silicone fluid phase. In such a system, the steady-state permeation rate (J) of the drug is given by

$$J = \frac{ADK\,\Delta C}{l}$$

where

A = surface area of the membrane

D = diffusivity of the drug in the membrane

K = reservoir/membrane distribution coefficient

ΔC = difference in concentrations of the drug in the reservoir and the fluid or medium adjacent to the membrane

l = thickness of the membrane

According to this equation, for a given device (i.e., A, D, K, and l are constant), the release rate of the drug would be constant as long as ΔC remains constant. That is, a zero-order release rate is obtained from such a device. Dissolution data (Fig. 34) for nitroglycerine from a Transderm Nitro device under sink conditions indeed show this profile. The very slight burst effect observed (relatively high initial release rate) is due to saturation of the silicone adhesive with nitroglycerine during storage. It may be noted that in systems in which the amount of drug in the adhesive at saturation or equilibrium is high, the magnitude of the burst effect will be proportionately greater. Such a burst effect may be advantageous for drugs with skin permeations that exhibit lag times, which sometimes occur because the drug binds to receptor sites within the skin. In these cases, the burst effect can be useful in reducing the duration of the lag time. On the other hand, for drugs with a narrow therapeutic index and a short or no lag time, a large burst effect could lead to toxic levels of the drug in the blood plasma.

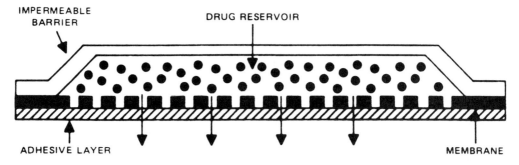

Figure 33 Schematic diagram of Transderm Nitro system.

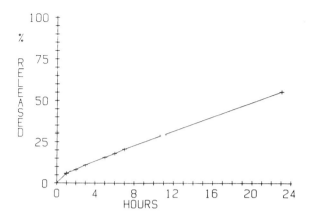

Figure 34 Dissolution profile of nitroglycerine from Transderm Nitro system. (From Ref. 175.)

The release profile from a reservoir device, in which the thermodynamic activity of the drug in a polymer or a solution formulation does not remain constant as the drug is released from the device, will follow first-order and not zero-order kinetics. If V is the volume of the drug reservoir, then the fractional release and the fractional release rate at any time t can be expressed by the following two equations:

$$\text{Fraction released:} \quad F = 1 - \exp\left(-\frac{ADKt}{Vl}\right)$$

$$\text{Release rate:} \quad J = \frac{M_0 ADK}{Vl} \exp\left(-\frac{ADKt}{Vl}\right)$$

where

M_0 = amount of drug at time zero

It can be seen that both F and J decline exponentially with time, although higher drug concentrations would result in greater release rates.

Monolithic Devices

There are two distinct categories of monolithic devices, one in which the drug is dissolved in the polymer matrix, and the other in which the drug is dispersed. The two categories differ basically in the extent of loading of the polymer matrix. In monolithic solutions the concentration levels of the drug are at or below saturation level, whereas in monolithic dispersions the drug levels are far in excess of its saturated concentration. The release profile in monolithic devices is a function of drug concentration, the chemical nature of the polymer matrix, and the device geometry.

A transdermal system generally has a slab geometry, with release occurring from only one side of the slab. For such a monolithic solution, as presented in Chapter 3, the fractional release and the release rate at any given time t are given by the following equations:

$$F = 2 \left(\frac{Dt}{\pi \ell^2}\right)^{1/2} \quad \text{for } F \le 0.6$$

$$J = M_0 \left(\frac{D}{\pi t \ell^2}\right)^{1/2}$$

The Nitroglycerine Transdermal System (NTS, Bolar Pharmaceuticals) exemplifies a monolithic solution. The drug matrix in the NTS (Fig. 35) is comprised of drug dissolved in plasticized vinyl chloride/vinyl acetate copolymer. The drug matrix is laminated to an impermeable polyester film, which in turn is attached to an adhesive-coated foam backing layer. Nitroglycerine release under sink conditions follows a typical pattern of declining release rates with time (Fig. 36a). When fractional release is plotted versus the square root of time, a straight-line relationship (Fig. 36b) is obtained, as would be indicated by the above equation.

In the case of monolithic dispersions, the release rate also follows a square-root-of-time relationship as expressed by the following equation:

$$J = A \times \left(\frac{DC_sC_0}{2t}\right)^{1/2} \quad \text{for } C_0 \gg C_s$$

where

C_0 = initial drug concentration

C_s = drug concentration at saturation

A = surface area of drug matrix

This proportionality of the release rate to the square root of time holds until the drug concentration in the matrix falls below the saturation value, C_s.

Example: A clonidine transdermal patch has the drug dispersed in polysiloxane matrix at a concentration of 100 mg/ml. The patch also has an EVA copolymer as control membrane. Saturated concentration of the drug in the polymer matrix was determined to be 10 mg/ml. Drug release from the patch under sink conditions was 5.6 μg/cm^2/h. The rate of drug transport across the skin from the patch was 3.85 μg/cm^2/h. Calculate the

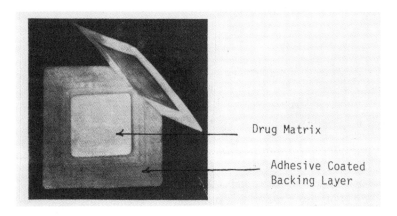

Figure 35 Monolithic NTS device construction.

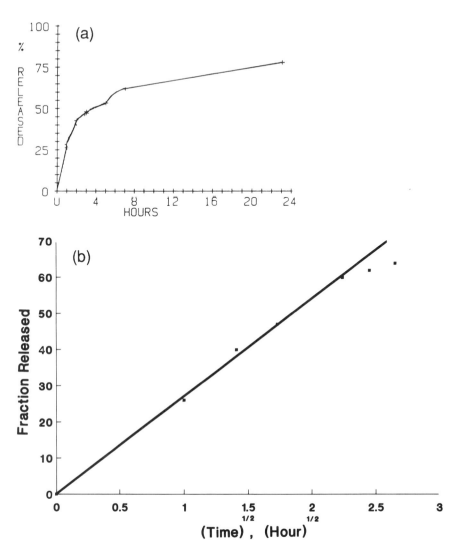

Figure 36 Nitroglycerine release from NTS as (a) a function of time and (b) a function of the square root of time. (Data from Ref. 175.)

skin resistance to clonidine and the extent of device control of transdermal delivery of the drug.

$$\textit{Solution:} \text{ Membrane resistance } R_m = \frac{\text{saturated drug concentration in polymer } (C_s)}{\text{drug release rate } (J_m)}$$

$$= \frac{10}{0.0056} = 1786 \frac{h}{cm}$$

$$\text{Total resistance } R = \frac{C_s}{\text{flux through skin}} = \frac{10}{0.00385}$$

$$= 2597 \text{ h/cm}$$

$$\text{Skin resistance } R_s = 2597 - 1786 = 811 \text{ h/cm}$$

$$\text{Device control of clonidine transport} = \left(\frac{1786}{2597}\right) \times 100 = 69\%$$

Preparation of Transdermal Devices

Polymers

The selection of a polymer for use in the drug matrix must be very carefully made. Analogous to pharmaceutical creams and ointments for dermatological applications, formulation of transdermal drug/polymer matrices is both a science and an art. Some of the criteria that must be considered in the drug/polymer matrix formulation include the following:

Drug solubility and diffusivity in the polymer
The desired drug loading and its effect on polymer integrity
Compatibility of the polymer with the necessary excipients, such as solvents and
    skin-permeation enhancers for the drug
Type of device construction—monolithic or reservoir
Skin compatibility—the effect of moisture occluded under the polymer formulation
Mechanical properties—softness, flexibility, conformability to skin, and mechanical
    integrity
Ease of fabrication
Toxicity and purity, i.e., compliance with safety requirements of the FDA
Cost and availability

Some of the polymeric materials that meet these criteria include polyisobutylene, polyacrylates, polysiloxanes, plasticized vinyl chloride polymers and copolymers, and hydrophilic polymer gels.

Polyisobutylene [176]. Polyisobutylene (PIB) is a highly paraffinic, nonpolar, and amorphous hydrocarbon polymer composed of essentially straight-chain macromolecules. Industrially, PIB is produced by low-temperature cationic polymerization:

$$
\underset{\text{Isobutylene}}{\underset{\underset{\displaystyle CH_3}{|}}{\overset{\overset{\displaystyle CH_3}{|}}{CH_2=C}}} \xrightarrow{\quad BF_3 \quad} \underset{\text{PIB}}{\underset{\underset{\displaystyle CH_3}{|}}{\overset{\overset{\displaystyle CH_3}{|}}{-[CH_2-C]-}}}
$$

Physical properties of PIB change gradually with increasing molecular weight, the lowest-molecular-weight polymers being viscous liquids. With increasing molecular weight, the liquids become more viscous, then change to balsamlike sticky masses, and finally form elastomeric solids. The precise properties required for a drug/polymer matrix may be attained through use of a mixture of different PIB molecular-weight grades and such additives as mineral oil.

PIB is soluble in hydrocarbon solvents and insoluble in polar solvents. Because of its amorphous character, lack of unsaturation, and low glass-transition temperature, PIB exhibits excellent low-temperature flexibility and oxidative stability.

Polyacrylates. Copolymers of acrylic and methacrylic acid esters can be most suitable for use as a drug/polymer matrix. They are prepared by free-radical copolymerization of the selected acrylic monomers by solution, emulsion, or suspension polymerization methods:

$$
\begin{array}{ccc}
CO_2R_1 & CO_2R_2 & \\
| & | & \\
CH_2{=}CH & + \ CH_2{=}CH & \longrightarrow
\end{array}
\quad
\begin{array}{cc}
CO_2R_1 & CO_2R_2 \\
| & | \\
-[CH_2{-}CH{-}CH_2{-}CH]{-}
\end{array}
$$

$$R^0$$

Acrylate copolymer

Acrylate polymers, which are moderately polar, colorless, and transparent, have excellent chemical resistance, plus thermal, light, and oxidative stability. For use as a drug/polymer matrix material, the copolymer must have its glass-transition temperature (T_g) considerably below the ambient-use temperature, as the diffusivity of a substance in glassy polymers $(T_g > 37°C)$ is very low. The effect on polymer T_g of chain substitution in the acrylate and methacrylate series is shown in Fig. 37.

The T_g of a polymer is a function of stiffness or rigidity of the chains. Thus, the methyl group on the main chain, in the case of methacrylates, contributes to increased chain stiffness as compared to the acrylates. There is an almost 100°C difference between the T_gs of methyl methacrylate and methyl acrylate homopolymers. However, increase in length of the side chain gives flexibility and therefore decreases T_g. After the number of carbon atoms in the alkyl group reaches 8 in the acrylate series and 12 in the methacrylate series, side-chain crystallization takes effect and restricts chain mobility, thus increasing the T_g. For example, poly(methyl acrylate) is a tough polymer which forms an extensible pliable film. Poly(ethyl acrylate) is soft and rubberlike. Poly(octyl acrylate) is soft and

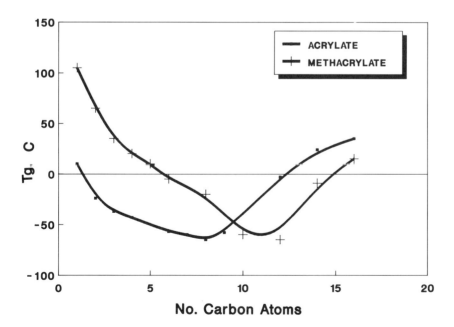

Figure 37 T_g of acrylic and methacrylic acid ester polymers.

tacky at room temperature, whole polymers of *n*-hexadecyl acrylate and *n*-octadecyl acrylate are hard, brittle, and waxlike solids. By copolymerization of selected acrylate and/or methacrylate monomers, a copolymer with the desired physical properties can be obtained.

Polysiloxanes. Polysiloxanes [177,178], also referred to as silicones, are organosilicon polymers having Si—O—Si bonds along the main chain and an alkyl group (R) attached to a significant proportion of silicon atoms by silicon-carbon bonds. Poly(dimethyl siloxanes) constitute the largest volume fraction of the commercially available silicone polymers. Unusual properties of silicones are their outstanding thermal and oxidative stability, chemical inertness, and very low surface tension. Their structural characteristics, such as very low rotational barrier about the silicon-oxygen bond of the main chain and relatively large intermolecular distances between the polysiloxane chains, contribute to their low modulus, low glass-transition temperature, and high solute permeability. Commercially important forms of silicones include fluids, elastomers, and resins. Polysiloxanes are formed by hydrolysis of chlorosilanes.

$$\begin{array}{ccc} R & R \\ | & | \\ -O-Si-O-Si-O- \\ | & | \\ R & R \end{array}$$

Polysiloxane

$$\begin{array}{ccc} R & & R \\ | & & | \\ Cl-Si-Cl + H_2O \longrightarrow [HO-Si-OH] + 2HCl \\ | & & | \\ R & & R \end{array}$$

$$\longrightarrow \begin{array}{cc} R & R \\ | & | \\ -[Si-O-Si-O- \\ | & | \\ R & R \end{array}$$

Silicone oils and elastomers can be used as inert carrier media for the preparation of controlled-release transdermal drug formulations. The drug/polymer matrix preparation is accomplished by dispersing the drug in low-molecular-weight silicone polymers (liquid or pastelike consistency) and then cross-linking to produce an elastomeric material in a sheet or slab form. Most compositions are based on poly(dimethyl siloxane), having a small amount of multiple reactive functionality. Room-temperature-curing elastomers can be one-component or two-component. Atmospheric moisture initiates the cure reactions of the one-component systems, which often contain acetoxysilane groups.

$$\begin{array}{cccc} R & R & & R & R \\ | & | & & | & | \\ -O-Si-O-Si-O- & + H_2O & -O-Si-O-Si-O- & +CH_3CO_2H \\ | & | & & | & | \\ R & O-C-CH_3 & & R & OH \\ & \parallel & & \\ & O & & \end{array}$$

$$-O-\overset{\overset{\displaystyle R}{|}}{\underset{\underset{\displaystyle R}{|}}{Si}}-O-\overset{\overset{\displaystyle R}{|}}{\underset{\underset{\displaystyle O-Si-O}{|}}{Si}}-O-$$

In the two-component systems, moisture is not required for cure. One of the two components contains cross-linkage functionality, such as vinyl groups, and the other component contains a hydrosilation catalyst, such as a solubilized platinum compound. The cure occurs upon mixing the two components.

$$-O-\overset{\overset{\displaystyle R}{|}}{\underset{\underset{\displaystyle R}{|}}{Si}}-CH{=}CH_2 + H-\overset{\overset{\displaystyle R}{|}}{\underset{\underset{\displaystyle R}{|}}{Si}}-O-\overset{\overset{\displaystyle R}{|}}{\underset{\underset{\displaystyle R}{|}}{Si}}- \xrightarrow{\text{catalyst}}$$

$$-O-\overset{\overset{\displaystyle R}{|}}{\underset{\underset{\displaystyle R}{|}}{Si}}-CH_2-CH_2-\overset{\overset{\displaystyle R}{|}}{\underset{\underset{\displaystyle R}{|}}{Si}}-O-\overset{\overset{\displaystyle R}{|}}{\underset{\underset{\displaystyle R}{|}}{Si}}-$$

At room temperature, a few hours are required for complete cure, which can be accelerated by heat. Other methods for curing polysiloxanes include peroxide or ultraviolet light-initiated reactions.

Hydrophilic polymer gels. A hydrophilic polymer gel matrix can be useful for some drugs to attain the desired drug permeability and thermodynamic activity in the diffusion matrix. A hydrophilic gel is particularly suitable when electrical conductivity is desired, as in the case of iontophoretic transdermal drug delivery.

Poly(vinyl pyrrolidone)/poly(vinyl alcohol) [179,180]: One of the convenient methods of preparing a gel diffusion matrix is to dissolve or disperse a drug in a mixture of PVP, PVA, a plasticizer such as glycerol, and water at somewhat elevated temperatures ($\sim$90°C) to produce a homogeneous mixture that is poured into a film form while still warm. Poly(vinyl alcohol) grades, which have a very high proportion (>95%) of hydroxyl groups per repeat unit in the chain, are insoluble in glycerol. Therefore, upon cooling the mixture, a swollen gel film or a slab containing the dissolved or dispersed drug is formed. In principle, other hydrophilic polymers exhibiting similar solubility behavior can also be used in the drug/polymer matrix formulation, as can other hydrophilic plasticizers, such as polyalkylene glycols.

Hydrogels [181–183]: Hydrogels are water-swollen but water-insoluble cross-linked networks of hydrophilic polymers. The chemical nature of cross-linkages can be covalent, ionic, or strong cooperative interpolymer interactions such as hydrogen bonding. The water content of synthetic hydrogels can range from 30% to 90 + %. Commonly known synthetic hydrogels are prepared by free-radical copolymerization of hydrophilic monomers such as 2-hydroxyethyl methacrylate, N-vinyl-2-pyrrolidone, or acrylamide with small amounts of a difunctional monomer such as ethylene glycol dimethacrylate or N,N′-methylenebisacrylamide. Such a polymerization is customarily carried out in aqueous medium in the absence of oxygen. The equilibrium water content of the hydrogel can be varied by the degree of cross-linking and by using a combination of hydrophilic and hydrophobic monomers in copolymerization. The drug/hydrogel matrix can be prepared by either incorporating the drug in the aqueous polymerization mixture or by a process of equilibration of the hydrogel in a concentrated aqueous solution of the drug. The chemical nature of the polymer network, the extent of cross-linking, and the water content determine the permeability of the hydrogel to a given drug. The drug diffusivity decreases with increasing cross-link density and decreasing water content.

Vinyl chloride polymers and copolymers (PVC).   Vinyl chloride polymers and co-polymers [184] useful for drug/polymer matrix preparation include homopolymers of vinyl chloride, CH_2=CH—Cl, and copolymers having a high vinyl chloride content. PVC by itself is a very stiff and rigid material, and therefore is seldom used alone; it needs plasticization in order to form a soft and flexible film suitable for use in a transdermal patch formulation. High-boiling liquids, such as dioctyl phthalate, adipate or sebacate esters and polyesters, epoxidized soybean oil, and citric acid esters, can be used as plasticizers for PVC. The proportion of the plasticizer in the formulation can range from 20% to as high as 75%. Although drug-containing plasticized PVC films can be prepared by conventional solvent-casting techniques, solventless methods involving PVC plastisol are more suitable. A colloidal, pastelike dispersion of PVC resin in a plasticizer is known as a plastisol. The drug can be easily dispersed in the plastisol, and the mixture coated to the desired thickness onto a substrate. Upon heating the coated mixture at 130–160° for 1 to 3 min, the PVC is solvated by the plasticizer to form a soft and flexible film. The chemical nature and the proportion of the plasticizer, and the drug loading in the plasticized polymer matrix, are three of the parameters affecting drug release and transport across skin.

Adhesives [185]

Adhesion can be defined in simple but adequate terms as wetting and setting. In order for a material to bond to another, it must first wet the surface of the other, and after doing so set in that state. It is the mode in which the wetting and setting is accomplished that characterizes the different adhesives. In the case of a pressure-sensitive adhesive (PSA) to be used to attach a transdermal device to the skin, the adhesive bond formation involves a liquidlike flow process resulting in adhesive wetting of the skin surface upon application of pressure. When the pressure is removed, the adhesive sets in that state. In order for the adhesive bond to have measurable strength, the storage of elastic energy is necessary during the bond-breaking process. Therefore, pressure-sensitive adhesion is a characteristic of a viscoelastic material, significantly above its glass-transition temperature, at which the viscous and the elastic processes are of a comparable order of magnitude. The balance of viscous flow and storage of elastic energy determines the usefulness of the material.

PSAs for transdermal systems can be selected from a range of polymeric materials exhibiting the above-described viscoelastic characteristics. Commonly employed PSAs in transdermal devices include acrylate copolymers, polyisobutylenes, and polysiloxanes. The adhesive selection is based on the results of screening for skin irritation or sensitization, adhesion of the device to skin at the intended site of application during the patient's normal activities, and ease of patch removal without leaving any residues of the adhesive mass on skin. In the case of a face adhesive patch, the transport of the drug and the excipients must occur through the adhesive layer. Therefore, the adhesive performance must not be affected by the drug or the excipients. In addition, the adhesive must not alter the overall rate of drug delivery.

Acrylic PSAs.   Conventional acrylic PSAs are generally copolymers of C_4—C_8 alkyl acrylates and polar monomers, such as acrylonitrile, acrylic acid, acrylamide, etc. Optional modifying monomers such as methyl or ethyl acrylate, methyl methacrylate, styrene, and vinyl acetate may also be incorporated in the copolymer structure. Optimum cohesive and adhesive properties of the copolymer are attained by a proper balance of its molecular weight (usually very high), covalent or ionic cross-linking, polarity, and a glass-transition temperature ranging from -25 to $+70°C$. These copolymers are prepared by free-radical solution or emulsion polymerization techniques to a high conversion and used as is. The copolymer solution or emulsion is applied to the desired substrate as a thin coating and subsequently dried.

Properties of the adhesive are tailored by copolymerization of appropriate monomers. The C_4—C_8 alkyl acrylates contribute to the copolymer chain flexibility. Small amounts of polar monomers enhance skin adhesion; and the modifying monomers alter the T_g to increase cohesive strength of the adhesive. All PSAs must possess a T_g substantially below their normal-use temperature. The greater the difference between the normal-use temperature and T_g, the greater will be viscous flow under the applied pressure, which will increase wetting but decrease cohesive strength. Therefore, an optimum T_g is required in order to provide a balance between wetting and cohesive strength. The effect of the T_g of an acrylate copolymer on its adhesive performance is illustrated in Fig. 38. Peel adhesion and tack decrease with increasing T_g, whereas shear strength, which is a measure of the cohesive strength, increases with increasing T_g. Likewise, increase in the copolymer molecular weight and cross-linking also decrease peel adhesion and tack while increasing cohesive strength.

One of the advantages of acrylic PSAs is that they are single-component systems. There are no low-molecular-weight additives, such as tackifiers, extenders, stabilizers, etc., which can adversely affect skin compatibility and adhesive performance. Acrylics have excellent thermal, oxidative, and ultraviolet light stability. They also exhibit good moisture-vapor permeability, which can help minimize the deleterious effects of skin occlusion in a transdermal system.

Polyisobutylene.   Polyisobutylene (PIB) pressure-sensitive adhesives are formulated by

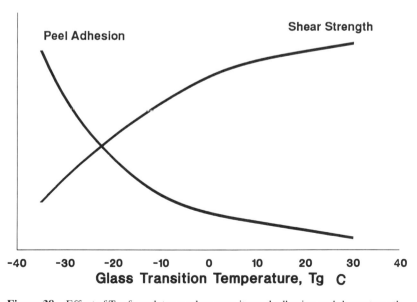

Figure 38   Effect of T_g of acrylate copolymer on its peel adhesion and shear strength.

mixing a combination of components that include both low- and high-molecular-weight fractions of PIB and, optionally, polyterpene-type tackifying resins and liquid plasticizers such as paraffinic oil and petrolatum. The high-molecular-weight PIBs provide elasticity and cohesive strength, while the low-molecular-weight PIB fractions impart tackiness. The plasticizers and tackifiers are used to further modify tackiness of the composition. The PSA film is prepared by a solution casting process employing a hydrocarbon or a chlorinated hydrocarbon solvent.

PIB adhesives have excellent flexibility due to the low T_g of PIB. However, PIB being nonpolar, its adhesive compositions do not exhibit the same aggressive tackiness as the acrylics. The PIB adhesives also have rather low moisture-vapor permeability.

Polysiloxanes. Polysiloxane PSA is made by condensation reaction of a linear poly(dimethyl siloxane) gum and a silicate resin. The gum and the resin are dissolved in a solvent and allowed to react chemically to effect the condensation of Si—OH groups to result in the formation of siloxane bonds. The desired PSA properties are attained by having an appropriate gum-to-resin ratio. Resin levels of 50 to 70% are typical in a PSA formulation. Tackiness of the composition decreases with increasing resin content.

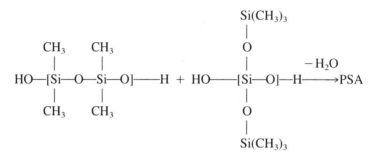

Polysiloxane PSAs have good skin compatibility, low toxicity, and high moisture-vapor permeability.

Control Membranes [186]

A membrane, in general, is a barrier that allows selective and/or greatly restricted transport of molecules. In a transdermal system, the membrane provides considerable resistance to the transport of drug from the reservoir or the drug/polymer matrix to the skin, the objective being to attain device-controlled transdermal drug delivery. Usually, the control membrane is a thin polymeric film. The kinds of polymeric materials that can be used in the preparation of membranes are quite varied and include olefinic polymers and copolymers, cellulose esters, polyamides, and poly(vinyl chloride). The membrane selected for a given transdermal system must be reasonably stable to the thermal and mechanical stresses encountered during fabrication and storage. In addition, it must be nontoxic and exhibit appropriate chemical resistance to the drug and excipients without altering its transport characteristics.

Membranes are classified according to their mode of transport as molecular-diffusion membranes or microporous membranes. A dense polymer film is an example of a molecular-diffusion membrane. Such a membrane does not have permanent pores. The permeant interacts chemically with the dense polymer matrix and dissolves in it, generating a diffusive mass transport along a chemical potential gradient. Permeability of a homogeneous polymer film is also a function of polymer film morphology, which reflects its polymer chain orientation, packing, and crystallinity. The polymer film preparation methods, such as solution or melt casting, extrusion, film blowing, and calendering, can have significant effects on morphology, and consequently on permeability characteristics.

Ethylene vinyl acetate (EVA) copolymers are ideally suited for the preparation of molecular diffusion-type membranes because their permeability properties can be varied over a wide range by judiciously changing their vinyl acetate content. Low-density polyethylene is a partially crystalline (50–60%) solid having a melting point of about 115°C. However, when ethylene is copolymerized with vinyl acetate, which is not isomorphous with ethylene (i.e., they are incapable of replacing one another in the crystal lattice), the degree of crystallinity and the crystalline melting point decrease. Solute diffusion in EVA copolymers occurs mainly through the amorphous regions. Therefore, as the degree of crystallinity decreases, the solute permeability increases. Further, copolymerization of ethylene with vinyl acetate (VA) also increases the polarity. Therefore, increasing the VA content of the copolymer also increases the solubility and hence diffusivity of polar compounds in the copolymer. However, at high vinyl acetate levels (>60 wt%), the T_g of the copolymer increases from about $-25°C$ to about 35°C. An increase in T_g is indicative of a decrease in the polymer chain mobility. Therefore, the solute diffusivity decreases.

This effect of changing membrane permeability by altering copolymer composition is well exemplified by measurements of camphor flux through a series of EVA copolymer membranes having different vinyl acetate contents (Fig. 39). Since the membrane flux is a function of both sorption and diffusion coefficients, the graph of camphor flux versus the EVA copolymer composition exhibits a maximum at about 60% vinyl acetate content.

The semipermeability of microporous membranes is based on the spatial cross section of the permeating species; that is, small molecules exhibit a higher permeation rate than large molecules, and linear-chain molecules permeate better than those with globular shapes. The effective pore size is at least several times the mean free path of the molecules, namely, from several micrometers to about 100 Å. They are generally opaque even when made of transparent material, because the microporous structure scatters visible light.

Microporous membranes are best prepared from polymeric materials that are not subject to cold flow, so that storage stability of the microporous structure is obtained. Cellulosic esters and polyolefins are two of the more commonly used materials of choice. The phase-separation process for preparing microporous membranes is well suited for materials such as cellulosic esters, which have good solubility in organic solvents. The phase-separation process consists of casting a solution of the polymer, followed by precipitation of the essentially polymer-rich phase from the solvent phase. Polymer precipitation can be accomplished by techniques such as solvent evaporation, cooling the solution to cause thermal gelation, and absorption of water by the solution.

Another important technique for the preparation of microporous membranes is uniaxial stretching of crystalline polymer films, produced by extrusion under conditions of low melt temperature and high melt stress. Isotactic polypropylene microporous membrane (Celgard, Hoechst-Celanese Corporation) is prepared commercially by this process.

The first of the critical steps in this process involves extrusion of the polymer into a thin film, generally at temperatures close to its melting point, coupled with a rapid drawdown of the melt, followed by annealing of the film in a relaxed or untensioned state. The film thus produced has a parallel-row, lamellar microcrystalline morphology. The oriented film is then elongated (up to 300% extension) at temperatures close to but below its crystalline melting point in the machine direction of the film to produce an interconnecting network of slitlike submicroscopic pores having sizes ranging from 200 to 2000

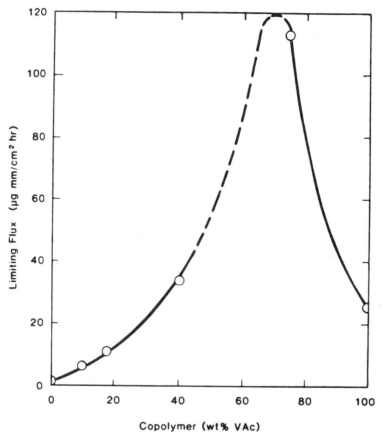

Figure 39   Flux of camphor through EVA copolymer membranes (From Ref. 186.)

Å. It has been suggested that the microvoid formation occurs as a result of spreading apart of the lamellae (Fig. 40).

Process Considerations

Transdermal therapeutic systems are relatively new and quite different from the traditional pharmaceutical dosage forms, such as solutions, tablets, capsules, etc. For this reason, the large-scale manufacturing processes for transdermal systems are very different from those used for the traditional dosage forms. As described earlier, there are basically two kinds of transdermal device constructions. There are the sealed "pouch" or "form-fill" type of construction, exemplified by Transderm-Nitro (GTN), and the multilaminate construction, exemplified by Catapres-TTS (clonidine). Each type of construction requires a different method of manufacturing.

Manufacturing of the "pouch"-type system involves first the preparation of a laminate of a release liner, pressure-sensitive adhesive layer, and a control membrane (or a non-rate-controlling drug and excipient permeable polymeric film). This laminate is usually prepared by casting a solution or an aqueous emulsion of the adhesive on a release linear,

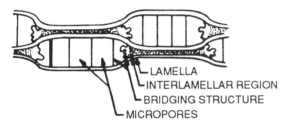

LAMELLA
INTERLAMELLAR REGION
BRIDGING STRUCTURE
MICROPORES

Figure 40 Spreading apart of lamellae to form micropores. (From Ref. 187.)

followed by air-drying in an oven, and then laminating or pressing the membrane onto the dry adhesive layer. In a commercial manufacturing operation, a continuous web of this laminate is fed into packaging-type equipment, where precisely metered, discrete amounts of the drug formulation are deposited onto the membrane surface of the laminate web, covered with the backing layer, and form-sealed using heat and pressure. Unit transdermal dosages are then die-cut from the continuous web and collected for packaging.

A multilaminate system consists of an assembly of three or more polymeric layers, which can include an impermeable backing layer, a drug/polymer layer, a rate-controlling membrane layer, and a pressure-sensitive adhesive layer, which is protected on one side by a release liner. In addition, a multilaminate assembly may also have a thin adhesive tie coat for bonding the membrane and/or the backing to the drug/polymer matrix. Incorporation of the drug into the polymer matrix can be done by dissolving or uniformly dispersing the drug into a liquid polymer formulation, which is then coated onto the impermeable backing layer. In the case of a polymer solution coating, the formation of the solid matrix layer involves solvent evaporation in an air-circulating oven line. In the case of solventless liquid polymer formulations, such as polysiloxane or PVC plastisol, the solid matrix formation involves heat curing or gelation. For transdermal systems in which the rate of drug transport through skin is controlled by the skin and not the device, a membrane may not be required. For such systems, it is possible to incorporate the drug into the pressure-sensitive adhesive formulation itself, which then functions as both a drug/polymer matrix and a means for attaching the device to the skin.

The assembly of a multilayered web is accomplished by pressing the separately manufactured individual layers together. This lamination is done using counterrotating roll systems having carefully controlled roll temperatures and pressure between the rolls. The control of temperature and pressure is very important to ensure interlayer bonding without deforming or destroying the integrity and or functionality of the drug/polymer matrix and the membrane layers. The multilaminate web is then slit longitudinally, if necessary, and die-cut into unit-dose sections.

For drugs that are very susceptible to thermal and oxidative degradation, the ''pouch''-type device is more suitable because the formulation is not subject to stressful conditions in the process of manufacturing the device. On the other hand, the multilaminate system has the advantage of being simple and amenable to high-speed manufacturing technologies.

In order to obtain the desired drug content uniformity in the drug/polymer matrix layer, precision coating techniques are required. The selection of the specific coating equipment (e.g., reverse-roll, gravure roll, or knife-over-roll coaters) depends on the viscosity and viscoelasticity of the liquid drug/polymer solution or dispersion, and the surface tension characteristics of the substrate on which the formulation is being coated. Attainment of precise coating weights requires accurate control of process variables. Continuous control and monitoring of oven drying temperatures, web speed, and air circulation are very critical toward obtaining defect-free curing or drying of drug/polymer matrix layers, that exhibit reproducible drug dissolution profiles. When solvents are employed in the coating process, the process parameters must be carefully adjusted to ensure absence of residual solvent in the finished laminate.

PROBLEMS

1. Drug permeation through the skin is influenced by the design of the transdermal delivery system. (a) What is the expression giving the permeation rate for a matrix-type device if the permeation is controlled not by the skin diffusion but by the device design (device-control system)? (b) What is the expression giving the permeation rate for a matrix-type device if the permeation is controlled not by the device design but by the skin permeability (skin-control system)? (c) Discuss the effect of the loading dose incorporated in the matrix system on the increase in the overall permeation rate across the skin. (*Ans.*: Ref. 174.)

2. Calculate the steady-state concentration in the plasma and the time lag to reach steady state following transdermal delivery of the following drugs: Stratum corneum solubility, 10 mg/ml; stratum corneum diffusion coefficient, 4×10^{-11} cm^2/s; viable skin diffusion coefficient, 6×10^{-8} cm^2/s; partition coefficient between stratum corneum and viable skin, 10; volume of distribution (one-compartment model), 2.1×10^5 ml; elimination half-life in plasma, 3 min or 10 h.

3. If binding takes place in the skin, the total concentration of drug in the skin is represented by the sum of the dissolved drug concentration and the bound-drug concentration as follows:

$$C_t = k_d C_d + \frac{v k_d P_t b C_d}{1 + k_d b C_d} \tag{a}$$

where k_d is the intrinsic partition coefficient between the skin and the donor solution, P_t is the total protein (enzymes), b is the association constant, v is the number of sites of binding, and C_d is the medium concentration. (a) Derive Eq. (16) by assuming a Langmuir-type binding. (b) Determine the individual parameters for the following experiments:

Drug A	Medium concentration (μg/ml)	Skin concentration (μg/g-skin)	Drug B	Medium concentration (μg/ml)	Skin concentration (μg/g-skin)
	0.5	15.0		0.5	7.1
	1.0	24.8		1.0	14.2
	2.0	39.9		2.0	23.2
	5.0	70.1		5.0	42.6
	10.0	116.6		10.0	72.1
	20.0	175.0		20.0	117.1
	30.0	237.4		30.0	143.0

Ans.: $k_d = 5.73$, $vP_t = 71.2$, $b = 0.060$ for drug A and $k_d = 1.75$, $vP_t = 132$, $b = 0.043$ for drug B.

4. Derive the following diffusion equation [Eq. (24)] for describing the simultaneous diffusion and binding in the skin by using Eq. (a) in Problem 3.

$$\left[1 + \frac{K}{(1 + bC)^2} \right] \frac{\partial C}{\partial t} = D \frac{\partial^2 C}{\partial x^2} \tag{24}$$

5. Derive the enhancement factor under iontophoretic transdermal delivery in the absence of convective flow. If the convective flow is appreciable, how does the enhancement factor deviate from the ideal case?
   Ans.: $E = (dQ/dt)_E/(dQ/dt)_0 = V/[1 - \exp(-V)]$; $V = zFE/RT$.

6. Summarize approaches for achieving a pulsed profile of the drug concentration in the plasma following transdermal delivery and discuss the mechanisms for such a temporaral increase in the skin permeability.

7. Human skin permeation rate of nitroglycerine from Transderm Nitro was found to be 14.6 $\mu g/cm^2/h$, and the resistance of the EVA membrane to nitroglycerine transport was determined to be 27 h/cm. Using the drug release data provided in the text, calculate the device control of nitroglycerine transport across skin.

8. For a transdermal fentanyl system having 15% drug dissolved in an acrylic copolymer, polyimide films were evaluated as membranes. Fentanyl flux values under sink conditions, from the drug/polymer matrix formulation, across 10-, 20-, 30-, and 40-μm-thick membranes were 7.2, 4.0, 3.1, and 2.5 $\mu g/cm^2/h$, respectively. Calculate the membrane permeability (DK) to fentanyl.

REFERENCES

1. A. F. Kydonieus, Fundamentals of transdermal drug delivery, in *Transdermal Delivery of drugs*, Vol. 1 (A. F. Kydonieus and B. Berner, Eds.), CRC Press, Boca Raton, Fla., p. 5 (1987).
2. G. B. Kasting, et al., Effect of lipid solubility and molecular size on percutaneous absorption, in *Skin Pharmacokinetics* (B. Shroot and H. Schaefer, Eds.), S. Karger, Basel, p. 138 (1987).
3. T. Winsor and G. E. Burch, *Arch. Intern. Med.*, *74*:428 (1944).
4. R. J. Scheuplein, *J. Invest. Dermatol.*, *45*:334 (1965).
5. P. M. Elias, *Arch. Dermatol. Res.*, *270*:95 (1981).
6. R. Eichner, T. Sun, and U. Aebi, *J. Cell Biol.*, *102*:1767 (1986).
7. J. Woodcock-Mitchell, R. Eichner, W. G. Nelson, and T. Sun, *J. Cell Biol.*, *95*:580 (1982).
8. C. A. Squier, *J. Ultrastruct. Res.*, *43*:160 (1983).
9. P. M. Elias, N. S. McNutt, and D. S. Friend, *Anat. Rec.*, *189*:577 (1977).
10. A. M. Kligman and E. Christophers, *Arch. Dermatol.*, *88*:702 (1964).
11. D. N. Menton, *Am. J. Anat. 145*:1 (1976)
12. W. L. Epstein and H. I. Maibach, *Arch Dermatol.*, *92*:462 (1965).
13. S. Gelfant, *Int. Rev. Cytol.*, *81*:145 (1983).
14. H. Baker and A. M. Kligman, *Arch. Dermatol.*, *95*:408 (1967).
15. S. Rothberg, R. G. Crounse, and J. L. Lee, *J. Invest. Dermatol.*, *37*:497 (1961).
16. M. A. Lampe, M. L. Williams, and P. M. Elias, *J. Lipid Res.*, *24*:131 (1983).
17. M. A. Lampe, A. L. Burlingame, J. Whitney, M. L. Williams, B. E. Brown, E. Roitman, and P. M. Elias, *J. Lipid Res.*, *24*:120 (1983).
18. R. J. Scheuplein and R. L. Bronaugh, in *Biochemistry and Physiology of Skin* (L. A. Goldsmith, Ed.), Oxford University Press, New York, p. 1255 (1983).
19. A. A. Fisher, *Contact Dermatitis*, 3rd ed., Lea & Febiger, Philadelphia (1986).
20. G. Von Krogh and H. I. Maibach, The contact urticaria syndrome, in *Dermatotoxicology* (F. N. Marzulli and H. I. Maibach, Eds.), Hemisphere, Washington, D.C., pp. 341–362 (1987).

21. R. M. Barr, S. Brain, R. D. R. Camp, J. Cilliers, M. W. Greaves, A. I. Mallet, and K. Misch, *Br. J. Dermatol.*, *111*:23 (1984).

22. B. W. Barry, Penetration enhancers, in *Skin Pharmacokinetics* (B. Shroot and H. Schaefer, Eds.), S. Karger, Basel, p. 121 (1987).

23. N. F. H. Ho, J. Y. Park, W. Morozowich, and W. I. Higuchi, Physical model approach to the design of drugs with improved intestinal absorption, in *Design of Biopharmaceutical Properties Through Prodrugs and Analogs* (E. B. Roche, Ed.), American Pharmaceutical Association, Washington, D.C., p. 136 (1977).

24. W. J. Dunn, Structural effects of partitioning behavior of drugs, in *Design of Biopharmaceutical Properties Through Prodrugs and Analogs* (E. B. Roche, Ed.), American Pharmaceutical Association, Washington, D.C., p. 47 (1977).

25. K. B. Sloan, M. Hashida, J. Alexander, N. Bodor, and T. Higuchi, Prodrugs of 6-thiopurines: Enhanced delivery through the skin, *J. Pharm. Sci.*, *72*:372 (1983).

26. K. H. Valia, K. Tojo, and Y. W. Chien, Long-term permeation kinetics of estradiol (III): Kinetic analysis of the simultaneous skin permeation and bioconversion of estradiol esters, *Drug Develop. & Ind. Pharm.*, *11*:1133 (1985).

27. W. I. Higuchi and C. Yu, Prodrugs in transdermal delivery, in *Transdermal Delivery of Drugs*, Vol. 3 (A. F. Kydonieus and B. Berner, Eds.), CRC Press, Boca Raton, Fla., p. 43 (1987).

28. A. S. Michaels, S. K. Chandrasekaran, and J. E. Shaw, Drug permeation through human skin. Theory and in vitro experimental measurement, *AIChE J.*, *21*:985 (1975).

29. H. E. Bodde, et al., The skin compliance of transdermal drug delivery systems, *Crit. Rev. Therapeutic Drug Carrier Systems*, *6*:94 (1989).

30. D. D. Munro and R. B. Stoughton, Dimethyl acetamide (DMAC) and dimethyl formamide (DMFA) effect on percutaneous absorption. *Arch. Dermatol.*, *92*:585–586 (1965).

31. S. K. Chandrasekaran, P. S. Campbell, and A. S. Michaels, Effect of dimethyl sulfoxide on drug permeation through human skin, *AIChE J.*, *23*: (1977).

32. D. Southwell and B. W. Barry, Penetration enhancers for human skin: Mode of action of 2-pyrrolidone and dimethyl formamide on partition and diffusion of model compounds water, *n*-alcohols and caffeine, *J. Invest. Dermatol.*, *80*:507–514 (1983).

33. R. B. Stroughton and W. O. McClure, Azone, a new non-toxic enhancer of cutaneous penetration, *Drug Develop. Ind. Pharm.*, *9*:725–744 (1983).

34. E. R. Cooper, Increased skin permeability for lipophilic molecules, *J. Pharm., Sci.*, *73*:1153–1156 (1984).

35. A. H. Ghanem, H. Mahmoud, W. I. Higuchi, U. D. Rohr, S. Borsadia, P. Liu, J. L. Fox, and W. R. Good, The effects of ethanol on the transport of β-estradiol and other permeants in hairless mouse skin. II. A new quantitative approach, *J. Controlled Release*, *6*:75–83 (1987).

36. E. S. Nuwayser, M. H. Gay, D. J. DeRoo, and P. D. Blaskovich, Transdermal nicotine— An aid to smoking cessation. *Proc. Int. Symp. Control. Rel. Bioact. Mater.*, *15*:213–214 (1988).

37. A. C. Williams and B. W. Barry, Urea analogues in propylene glycol as penetration enhancers in human skin, *Int. J. Pharm.*, *36*:43 (1989).

38. Y. L. Cheng, et al., Skin penetration enhancer compounds using sucrose esters, U.S. Patent 4,900,555 (1990).

39. G. Cleary and S. Roy, Transdermal drug delivery composition, U.S. Patent 4,906,463 (1990).

40. T. W. Leonard, et al., Carvone enhancement of transdermal drug delivery, U.S. Patent 4,888,360 (1989).

41. T. W. Leonard, et al., Eugenol enhancement of transdermal drug delivery, U.S. Patent 4,888,362 (1989).

42. N. S. Bodor and T. Loftsson, Composition and method for enhancing permeability of topical drugs, U.S. Patent 4,892,737 (1990).

43. M. S. Wu, Determination of concentration-dependent water diffusivity in a keratinous membrane, *J. Pharm. Sci.*, *72*:1421 (1983).

44. D. E. Wurster and S. F. Kramer, Investigation of some factors influencing percutaneous absorption, *J. Pharm. Sci.*, *50*:288 (1961).

45. E. Cronin and R. B. Stroughton, Percutaneous absorption: Regional variations and the effect of hydration and epidermal stripping, *Br. J. Dermatol.*, *74*:265 (1962).

46. C. F. H. Vickers and W. C. Fritsch, A hazard of plastic film therapy, *Arch. Dermatol.*, *87*:633 (1963).

47. D. E. Wurster and R. Munies, Factors influencing percutaneous absorption II: Absorption of methyl ethyl ketone, *J. Pharm. Sci.*, *54*:554 (1965).

48. B. Idson, Percutaneous absorption, *J. Pharm. Sci.*, *64*·901 (1975).

49. D. Southwell and B. W. Barry, Penetration enhancement in human skin: Effect of 2-pyrrolidone, demethylformamide and increased hydration on finite dose permeation of aspirin and caffeine, *Int. J. Pharm.*, *22*:291 (1984).

50. K. Knutson, R. O. Potts, D. B. Guzek, G. M. Golden, J. E. MeKie, W. J. Lambert, and W. I. Higuchi, Macro- and molecular physical-chemical considerations in understanding drug transport in the stratum corneum, *J. Control. Rel.*, *2*:67 (1985).

51. E. R. Cooper, Effect of decylmethyl sulfoxide on skin penetration, in *Solution Behavior of Surfactants* (K. L. Mittal and E. J. Fendler, Eds.), Plenum Press, New York, p. 1505 (1982).

52. E. R. Cooper and B. Berner, Penetration enhancers, in *Transdermal Delivery of Drugs*, Vol. 2 (A. F. Kydonieus and B. Berner, Eds.), CRC Press, Boca Raton, Fla., p. 57 (1987).

53. G. M. Golden, et al., *J. Pharm., Sci.*, *76*:25 (1987).

54. A. H. Ghanem, et al., *J. Control. Rel.*, *6*:75 (1987).

55. K. Knutsen, et al., Solvent-mediated alterations of the stratum corneum, *J. Control. Rel.*, *11*:94 (1990).

56. General Hospital Corp., Controlled removal of human stratum corneum by pulsed laser to enhance percutaneous transport, U.S. Patent 4,775,361.

57. J. S. Nelson, Mid-infrared laser ablation of stratum corneum enhances topical delivery of drugs, *J. Invest. Derm.*, *94*:559 (1990).

58. A. K. Banga and Y. W. Chien, *J. Control. Rel.*, *7*:1–24 (1988).

59. L. E. Gibson and R. E. Cooke, *Pediatrics*, *23*:545 (1959).

60. S. Bellantone, et al., *Int. J. Pharm.*, *30*:63–72 (1986).

61. M. Comeau, et al., *Arch. Otolaryngol.*, *98*:114 (1973).

62. J. B. Sloan and K. J. Solanti, *Am. Acad. Dermatol.*, *15*:671–684 (1986).

63. B. Kari, *Diabetes*, *35*:217–221 (1986).

64. O. Siddiqui, et al., *J. Pharm. Sci.*, *76*:341–345 (1987).

65. L. P. Gangarosa, et al., *Proc. Soc. Exp. Biol. Med.*, *154*:439–443 (1977).

66. A. Pikal, *Pharm. Res.*, *7*:118 (1990).

67. R. R. Burnette and B. Ongipasttankul, *J. Pharm. Sci.*, *77*:132 (1988).

68. G. B. Kasting and A. Bowman, *Pharm. Res.*, *7*:134 (1990).

69. S. Riar, et al., *Antrologia*, *14*:481 (1982).

70. A. Satyanand, et al., *J. Indian Med. Assoc.*, *84*:134 (1986).

71. M. W. Cornwall, *Phys. Ther.*, *61*:359 (1981).

72. M. Tannenbaum, *Phys. Ther.*, *60*:792 (1980).

73. F. Soan, *Int. J. Dermatol.*, *19*:519 (1980).

74. B. Chantraine, et al., *Arch. Phys. Med. Rehabil.*, *67*:38 (1980).

75. W. B. Shelly, et al., *J. Invest. Dermatol.*, *11*:275 (1948).

76. M. L. Elgart and G. Fuchs, *Int. J. Dermatol.*, *26*:194 (1987).

77. B. C. Palombini and P. R. K. Desouza, *Rev. Bras. Biol.*, *32*:197 (1984).

78. H. S. Schwarz, *Arch. Intern. Med.*, *95*:662 (1955).

79. N. H. Gordon and M. V. Weinstein, *Phys. Ther.*, *49*:869 (1969).

80. B. S. Kwon, et al., *J. Infect. Dis.*, *140*:1014 (1979).

81. A. S. Rapperport, et al., *Plastic Reconstruct. Surg.*, *36*:547 (1965).

82. Y. W. Chien, et al., *Ann. N.Y. Acad. Sci.*, *507*:32–51 (1987).

83. R. R. Burnette and D. Marrero, *J. Pharm. Sci.*, *75*:738 (1986).

84. D. M. Skauen and G. M. Zentner, *Int. J. Pharm.*, *20*:235 (1984).

85. J. Kost, K. Leong, and R. Langer, *Polym. Sci. Technol.*, *34*:387 (1986).

86. S. Miyasaki, C. Yokouchi, and M. Takada, *J. Pharm. Pharmacol.*, *40*(10):716 (1988).

87. T. N. Julian and G. M. Zentner, *J. Pharm. Pharmacol.*, *38*:871 (1986).

88. N. T. Julian and G. M. Zentner, Mechanism for ultrasonically enhanced transmembrane solute permeation. *J. Control. Rel.*, *12*:77 (1990).

89. J. Griffin and J. Touchstone, *Proc. Soc. Exp. Biol. Med.*, *109*:461–465 (1962).

90. J. McElnay, T. Kennedy, and R. Harland, *Int. J. Pharm.*, *40*:105–110 (1987).

91. H. Benson, J. McElnay, and R. Harland, *Int. J. Pharm.*, *44*:65–69 (1988).

92. J. Kost, D. Levy, and R. Langer, *Abstracts, 3rd Ann. Mtg. Am. Assoc. of Pharmaceut. Sci.*, Orlando, Fla., October 1988.

93. R. Bricks, et al., The effect of ultrasound on the in vitro penetration of ibuprofen through human epidermis, *Pharm. Res.*, *6*(8):697 (1989).

94. J. Kost, et al., Experimental approaches to eluadate the mechanism of ultrasonically enhanced transdermal drug delivery, *Program and Abstracts of the 17th Int. Symp. on Controlled Release of Bioactive Materials* (V. H. L. Lee, Ed.), p. 53 (1990).

95. D. Bommannan, et al., Sonophoresis: Enhancement of transdermal drug delivery using ultrasound, *Program and Abstracts of the 17th International Symposium on Controlled Release of Bioactive Materials* (V. H. L. Lee, Ed.), p. 53 (1990).

96. A. Obermayer, et al., Transdermal pharmaceuticals containing amines and nonpolar, non-volatile solvents, European Patent Appl. EP267051 (1989).

97. D. C. Patel and C. D. Ebert, Method for reducing skin irritation associated with drug/penetration enhancer compositions, U.S. Patent 4,855,294 (1989).

98. S. K. Govil and P. Kohlman, Transdermal delivery of nicotine, U.S. Patent 4,908,213 (1990).

99. A. Amkraut and J. E. Shaw, Prevention of contact allergy by coadministration of a corti-costeroid with a sensitizing drug, European Patent Appl. 88300771.8 (1988).

100. Rockefeller University, Glucocorticoid carboxylic acid ester for suppression of cutaneous delayed hypersensitivity, PCT Inst. App. WO8809175.

101. M. J. N. Cormier, Method for reducing sensitization or irritation in transdermal drug delivery and means thereof, U.S. Patent 4,885,154 (1989).

102. H. Schaefer, A. Zesch, and G. Stuttgen, *Skin Permeability*, Springer-Verlag, New York (1982).

103. A. M. Kligman, A biological brief on percutaneous absorption, *Drug Dev. Ind. Pharm.*, *9*:521–560 (1983).

104. R. J. Scheuplein, Mechanism of percutaneous absorption II, *J. Invest. Derm.*, *48*:79–88 (1967).

105. K. Tojo, C. C. Chiang, and Y. W. Chien, Drug permeation across the skin, *J. Pharm. Sci.*, *76*:123–126 (1987).

106. H. Okamoto, M. Hashida, and H. Sezaki, *J. Pharm. Sci.*, *77*:418–424 (1988).

107. R. J. Scheuplein and I. H. Blank, Permeability of the skin, *Physiol. Rev.*, *51*:702–747 (1971).

108. S. D. Roy and G. L. Flynn, Transdermal delivery of narcotic analgesics, *Pharm. Res.*, *6*:825–832 (1989).

109. W. J. Albery and J. Hadgraft, Percutaneous absorption: Theoretical description, *J. Pharm. Pharmacol.*, *31*:129–139 (1979).

110. W. J. Albery and J. Hadgraft, Percutaneous absorption: In vivo experiment, *J. Pharm. Pharmacol.*, *31*:140–147 (1979).

111. B. Berner and E. Cooper, Models of skin permeability, in *Transdermal Delivery of Drugs*, Vol. 2 (A. F. Kydonieus and B. Berner, Eds.), CRC Press, Boca Raton, Fla., chap. 3 (1987).

112. C. A. Kumins and T. K. Kwei, Free volume and other theories, in *Diffusion in Polymers* (J. Crank and G. S. Park, Eds.), Academic Press, New York, p. 107 (1968).

113. S. H. Yalkowsky, Solubility and solubilization of nonelectrolytes, in *Techniques of Solubilization of Drugs* (S. H. Yalkowsky, Ed.), Marcel Dekker, New York, p. 1 (1981).

114. D. W. Osborne, Computational methods for predicting skin permeability, *Pharmaceut. Manufact.*, 41–48 (1986).

115. R. H. Guy and J. Hadgraft, Prediction of drug disposition kinetics in skin and plasma following topical administration, *J. Pharm. Sci.*, *73*:883–887 (1985).

116. B. Anderson, Presentation at NICHD workshop on transdermal delivery of drugs, Bethesda, Md. (May 1988).

117. W. I. Higuchi, Simultaneous transport and metabolism of β-estradiol in hairless mouse skin, *Therapeut. Res.*, *10*:149–166 (1989).

118. A. S. Michaels, S. K. Chandrasekaran, and J. E. Shaw, Drug permeation through human skin, *AIChE J.*, *21*:985–996 (1975).

119. K. Tojo, Random brick model for drug transport across stratum corneum, *J. Pharm. Sci.*, *76*:889–996 (1987).

120. H. Schaefer, A. Zesch, and G. Stuttgen, *Skin Permeability*, Springer-Verlag, New York (1982).

121. K. Tojo, C. C. Chiang, U. Doshi, and Y. W. Chien, Stratum corneum reservoir capacity affecting dynamic of transdermal drug delivery, *Drug Dev. Ind. Pharm.*, *14*:561–572 (1988).

122. H. Bundgaard, N. Mork, and A. Hoelgaard, Enhanced delivery of nalidixic acid through human skin via lacyloxymethyl ester prodrugs, *Int. J. Pharm.*, *55*:91–97 (1989).

123. F. Geneser, *Textbook of Histology*, Lea & Febiger, Philadelphia, chap. 17 (1986).

124. R. J. Scheuplein, Mechanism of percutaneous absorption, *J. Invest. Derm.*, *48*:79–88 (1967).

125. E. Christophers, C. Schubert, and M. Good, The epidermis, in *Pharmacology of the Skin* (M. W. Greavea and S. Shuster, Eds.), Springer-Verlag, Berlin, chap. 1 (1989).

126. K. Tojo, Random brick model for drug transport across stratum corneum, *J. Pharm. Sci.*, *76*:889—891 (1987).

127. J. Crank, *The Mathematics of Diffusion*, Clarendon Press, Oxford, chap. 4 (1975).

128. J. Shaw, Development of transdermal therapeutic systems, *Drug. Dev. Ind. Pharm.*, *9*:579–603 (1983).

129. R. Scheuplein, Skin permeation, in *The Physiology and Pathophysiology of the Skin* (A. Jarrett et al., Eds.), Academic Press, New York, chap. 55 (1978).

130. K. Tojo, C. C. Chiang, and Y. W. Chien, Drug permeation across the skin: Effect of penetrant hydrophilicity, *J. Pharm. Sci.*, *76*: 123–126 (1987).

131. H. Okamoto, Development of polypronylazacycloalkanone derivatives as percutaneous penetration enhancers and their mechanism of action based on diffusion model, Ph.D. thesis, Kyoto University, 1989.

132. K. Tojo, In vivo/in vitro/in numero correlation for transdermal drug delivery, *Pharm. Tech.*, *Japan*, *5*:19–27 (1989).

133. B. Anderson, NICHD workshop on transdermal delivery of drugs, Bethesda, Md. (May 1988).

134. H. Okamoto, M. Hashida, and H. Sezaki, Structure-activity relationship of 1-alkyl- or 1-alkynylazacycloalkanone derivatives as percutaneous penetration enhancers, *J. Pharm. Sci.*, *77*:418–424 (1988).

135. E. Cooper, Pharmacokinetics of skin penetration, *J. Pharm. Sci.*, *65*:1396–1397 (1976).

136. J. Swarbrick, G. Lee, and J. Brom, Drug permeation through human skin:, I. Effect of storage conditions of skin, *J. Invest. Derm.*, *78*:63–66 (1982).

137. B. Mollgaard, and A. Hoelgaard, Permeation of estradiol through the skin—Effect of vehicles, *Int. J. Pharm.*, *15*:185–197 (1983).

138. W. R. Vieth, *Membrane Systems: Analysis and Design*, Oxford University Press, New York, chap. 2 (1988).

139. J. Crank, *The Mathematics of Diffusion*, Clarendon Press, Oxford, chap. 10 (1975).
140. A. Martin, J. Swarbrick, and A. Cammarata, *Physical Pharmacy*, 3rd ed., Lee & Febiger, Philadelphia, chap. 13 (1983).
141. K. Tojo, Y. Sun, M. Ghannam, et al., Simple evaluation method for intrinsic diffusivity for membrane-moderated controlled release, *Drug Dev. Ind. Pharm.*, *11*:1363–1372 (1985).
142. S. K. Chandrasekeran, A. S. Mihaels, P. S. Cambell, et al., Scopolamine permeation through human skin in vitro, *AIChE J.*, *22*:828–832 (1976).
143. L. L. Wearley, K. Tojo, and Y. W. Chien, A numerical approach to study the effect of binding on the iontophoretic transport of a series of amino acids, *J. Pharm. Sci.*, *79*:992–998 (1990).
binding on the iontophoretic transport of a series of amino acids, *J. Pharm. Sci.*, in press.
144. L. L. Wearley, Factors affecting reversibility of iontophoretic transdermal transport, Ph.D. thesis, Rutgers University, New Brunswick, N.J. (1989).
145. A. C. Lee, Percutaneous absorption and bioconversion of vitamins C, E and its pro-vitamin, Ph.D. thesis, Rutgers University, New Brunswick, N.J. (1990).
146. W. Montagna, Histology and cytochemistry of human skin: IX. The distribution of non-specific esterases, *J. Biophys. Biochem.*, *1*:13–16 (1955).
147. Y. Miura, Histochemical studies on enzyme activities in the skin, *Jpn. J. Dermatol.*, *74*:556–570 (1964) (in Japanese).
148. J. K. Leypolt and D. A. Gough, Comments on the penetrant time-lag in a diffusion-reaction system, *J. Phys. Chem.*, *84*:1058–1059 (1980).
149. K. Tojo, Mathematical modeling of skin permeation of drugs, *J. Chem. Eng. Jpn.*, *20*:300–308 (1987).
150. K. Tojo and A. R. C. Lee, A method for predicting steady-state rate of skin penetration in vivo, *J. Invest. Derm.*, *92*:105–108 (1989).
151. C. D. Yu, J. L. Fox, N. F. H. Ho, et al., Physical model evaluation of topical prodrug delivery I, *J. Pharm. Sci.*, *68*:1341–1346 (1979).
152. C. D. Yu, J. L. Fox, N. F. H. Ho, et al., Physical model evaluation of topical prodrug delivery II, *J. Pharm. Sci.*, *68*:1347–1357 (1979).
153. K. Tojo, Concentration profile in plasma after transdermal drug delivery, *Int. J. Pharm.*, *43*:201–205 (1988).
154. O. Okamoto, F. Yamashita, K. Saito, et al., Analysis of drug penetration through the skin by the two-layer skin model, *Pharm. Res.*, *6*:931–937 (1989).
155. W. I. Higuchi, Simultaneous transport and metabolism of β-estradiol in hairless mouse skin, *Therap. Res.*, *10*:149–166 (1989).
156. K. Tojo, K. H. Valia, and Y. W. Chien, Bioconversion of estradiol esters in skin, *Biochem. Eng. J.*, *33*:B63–B67 (1986).
157. M. Gibaldi and D. Perrier, *Pharmacokinetics*, Marcel Dekker, New York, chap. 2 (1982).
158. For example, A. G. Gilman, L. S. Goodman, and A. Gilman, *The Pharmacological Basis of Therapeutics*, 6th ed., Macmillan, New York (1980).
159. B. N. Singh, G. Ellrodt, and C. T. Peter, Verapamil: A review of its pharmacological properties and therapeutic case, *Drugs*, *15*:167–197 (1978).
160. K. Tojo, A dynamic model for iontophoretic transdermal delivery of peptides, *Pharm. Tech. Jpn.*, *5*:971–976 (1989).
161. C. Vickers, Existence of reservoir in the stratum corneum, *Arch. Dermatol.*, *88*:20–25 (1963).
162. C. F. H. Vickers, Reservoir effect of human skin: Pharmacological speculation, in *Percutaneous Absorption of Steroids* (P. Mauvais-Jarvis, C. F. H. Vickers, and J. Wepierre, Eds.), Academic Press, London (1980).
163. D. Dupium, A. Rougier, R. Roguet, et al., In vivo relationship between horny layer reservoir effect and percutaneous absorption in human and rat, *J. Invest. Derm.*, *82*:353–356 (1984).
164. N. H. Bellantone, S. Rim, M. L. Franceur, et al., *Int. J. Pharm.*, *30*:63 (1986).
165. H. O. Calvery, J. H. Drajze, and E. P. Lang, *Physiol. Rev.*, *26*:495 (1946).

166. R. R. Burnette and D. Marrero, *J. Pharm. Sci.*, *75*:738 (1986).

167. B. R. Meyer, W. Kreis, J. Eschbach, et al., Successful transdermal administration of therapeutic doses of a polypeptide to normal human volunteers, *Clin. Pharmacol. Therap.*, *44*:607–612 (1988).

168. R. R. Burnette and B. Ongpipattanakul, Characterization of the permselective properties of excised human skin during iontophoresis, *J. Pharm. Sci.*, *76*:765–773 (1987).

169. P. Lelawongs, Transdermal iontophoretic delivery of arginine vasopressin, Ph.D. thesis, Rutgers University, New Brunswick, N.J. (1990).

170. V. G. Levich, *Physicochemical Hydrodynamics*, Prentice-Hall, Englewood Cliffs, N.J., chap. VI (1962).

171. K. Tojo, Mathematical model of iontophoretic transdermal drug delivery, *J. Chem. Eng. Jpn.*, *22*:512–518 (1989).

172. K. Tojo, P. Lelawongs, and Y. W. Chien, A dynamic model for iontophoretic transdermal delivery of peptides, *Proc. Int. Symp. Control. Rel. Bioact. Mater.*, *16*:452–453 (1989).

173. W. R. Good, Transderm-Nitro controlled delivery of nitroglycerine via the transdermal route, *Drug Dev. Ind. Pharm.*, *9*:647–670 (1983).

174. K. Tojo, P. Keshary, and Y. W. Chien, Drug permeation through skin from matrix-type drug delivery systems, *Chem. Eng. J. Biochem. Eng.*, Sec. 32:B57–B64 (1986).

175. V. P. Shah, N. W. Tymes, and J. P. Skelley, *J. Control. Rel.*, *7*:79 (1988).

176. D. J. Enscore and R. M. Gale, U.S. Patent 4,559,222 (1985).

177. B. Hardman and A. Torkelson, Silicones, in *Encyclopedia of Polymer Science and Engineering*, 2nd ed., John Wiley, New York, p. 204 (1989).

178. Y. W. Chien and H. J. Lambert, U.S. Patent 4,053,580 (1977).

179. A. D. Keith, *Drug Dev. & Ind. Pharm.*, *9*:605 (1983).

180. A. D. Keith and W. Snipes, U.S. Patent 4,292,303 (1981).

181. S. Hosaka, H. Ozawa, and H. Tanzawa, *J. Appl. Poly. Sci.*, *23*:2089 (1979).

182. N. B. Graham and D. A. Wood, *Polymer News*, *8*:230 (1982).

183. K. R. Shah, U.S. Patent 4,693,887 (1987).

184. A. F. Kydonieus, K. R. Shah, and B. Berner, U.S. Patent 4,758,434 (1988).

185. D. Satas (Ed.), *Handbook of Pressure-Sensitive Adhesive Technology*, 2nd ed., Van Nostrand Reinhold, New York (1989).

186. R. W. Baker, *Controlled Release of Biologically Active Agents*, John Wiley, New York, pp. 161–162 (1987).

187. R. Gale and L. A. Spitze, Pemeability of camphor in ethylene-vinyl acetate copolymers, *Proc. Int. Symp. Control. Rel. Bioact. Mater.*, *8*:183 (1981).

188. H. S. Bierenbaum, R. B. Isaacson, M. L. Druin, and S. G. Ploven, *Ind. Eng. Chem.*, *Prod. Res. Dev.*, *13*:2 (1974).

9

Nasal Drug Delivery

Shyi-Feu Chang   *Amgen, Inc., Thousand Oaks, California*

Yie W. Chien   *Rutgers—The State University of New Jersey, Piscataway, New Jersey*

INTRODUCTION

Historically, the nasal route has been used for drug delivery since ancient times [1–4]. In the Ayuredic system of Indian medicine, nasal therapy was a recognized form of treatment. Psychotropic drugs and hallucinogens were inhaled via nose snuff by the Indians of South America. More recently, better systemic bioavailability has been achieved with many drugs by self-medication through the nasal route than by oral administration. For some peptide/protein drugs with low nasal absorption, absorption enhancers (surfactants or bile salts), enzyme inhibitors, bioadhesives, and microspheres (manufactured from biodegradable polymers) have been used to promote nasal absorption.

For systemic drug administration, the nasal route appears to be an ideal alternative to use of the parenterals, especially in view of the rich vascularity of the nasal mucosa and the ease of intranasal drug delivery [1–4]. Advantages of nasal administration include the following: (1) the nasal route avoids hepatic "first-pass" elimination, gut wall metabolism, and/or destruction in the gastrointestinal tract; (2) the rate and extent of absorption as well as the plasma concentration-versus-time profiles are comparable to similar data obtained by intravenous medication; and (3) as noted, the rich vasculature and highly permeable structure of the nasal mucosa greatly enhance drug absorption.

The nasal passage, which runs from the nasal vestibule (i.e., nasal valve) to the nasopharynx, has a depth of approximately 12–14 cm [5]. The lining is ciliated, highly vascular, and rich in mucous glands and goblet cells. The blanket of nasal mucus is transported in a posterior direction by the synchronized beat of the cilia. An individual cilium, which is approximately 5 μm in length and 0.2 μm in diameter, moves at a frequency of about 20 beats/s [6,7]. The rate of diffusion of a nasal preparation through the mucus blanket and its rate of clearance may be influenced by the vehicle, drug particle size and surface charge, and the surfactants incorporated (to reduce diffusional resistance of the mucus blanket).

The normal pH of adult nasal secretions ranges from 5.5 to 6.5, and many enzymes exist in the secretions [1,8]. Insulin (zinc-free) was found to be hydrolyzed slowly by leucine aminopeptidase [9]; prostaglandin E, progesterone and testosterone were similarly inactivated by nasal enzymes [1]. An important aspect of nasal drug delivery is the effect of drugs and additives on nasal ciliary functions; indeed, among drugs and additives found to have a negative effect are preservatives, antihistamines, propranolol, bile salts, cocaine, xylometazoline, atropine, and hairspray [1].

Nasal polyposis, atrophic rhinitis, and severe vasomotor rhinitis can reduce the capacity for nasal absorption of drugs such as caerulein [10]. The common cold or any pathological condition involving mucociliary dysfunction can greatly affect the rate of nasal clearance and subsequently the therapeutic efficacy of drugs administered nasally.

The vehicle for nasal formulations and mode of application may be optimized to deliver drugs to the absoptive turbinate region. Toward this end, several factors should be considered: (1) methods and techniques of administration; (2) the site of deposition; (3) the rate of clearance; and (4) the existence of pathological conditions.

Absorption enhancers have long been used to achieve better systemic bioavailability of nasally administered drugs. However, long-term use of an absorption enhancer and chronic use of an enhancer-containing nasal formulation will affect nasal functions and thus the efficiency of nasal absorption. Accordingly, local toxic effects, antibody formation, and tolerance of nasal formulations must be evaluated.

ANIMAL MODELS

In-Vivo Nasal Absorption Models

Rat Model

The rat model used for studying the nasal delivery of drugs was first reported in the late 1970s [1,11]. The surgical preparation for this in-vivo procedure is outlined as follows: The rat is anesthetized by intraperitoneal injection of sodium pentobarbital. After an incision is made in the neck, the trachea is cannulated with a polyethylene tube. Another tube is inserted through the esophagus toward the posterior part of the nasal cavity. The passage of the nasopalatine tract is sealed to prevent drainage of drug solution from the nasal cavity into the mouth. The drug solution is delivered to the nasal cavity through the nostril or through the esophagus cannulation tubing. Blood samples are then collected from the femoral vein.

With all possible outlets in the rat blocked following surgical preparation, the only possible passage for the drug to be absorbed and transported into the systemic circulation is by penetration and/or diffusion through the nasal mucosa.

Rabbit Model

A rabbit weighing about 3 kg is either anesthetized or maintained in the conscious state, depending on the purpose of experiment. (The rabbit is usually anesthetized by an intra-muscular injection of a combination of ketamine and xylazine.) The drug solution is delivered by nasal spray into each nostril while the rabbit's head is held upright. During the study, the rabbit is permitted to breathe normally through its nostrils, and its body temperature is maintained at 37°C by a heating pad. Blood samples are collected via an indwelling catheter in the marginal ear vein [1,12].

Rabbits are generally inexpensive, readily available, and easily maintained in a laboratory setting. The blood volume of the rabbit is large enough (approx. 300 ml) to permit multiple blood samplings (1–2 ml each) at a frequency that permits full characterization of the pharmacokinetic profile of the drug being evaluated.

The rabbit model described here has been used to study nasal absorption and controlled delivery of progesterone and its hydroxy derivatives [12,13].

Dog Model

A dog is either anesthetized or maintained in the conscious state, depending on the purpose of the study and the characteristics of the drug. For anesthetization, the dog is injected intravenously with sodium thiopental and maintained with sodium phenobarbital. A positive-pressure pump provides ventilation through a cuffed endotracheal tube, and a heating pad keeps the body temperature at 37–38°C. Blood samples are collected from the jugular vein [1].

Sheep Model

The sheep model is basically similar to that described for the dog model [1]. Male in-house-bred sheep are selected for their lack of nasal infectious diseases.

Because of their larger nostrils and body size compared to those of the rat, rabbits, and dogs, sheep are more suitable and practical for evaluating parameters of more sophisticated formulations.

Monkey Model

A monkey weighing about 8 kg is either tranquilized, anesthetized, or maintained in the conscious state, depending on the purpose of the experiment. The monkey is tranquilized by intramuscular injection of ketamine hydrochloride, or it is anesthetized by intravenous injection of sodium phenobarbital. With the head of the monkey upright, the drug solution is delivered into each nostril. The monkey is then placed in a supine position in a metabolism chair for 5–10 min. During the entire course of the study, the monkey breathes normally through its nostrils. Blood samples are collected via an indwelling catheter in the vein [1].

In summary, the rat is small, low in cost, easy to handle, and inexpensive to maintain. Unfortunately, its applications are limited to the biopharmaceutics and pharmacological studies of nasal drug absorption. The primate model continues to be very useful, even though it is expensive and, increasingly, its use is being opposed by animal rights groups. Alternatively, rabbit, dog and sheep models are excellent and particularly useful for formulation studies. However, the constraints in selection of an animal model and the differences in physiology/anatomy among the models are unavoidable. Animal models, nevertheless, play an important role in evaluating the nasal delivery of drugs.

Ex-Vivo Nasal Perfusion Model

Figure 1 shows the experimental setup for in-situ nasal perfusion studies [1]. The same surgical preparation as that described for the in-vivo rat model may be followed. During perfusion, a funnel is placed between the nose and reservoir to minimize loss of drug solution. The drug solution, which is held in a reservoir maintained at 37°C, is circulated through the nasal cavity of the rat by means of a peristaltic pump. The perfusion solution passes out of the nostril, through the funnel, and returns to the reservoir, where the drug solution is stirred constantly. The amount of drug absorbed is determined by measuring the initial and final drug concentrations of the perfusing solution. Possible loss of drug

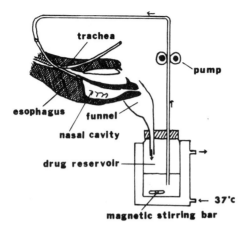

Figure 1 Experimental setup for ex-vivo nasal
perfusion. (From Ref. 1.)

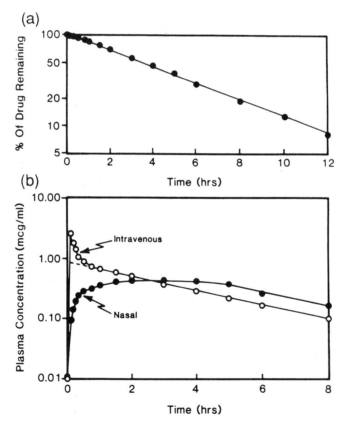

Figure 2 The ex-vivo nasal absorption study of hydromorphone
in the rabbits (n = 3): (a) first-order disappearance of drug from
the perfusion solution; (b) comparative plasma profiles of drug
following nasal delivery and intravenous administration at the same
dose (5 mg/kg). (From Ref. 14.)

activity due to stability, such as loss of peptides and proteins by proteolysis, aggregation, etc., must be considered.

The rabbit can also be used with the ex-vivo nasal perfusion procedure to study the pharmacokinetics of drugs. In this case, the kinetics of nasal absorption can also be monitored by simultaneous measurements of the drug circulating systemically (Fig. 2) [14].

NASAL ABSORPTION

Physicochemical Parameters

The following physical and chemical properties of drug molecules should be evaluated before development of a nasal drug delivery system [1].

Molecular Size

McMartin et al. [15] found that nasal absorption falls off sharply as molecular weight increases above 1000 Da, while oral absorption declines even more steeply as molecular weight falls below 200 Da (Fig. 3). The nasal absorption of a wide range of water-soluble

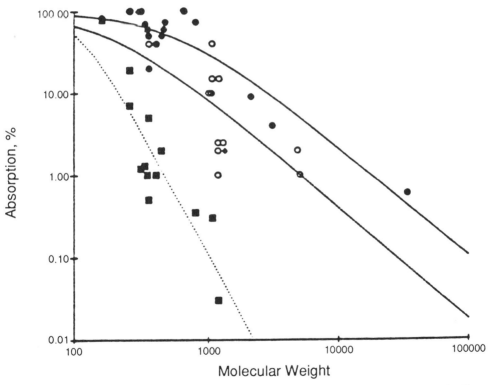

Figure 3 Log-log plot of the absorption after nasal and oral administration of the compounds. Key: (●) rat, nasal; (○) human, nasal; (◆) other species, nasal; (■) all species, oral. The curves were generated by fitting the least squares of the function % absorption = $100[1 + a(MW)^b]$. Curves: continuous, rat, nasal (a = 0.003, b = 1.3); dashed, human, nasal (a = 0.001, b = 1.35); dotted, all species, oral (a = 8.4 × 10^{-7}, b = 3.01). (From Ref. 15.)

compounds with different molecular weights (e.g., insulin, dextran, p-aminohippuric acid) has been studied in the male Wistar rat [16]. The results indicated that a good linear correlation exists between the log percentage of drug absorbed nasally and the log of molecular weight; this finding suggests participation of aqueous channels in the nasal absorption of water-soluble compounds.

Effect of Perfusion Rate

Using the in-situ nasal perfusion procedure, the nasal absorption of phenobarbital was found to increase as the perfusion rate increased. The absorption then reached a plateau level which became independent of the rate of perfusion (> 2 ml/min) [17].

Effect of Perfusate Volume

As the volume of perfusion solution increases, the first-order disappearance rate of phenobarbital has been observed to decrease (Figure 4) [17]. The results from studies using

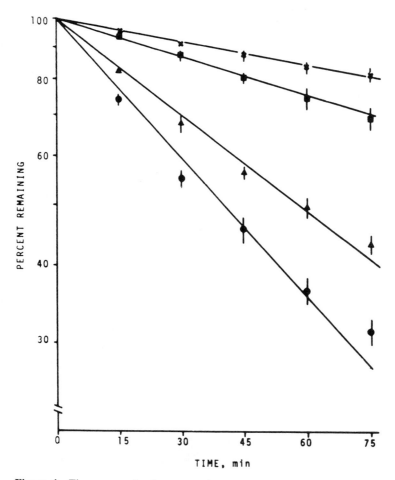

Figure 4 Time course for the percent (mean $\pm$ SEM, n = 3–5) of phenobarbital remaining in the in-situ perfusate solution (at pH 6, 37°C) as a function of perfusion volume. ($\times$) 20 ml; (■) 10 ml; (▲) 5 ml; (●) 3 ml. (From Ref. 17.)

drugs with various structures suggest that the intrinsic rate constant varies from one drug to another.

Effect of pH

The effect of pH of the perfusion solution on nasal absorption was examined using a water-soluble ionizable compound, specifically, benzoic acid ($pK_a = 4.19$), over the pH range 2.0 to 7.1. Nasal absorption was found to be pH-dependent, with absorption being greater at pH levels below the pK_a (Fig. 5) [17]. The rate of nasal absorption decreased as pH increased, owing to ionization of the penetrant molecule. Varying the pH in nasal insulin solutions also reduced the plasma glucose levels of dogs and rats, as shown in Figs. 6 and 7, respectively [18,19]. At pH 6.1 only slight hypoglycemia occurred, whereas at pH 3.1 the plasma glucose level was reduced about 55% (Fig. 6). Figure 7 shows a dose-dependent decrease in the plasma glucose level at pH 3.1; but at pH 5.5 and 7.4, no hypoglycemic effect was observed.

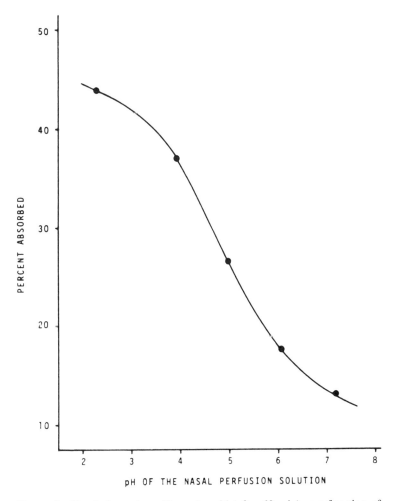

Figure 5 Nasal absorption of benzoic acid (after 60 min) as a function of pH. (From Ref. 17.)

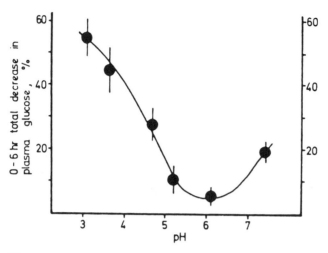

Figure 6 Effect of pH of nasal insulin solution on reduction of
plasma glucose levels in dogs (dose: 50 IU/dog). The data are
expressed as mean ± SEM (n = 4). (From Ref. 18.)

A good linear relationship was found to exist between the absorption rate constant
for hydralazine and the fraction of its undissociated species calculated from the pK_a value
(Fig. 8) [20]. The nasal absorption of hydralazine was facilitated, with the peak plasma
level being achieved within 10 min when the pH of the drug solution was increased from
3.0 to 6.5 or when 0.5% BL-9 (polyoxyethylene 9-lauryl ether) was added.

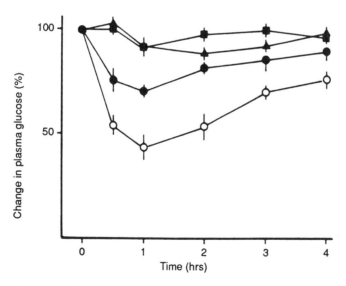

Figure 7 Effect of pH on plasma glucose levels during the 4 h
following intranasal administration of insulin in rats. Key: (○), pH
3.1 (20 U/kg); (●), pH 3.1 (10 U/kg); (■), pH 5.5 (10 U/kg); (▲),
pH 7.4 (10 U/kg). The data are expressed as mean ± SEM (n =
4). (From Ref. 19.)

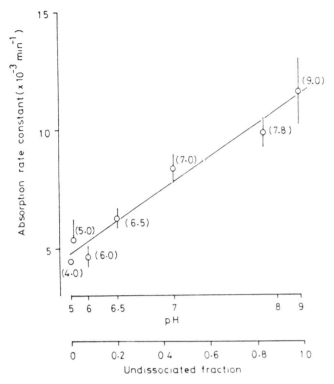

Figure 8 Relationship between the rate constant for the absorption of hydralazine (1 mM) in rats and its undissociated fraction at various pHs. The number in parentheses indicates the pH value. The data are expressed as mean ± SEM (n = 3). (From Ref. 20.)

The nasal absorption rate of decanoic, octanoic, and hexanoic acids is pH-dependent, with the maximum at pH 4.5; absorption decreases steadily at higher or lower pH [21].

Effect of Drug Lipophilicity

The effect of lipophilicity on nasal absorption was studied using a series of barbiturates at pH 6.0, a point at which the barbiturates (pH = 7.6) exist entirely in the nonionized form [17]. The nasal absorption of pentobarbital and barbital differed by only fourfold, even though their partition coefficients differed by 40-fold. Similarly, despite the great difference in partition coefficients of propranolol and l-tyrosine, the variation in their rate constants of nasal absorption was very small [17]. Results from nasal delivery studies of a series of progestational steroids in ovariectomized rabbits demonstrated that the (octanol/water) partition coefficient cannot be used to predict the penetration of nasal mucosa by progesterone and its hydroxy derivatives (Fig. 9) [13]. The results, however, show that the systemic bioavailability of progesterone and its hydroxy derivatives correlate well with their (nasal mucosa/buffer) partition coefficients.

Effect of Initial Drug Concentration

The effect of variation in the initial concentration of drug in the nasal perfusion solution was studied by monitoring the disappearance of l-tyrosyl-l-tyrosine and the formation of l-tyrosine using the rat in-situ perfusion technique. Nasal absorption of l-tyrosine was

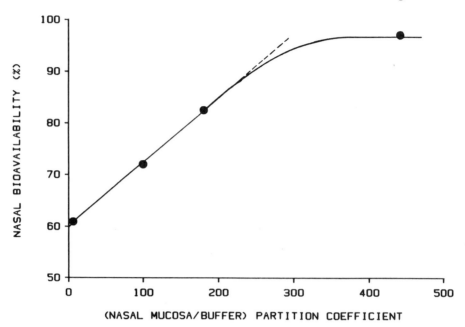

Figure 9 Relationship between the systemic bioavailability of progesterone and its hydroxy derivatives in rabbits and (nasal mucosa/buffer) partition coefficient. (From Ref. 13.)

found to depend on the initial concentration, because the concentration of l-tyrosine formed depended on the initial concentration of l-tyrosyl-l-tyrosine (Fig. 10) [17].

Mechanisms and Pathways

Mechanisms

McMartin et al. [15] studied the transport of SS-6 (an octapeptide) and horseradish peroxidase through the rat's nasal cavity. They found that two mechanisms of transport were involved, a fast rate that depended on lipophilicity, and a slower rate that depended on molecular weight. Their absorption studies are consistent with nonspecific diffusion of the penetrant molecules through aqueous channels located between the nasal mucosal cells, which impose a size restriction on nasal permeability. Along with other literature results, their data indicate that good bioavailabilities can be achieved for molecules up to 1000 Da without use of enhancers and good availability can be extended to at least 6000 Da with enhancers.

Wheatly et al. [22] used Ussing chambers to study transport mechanisms of substances across nasal mucosal tissue. They found that the transport of insulin, mannitol, or pro-pranolol occurs by a passive transport mechanism. The addition of deoxycholate (0.1%) to the mucosal bathing solution reversibly increased the transepithelia conductance across the nasal membrane and enhanced the transport of mannitol and insulin 10- to 20-fold.

The transport of tyrosine (Tyr) and phenylalanine (Phe) across rat mucosa was studied by Tengamnuay and Mitra using an in-situ perfusion technique [23]. They found that both amino acids were absorbed by an active saturable transport process, which appeared to be Na^+-dependent, and that transport may have required metabolic energy as a driving force, based on the inhibition of the l-Phe uptake by Ouabain and 2,4-dinitrophenol. When

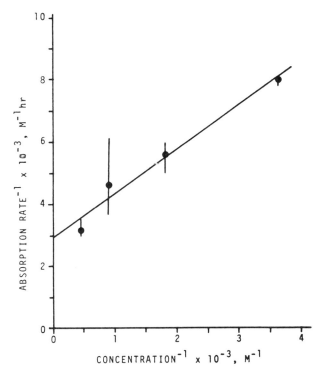

Figure 10 Linear dependency of in-situ nasal absorption rate of l-tyrosine on its concentration (at pH 7.4 and 37°C). Values are mean ± SEM (n = 4–5). (From Ref. 17.)

d-Tyr and d-Phe were used as substrates, the extent of nasal absorption was significantly reduced, indicating a specific affinity of the carrier for the l-amino acids. When mixtures of l-Tyr and l-Phe were used as the perfusates, both amino acids were found to be concomitantly absorbed, but in a competitive manner.

Water-soluble compounds such as sodium cromoglycate are absorbed well, so their nasal absorption is likely to be dependent on aqueous channel diffusion (pores) [24]. For such channel, the molecular size of the compound will be a determinant in the rate of absorption.

Pathways

The olfactory epithelium is known to be a portal of entry for substances into the central nervous system (CNS) and the peripheral circulation [1]. In addition, the nasopharynx has been shown to act as the portal of entry for such common viral diseases as measles, the common cold, smallpox, chicken pox, and poliomyelitis. Communication appears to exist between the subarachnoid space and the nasal cavity, between the lymphatic plexus in the nasal mucosa and the subarachnoid space, as well as between the perineural sheaths in the olfactory nerve filaments and the subarachnoid space. The transport of drugs across the nasal membrane and into the bloodstream may involve either passive diffusion of drug molecules through the pores in the nasal mucosa or some form of nonpassive transport [19].

It is important to investigate the pathways for nasal absorption before nasal delivery rate and transnasal bioavailability of drugs can be improved and controlled. The potential pathways involved in the nasal absorption of drug reported in the literature are summarized in Table 1 [1].

Drug Distribution and Deposition

Drug distribution in the nasal cavity is an important factor that affects the efficiency of nasal absorption [1]. The mode of drug administration may affect this distribution, which in turn can help determine the absorption efficiency of a drug. Using a cast of a human nose, significant differences in drug distribution were demonstrated with such different nasal delivery systems as nose drops, plastic bottle nebulizers, atomized pumps, and metered-dose pressurized aerosols. Among the systems evaluated, the atomized pump was best because it gave a constant dose and a very good mucosal distribution. Results also suggested that the use of a large volume of weak solution was preferable to use of a small volume of concentrated solution. A simulated nasal cavity made of acrylic resin has also been developed for studying distribution of beclomethasone dipropionate aerosol particles in the nasal cavity [25]. No significant differences were found among the gas-, liquid-, and powder-type preparations. The highest concentration of beclomethasone dipropionate was usually found at the anterior portion of the middle turbinate.

Nasal deposition of particles is related to the individual's nasal resistance to air flow [26]. With nasal breathing, nearly all the particles having an aerodynamic size of 10–20 μm are deposited on the nasal mucosa [5]. Deposition in both the poorly absorptive stratified epithelium of the anterior atrium and in the posterior nasopharyngeal region should be avoided, as such deposition leads to drug loss to the stomach by swallowing. Insoluble particles deposited in the main nasal passage are likely to be carried back by the ciliary movement and dispatched to the stomach. If the drug is introduced as a vapor or a soluble particle, it may readily pass into the lining secretions and then be absorbed into the blood.

The deposition of aerosols in the respiratory tract is a function of particle size and respiratory patterns [27]. The density, shape, and hygroscopicity of the particles, and the pathological conditions in the nasal passage, will influence the deposition of particles, whereas the particle-size distribution will determine the site of deposition and affect the subsequent biological response in experimental animals and humans. A uniform distribution of particles throughout the nasal mucosa has been achieved by delivering the particles from a new nasal spray that uses a pressurized gas propellant; the system, which delivers a metered dose of flunisolide, a synthetic fluorinated corticosteroid, provides a consistent dose delivery and spray pattern [28].

The particle or droplet size of an aerosol is important for both efficacy and toxicity. For example, the metered-dose flunisolide solution discussed above required that the majority of particles have a diameter greater than 10 μm to achieve localized delivery in the nasal cavity and to avoid any potential undesired effects resulting from deposition of flunisolide aerosol in the lung [29]. Beclomethasone dipropionate (BD) particles were sprayed into each nostril by Freon propellant. Results indicated that the shape of the nasal cavity had a greater effect on the deposition of BD from the gas spray than on that from the powder spray. This difference may be due to the spray angle as well as the size and speed of the aerosol particles. The powder spray was preferable with regard to the deposition and distribution of drug particles in the nasal cavity. Furthermore, improvement of the

Table 1 Pathways for Nasal Absorption

Substances	Possible pathways
Albumin	
Albumin (labeled with Evans blue and horseradish peroxidase)	Nasal mucosa → sensory nerve cells of olfactory epithelium → subarachnoid space → bloodstream
Egg albumin	Nasal mucosa → lymphatic stream
Serum albumin	Nasal mucosa → lymphatic stream
Amino acids	
Arginine, glutamic acid, glycine, γ-aminobutyric acid, proline, serine, tritiated leucine	Nasal mucosa → blood vessel Nasal mucosa → olfactory nerve fiber → CNS
Bacteria	
Rabbit virulent Type III pneumococci	Nasopharyngeal epithelium → lymphatics → cervical lymphatic vessel → blood vessel
E. coli and staphylococci	Nasal mucosa → lymphatics → blood
Clofilium tosylate	Nasal mucosa → epithelial cells → systemic circulation
Dopamine	Nasal mucosa → CSF and serum
Hormones	
Estradiol	(1) Nasal membrane → CSF (<1 min) (2) Nasal membrane → olfactory neurons → brain and CSF
Norethisterone	(1) Nasal membrane → olfactory dendrites → nervous system → supporting cells in the olfactory mucosa → submucosal blood vascular system (2) Nasal membrane → peripheral circulation (high levels) and CSF (low levels) → CNS
Progesterone	Nasal membrane → olfactory dendrites → nervous system → supporting cells in the olfactory mucosa → submucosal blood vascular system → CSF nasal membrane → CSF (<1 min)
Penicillines	Nasal membrane → bloodstream
Viruses	
Herpes virus encephalitis	Nasal mucosa → peripheral and cranial nerves → CNS
Herpes virus simplex	Nasal mucosa → cranial nerve → CNS
Mouse-passage strains of herpes virus	Nasal mucosa → trigeminal and olfactory pathways → CNS
Neurotropic virus and poliomyelitis virus	Nasal mucosa → olfactory nerve → CNS
Vaccina virus	Nasal mucosa → submucous lymphatics → cervical lymphatic pathways → CNS
Water	
Distilled water	Nasopharynx → cervical lymph

delivery system and drug formulation is necessary to achieve a better clinical effect and easier handling by patients.

Of the three mechanisms usually considered in assessing particle deposition in the respiratory tract, i.e., inertia, sedimentation, and diffusion, inertial deposition is the dominant one in nasal deposition [30]. Any particles with an aerodynamic diameter of 50 μm or greater do not enter the nasal passage. Fry and Black demonstrated that 60% of aerosolized particles of 2–20 μm are deposited in the anterior regions of the nostrils, 2–3 μm from the external nares [31]. The site of drug deposition within the nasal cavity depends on the type of delivery system and the technique used in application [32]. It was found that greater deposition coverage of the nasal walls was achieved following administration by nose drops than by other delivery systems. The area of deposition was independent of the volume administered (over the range 0.1–0.75 ml) [33]. The particles, once deposited at the anterior region of the nasal cavity, may again be conveyed posteriorly by inhaled air, ciliary movement, and/or diffusion in the mucous layer.

The pattern of nasal deposition and the rate of clearance were studied in normal subjects using nasal spray and nose drops of 99mTc-labeled human serum albumin (HSA) [34]. The nasal spray deposited HSA anteriorly in the nasal cavity, with little of the dose reaching the turbinates. In contrast, the nose drops dispersed the dose throughout the length of the nasal cavity, from the atrium to the nasopharynx; also noted, the dosing with three drops resulted in a greater coverage of the nasal walls compared with that of a single drop. The solution deposited anteriorly in the nasal cavity was slow to clear, especially with spray administration. The nose drops cleared more rapidly than the dose administered as a spray.

Hallworth and Padfield compared, in a model nose, the regional deposition of drug discharged from a pressurized aerosol product and a metered-pump spray product [35]. No significant difference in regional deposition was detected between these two products. In addition, it was observed that most of the drug in each case was deposited in the anterior region of the nose by inertial impaction, and that there was little nasal penetration of the drug. The anteriorly deposited drug can be spread backward by mucociliary and general-surface flow. In another study of initial distribution and subsequent clearance of aerosol from a nasal pump spray, Newman et al. [36] showed that the aerosol is concentrated mainly in the anterior part of the nose, but the area of deposition varies from one subject to another. On the average, 56% of the dose remained at the initial site of deposition 30 min after application; the remaining 44% cleared to the nasopharynx. Whaley and Renken [37] developed a method for exposing the nasal cavity of beagle dogs to a radiolabeled aerosol without exposure of the remainder of the respiratory tract. They reported the deposition to be $15(\pm 2)\%$ of inhaled activity, with the maximum deposition occurring in the anterior third of the nasal cavity, which contained $78(\pm 4)\%$ of the total radioactivity deposited. The middle third of the nasal cavity received $13(\pm 3)\%$ and the posterior third $9(\pm 2)\%$ of the radioactivity deposited.

A mathematical model was developed by Gonda and Gipps [38] to describe the rate processes involved in the disposition of drug in the human nasal cavity by their delivery systems. Their model contains a series of parallel first-order rate processes consisting of the convective drug and carrier transport by fluid flow, mucociliary clearance and peristalsis, and drug decomposition, as well as a series of sequential irreversible first-order rate processes consisting of the release and absorption of the drug prior to its appearance in the systemic circulation. The simulation using this model showed that bioadhesion could improve bioavailability and reduce the variability in absorption because it would lessen removal of the drug from the nasal cavity by sniffing, blowing, or wiping the nose.

Enhancement of Drug Absorption

Several methods are used to facilitate nasal absorption of drugs [1].

1. Chemical modification: A drug may be chemically modified to alter its physicochemical properties and thereby enhance its nasal absorption.
2. Salt or ester formation: A drug may be converted to a salt or an ester for better transnasal permeability. Such a salt may be more soluble; the ester may provide better uptake by the nasal epithelium.
3. Formulation design: Proper selection of formulation excipients may enhance the stability or nasal absorption of drugs.
4. Surfactants: Incorporation of surfactants into the nasal formulations may modify the permeability of the nasal mucosa and facilitate drug absorption.

Several approaches have been applied to enhance the nasal absorption of drugs [39]. In the first approach, various surfactants or bile salts are used to promote absorption [1,23,39–52]. Table 3 lists the surfactants and bile salts that have been used in this approach. The systemic bioavailabilities of drugs that contain enhancers are dramatically increased, as shown in Tables 2A and 2B, often to a level sufficient to achieve the desired systemic effects. Mild surfactants at low concentrations may only alter structure and permeability of membranes; whereas certain surfactants, at high concentrations, may disrupt and even dissolve nasal membranes. The mechanisms of action and toxic effects of some surfactants or bile salts are shown in Tables 4A and 4B.

In the second approach to enhancement of transnasal delivery of drugs, various bioadhesive agents, such as methyl celluose, carboxymethyl cellulose, hydroxypropyl cellulose, and polyacrylic acid, have been used [39,53,54]. This enhancement is pre-

Table 2A Biopharmaceutics Data on Nasal Delivery of Some Peptide-Base Pharmaceuticals

Drugs	Animal model	T_{max}	Bioavailability (%)
Alsactide (ACTH-17)	Rat	1 h	12
alpha$^{1-18}$-ACTH	Human	<4 h	12
Buserelin	Human	1–6 h (LH) 2 h (FSH)	
Calcitonin (with sodium glycocholate)	Human	15 min	
[Asu^{1,7}]-eel calcitonin (with polyacrylic acid, pH 6.5)	Rat	30 min	
Cerulein	Human		1
DDAVP	Human		10–20
beta-Endorphin	Bonnet monkey	30 min	
Enkephalin analogs:			
(1) Leucine enkephalin	Rat		<10
(2) DADLE (in saline)	Rat		59
(3) DADLE (with 1% Na glycocholate)	Rat		94

Table 2A   Continued

Drugs	Animal model	T_{max}	Bioavailability (%)
(4) Mekephamid	Rat	10 min	102
GnRh	Human	<30–90 min (LH)	
		<90–120 min (FSH)	
Glucagon (with Na glycocholate)	Human	10 min	50
Growth hormone (hpGRF-40)	Human	<30 min	1–2
Met-hGH			
No enhancer	Rat	30 min	<1
With enhancer			
(1) BL-9	Rat	10–15 min	79 (0.2 mg/kg)
			65 (0.5 mg/Kg)
			57 (1 mg/Kg)
(2) Na glycocholate	Rat	10–15 min	7–8
Horseradish peroxidase	Rat	5–10 min	0.6
Insulin			
no promoter	Rat		5
with promoter	Rat		10
with saponin (1%)	Dog		30
with Na glycocholate (1%)	Dog		25–33
	Human	13.5 min	12.5
with Na deoxycholate	Human	10 min	10–20
with BL-9 (1%)	Human	<15 min	7–10
with Na carbenoxolone	Rat	15 min	15
with Na caprate	Rat	5 min	98
with Na caprylate	Rat	5 min	27
with K glycyrrhizinate	Rat	15 min	13
with Na glycyrrhizinate	Rat	5 min	27
with Na laurate	Rat	5 min	55
with STDHF	Rat	10 min	18
	Rabbit	10 min	5
	Sheep	10 min	16
Freezed-dried powder with car-bopol 934p	Dog	30 min	33
beta-Interferone	Rabbit	15–60 min	22
[D-Arg^2]Kyotorphin	Rat		47
LHRH	Human	1 h	1
Lypressin	Human		14
Nafarelin acetate	Rhesus monkey	<15 min	2
		30 min	5
Oxytocin	Human	<10–20 min	
	Human		<1–2
	Rabbit		1–10
	Human		1
Pentagastrin	Human		20–33
Secretin	Rat		10
SS-6	Rat	5–10 min	73
Substance P	Rat	2–5 min	
Thyroxine-releasing hormone (TRH)	Rat	15 min	20

Table 2B Biopharmaceutics Data for Some Non-Peptide-Based Pharmaceuticals

Drugs	Animal model	T_{max}	Bioavailability (%)	
			Nasal	Other
Buprenorphine	Rat	2–5 min	95	9.7 (i.d.)
	Human	5 min	48	
Clofilium tosylate	Rat	<10 min	69.6	1.3 (p.o.)
Cocaine	Human	15–60 min		
	Human	60–120 min		
	Human	30 min		
	Human	58 min (solution)		
		35 min (crystalline)		
	Human	37 min (64 mg)		
	Human	41 min (96 mg)		
Cromoglycate disodium	Rat	20 min	60 (plasma)	
		15–30 min	53 (bile)	
Dopamin	Rhesus monkey	15 min		
Diazepam	Humans	1 h	72–84	
Ergotamine tartrate	Rat	20 min	62	12.7 (i.d.)
Ergotamine tartrate	Rat	20 min	62	12.7 (i.d.)
(with caffeine)	Rat	20 min	65.4	5.1 (i.d.)
Gentamycin (with Na	Human	30 min (n = 4)		
glycocholate)		60 min (n = 3)		
Hydralazine				
At pH 3.0	Rat	30 min	127	
At pH 6.5	Rat	<10 min	83	
At pH 3.0 (+BL-9,	Rat	<10 min	113	
0.5%)				
Lorazepam	Humans	0.5–4 h	51	
Meclizine	Rat	8.5 min	51	8 (p.o.)
	Dog	12 min	89	22 (p.o.)
Naloxone	Rat	20 min	101	1.5 (i.d.)
Nitroglycerine	Human	1–2 min		
Norethisterone	Rhesus monkey	30 min		
	Rhesus monkey	5 min		
Phenol red				
With Na deoxycholate	Rabbit		98	
With BL-9	Rabbit		87	
With STDHF	Rabbit		83	
Progesterone	Rhesus monkey	<10 min		
	Rat	6 min	100	1.2 (i.d.)
	Rhesus monkey	5.5 min	91	
	Rabbit	15 min	88	10 (p.o.)
	Rabbit (spray)	5 min	82.5	7.9 (i.d.)
	Rabbit (device)	20–30 min	72.4	
Propranolol	Rat	5 min	100	15 (p.o.)
	Rat	6.3 min	99.9	19 (p.o.)
	Dog	5 min	103	7 (p.o.)
	Human	5 min	109	
Testosterone (25 mg)	Rat	<2 min	99	1 (i.d.)
(50 mg)			90	1 (i.d.)
Verapamil	Dog	5 min	37	13 (p.o.)

Table 3 Some Representative Surfactants and Bile Acid Salts Used to Enhance Nasal Drug Absorption

Drugs	Enhancers Used
Atropine	Sodium lauryl sulfate
Buserelin	Bacitracin
[Asu^{1,7}]-eel Calcitonin	Polyacrylic acid
Calcitonin	Sodium glycocholate
	Polyacrylic acid
Cholecystokinin	Sodium deoxycholate
Enkephalin analog (DADL)	Sodium glycocholate
Gentamycin	Sodium glycocholate
	Lysophosphatidyl choline
Glucagon	Sodium glycocholate
Met-hGH	Sodium glycocholate
	BL-9
Hydralazine	Sodium glycocholate
	BL-9
Insulin	Bl-9
	l-Alpha-lysophosphatidyl choline
	Disodium carbenoxolone
	Dipotassium glycyrrhizinate
	Saponin
	Sodium caprate
	Sodium caprylate
	Sodium deoxycholate
	Sodium glycocholate
	Sodium glycyrrhizinate
	Sodium laurate
	Sodium taurodihydrofusidate (STDHF)
Interferon	Azone
	Sodium cholate
	Sodium glycocholate
[D-Arg^2] Kyotorphin	Sodium glycocholate and oleic acid (or linoleic acid) mixed micelle
LHRH	STDHF
Phenol red	BL-9
	Sodium deoxycholate
	STDHF
Progesterone	Polysorbate 80
Scarlet fever toxin	Sodium taurocholate
Testosterone	Polysorbate 80

sumably due to the increase in drug residence time and higher local drug concentrations in the mucus lining on the mucosal surface. The bioadhesive polymers swell by absorbing water from the mucous layer in the nasal cavity, thereby forming a gel-like layer in which the polymer bonds with the glycoprotein in the mucus. Methyl cellulose and polyacrylic acid gel have been shown, in rat studies, to enhance nasal bioavailability of propranolol as well as irritate nasal mucosa with some morphological change [55]. A nasal formulation of meclizine (Mecnazone), prepared at a concentration of 50 mg/ml in 85% propylene

Table 4A Mechanisms of Action for Some Nasal Absorption Enhancers

Enhancers	Mechanisms of action
Bile salts (sodium glycocholate, sodium deoxycholate, etc.)	(1) Inhibit aminopeptidase activity in nasal mucosa. (2) Form transient hydrophilic pores in membrane bilayer. (3) Reduce viscosity of mucus. (4) Remove epithelial cells which constitute a major permeability barrier. (5) Drug is solubilized in bile salts micelles, thus creating a transmembrane concentration gradient. Reverse micellar structures may form within the mucosal lipid membranes, which act as temporary aqueous pores (Fig. 11)
Bile salts unsaturated fatty acid mixed micelle	Bile salts can solubilize fatty acid in mixed micelles, thus making it more available at the mucosal surface for absorption.
BL-9	Irreversibly removes membrane proteins or lipids.
EDTA	Increases paracellular transport by removal of luminal calcium, thus affecting the permeability of the tight junctions.
Fatty acid salts (sodium caprate, sodium laurate, etc.)	(1) Inhibit leucine aminopeptidase activity. (2) Act to create intercellular space by temporarily extracting calcium ions from the nasal mucosa.
Glycyrrhetic acid derivatives	Inhibit leucine amino-peptidase in nasal mucosa.
Lysophosphatidyl choline	Affects enzyme activity and morphological changes of nasal membranes.
STDHF	(1) Inhibits leucine aminopeptidase activity. (2) Forms peptide/adjuvant complex. (3) Facilitates paracellular transport from nasal cavity to capillary bed.

glycol and 10% glycerol, was administered to rats [56]. About 50% of the drug was absorbed, which was as effective as intravenous injection, but was about six times more effective than oral administration (8%). The mean times to peak plasma levels were about 8.5 min for the nasal administration and 49.0 min for oral delivery. Nagai et al. [53] demonstrated enhancement in the absorption of insulin after intranasal administration in dogs of a bioadhesive powder formulation containing hydroxypropyl cellulose and neu-

Table 4B Toxic Effects of Some Enhancers of Nasal Absorption

Enhancers	Toxic effects
Bile salts	(1) Have irritation and congestion effects on nasal mucosa (2) Decrease ciliary function (3) Cause ultrastructural abnormalities
BL-9	Has hemolytic and protein-releasing actions on nasal mucosa
EDTA	Has ciliostatic action on nasal mucosa
Fatty acid salts	Have hemolytic action on nasal mucosa
STDHF	Has ciliostatic action on nasal mucosa

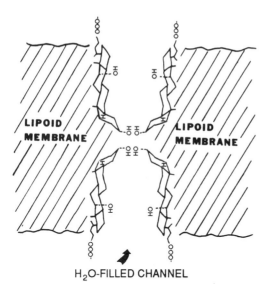

H$_2$O-FILLED CHANNEL

Figure 11 Schematic illustration of the molecule model for reverse micelle formation on cell membrane. Two pairs of sodium deoxycholate molecules are shown stacked end-to-end, spanning a lipid bilayer and forming an aqueous pore for the transport of insulin monomers from the extracellular space, where high concentrations of insulin monomers are solubilized in mixed bile salt and insulin micelles. (From Ref. 50.)

tralized cross-linked polyacrylic acid (Carbopol 934p). Similarly, Morimoto et al. [54] used a polyacrylic acid (0.1%) gel base to enhance the absorption of both insulin and calcitonin in rats by intranasal administration. Formulation pH in the range 4.5 to 7.5 did not appear to affect nasal absorption of insulin from the gel formulation; however, lower-viscosity gels provided more rapid onset of insulin activity, as measured by hypoglycemic response.

In the third approach to enhancement of transnasal drug delivery, microspheres that have good bioadhesive characteristics and swell easily in contact with the nasal mucosa are used [39,57]. This delivery system can control the rate of drug clearance from the nose and, by protecting the drugs from enzymatic degradation in the nasal cavity, increase the systemic bioavailability of drugs. The microspheres, made from bioadhesive polymers such as starch, albumin, gelatin, and dextran, have been retained in the nasal cavity with clearance half-lives increased to 3 h and longer [58]. Bioadhesive microsphere delivery systems have now been investigated for improving nasal bioavailability of several drugs, e.g., gentamycin, insulin, rose bengal, and cromoglycate disodium [58–61].

The fourth approach to enhancing nasal drug absorption involves the use of "physiological modifying agents" [39]. These agents have vasoactive properties and exert their action by increasing nasal blood flow. Agents known to increase nasal blood flow include histamine, leukotriene D$_4$, prostaglandin E$_1$, and the beta-adrenergic agents, isoprenaline and terbutaline. These "physiological modifying agents" represent a new class of transnasal absorption promoters.

Delivery Systems

A number of systems have been used to deliver drugs to the nasal cavity, e.g., nasal sprays, nose drops, saturated cotton pledgets, aerosol sprays, and insufflators [1].

The metered-dose nebulizer has recently been introduced. Activated mechanically, it delivers a predetermined volume into the nasal cavity. The dose of active ingredient delivered depends on the volume of solution released at each actuation and the concentration of drug in the formulation. This nebulizer has already been utilized successfully in the delivery of several topical drugs: corticosteroid (Extracort), beclomethasone dipropionate (Aldecin, Beconase, and Becotide), flunisolide (Nasalide), tramazoline (Tobispray), and nasal decongestant (Rhinospray). Also, this nebulizer has been explored for use with the following drugs: DDAVP, enviroxime, insulin, and nitroglycerine [1].

An inflatable nasal device having a microporous membrane wall was developed to provide controlled release of drugs from suspension formulations. Following insertion into the rabbit nasal passage, the device is inflated by filling with donor drug suspension (Fig. 12) [12]. The device, which conforms to the contour of the nasal passage, is left in place for a prolonged period. With progesterone as the model drug and ovariectomized rabbits as the animal model, the plasma profile of progesterone delivered by this controlled-release device was compared with that obtained from an immediate-release nasal spray. The nasal spray produced a peak plasma level of progesterone within 2 min, which indicated rapid absorption of progesterone by the nasal mucosa (Fig. 13) [12]. On the other hand, the controlled-release device produced a gradual increase in the plasma progesterone

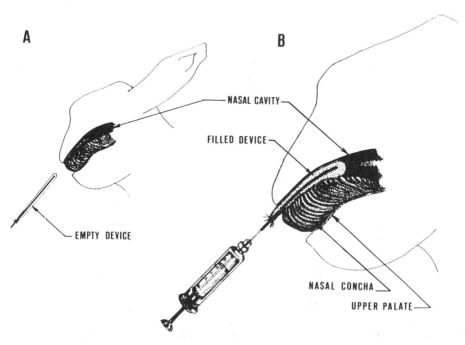

Figure 12 Illustration of controlled-release nasal delivery device (A) prior to insertion into the rabbit nasal passage, and (B) the in-situ filling of the nasal delivery device with the progesterone formulation in the nasal cavity. (From Ref. 12.)

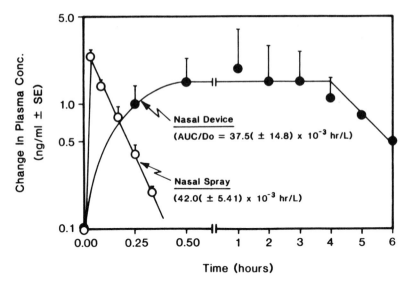

Figure 13 Change in plasma concentration of progesterone with time after intranasal administration of progesterone in the ovariectomized rabbits (n = 3) by: (○) immediate-release nasal spray (2 μg/kg) and (●) controlled-release nasal device (60 μg/kg). (From Ref. 12.)

concentration, which reached a plateau level within 20–30 min and remained at an elevated level throughout the 6-h insertion period (Fig. 13). The systemic bioavailabilities of progesterone following intranasal administration by the immediate-release spray formulation (82.5 ± 13.5%) and the controlled-release nasal device (72.4 ± 25.7%) did not differ statistically, but both were significantly greater than that resulting from oral administration (7.9 ± 1.6). The nasal data suggested a 9- to 10-fold reduction in the extent of hepatic "first-pass" metabolism (compared to oral administration), as well as substantial prolongation of plasma level (from 10 min to 300 min) with the inflatable device.

Sustained-release formulations for propranolol have been developed using methyl cellulose, and studies carried out on rats have achieved a low initial but prolonged blood level along with bioavailability equal to that following intravenous administration [11]. Hydrogel, such as carbopol 934p, might be useful for preparing a sustained-release nasal delivery system, since it provides long-term intimate contact with the nasal mucosa and thus a prolonged therapeutic effect without suppressing the mucociliary functions. Microspheres fabricated from various biodegradable materials have the potential to become controlled-release drug delivery systems for nasal administration [58–61]. Microspheres prepared in suitable sizes can be loaded with a variety of drugs, and the drugs can be released at a controlled rate at the desired site in the nasal cavity. Efficient trapping of the microspheres in the nose depends very much on the size and surface characteristics of the microspheres. Microspheres with particle sizes larger than 10–15 μm should be suitable for targetting deposition in the nasal cavity.

Nagai et al. reported the use of an adhesive powder dosage form and adhesive powder spray for the delivery of insulin nasally to rabbits and dogs [53,62]. The bioadhesive powder dosage form provided drug absorption that was more effective and less irritating than liquid forms. The adhesive powder spray gave a large reduction in plasma glucose

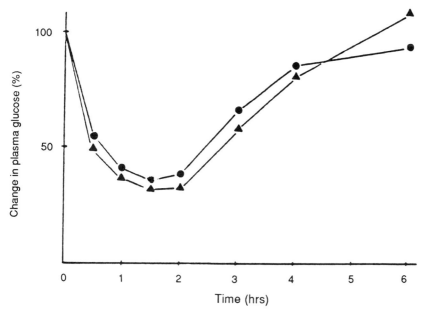

Figure 14 Plasma glucose level after intranasal administration of insulin preparation with (-●-) or without (-▲-) tracheal cannulization. Data are expressed as the mean of four determinations. (From Ref. 26.)

level, to almost the same extent, but without a great pH dependency (Fig. 14) [26,53]. Their final powder dosage form successfully produced a hypoglycemia one-third of that reached by intravenous injection. The nasal absorption of powdered insulin became fast and sustained in the preparation with the addition of crystalline cellulose, starch, and neutralized CM-Na. For the bioadhesive powder spray, the powder mixture was prepared with hydroxypropyl cellulose (HPC) as the bioadhesive base and it was applied by a special applicator (Pulizer). The advantages of HPC were as follows: (1) Drug dose may be reduced; (2) side effects may be lowered; and (3) the effect lasts longer. The powder swells and adheres to the mucosal membrane for as long as 6 h. Drug (beclomethasone) remained in the body very much longer than the drug formulation in regular nose drops.

Another means of nasal drug delivery uses a dosage form with better adhesion and less irritation, but which contains only a small amount or no absorption enhancer. Illum et al. [60] reported on bioadhesive microspheres that could not be cleared easily from the nasal cavity because of their intimate contact with the mucosa. In this system, albumin, starch, and DEAE-dextran could be used as the formulation base. Using Rose bengal and sodium cromoglycate as model drugs, they found it possible to control release of the drugs from the bioadhesive microsphere system. Studies have been conducted in human volunteers using a variety of systems containing microspheres with good bioadhesive properties. The microspheres were labeled with a technetium complex, using powder and solution forms as the controls, and nasal clearance was monitored using gamma scintigraphy. The microspheres prepared from starch and DEAE-Sephadex as the formulation bases were found to be effective in delaying nasal mucociliary clearance. The half-life of clearance for starch microspheres was on the order of 240 min as compared to 15 min for the control (liquid and powder) formulations. The albumin base was less effective than either of these two, but more effective than the powder and solution dosage forms. Not only could the drug be incorporated into such a system, but absorption enhancers could also be incorporated. Retention of drug and enhancer on the nasal mucosa for an extended period of time could lead to an improvement in systemic bioavailability of peptides and proteins.

Illum et al. [58] investigated the possibility of improving the bioavailability of gentamycin administered intranasally by means of a gelling microsphere delivery system in rats and sheep. They found that the combination of gentamycin with the microsphere delivery system increased uptake of the drug across the nasal membrane. Uptake was furthered enhanced by incorporation of an absorption enhancer, lysophosphatidyl choline (LPC), into the microspheres. Administration of gentamycin solution alone produced poor bioavilability, whereas addition of an enhancer gave rise to a fivefold increase in peak level (Fig. 15) [58]. If gentamycin was administered in combination with the starch microspheres, a significant increase in bioavailability was obtained (Fig. 16) [58]. An even more dramatic effect was observed when the gentamycin and enhancer were administered together in the starch microsphere formulation; the blood level peaked at 6.3 μg/ml compared to only 0.4 μg/ml for gentamycin solution. The combination of microspheres with the enhancer, LPC, produced a plasma concentration-versus-time profile very similar to that obtained when gentamycin was given intravenously (Fig. 16) [58]. The bioavailability for nasally administered gentamycin was increased to about the same as that of the intravenous dose, whereas bioavailability was less than 1% for a simple nasal gentamycin solution.

Björk and Edman [59] investigated a nasal delivery system for insulin in the rat which used degradable starch microspheres (DSM). Administered nasally as a dry powder, their preparations gave a dose-dependent reduction in blood glucose and a concomitant increase in serum insulin (Figs. 17 and 18) [59]. Blood glucose was reduced by 40% and 64%,

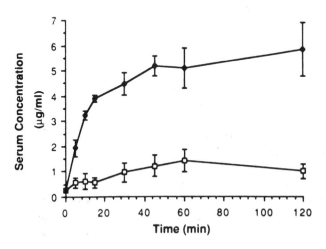

Figure 15 Effect of lysophosphatidylcholine on serum concentration (mean $\pm$ SEM) of gentamycin in rats following intranasal administration of gentamycin (5 mg/kg) (in-situ model) in a phosphate buffer solution (20 mg/ml) ($\square$) and with addition of lysophosphatidylcholine (2 mg/ml) ($\blacklozenge$). (From Ref. 58.)

combined system, the bioavailability of insulin was 31.5% of that obtained by subcutaneous administration. Thus, this system has potential for the nasal delivery of insulin.

Pharmacokinetics and Bioavailability

Factors reported to affect the pharmacokinetic parameters following intranasal administration include [1,63]:

1. Physiology-related factors
   (a) Speed of mucus flow
   (b) Presence of infection
   (c) Atmospheric condition
2. Dosage form-related factors
   (a) Concentration of drug
   (b) Physicochemical properties of drug
   (c) Density/viscosity properties of formulation
   (d) pH/toxicity of dosage form
   (e) Pharmaceutical excipients
3. Administration-related factors
   (a) Size of droplets
   (b) Site of deposition
   (c) Mechanical loss to the esophagus
   (d) Mechanical loss to other regions in the nose
   (e) Mechanical loss anteriorly from nose

The bioavailability of a drug after intranasal administration may be expressed in terms of absolute nasal absorption, Ae, determined from the area under the plasma concentration curve (AUC) following an intravenous (i.v.) and an intranasal (i.n.) dose:

$$Ae = \frac{(AUC)i.n. \ (Dose)i.v.}{(AUC)i.v. \ (Dose)i.n.} \tag{1}$$

where the AUC is extrapolated to an infinite time following the administration of a single intravenous or intranasal dose.

Zero-Order Transnasal Permeation Kinetics

In cases where the absorption of drugs from the nasal site of administration follows zero-order kinetics, e.g., a controlled delivery of drug at constant rate of absorption, the plasma profile of the drug may be described by

$$\frac{dX_B}{dt} = K_0 - K_e X_B \tag{2}$$

in which K_0 is the zero-order absorption rate constant, K_e is the overall rate constant for plasma elimination, and X_B is the amount of drug absorbed into the body or into the blood circulation (i.e., central compartment). The plasma concentration (C_p) of the drug may then be expressed as

$$C_p = \frac{K_0}{Cl}(1 - e^{-K_e t}) \tag{3}$$

where Cl is total body clearance and t is any specified time interval following drug administration. The plasma drug level following zero-order transnasal permeation of the drug increases to a steady-state plateau level [$C_{p(ss)}$] and then begins to decline exponentially after time t_p, the time when there is no more absorption of the drug from the nasal cavity.

First-Order Transnasal Permeation Kinetics

In cases where the absorption of drugs from the nasal site of administration follows first-order kinetics, the plasma profile of the drug can be described by

$$\frac{dX_B}{dt} = FX_{in}K_a - K_eX_B \tag{4}$$

where K_a is the first-order absorption rate constant, F is the fraction of applied dose absorbed, and X_{in} is the amount of drug administered intransally to the absorption site. Then, the plasma concentration of drug (C_p) can be expressed as

$$C_p = \frac{FX_{in}{}^0K_a}{V_d(K_a - K_e)}(e^{-K_et} - e^{-k_at}) \tag{5}$$

where X^0_{in} is the initial drug dose delivered intranasally at time zero and V_d is the volume of distribution.

The systemic bioavailability of drug following intranasal or oral administration can be calculated using the following example: The areas under the plasma concentration-versus-time curve (AUC) for naloxone following intravenous, intranasal, and oral administration of 2.5 mg/kg to three rats are 1498.7 ($\pm$121.9), 1517.5 ($\pm$193.5), and 22.0 ($\pm$7.1) ng/ml-min, respectively. The systemic bioavailability for nasal or oral administration can be calculated using Eq. (1) as follows:

$$F(\%)\text{i.n.} = \frac{1517.5}{1498.7} * \frac{2.5}{2.5} = 101$$

$$F(\%)\text{oral} = \frac{22.0}{1498.7} * \frac{2.5}{2.5} = 1.5$$

The pharmacokinetic data for the nasal absorption of drugs are summarized in Tables 2A and 2B. The plasma concentration following intranasal administration shows a concentration profile more similar to the intravenous bolus injection than to those following other nonparenteral routes. Drugs with poor oral absorption (e.g., sulbenicillin, cephacetrile, cefazoline, phenol red, and disodium cromoglycate) and drugs with extensive "first-pass" heptic metabolism (e.g., progesterone, estradiol, testosterone, insulin, hydralazine, propranolol, cocaine, buprenorphine, naloxone, and nitroglycerine) can be absorbed rapidly through the nasal mucosa with near 100% systemic bioavailability [1].

Several factors affect the systemic bioavailability of drugs, including the deposition and clearance of drug in the nasal cavity, and the rate of movement of drug into the nasal blood supply. Some compounds could have very satisfactory absorption characteristics if sufficient time were allowed for contact with the absorptive regions of the nasal mucosa. Therefore, poor nasal absorption of some drugs may not necessarily be related to insufficient permeability. The rapid clearance of drugs by the mucociliary clearance mechanism in the nasal cavity may be the factor affecting nasal bioavailability.

Fisher et al. [16] studied the relationship between the systemic bioavailability of hydrophilic compounds delivered nasally and their molecular weights. They found that

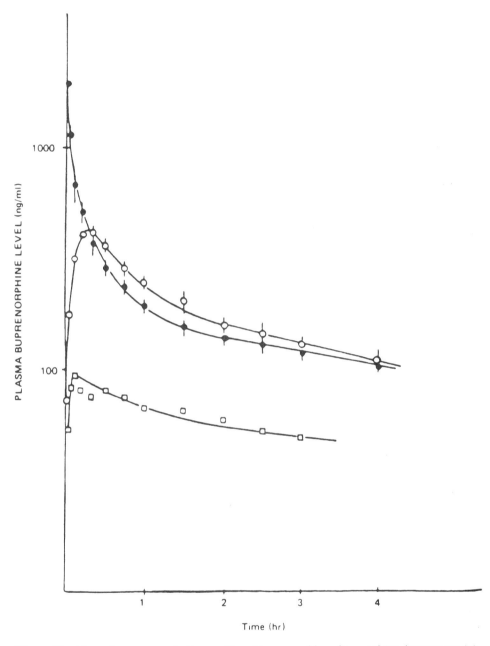

Figure 19 Mean plasma concentration profiles of buprenorphine after nasal (○), intravenous (●), and intraduodenal (□) administration of 30 μg of buprenorphine. Points represent mean (± SEM) of three animals. (From Ref. 72.)

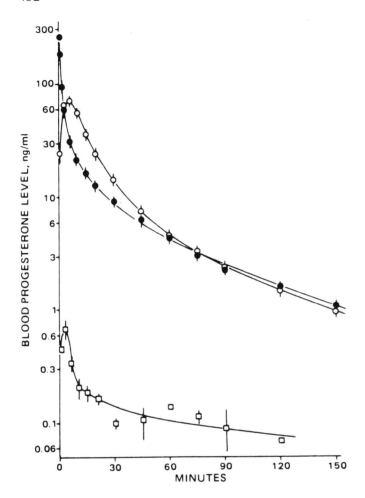

Figure 20   Mean blood progesterone concentration profiles in rats
after nasal (○), intravenous (●), and intraduodenal (□) administration
of progesterone (50 μg) in rats. The vertical bars give the standard
error. (From Ref. 73.)

uptake of the drugs decreased with increasing molecular weights; however, polar molecules
with relatively high molecular weights could still be taken up to a significant extent.
McMartin et al. [15] observed that the transport rate across the nasal membrane falls off
sharply for drugs with molecular weights above 1000. Low-molecular-weight drugs, such
as propranolol, buprenorphine, and steroids were well absorbed following administration
from a nasal spray formulation; the resultant plasma concentration-versus-time profiles
are almost indistinguishable from intravenous injections (Figs. 19–22).

Irritation and Toxicological Evaluations

The assessment of irritation and toxicity of drugs or formulations to the nasal mucosa is
difficult. Anatomical differences in nasal tissues between small laboratory animals and
humans are great, and therefore different results in disposition and concentration of drugs

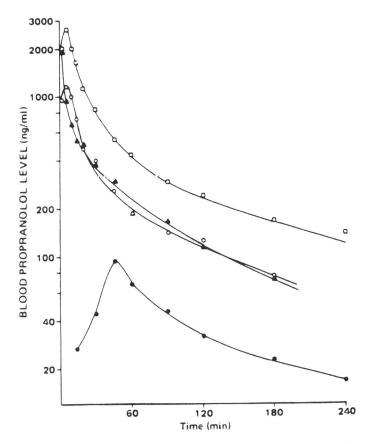

Figure 21 Average blood propranolol levels versus time in rats after intranasal administration of 1 mg/rat (○) and 2 mg/rat (□), intravenous administration of 1 mg/rat (▲), oral administration of 1 mg/rat (●). (From Ref. 55.)

administered are expected. Nevertheless, the nasal mucosa of experimental animals should be examined histologically to be sure that drugs administered do not do any damage to the absorption cells of the nasal mucosa [64]

Toxicological evaluations of intranasal drug administration present a number of unique challenges. In addition to evaluation of adverse local and systemic effects of the drug, other considerations deserving special attention are administration of an appropriate dose for clinical use, selection of concentration and volume of the test drug, and method of administration and formulation to be used in the animal models, in which nasal anatomy may differ greatly from that of humans [1,65].

Both the drug and the final nasal dosage form for clinical use must be submitted to both acute and subchronic toxicity studies. To avoid toxicity problems, ingredients used in nasal formulations are preferably GRAS (generally regarded as safe) materials. Should the toxicity of an excipient or an absorption enhancer that is added to the formulation be unknown, an extensive program designed to define the absorption, distribution, metabolism, excretion, and toxicity/carcinogenicity of the agent may be in order [1].

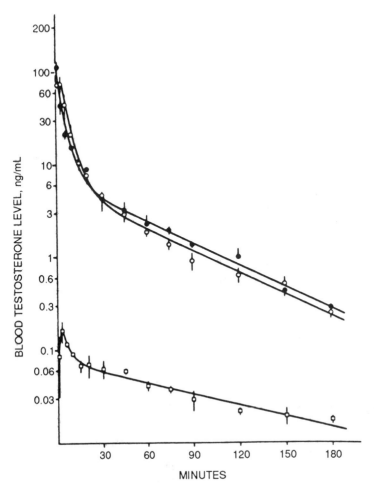

Figure 22 Mean blood levels of testosterone in rats following nasal (○), intravenous (●), and intraduodenal (□) administration of testosterone (25 μg/rat). Points are mean values of three animals (± SEM). (From Ref. 74.)

Subchronic toxicity studies usually include a range of doses (or concentrations) and require testing in two different animal species. Other studies, including fertility, teratology, and carcinogenicity, should also be considered. In general, a 30-day toxicity test on the finished product is sufficient to support a single-dose clinical trial, whereas a 90-day toxicity test will be needed to substantiate multiple-dose clinical trials. The nasal mucosa of the experimental animals should be examined histologically to ensure the integrity of the nasal mucosa. Local effects, such as irritation, cell damage, and mucociliary clearance, must also be evaluated. During a clinical study, gross examination of the nasal cavity of subjects should be conducted by an ENT specialist [1].

Long-term local toxic effects are even more important for the formulations where enhancers are used to increase nasal membrane permeability. Chronic erosion of the mucous membrane by certain surfactants could result in inflammation, hyperplasia, metaplasia, and deterioration of normal nasal function [52]. Certain bile salts are known to break down mucous membrane structures, accelerate phospholipid and protein release from the membranes, and damage interstitial mucosa.

INTRANASAL DELIVERY OF PEPTIDE/PROTEIN DRUG

Because of their physicochemical instability and susceptibility to hepato-gastrointestinal ''first-pass'' elimination, peptide/protein drugs are generally administered parenterally. Most nasal formulations of peptide/protein drugs have been made up in simple aqueous or saline solution with preservatives. Recently, more R&D work has been directed toward development of nasal drug delivery systems for peptides/proteins.

Currently, in the United States only four intranasal pharmaceutical products for systemic delivery have been marketed, i.e., desmopressin (DDAVP), lypressin (Diapid), oxytocin (Syntocinon), and nafarelin acetate (Synarel). If the therapeutic index of the peptide/protein is broad and its cost low, nasal delivery will be a viable alternative to parenteral administration. For peptides/proteins with a narrow therapeutic index and high bulk material cost, nasal delivery is not acceptable owing to their low bioavailabilities.

Table 5 Amino Acids, Peptides, Proteins, and Biological Products Being Studied for Nasal Delivery

A. Amino acids
B. Peptides
    1. Calcitonin
    2. Cerulein
    3. Cholecystokinin
    4. Enkephalins
    5. Kyotorphin
    6. Pentagastrin
    7. Secretin
    8. SS-6
    9. Substance P
    10. Thyrotropin-releasing hormone
C. Polypeptides and proteins
    1. Albumins
    2. Anterior pituitary hormones
        (a) Adrenal corticotropic hormone
        (b) Gonadotropin-releasing hormone
        (c) Growth hormones
    3. Biological products
        (a) Interferon
        (b) Vaccines
    4. Horseradish peroxidase
    5. Pancreatic hormones
        (a) Insulin
        (b) Glucagon
    6. Posterior pituitary hormones
        (a) Oxytocin
        (b) Vasopressin

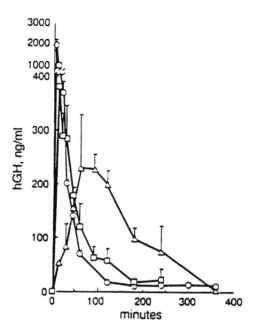

Figure 23 Plasma hGH concentration (ng/ml)
after i.n., i.m., and i.v. administration of Met-
hGH. Each point is the mean value (± SD) of four
determinations for i.v. and six determinations for
i.m. and i.n.: (□) 0.2 mg/kg with 1% BL-9 (i.n.);
(○), 0.2 mg/kg (i.v.); (△), 0.5 mg/kg (i.m.).
(From Ref. 67.)

The efficacy of peptide/protein delivered transnasally is highly dependent on the
molecular structure of the drugs and their size. Respiratory epithelial cells are capable of
absorbing peptide/protein by a vesicular transport mechanism, followed by transfer to the
extracellular spaces, and subsequent uptake by the submucosal vascular network [8,66].

The extent of systemic delivery of peptides/proteins by transnasal permeation may
depend on (1) structure and size of molecules, (2) partition coefficient, (3) susceptibility
to proteolysis by nasal enzymes, (4) particle residence time, and (5) formulation variability
(pH, viscosity, and osmolarity).

High-molecular-weight peptides, i.e., insulin, calcitonin, and growth hormone-releasing factor, all have poor stability in the nasal cavity and low systemic bioavailability.
Investigations have been initiated to concentrate on enhancement of the nasal absorption
of peptides using the following agents:

Viscosity-enhancing agents: hydroxymethyl cellulose, methyl cellulose, hydroxy-
    propyl cellulose, polyethylene glycol, propylene glycol, polyacrylic acid
Bile salts: sodium deoxycholate, sodium glycocholate
Surfactants: polyoxyethylene-9-lauryl ether
Enzyme inhibitors: aprotinin
Bioadhesive polymers: starch, albumin, gelatin

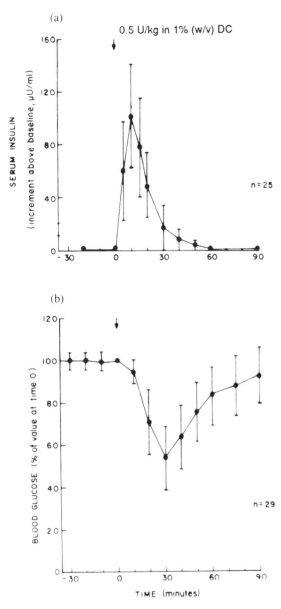

Figure 24 (a) Serum insulin concentration profile (mean ± SD) for 25 normal subjects who received a nasal spray of 0.5 IU/kg insulin in 1% (w/v) deoxycholate at time 0. (b) Blood glucose as percentage of the value at time 0 (mean ± SD) versus time for 29 normal subjects who received 0.5 IU/kg insulin in 1% (w/v) deoxycholate aerosol at time 0. (From Ref. 68.)

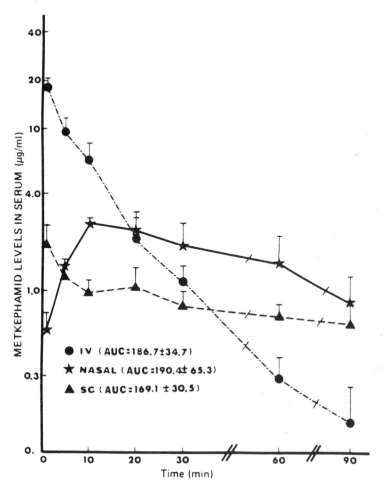

Figure 25 Serum concentration profiles of metkephamid after administration of a 25-mg/kg dose in rat (n = 3–4). Key: (●) intravenous administration; (▲) subcutaneous administration; (★) intranasal administration. No compound was detected after oral administration. (From Ref. 69.)

Peptide/proteins that have been studied for nasal delivery are listed in Table 5; their pharmacokinetic parameters and bioavailabilities are shown in Table 2A [1,39–52,67–71]. Several examples of the increase in nasal systemic bioavailability of peptides/proteins using absorption enhancers are shown in Figs. 23–25 [67–69].

INTRANASAL DELIVERY OF NONPEPTIDE DRUGS

Nonpeptide drugs investigated for intranasal delivery are listed in Table 6, and their biopharmaceutics data are shown in Table 2B. Drugs with extensive heptic "first-pass" metabolism, such as progesterone, estradiol, testosterone, hydralazine, propranolol, cocaine, naloxone, and nitroglycerine, can be rapidly absorbed through the nasal mocosa with systemic bioavailability of approximately 100% (Figs. 19–22) [55,72–74]. Non-

Table 6 Nonpeptide Drugs Being Studied
for Nasal Delivery

A. Adrenal corticosteroids
B. Antibiotics
    1. Aminoglycosides
        (a) Gentamycin
        (b) Streptomycin
    2. Cephalosporins
    3. Penicillins
    4. Tyrothricin
C. Antimigraine drugs
    1. Diergotamine
    2. Ergotamine tartrate
D. Antiviral agents
    1. Enviroxime
    2. Phenyl-*p*-guanidino benzoate (PGB)
E. Autonomic nervous system drugs
    1. Sympathomimetics
        (a) Dobutamine
        (b) Dopamine
        (c) Ephedrine
        (d) Epinephrine
        (e) Phenylephrine
        (f) Tramazoline
        (g) Xylometazoline
    2. Parasympathomimetics
        (a) Methacholine
        (b) Nicotine
    3. Parasympatholytics
        (a) Atropine
        (b) Ipratropium
        (c) Prostaglandins
        (d) Scopolamine
F. Cardiovascular drugs
    1. Angiotensin II antagonist
    2. Clofilium tosylate
    3. Hydralazine
    4. Isosorbide dinitrate
    5. Nitroglycerine
    6. Propranolol
    7. Verapamil
G. Central nervous system drugs
    1. Stimulants
        (a) Cocaine
        (b) Lidocaine
    2. Depressants
        (a) Diazepam
        (b) Lorazepam
H. Diagnostic drugs
    1. Dye T-1824
    2. Phenolsulfonphthalein
    3. Potassium ferrocyanide
    4. Vital dyes

Table 6   Continued

I.   Histamines and antihistamines
    1.   Histamines
    2.   Antihistamines
        (a)   Disodium cromoglycate
        (b)   Meclizine
J.   Inorganic compounds
    1.   Colloidal carbon
    2.   Colloidal gold
    3.   Colloidal silver
    4.   Inorganic salts
    5.   Lead carbonate
    6.   P and thorium B
K.   Narcotics and its antagonists
    1.   Buprenorphine
    2.   Naloxone
L.   Sex Hormones
    1.   Estradiol
    2.   Progesterone
    3.   Norethindrone
    4.   Testosterone
M.   Vitamins

---

peptide, water-soluble compounds such as sodium cromoglycate were well absorbed, their nasal absorption likely to be dependent on diffusion in aqueous channels.

PROBLEMS

1.   If you are developing a nasal formulation for a peptide/protein drug with a short half-life and high molecular weight, what will be your approaches to achieving a high systemic bioavailability for this peptide/protein?

2.   If you are developing a nasal formulation for a peptide drug with a low molecular weight and water-soluble properties, what are the factors that will affect systemic bioavailability of this peptide following nasal administration?

3.   What are the mechanisms of action for the following compounds? (a) sodium glycocholate, (b) lysophophatidyl choline (LPC), (c) BL-9, (d) STDHF, and (e) sodium caprylate.

4.   What will be the FDA's concern for nasal formulations containing a high-molecular-weight protein?

5.   The nasal absorption of propranolol has been evaluated and compared with intravenous and oral administration in humans. Using the following table, calculate the systemic bioavailability of propranolol following nasal and oral administration (using intravenous administration as the control), and explain your results.

Route	Dose (mg)	AUC (ng/h)/ml	F (%)
Intravenous	10	175.4	100
Nasal	10	190.3	
Oral	80	349.5	

REFERENCES

1. Y. W. Chien, K. S. E. Su, and S.-F. Chang, *Nasal Systemic Drug Delivery*, Marcel Dekker, New York/Basel (1989).
2. Y. W. Chien and S. F. Chang, *Crit. Rev. Ther. Drug Carr. Sys.*, *4*:67–194 (1987).
3. Y. W. Chien and S. F. Chang, in *Transnasal Systemic Medications: Fundamental Concepts and Biomedical Assessments* (Y. W. Chien, Ed.), Elsevier, Amsterdam, pp. 1–99 (1985).
4. S. F. Chang and Y. W. Chien, *Pharm. Int.*, *5*:287 (1984).
5. D. F. Proctor, *Am. Rev. Respir. Dis.*, *115*:97 (1977).
6. G. Ewert, *Acta Otolaryngol. (Stockh.)*, (Suppl.), *200*:1 (1965).
7. G. A. Laurenzi, *J. Occup. Med.*, *15*:174 (1973).
8. R. E. Stratford, Jr. and V. H. L. Lee, *Int. J. Pharm.*, *30*:73 (1986).
9. E. L. Smith, R. L. Hill, and A. Borman, *Biochim. Biophys. Acta*, *29*:207 (1958).
10. D. F. Proctor, in *Transnasal Systemic Medications* (Y. W. Chien, Ed.), Elsevier, Amsterdam, pp. 101–106 (1985).
11. A. Hussain and R. Bawarshi, *J. Pharm. Sci.*, *69*:1411 (1980).
12. D. C. Corbo, Y. C. Huang, and Y. W. Chien, *Int. J. Pharm.*, *46*:133 (1988).
13. D. C. Corbo, Y. C. Huang, and Y. W. Chien, *Int. J. Pharm.*, *50*:253 (1989).
14. S.-F. Chang, L. C. Moore, and Y. W. Chien, *Pharm. Res.*, *5*:718 (1988).
15. C. McMartin, L. E. F. Hutchinson, R. Hyde, and G. E. Peters, *J. Pharm. Sci.*, *76*:535–540 (1987).
16. A. N. Fisher, K. Brown, S. S. Davis, G. D. Parr, and D. A. Smith, *J. Pharm. Pharmacol.*, *39*:357–362 (1987).
17. A. A. Hussain, R. Bawarshi-Nassar, and C. H. Huang, in *Transnasal Systemic Medications* (Y. W. Chien, Ed.), Elsevier, Amsterdam, pp. 121–137 (1985).
18. S. Hirai, T. Ikenaga, and T. Matsuzawa, *Diabetes*, *27*:296 (1978).
19. S. Hirai, T. Yashiki, T. Matsuzawa, and H. Mima, *Int. J. Pharm.*, *7*:317 (1981).
20. Y. Kaneo, *Acta Pharm. Suec.*, *20*:379 (1983).
21. R. E. Gibson and L. S. Olanoff, *J. Control. Rel.*, *6*:361–366 (1987).
22. M. A. Wheatly, J. Dent, E. B. Wheeldon, and P. L. Smith, *J. Control. Rel.*, *8*:167 (1988).
23. P. Tengamnuay and A. K. Mitra, *Life Sci.*, *43*.585 (1988).
24. A. N. Fisher, K. Brown, S. S. Davis, G. D. Parr, and D. A. Smith, *J. Pharm. Pharmacol.*, *37*:38–41 (1985).
25. T. Unno, Y. Okude, O. Yanai, and S. Onodera, *Jpn. J. Otol. (Tokyo)*, *85*:277 (1982).
26. R. F. Hounan, A. Black, and M. Walsh, *Aerosol Sci.*, *2*:47 (1971).
27. B. O. Stuart, *Arch. Intern. Med.*, *131*:60 (1973).
28. C. D. Yu, R. E. Jones, J. Wright, and M. Henesian, *Drug. Devel. Ind. Pharm.*, *9*:473 (1983).
29. C. D. Yu, R. E. Jones, and M. Henesian, *J. Pharm. Sci.*, *73*:344 (1984).
30. J. Wolfsdorf, D. L. Swift, and M. E. Avery, *Pediatrics*, *43*:799 (1969).
31. F. A. Fry and A. Black, *J. Aerosol. Sci.*, *4*:113 (1973).
32. N. Mygind, *Nasal Allergy*, 2nd ed., Blackwell, Oxford, pp. 260–262 (1979).
33. J. G. Hardy, S. W. Lee, and C. G. Wilson, *J. Pharm. Pharmacol.*, *37*:294 (1985).
34. F. Y. Aoki and J. C. W. Crawley, *Br. J. Clin. Pharmacol.*, *3*:869 (1976).

35. G. W. Hallworth and J. M. Padfield, *J. Allergy Clin. Immunol.*, 77:348–353 (1986).
36. S. P. Newman, F. Moren, and S. W. Clarke, *Rhinology*, 25:77–82 (1987).
37. S. L. Whaley and S. Renken, B. A. Muggenburg, and R. K. Wolff, *J. Toxicol. Environ. Hlth.*, 23:519–525 (1988).
38. I. Gonda and E. Gipps, *Pharm. Res.*, 7:69–75 (1990).
39. L. S. Olanoff and R. E. Gibson, in *Conrolled-Release Technology, Pharmaceutical Applications* (P. I. Lee and W. R. Good, Eds.), American Chemical Society, Washington, D.C., pp. 301–308 (1987).
40. M. Mishma, Y. Wakita, and M. Nakano, *J. Pharmacobio-Dyn.*, 10:624–631 (1987).
41. J. P. Longnecker, A. C. Moses, J. S. Flier, R. D. Silver, M. C. Carey, and E. J. Dubov, *J. Pharm. Sci.*, 76:351–355 (1987).
42. M. J. M. Deurloo, W. A. J. J. Hermens, S. G. Romeyn, J. C. Verhoef, and F. W. H. M. Merkus, *Pharm. Res.*, 6:853–856 (1989).
43. B. J. Aungst and N. J. Rogers, *Pharm. Res.*, 5:305–308 (1988).
44. L. Illum, N. F. Farraj, H. Critchley, B. R. Johanson, and S. S. Davis, *Int. J. Pharm.*, 46:261 (1988).
45. M. Mishima, S. Okada, Y. Wakita, and M. Nakano, *J. Pharmacobio-Dyn.*, 12:31–36 (1989).
46. B. J. Aungst, N. J. Rogers, and E. Shefter, *J. Pharmacol. Exp. Therap.*, 244:23–27 (1988).
47. E. Hayakawa, A. Yamamoto, Y. Shoji, and V. H. L. Lee, *Life Sci.*, 45:167–174 (1989).
48. G. S. M. J. E. Duchateau, J. Zuidema, and S. W. J. Basseleur, *Int. J. Pharm.*, 39:87–92 (1987).
49. S. C. Raehs, J. Sandow, K. Wirth, and H. P. Merkle, *Pharm. Res.*, 5:689–693 (1988).
50. G. S. Gordon, A. C. Moses, R. D. Silver, J. S. Flier, and M. C. Carey, *Proc. Natl. Acad. Sci. (U.S.A.)*, 82:7419 (1985).
51. W. A. J. J. Hermens, P. M. Hooymans, J. C. Verhoef, and F. W. H. M. Merkus, *Pharm. Res.*, 7:144–146 (1990).
52. W. A. Lee and J. P. Longneckcr, *Biopharm*, 1:30–37 (1988).
53. T. Nagai, Y. Nishimoto, N. Nambu, Y. Suzuki, and K. Sekine, *J. Control. Rel.*, 1:15 (1984).
54. K. Morimoto, K. Morisaka, and A. Kamada, *J. Pharm. Pharmacol.*, 37:134 (1985).
55. A. A. Hussain, S. Hirai, and R. Bawarshi, *J. Pharm. Sci.*, 68:1196 (1979).
56. Y. Kaneo, *Acta Pharm. Suec.*, 20:379 (1983).
57. S. S. Davis, L. Illum, D. Burges, J. Ratcliffe, and S. N. Mills, in *Controlled-Release Technology, Pharmaceutical Applications* (P. I. Lee and W. R. Good, Eds.), American Chemical Society, Washington, D.C., pp. 201–213 (1987).
58. L. Illum, N. Farraj, H. Critchley, and S. S. Davis, *Int. J. Pharm.*, 46:261–265 (1988).
59. E. Björk and P. Edman, *Int. J. Pharm.*, 47:233–238 (1988).
60. L. Illum, H. Jorgensen, H. Bisgaard, O. Krogsgaard, and N. Rossing, *Int. J. Pharm.*, 39:189–199 (1987).
61. N. F. Farraj, L. Illum, S. S. Davis, and B. R. Johansen, *Diabetologia*, 32:486A (1989).
62. T. Nagai and Y. Machida, in *CRC Bioadhesive Drug Delivery Systems* (V. Lenaerts and R. Gurney, Eds.), CRC Press, Boca Raton, Fla., p. 170–178 (1990).
63. J. L. Colaizzi, in *Transnasal Systemic Medications* (Y. W. Chien, Ed.), Elsevier, Amsterdam, pp. 107–119 (1985).
64. K. S. E. Su, K. M. Campanale, and C. L. Gries, *J. Pharm. Sci.*, 73:1251 (1984).
65. M. A. Dorato, Nasal delivery of peptides and proteins: Toxicologic considerations, presented at a symposium on Nasal Administration of Peptide and Protein Drugs, Princeton, New Jersey (October 1987).
66. J. Richardson, T. Bouchard, and C. C. Ferguson, *Lab. Invest.*, 35:307–314 (1976).
67. A. L. Daugherty, H. D. Liggitt, J. G. McCabe, J. A. Moore, and J. S. Patton, *Int. J. Pharm.*, 45:197 (1988).
68. A. C. Moses, G. S. Gordon, M. C. Carey, and J. S. Flier, *Diabetes*, 32:1040 (1983).

69. K. S. E. Su, K. M. Campanale, L. G. Mendelsohn, G. A. Kerchner, and C. L. Gries, *J. Pharm. Sci.*, *74*:394 (1985).

70. B. Tarquini, V. Cavallini, A. Cariddi, M. Checchi, V. Sorice, and M. Cecchettin, *Chronobio. Int.*, *5*:149–152 (1988).

71. Y. Maitani, T. Igawa, Y. Machida, and T. Naigi, *Drug Design and Delivery*, *1*:65–70 (1986).

72. A. A. Hussain, R. Kimura, C. H. Huang, and T. Kashihara, *Int. J. Pharm.*, *21*:233 (1984).

73. A. A. Hussain, S. Hirai, and R. Bawarshi, *J. Pharm. Sci.*, *70*:466 (1981).

74. A. A. Hussain, R. Kimura, and C. H. Huang, *J. Pharm. Sci.*, *73*:1300 (1984).

10

Drug Delivery in Veterinary Medicine

John R. Cardinal and Leonore C. Witchey-Lakshmanan   *Merck Sharp & Dohme Research Laboratories, Rahway, New Jersey*

INTRODUCTION

This chapter reviews a number of controlled drug delivery systems in use or under development in the veterinary field. The information covers both food-producing animals and companion pets. This summary should provide an understanding of the state of the art for controlled drug delivery in animals, and an appreciation for the unique challenges in drug delivery for those interested in this field.

The veterinary field is probably not familiar to the average reader, but it is one for which controlled-release drug delivery system technology offers numerous benefits and opportunities to the end user. Tables 1 and 2 are a summary of the primary markets and formulations utilized in the veterinary area. The markets can be broken down into two separate sectors: "food-chain" animals (e.g., cattle, sheep, swine, and poultry), and "companion" animals (e.g., dogs and cats). Therapeutic approaches include products for growth promotion, such as antibiotics, mineral supplements, and the like, as well as products for disease prevention. Therapeutic antibacterials dominate the disease-prevention market, but products for parasitic diseases, such as coccidiosis in chickens and various ecto- and endoparasitic diseases in cattle, sheep, and companion animals, also play a major role. Overall, the food-producing animal is the major sector of the total market, with approximately 90% of total sales.

The reader with a background in human pharmaceuticals will recognize that this product mix is vastly different from that seen in the human field. The area of disease prevention, particularly antibiotics, is common to both human and veterinary practices. However, products for the health and well-being of humans, such as cardiovasculars, anti-inflammatories, antipsychotics, antitumor agents, etc., dominate the human market but occupy only a minor role in the veterinary field. Over the next decade, it is likely

Table 1 World Animal Health Product Market,
1986

Category	Sales ($ million)	Percent of total sales
Anthelmintics	706	10.7
Ectoparasiticides	416	6.3
Antibacterials	1,069	16.2
Anticoccidials	485	7.3
Biologicals	825	12.5
Feed additives	2,155	32.7
Others	944	14.3
Total	6,600	100.0

that products from many of these therapeutic areas will be introduced in the veterinary field, especially in the companion-animal sector; however, it is unlikely that the overall dominance of products for growth and disease prevention will change.

A summary of the animal health market by formulation approach is presented in Table 2. Given that products for growth promotion dominate the market and that the target species are food animals, it is not surprising that feed premixes play a dominant role in the formulation work. However, that injectables have such a dominant role may be surprising. The reason for this is that injectables are the most convenient route of administration. If you doubt that, consider the challenges presented in getting a cow (or a dog) to swallow a tablet or capsule! After injectables, oral and topical products provide the most important means of drug administration to animals.

A number of important issues (Table 3) arise in animal health that differentiate this field from that of human health. First and foremost is product cost. With the possible exception of the companion-animal market, the cost/benefit ratio drives the animal health market. In particular, the producer will critically determine the potential economic benefit from the use of, for example, a mineral supplement for the weight gain of the animal. The product will be used only if the economic gain from use of the product is greater

Table 2 U.S. Animal Health Market—By Formulation (FDA-Approved Pharmaceuticals and USDA-Approved Biologicals)

Category	1987 Sales, $mm, manufacturer's level	Percent
Feed premixes	510	35
Injectables	490	33
Oral, tab, cap, boluses	137	9
Oral liquids/powders	123	8
Topicals	77	5
Implants	52	4
Paste/gels	42	3
Intramammary	31	2
Other	10	1
Total	1,417	100

Table 3 Animal Health Drug Delivery Regulatory/Husbandry Issues

Cost
Ease/speed of administration
Frequency of dosing (handling)
Tissue residue
Side effects
Environmental/handler safety
Needs based on slaughter procedures

than its initial cost. An important component of this decision will be the ease of administration, and the resulting impact on the overall cost for the use of the product. For example, if grazing animals must be specifically rounded up for drug administration, then significant labor costs are added to the total product cost. Such factors must be critically examined in the design of any product for use in animals.

Another issue unique to food-producing animals is the question of tissue residues. These are the levels of drug remaining in edible tissues at the time of slaughter. To ensure that these levels cause no danger to humans, the time required post-administration for drug tissue levels to decline to nontoxic levels must be ascertained. Slaughter of the animal prior to that date is not permitted. This issue, therefore, strongly affects the economics of the product, and, in particular, it influences the formulation approaches that can be considered for a given application.

Other issues include side effects, environmental/handler safety, and slaughter procedures. Side effects are certainly not unique to animals, but the spectrum of these effects may be vastly different from those in humans. Environmental/handler safety issues reflect the impact of, for example, product run-off onto fields or into streams from a topical solution, or accidental contact by the person administering the dosage form. All of these possibilities must be evaluated and considered in the design of delivery systems for animals. The final point in Table 3 is the issue of slaughter procedures and how this reflects, for example, the need to ensure that the product will not harm equipment used in the slaughter of animals. The metals used in some ruminal boluses have been of particular concern in this regard.

One of the major advantages that have been extolled for "novel" drug delivery systems has been that of extending product life through formulation. Nowhere has this concept been more visible than in the animal market. Minor advances in terms of product duration, which decrease the number of times that a product is administered, can have a major impact in the marketplace. Perhaps the best example of this is Terra LA, first marketed by Pfizer. This is an injectable form of oxytetracycline for cattle. The conventional form provided efficacious blood levels for 1 day. The sustained-release product, Terra LA, extended the duration of efficacious blood levels to approximately 3 to 5 days. This enhanced duration was achieved through the use of various solvents that control drug disposition at the injection site. Given that a common usage of this produce was for the prevention of shipping fever, it is apparent that this prolonged duration minimized the number of times that animals had to be treated during this stressful period. This change provided a major economic benefit and helped establish Terra LA as the dominant product in the field.

Numerous other products have been developed that further establish the benefits of controlled drug delivery to animals. Examples include once-a-season oral boluses for cattle, topical products for cattle and sheep, flea collars for dogs and cats, and biodegradable implants for various applications in both food-producing and companion-animal markets. Products from most of these areas will be described in this chapter. It will be demonstrated that the economic benefit, animal husbandry practices, and the unique physiological features of animals provide a marketplace wherein controlled-release dosage forms are often the preferred dosage form. This is in contrast to the field of human pharmaceuticals, where the controlled-release form is generally a product-line extension of conventional oral tablets and capsules.

In this chapter, several approaches that have been utilized to achieve controlled drug delivery in the veterinary field will be described. Focus will be given to applications in the oral and parenteral areas. Delivery systems for vaccines and feed additives are of broad utility in the field, but of little academic interest; they will not be discussed specifically in this chapter. Topicals utilize many concepts common to the transdermal field, as discussed in Chapter 8; these discussions will not be repeated. Canine and feline flea collars represent a topical system of importance in the companion-animal market; they will be discussed in detail.

ORAL FORMULATIONS

Ruminal Boluses

Ruminants, including sheep and cattle, are a class of animals of great economic importance that spend much of their life grazing in open fields. As such, they present significant challenges to the drug delivery scientist, who must develop methods of administration that assure disease protection for extended periods and minimize handling during the grazing season.

These animals have several physiological features that provide important opportunities for drug delivery. First and foremost is the physiology of the stomach. Ruminants have a stomach composed of four compartments, namely, the rumen, reticulum, omasum, and abomasum, as illustrated in Fig. 1. The largest of these compartments is the rumen, which can be viewed as a large fermentation tank. Food entering through the esophagus is broken down through the action of bacteria to low-molecular-weight compounds capable of being absorbed in the lower portions of the gastrointestinal (GI) tract. To aid in this process, food is regurgitated by the animal for additional mechanical breakdown by chewing. Devices placed in the rumen can be easily regurgitated along with the food. A means to prevent regurgitation is critical to the design of systems intended for long-term delivery in the rumen. This prevention is normally achieved by producing devices with high density (>2.75 g/ml), [1], or by producing devices that expand in some physical dimension upon entry in the rumen. These measures ensure that the device cannot reenter the esophagus [2]. Approaches that utilize both options will be described in this chapter.

A second feature is that ruminants are, in general, large animals. As illustrated in Fig. 2, drug delivery systems intended for use in these animals also tend to be large. Depicted in this figure are two ruminal boluses, the Paratect Flex, which is a controlled-release bolus for cattle, and a conventional capsule for humans. The size differences are striking. The Paratect Flex bolus utilizes a polymeric sheet which unrolls following administration to increase its dimensions and thereby prevent its regurgitation. Heavy-metal salts are often incorporated into the conventional tablets to ensure retention in animals.

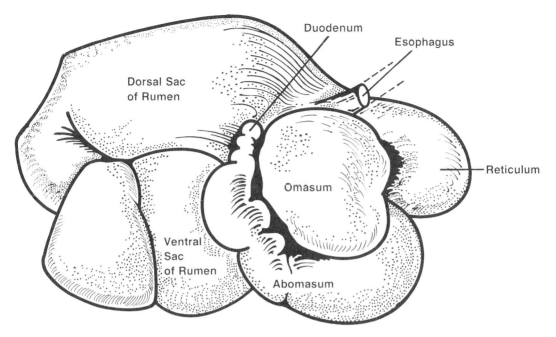

Figure 1 Diagram showing the stomach of ruminants.

An example is the use of $BaSO_4$ for the tetracycline ruminal bolus described by Hare et al. [3,4].

Paratect Bolus

The Paratect bolus is composed of a stainless steel cylinder encompassing a central reservoir containing a paste of morantel tartrate, an anthelmintic for control of gastrointestinal worms, and polyethylene glycol 400 [5]. The ends of the cylinder are capped with rate-controlling membranes prepared from a porous polyethylene sintered disk filled with cellulose triacetate. In use, these porous disks become impregnated with water and form the rate-limiting barrier to drug transport. As such, the Paratect bolus is a simple reservoir

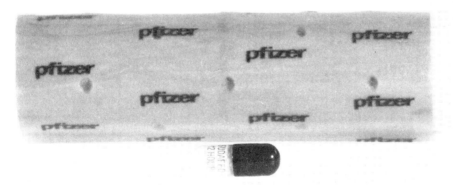

Figure 2 Comparison showing relative size of drug delivery devices for cattle (PARATECT Flex bolus) and for humans (Sudafed capsule).

device with two rate-limiting porous membranes. Transport of the highly water-soluble salt, morantel tartrate, occurs within the water filled-channels of the membrane.

The Paratect bolus was designed to treat diseases arising from exposure of grazing cattle to gastrointestinal nematodes. These parasites, endemic to pastures in nonarid climates, enter the animal through feeding. The parasites have various stages in their life cycle and are normally present on the fields throughout the grazing season. The severity of disease is related to the level of exposure and infection in cattle. Conventional treatment is to treat the herd with therapeutic doses of an anthelmintic at the first sign of clinical disease. The clinical sign associated with an infestation with these parasites is significant weight loss, or death in severe cases, if left untreated.

The Paratect bolus is designed to treat cattle with prophylactic doses of the anthelmintic throughout the grazing season. The device delivers at least 75 mg of drug on a daily basis for approximately 90 days [5]. The clinical performance of this device in terms of efficacy against gastrointestinal nematodes is shown in Fig. 3 [6,7], where the eggs per gram of feces are plotted as a function of time for treated and control animals. These values provide a measure of the level of parasite infestation in cattle. The pathology of this disease is a function of the number of larvae on the field. When levels of larvae are high, significant numbers of eggs are found in the feces of untreated controls (group 1). The corresponding treated animals from the same study (group 2) have limited numbers of eggs in the feces. In a second study, both control (group 3) and treated animals were found to have limited egg burdens during the early portion of the grazing season. Subsequently, larval burdens in the field increased, and significant increases in fecal egg counts were found in untreated controls. Treated animals had low egg burdens throughout the grazing season. This exposure is reflected in the weight gain observed with these animals, as shown in Fig. 4. Note that the untreated controls from group 1 show significantly lower weight gains than the other groups. The second control group did not show significant loss in performance suggesting a buildup in immunity, which can occur upon exposure to low levels of these parasites. The development of immunity is also a characteristic of the performance of the Paratect bolus per se in that the exposure of the animal to larvae in treated animals is

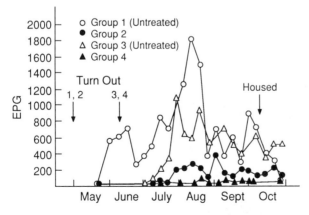

Figure 3   Egg output per gram of feces (EPG) versus time for animals treated with the PARATECT bolus at turnout for treated and control (untreated) animals. Results of two studies are summarized. (From Ref. 7.)

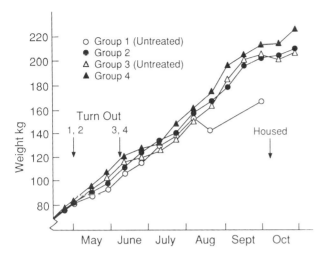

Figure 4 Plots of live weight gain versus time for treated and control animals described in Fig. 3. (From Ref. 7.)

sufficient for immunity to develop during the first grazing season. Therefore, through the combination of drug treatment and the natural development of immunity, a single administration of the Paratect bolus is sufficient to provide lifetime protection from gastrointestinal parasites.

Paratect Flex Bolus

A second bolus designed to deliver morantel tartrate to grazing cattle, termed the Paratect Flex Bolus, is illustrated in Fig. 5 [8]. This device is a trilaminate design wherein a core

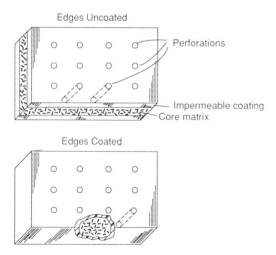

Figure 5 Diagrams of the Paratect Flex bolus showing options with either coated or uncoated edges. For administration to animals, the device is coiled in the form of a cylinder. (From Ref. 8.)

matrix, containing a 50:50 mixture of morantel tartrate and ethylene vinyl acetate (EVA), is coated on its outer surfaces (but not the edges) with a layer of pure EVA. Morantel tartrate is a water-soluble salt and cannot permeate EVA. Release of drug occurs via diffusion within water-filled channels of the porous core matrix to the uncoated edges of the device, or to the uncoated edges of a series of holes punched through the matrix. For administration, the device is rolled along its long axis to produce a cylinder similar in dimension to the Paratect bolus. A piece of tape with a water-dissolvable adhesive is utilized to hold this shape during passage down the esophagus. Upon entry into the rumen, the adhesive dissolves and the device unrolls, thereby preventing its regurgitation.

For a trilaminate whose edges are coated so that release occurs only through the holes, the release profile can be described by [8]

$$Mr = N(\lambda^2 - n^2)L\pi\epsilon\left(\rho - Cs + \frac{Cs}{2[h\epsilon/n\tau + \ln(\lambda/n)]}\right) - \frac{NL\pi\epsilon Csn^2}{(h\epsilon/n\tau) + \ln(\lambda/n)}\ln\frac{\lambda}{n} \quad (1)$$

and

$$t = \frac{[(h\epsilon/n\tau) - 1/2](\lambda^2 - n^2)\,p\tau + \lambda^2\tau p\,\ln(\lambda/n)}{2DCs} \quad (2)$$

where

Mr = total amount of drug released
N = number of perforations
t = time
λ = radius of depletion zone
n = radius of perforation
L = thickness of the core matrix
ϵ = porosity of the core matrix
τ = tortuosity of the core matrix
p = density of the drug
Cs = saturation solubility of the drug in the dissolution medium
D = diffusivity of the drug in the dissolution medium
h = effective diffusion-layer thickness

In spite of the apparent complexity of release, this design offers several advantages in the control of the rate and duration of drug delivery [9]. Similar to the hemispherical device proposed by Rhine et al. [10], or the various inward-releasing cylinder options described in the literature over the past few years [11–14], this design offers the advantage of a relatively constant release rate over time due to the compensating effects of increasing surface area at the zone of depletion with increasing diffusional distance. The extent of this effect and the relative constancy of the release rate depends on the diameter of the hole(s), as shown in Figs. 6 and 7. As shown (Fig. 6), the amount released increases with an increase in the perforation diameter, n, due to an increase in the average surface area at the zone of depletion at any given time t. As illustrated in Fig. 7, the apparent steady-state release rate at any time t also increases with increasing perforation diameter.

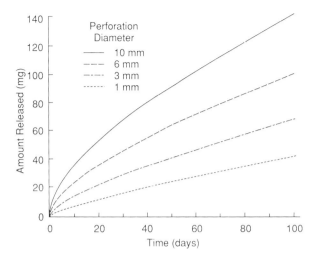

Figure 6 Plots of calculated amount released verses time for the trilaminate device as a function of the diameter of the perforation. (From Ref. 8.)

An additional advantage of the trilaminate design is that the total release from a device of a given thickness and drug load can be controlled by the number of holes, as demonstrated in Fig. 8. In this graph, the symbols represent in-vitro data on the release of morantel tartrate and the solid lines represent the model calculations developed with Eqs. (1) and (2). As shown, when the number of holes, N, of a given diameter increases, the overall output increases. This greater output is due to the increase in the total surface area at the zone of depletion with an increase in the number of holes. Hence, this device is very

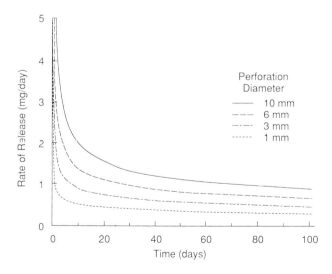

Figure 7 Plots of the calculated rates of release versus time for the trilaminate device as a function of the perforation diameter. (From Ref. 8.)

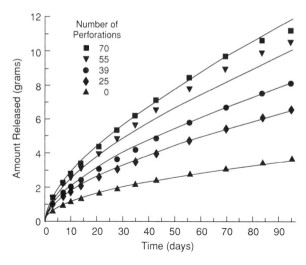

Figure 8 Plots of the amount of morantel tartrate released versus time as a function of the number of perforations in the Paratect Flex Bolus. Solid lines are calculated based on Eqs. (1) and (2). (From Ref. 8.)

versatile, because various methods can be used to control the release profile by controlling the surface area available for diffusion.

Another means of varying the overall release rate from these devices is drug solubility. When the variations in the number of holes is combined with drug solubility, wide differences in the rate and duration of drug delivery can be achieved. Predicted release rates are presented in Table 4 for devices with 196 perforations, which is the practical limit for the number of 2.7-mm-diameter perforations in a 20.8-mm × 9.5-mm × 1.9-mm device. From this table it can be seen that the release rate decreases from 802 to 2.1 mg/day and the duration increases from 8.5 to 4152 days when drug solubility decreases

Table 4 Predicted Values for Drug Release from Sustained-Release Trilaminate Devices

Saturation solubility (mg/ml)	Time for depletion zones to overlap (days)	Release rate at overlap (mg/day)
500	8.5	802
200	21	387
100	42	204
50	85	91
25	169	53
10	423	21
5	847	10
2	2075	4.3
1	4152	2.1

Source: Ref. 8.

from 500 to 1 mg/ml. These calculations demonstrate the wide range of release characteristics that can be achieved with the device.

Extensive work has been completed on the in-vitro/in-vivo performance of this device. As noted, the in-vitro release rates for morantel tartrate from a series of devices with a varying number of holes are presented in Fig. 8. Also shown are the predicted release rates using Eqs. (1) and (2), where the porosity was taken as an adjustable parameter to fit one of the curves. All other curves were calculated with the same value. The fit is excellent with the exception of the device with 55 holes. The reasons for the discrepancy with the 55-hole device, although not great, are not known.

These devices were prepared with the stated number of holes and with open edges so that release occurs from both holes and the uncoated edges of the device. Such devices are easier to prepare than are devices with coated edges, and they were utilized for all work on this device. Theoretical predictions of the release of drug from devices with uncoated edges were developed by the adding a term to Eq. (1) to account for the release from a flat sheet [8]. Calculated curves for devices with and without the uncoated edges are presented in Fig. 9. Device dimensions were taken to be typical of the other devices described. Figure 9 shows that the total drug release at day 90 is the same for a device with 40 perforations and an uncoated edge, and a device with 69 perforations and a coated edge. Since drug is released from the uncoated edges, fewer holes are required to achieve the same total release. During the test period, the shapes of the two curves are not markedly different.

Figure 10 is a comparison of in-vitro data on the release rate per day of devices with uncoated edges with in-vivo data obtained from similar devices placed in fistulated cattle. (Fistulated cattle are fitted with a port to allow placement and withdrawal of objects in

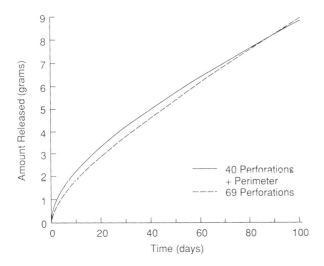

Figure 9   A plot of the amount released from the trilaminate device versus time for devices with and without coated edges. The device with uncoated edges is assumed to have 69 perforations, whereas the device with coated edges is assumed to have 40 perforations and drug is assumed to be released from the uncoated edge. (From Ref. 8.)

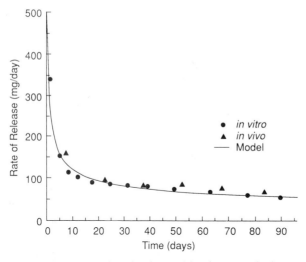

Figure 10   A plot of the in-vitro and in-vivo rate of release for morantel tartrate versus time for the Paratect Flex bolus compared with calculated values. The in-vivo data were obtained in fistulated cattle. (From Ref. 8.)

the rumen.) Agreement is excellent, suggesting that the harsh environment of the rumen does not dramatically alter the overall release pattern.

Figure 11 [15] is a plot of the mean cumulative weight gain for animals treated with this device compared to controls. Over the grazing season, the treated animals gained about twice as much weight as the untreated controls.

Rumisert Bolus

Another ruminal bolus currently under development is the Rumisert bolus, illustrated in Fig. 12 [16]. The device described here is being developed to deliver ivermectin for the

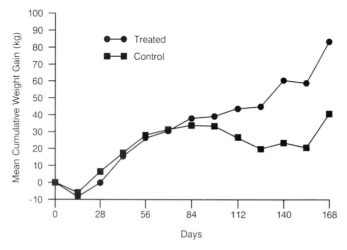

Figure 11   A plot of the mean weight gain for treated and control animals during the grazing season. Note that control animals lose weight due to the onset of disease. (From Ref. 15.)

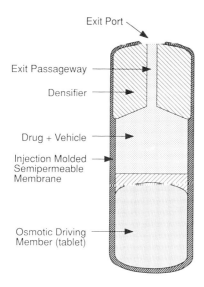

Figure 12 Cutaway view showing the components of the Rumisert bolus. (From Ref. 16.)

control of ecto- and endoparasites in cattle. This device utilizes the osmotic technology originally developed by Alza for the Oros and Alzet Minipump (discussed in Chapter 5). The drug delivery rate is controlled by a cellulose acetate cup of sufficient thickness to give the overall device mechanical integrity. When placed in the rumen, water from the ruminal fluid diffuses through the cup into a tablet composed of a salt and a swelling hydrogel. The tablet expands as the hydrogel absorbs water. Expansion of the tablet forces a drug-containing vehicle through an exit port in the top of the bolus. This exit port is a passageway through a metal element placed at the top of the bolus. This element, called a densifier, is included to increase the density of the bolus and thereby prevent regurgitation from the rumen. In-vitro release rates through 110 days are presented in Fig. 13 [16]. The slight increase in the release rate over time arises from the increasing membrane area in contact with swelling hydrogel over time.

In Figure 14 [16], the in-vivo release rates are shown for devices soaked in water

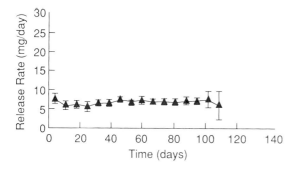

Figure 13 In-vitro release of ivermectin versus time from the Rumisert bolus.

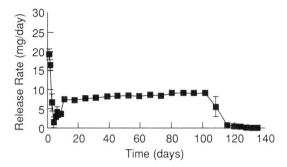

Figure 14 In-vivo release of ivermectin versus time for the Rumisert bolus. Data obtained using boluses placed in collection vessels in fistulated cattle. (From Ref. 16.)

for 2 weeks prior to administration. This initial hydration is performed to assure immediate startup upon placement of the device in the animal. Again, the slight increase in release rate over time is seen. However, an initial burst is observed. This burst arises from a combination of a pressure buildup due to hydration, and the thermal expansion of the drug vehicle upon administration.

A similar device has been developed by Alza for Schering for the release of selenium to grazing cattle [17]. Selenium is a critical growth component for cattle that is often missing from natural food sources.

Other Ruminal Devices

The Rumensin RDD is illustrated in Fig. 15 [18,19]. This ruminal delivery device was developed by Lilly for the long-term delivery of monensin sodium (an antibacterial agent) to promote growth in cattle. The core matrix is a mixture of drug (40%) and a biodegradable copolymer prepared from lactic and glycolic acids (80/20 w/w). To achieve the desired release rate, a relatively low-molecular-weight copolymer is utilized. Surface hydrolytic degradation of this copolymer matrix controls the overall release rate. Bulk hydration and erosion of the matrix is minimized by the hydrophobicity of monensin sodium and the copolymer. Hot-melt adhesives are used to secure the drug/polymer mixture to the interior surface of the cylinder. The ends of the cylinder are covered with a polymer shield to prevent metals and other abrasive ruminal contents from entering the device and physically

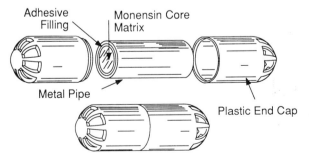

Figure 15 Sketch of the Rumensin RDD. (From Ref. 19.)

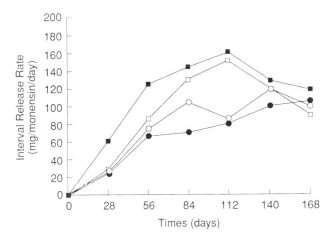

Figure 16 Interval release rates for monensin from the RDD in fistulated cattle. Each device contained 40% by weight of monensin and a copolymer of lactic and glycolic acids with different number average molecular weights (Mn). Key: Mn = 4480 g/mole (●), 3720 g/mole (○), 3000 g/mole (□), and 2320 g/mole (■). (From Ref. 18.)

abrading the core matrix. Drug release rates from this device increase over the first 50 days and are then at various steady-state rates which depend on the average molecular weight of the copolymer used in the preparation of the matrix (Fig. 16) [19].

The device of Laby [20], illustrated in Fig. 17, utilizes wings that remain folded during administration but spread out after administration for retention in the rumen. Drug is incorporated into a bioerodible matrix to control the rate and duration of drug delivery. A spring is placed inside the device to push the matrix firmly to the opening, thereby maintaining a constant exposed surface area, and, therefore, a constant release rate. This device has been utilized for the delivery of oxfendazole (an anthelmintic) for the control of gastrointestinal disease in sheep [20]. It has also been evaluated in Australia by Lilly for the delivery of monensin to control bloat in cattle [21].

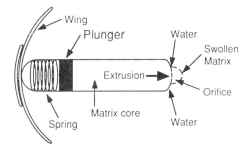

Figure 17 Sketch of a ruminal bolus for sheep. During administration, the plastic wings are held against the device by a water-dissolvable tape. (From Ref. 20.)

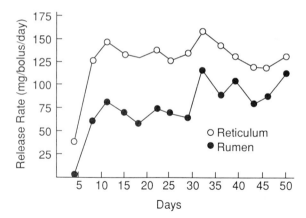

Figure 18 Plots of the release rate of oxytetracycline from erodible boluses placed in fistulated cattle. (From Ref. 3.)

Numerous "conventional" boluses provide sustained action via slow erosion of the core matrix. J. A. Hair and co-workers [3,4] described such systems for the long-term delivery of tetracycline. These boluses are made from drug, carnauba wax, barium sulfate, polyethylene glycol, and iron powder. With oxytetracycline, release for approximately 50 days has been achieved; however, as shown in Fig. 18, the release rate appears to be dependent on bolus location.

Pulsatile Boluses

For treatment of cattle or sheep with anthelmintics, a significant concern is the development of resistance due to the administration of low levels of drug. One method of minimizing this problem is through pulsed dosing, wherein therapeutic levels are administered on a regular basis, followed by periods of no drug delivery. One early attempt at the design of such a system is the device of Holloway, illustrated in Fig. 19 [22]. This device has a series of drug compartments separated via degradable cellulosic partitions. Each partition degrades upon exposure to ruminal contents. As successive partitions degrade, drug is released periodically.

Another pulsed delivery system is the ruminal device shown in Fig. 20 [23]. This device provides five doses of the anthelmintic, oxfendazole, at 23-day intervals. It was

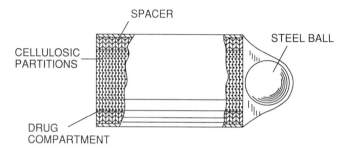

Figure 19 Sketch of a pulsatile bolus for cattle using degradable partitions to time the release of drug. (From Ref. 22.)

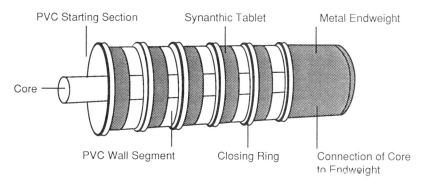

Figure 20 Diagram of a pulsatile bolus for grazing cattle. In the product the tablet is completely encased by the wall segments and is not exposed to fluid until the preceding wall segment and closing ring separate from the core. (From Ref. 23.)

designed by Coopers in the U.K. and is marketed as the Autoworm by Coopers and the Multidose 130 by Syntex. The bolus is made up of a steel endweight connected to a magnesium-alloy spindle. Around the spindle are placed five oxfendazole ''lifesaver''-shaped tablets, which are encased in PVC caps and are separated by silicon-rubber sealing washers. Delivery of the tablets is based on the well-established principle of corrosion called a galvanic couple. A galvanic couple is produced when two dissimilar metals are touching while in the presence of conductive liquid. A voltage potential develops between the two metals, causing one metal, the anode, to corrode. The other, more noble metal, the cathode, remains intact. The anode corrodes at a relatively constant rate as long as the ratios of the surface areas of the anode and cathode remain the same. Eventually, the anodic metal corrodes completely. Only then does the cathodic metal begin to corrode appreciably.

In this same way, the conductive ruminal fluid allows a voltage potential to develop between the magnesium central spindle and the mild-steel retaining weight. This galvanic couple causes the magnesium spindle to corrode preferentially at relatively constant rate, while the endweight does not corrode. As the spindle disintegrates, the PVC caps containing the drug in ''lifesaver''-shaped tablets fall off, exposing the next tablet of drug. In-vivo trials of this device confirm that the drug is delivered approximately every 19 to 21 days, as predicted by the corrosion rate of the magnesium spindle [23,24].

A significant advance would be the development of pulsatile devices for sheep, since the development of resistance by gastrointestinal helminths is an especially difficult problem in these animals. Unfortunately, high-density devices are not retained by sheep, so that currently devices with expanding geometry for retention are the only option.

Systems to Bypass the Rumen

The final device to be discussed here resulted from the work of Stephen Wu and his co-workers [25,26]. They developed an oral delivery system for sheep or cattle that permits the system (usually a multiparticulate) to pass through the rumen without releasing drug, and then to release drug in the lower GI tract in areas such as the abomasum. One application for this technology is in the delivery of methionine to ruminants. Methionine is degraded by exposure to the microbial flora of the rumen. The problem is essentially the reverse of the preparation of enteric-coated dosage forms for human applications. To prevent drug release in the rumen, Wu et al. [25] use amine-containing polymers with a pK_a of about 4.0, so that the polymer is nonionized at rumen pH, approximately 5.5 to 7.0, but ionized at gastrointestinal pH, approximately 3.0. These polymers, when used alone or with such waxes as AL/(OH) (oleate)2, prevent the dissolution of the coating in rumen fluids. The waxes decrease swelling of the polymer and/or drug core in the aqueous environment. Polymers such as cellulose propionate 3-morpholinobutyrate (CPMB) or poly(2-methyl-5-vinylpyridine/styrene, 80/20) are useful in these applications. With methionine and phenylalanine, CPMB alone is a useful coating material, but waxes are needed with more water-soluble amino acids such as lysine. Such coatings lead to significantly higher blood levels of methionine compared to those in control animals, which lead to significant increases in wool production from treated animals.

INJECTABLE/IMPLANTABLE SYSTEMS

As noted earlier, injectable and/or implantable systems represent a practical dosage form for use in animals. This approach requires less animal handling and does not suffer from the cosmetic and/or psychological factors that tend to minimize the use of this route of administration for human dosage forms. By using relatively simple technologies, a number of successful products have been launched which provide sustained release via injectable/implantable systems.

Injectable Systems

Perhaps the simplest of injectable system is the approach used by Pfizer (and other companies) in the development of long-acting formulations for oxytetracycline. The approach is essentially a variation on the idea utilized for procaine or benzathine penicillin wherein materials utilized are insoluble at the injection site and are slowly leached from that site. An example of the comparative blood-level profiles of conventional and long-acting formulations of oxytetracycline are presented in Fig. 21 [27]. Efficacious plasma levels are achieved for at least 72 h with the long-acting formulation. Table 5 provides

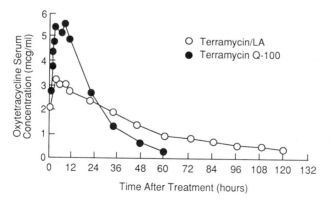

Figure 21 Serum levels of oxytetracycline following an intramuscular injection of a conventional and a long-acting form at a dosage of 20 mg/kg body weight. (From Ref. 27.)

Table 5 Effects of Repeated Intramuscular Doses of a Conventional Solution Versus One Dose of a Long-Acting Solution of Oxytetracycline for Treatment of Pasteurella Pneumonia in Cattle—Summary of Five Experiments

Treatment	No. of calves	Average daily gain (kg)	Lung lesion score[a]	Mortality (%)
Infected, nonmedicated	23	−0.07	2.6	35
Infected, OTC-C Solution (6.6 mg/kg × 3 or mg/kg × 2)	24	1.83	1.3	0
Infected, OTC-LA Solution (20 mg/kg × 1)	24	2.77	0.9	0

[a] At necropsy: Scale 0 = normal, 4 = extensive lesions.
Source: Ref. 27.

a summary of performance data from cattle treated with the long-acting formulation compared with controls and animals treated with repeat daily injections of the conventional formulation for the treatment of pneumonia. Note the efficacy of treated verses untreated animals, and the finding that the average daily weight gain of animals treated with the long-acting formulation exceeded that of animals treated with the conventional formulation.

Along the same lines, prodrugs are utilized in animal health applications to achieve prolonged release [28 & 29]. Presented in Fig. 22 and Table 6 are the variation of the duration of release of steroid esters using either different prodrugs or different formulations of the same prodrug. With norgesterol esters (Fig. 22), plasma concentrations with the C11 and C16 esters are relatively constant through 15 weeks. With the C6 ester, high plasma concentrations are found at early times, with a rapid decline to low plasma concentrations by week 6. The data on testosterone esters presented in Table 6 demonstrate that crystalline materials provide a longer duration of activity than solutions in sesame

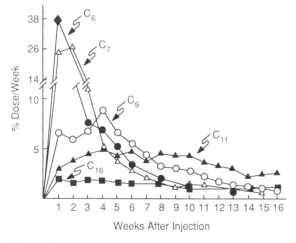

Figure 22 Plasma concentrations of fatty acid esters of norgesterol following intramuscular injection. (From Ref. 28.)

Table 6 Effect of Formulation Variables on the Duration of Effect
of Testosterone Esters

Ester/formation	Route of administration	Duration of effect
T propionate in oil solution	i.m.	3–4 days
T *n*-valerate in oil solution	i.m.	9 days
T propionate in emulsion	i.m.	3–4 days
T propionate small crystals	i.m.	8 days
T propionate large crystals	i.m.	12 days
T propionate pellet	s.c.	4–5 weeks

Source: Ref. 29.

oil for the same compound. A number of animal health products utilize the sesame oil approach to attain prolonged release of either anabolic steroids or corticosteroids.

Implantable Systems

A number of highly successful implantable products have been utilized for promoting growth in cattle. These systems employ conventional technology that is proprietary to deliver hormones for growth promotion. Synovex S (marketed by Syntex) releases estradiol benzoate and progesterone for steers, while Synovex H (Syntex) releases estradiol benzoate and testosterone propionate. Ralgro (marketed by IMC) releases the synthetic hormone zeranol, which is reported to induce the release of somatotropin (growth hormone) from the pituitary. Each of these devices is highly effective in that growth increases are observed relative to controls [30].

A nonerodible system for similar applications is the Compudose [31]. This product, marketed by Lilly, is a silastic implant that releases estradiol for up to 1 year. A drug-free silastic cylinder is coated with a layer of silastic containing 20% estradiol (Fig. 23).

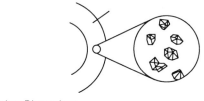

Implant Dimensions

Surface Area 4.84 sq. cm

Figure 23 Sketch of the Compudose implantable system for cattle. Drug is dispersed in a layer at the outer edge of the device. The core contains only that amount of drug that is soluble in the polymer, which is a small percentage of the total drug load. (From Ref. 31.)

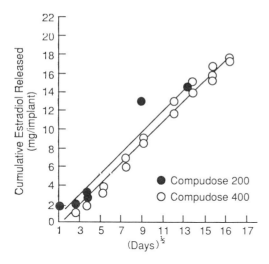

Figure 24 Plots of the cumulative amount of drug released over time for the Compudose 200 and 400. Note that the plots are linear with the square root of time, as expected for a dispersed matrix device.

Release profiles for the Compudose 200 and 400 plotted verses $t^{1/2}$ are shown in Fig. 24. From the design, square-root-of-time release is expected. The advantage of this particular design is that the interior of the device contains no dispersed drug. This minimizes tailing in the release profile when the release rate falls below desired levels. Furthermore, since the coating layer is relatively thin and the loading high, the overall change in release rate is not severe, at least compared with that observed if drug were dispersed throughout the device.

Recently, a number of papers describing relative clinical effectiveness of these three devices have been presented. As an example, Table 7 is a summary of the data of Lambert and Davis [31] which compares the relative rates of growth of cattle treated with Compudose, Ralgro, and Synovex. All treatments increase the rate of growth compared with controls. In addition, this increased growth occurs with a lower feed-to-gain ratio than with the controls, leading to a decrease in total feed cost per kilogram of body weight. Of the treatments listed, a double implantation with Ralgro appears to be the best overall treatment.

Table 7 Comparison of Ralgro, Synovex, and Compudose

	Treatment group				
	Control	Ralgro	Ralgro + Ralgro	Synovex + Synovex	Compudose
Average daily gain (kg)	1.25	1.33	1.46	1.50	1.4
Feed intake (kg)	8.80	9.09	9.29	9.72	9.08
Feed/gain	7.04	6.83	6.34	6.48	6.49

Source: Ref. 32.

COLLAR TECHNOLOGY FOR COMPANION ANIMALS

The concept of a sustained-release pesticide system for companion animals stemmed from the No-Pest Strip produced by Shell in the early 1960s. This device is a polymer resin impregnated with the pesticide dichlorvos (dimethyl 2,2-dichlorovinyl phosphate or DDVP) [33]. In this device, DDVP diffuses to the surface of the polymer resin and evaporates due to its high vapor pressure. The pesticide vapors kill any offending insect pests. DDVP is one example of the class of compounds known as dialkyl beta-halogen-substituted vinyl phosphates. In general, these compounds are effective against fleas and ticks. The application of the No-Pest Strip technology to flea and tick collars was apparent, and several products have been developed [33–37].

Several types of polymeric resins can be used as carriers of these materials. The polymer used depends on the compatibility of the chemical and the polymer, and the physical properties required of the polymer. Polyvinylchloride is usually employed to carry dichlorvos, because PVC has an excellent compatibility with the pesticide and because it affords excellent physical properties for use as a collar [37]. In general, at least 0.1% w/w of the dichlorvos must be present to be effective against insects; however, the amount of dichlorvos in the system can be as high as 70%, to ensure that enough pesticide is available for controlled delivery. A typical loading of dichlorvos would be ~5–20% w/w [33,35].

Because the active ingredient continually evaporates from the surface of the collar, even before application, one might anticipate losses of the active ingredient during storage, which could reduce the lifetime of the device. This problem can be averted if the device is packaged in an air-tight container. In this sealed environment, the pesticide evaporates until the surrounding atmosphere is saturated, at which point no more active is lost from the collar. Moreover, since these devices release drug via a square-root-of-time profile, the initial release rate is high, providing a large initial dose, resulting in a very effective product at startup [33]. Although this high initial dose might be beneficial in many instances, the excess material often irritates the animal's skin.

This irritation, and several other problems, has raised questions about the use of the dialkyl phosphonate compounds as pesticides. For example, while the volatile materials work well against fleas over the whole body of the animal, they are not as effective over the whole body against ticks [34]. As a consequence, an alternative approach to controlled-release collars was developed using a nonvolatile drug. The concept has been described as "blooming." This technique relies on a basic incompatibility between the polymer matrix and the active ingredient, which typically is a solid and nonvolatile. Because the two materials are incompatible, the nonvolatile pesticide preferentially diffuses to the surface, where it collects and rubs onto the fur of the animal. As the animal grooms itself, the pesticide is distributed across the coat.

One of the first nonvolatile pesticides was propoxur, or O-isopropoxyphenyl methylcarbamate, initially marketed by Thuron. In 1980, A. H. Robins began to market the Sergeant's Sentry V collar, which combined a vaporizing agent with a nonvolatile agent [33]. The nonvolatile agent was propoxur and the vaporizing agent was Naled, a material developed in response to the toxicity of dichlorvos. Other nonvolatile active ingredients included carbaryl (1-naphthyl-N-methylcarbamate), made by Union Carbide [38], and rabon [B-2-chloro-1-(2,4,5-trichlorophenyl)-vinyl dimethyl phosphate], developed by Shell [39].

As with the collars containing volatile pesticides, these nonvolatile pesticides diffuse to the surface of the collar during storage. But unlike the collars with volatile pesticides, the convenience of establishing a vapor/liquid equilibrium in the package does not exist. The carbaryl and rabon collars, however, can prevent premature release by taking advantage of the particular crystal structure, or polymorphs, of the pesticide [34,38]. In these collars, the pesticide is in the form of the needlelike "polymorph II" when the collar is first extruded. In this form, the pesticide diffuses to the surface. However, to prevent this phenomenon from occurring in the package, the newly extruded collar is heated before packaging so that the surface layer of needlelike crystals of polymorph II become the plateletlike crystals of "polymorph I." The flat crystals coat the surface of the collar and prevent polymorph II from diffusing to the surface. The consumer breaks this layer of flat crystals by stretching the collar before putting it on the animal. Once the layer of polymorph I is broken, fresh collar surface becomes available and polymorph II crystals from the interior of the device begin to diffuse to the surface [34,38].

As more potent drug products for flea and tick control are developed, delivery of these compounds will require more precise control over the delivery period. Therefore, future products are apt to evolve from membrane reservoir devices and the like, to provide release profiles that can be more accurately characterized and controlled.

FUTURE FRONTIERS

In the future, we can expect to see a number of new veterinary products that utilize controlled-release technology. An area of clear importance in the future is that of growth-hormone and low-molecular-weight peptides for growth promotion and increased milk yield. Because of the high doses suggested in the literature, it seems likely that any delivery system will be of relatively short duration, e.g., 1 week. This situation could change, however, if more potent low-molecular-weight analogs are developed. In either area, it is likely that microencapsulation will be a first choice as a method of delivery. As a subclass, microparticulate dispersed or porous matrix devices will also be considered as well.

PROBLEMS

1. For the Rumisert bolus, assume that the osmotic tablet contains 25% NaCl (MW = 58.5) and that the tablet can be approximated as the union of a cylinder (r = $\frac{1}{2}$ in., h = $\frac{1}{2}$ in.) and a hemisphere (r = $\frac{1}{2}$ in.).

   (a) Calculate the change in the osmotic pressure of the tablet at a time where the height of the cylindrical portion is 3 in. Neglect changes that arise from the hydrogel portion of the tablet.

   (b) At this same point, determine the overall percent increase in the rate of water transport due to the increase in surface area for transport.

   (c) Assuming that the initial rate of release is 60 mg/day, what is the rate at the point where the tablet height is 3 in.?

2. For the Rumisert bolus, solve the following problems, which relate to practical questions involved in the development of this device. Assume that the volume flow of water can be described by the following equation:

$$\frac{dV}{dt} = \frac{A\sigma Lp}{l}(\pi_2 - \pi_1)$$

(a) Assuming that $A = 20$ cm^2, $l = 0.5$ cm, $\sigma = 1$, and $\Delta\pi = 200$ atm, determine the value of Lp required to deliver 120 µg/day.

(b) If Lp is 1 µg-cm/(day atm-cm^2), what is the formulation delivery rate in micrograms per day?

(c) Manufacturing variables and experimental errors of measurement usually provide a range for a given value. If Lp $= 1.0$ µg-cm/(day atm-cm^2) $\pm$ 5%, what is the range of resulting delivery rates?

(d) Assume that you must deliver 80 µg of drug $\pm$ 10% per day and that the concentration of drug in the formulation is 1%. Can the cups from part (c) be used to achieve this performance?

(e) Changes in membrane transport rates also arise from variations in membrane thickness. Assume that the membrane thickness is $0.5 \pm 5\%$. Will this lot of membrane cups achieve the desired performance?

3. A sustained-release topical delivery system is required for sheep which must achieve sustained plasma levels for 20 days. One approach to this problem is utilized a solvent containing drug and a dissolved polymer. Upon evaporation of the solvent, a film composed of drug and polymer is deposited on the skin/wool of the animal. Drug will be released from this polymer film and permeate the skin to achieve the desired plasma level.

(a) Discuss the role of each of the following in the formulation of this product:

(i) Drug solubility in polymer, solvent, and skin

(ii) Drug transport in the polymer and skin

(iii) Factors influencing drug stability

(iv) Formulation variables influencing skin irritation

(b) Discuss formulation variables that must change if this formulation is to be used for either pigs or cattle.

REFERENCES

1. H. R. Marsten, Therapeutic pellet for ruminants, U.S. Patent 3,056,724 (1962).
2. R. H. Laby, Device for administration to ruminants, U.S. Patent 3,844,285 (1974).
3. J. L. Riner, R. L. Byford, L. G. Stratton, and J. A. Hair, *Am. J. Vet. Res.*, *43*:2023 (1982).
4. R. L. Byford, J. L. Riner, and J. A. Hair, *Bovine Pract.*, *15*:91 (1981).
5. R. M. Jones, *Vet. Parasitol.*, *12*:223 (1983).
6. H. Prosl, R. Supperer, R. M. Jones, P. W. Lockwood, and D. H. Bliss, *Vet. Parasitol.*, *12*:239 (1983).
7. F. H. M. Borgsteed, *Vet. Parasitol.*, *12*:251 (1983).
8. W. A. Boettner, A. J. Aguiar, J. R. Cardinal, A. C. Curtiss, G. R. Ranade, J. A. Richards, and W. F. Sokol, *J. Control. Rel.*, *8*:23 (1988).
9. J. R. Cardinal, in *Recent Advances in Drug Delivery Systems*, (J. M. Anderson and S. W. Kim, Eds.), Plenum Press, New York, p. 229 (1984).
10. W. D. Rhine, V. Sukhatme, D. S. T. Hsieh, and R. S. Langer, in *Controlled Release of Bioactive Materials*, (R. Baker, Ed.), Academic Press, New York, p. 177 (1980).
11. D. Brooke and R. J. Washkuhn, *J. Pharm. Sci.*, *66*:159 (1977).
12. M. Vadnere, S. Borodkin, D. Hoffman, T. Hale, D. Davidson, and J. Chu, Linear release kinetics from an inwardly releasing tablet matrix tablet, *Proc. Int. Symp. Control. Rel. Bioact. Mater.*, *14*:196, 1987.
13. W.-Y. Kuu and S. H. Yalkowski, *J. Pharm. Sci.*, *74*:926 (1985).
14. A. G. Hansson, A. Giardino, J. R. Cardinal, and W. Curatolo, *J. Pharm. Sci.*, *77*:322 (1988).

15. W. T. R. Grimshaw, A. J. Weatherly, and R. M. Jones, *Vet. Record*, *124*:453 (1989).
16. P. K. Wilkinson, J. B. Eckenhoff, presented at the 14th Int'l Symposium on the Controlled Release of Bioactive Materials, Aug 2–5, 1987, Toronto, Canada.
17. D. T. Campbell, J. Maas, D. W. Weber, O. R. Hedstrom, and B. B. Norman, *Am. J. Vet. Res.*, *51*:813 (1990).
18. J. M. Conrad and D. S. Skinner, *J. Control. Rel.*, *9*:133 (1988).
19. J. C. Parrott, J. M. Conrad, R. P. Basson, and L. C. Pendlum, *J. Animal Sci.* 68:2614 (1990).
20. *The Farmers Annual*, p. 15 (1985).
21. J. M. Conrad, personal communication (1989).
22. J. W. Holloway, Delivery system for the sustained release of a substance in the reticulorumen, European Patent 0,062,391 (1982).
23. M. Pringle, *Animal Pharm*, Nov. (1986).
24. D. Rowlands, M. Shepherd, and K. Collins, *J. Vet. Pharmacol. Therapy*, *11*:405 (1988).
25. S. H. Wu, C. C. Dannelly, and R. J. Komarek, in *Controlled Release of Pesticides and Pharmaceuticals* (D. A. Lewis, Ed.), Plenum Press, New York, 319 (1981).
26. S. H. Wu and S. A. Sandhu, Rumen-stable pellets, International Patent W.O. 84/00282.
27. A. J. Aguiar, W. A. Armstrong, and S. J. Desai, *J. Control. Rel.*, *6*:375 (1987).
28. Y. W. Chien, *Novel Drug Delivery Systems* Marcel Dekker, New York, p. 291 (1982).
29. A. A. Sinkub, in *Sustained and Controlled Release Drug Delivery Systems* (J. R. Robinson, Ed.), Marcel Dekker, New York, p. 437 (1978).
30. B. D. Schanbacker and J. R. Brethour, *J. Animal Sci.*, *57* (Suppl. 1):468 (1983).
31. I. H. Ferguson, G. F. Needham, R. R. Pfeiffer, and J. F. Wagner, Compudose: An implant system for cattle, *Proc. Int. Symp. Control. Rel. Bioact. Mater.*, *14*:51, 1987.
32. S. B. Lambert and G. V. Davis, *J. Animal Sci.*, *57*(Suppl. 1):468 (1983).
33. J. Greenberg, Modern insecticide/resin collars for the control of ecto-parasites on pet dogs, Abstracts 7th Int. Symp. Control. Rel. Bioact. Mater., 103, Controlled Release Society (1980).
34. D. G. Pope, in *Animal Health Specialized Delivery Systems* (D. C. Monkhouse, Ed.), Academy of Pharmaceutical Sciences, p. 78 (1978).
35. H. L. Quick, *Vet. Med. Small Animal Clin.*, 66:773 (1971).
36. I. Fox, I. G. Bayona, and J. L. Armstrong, *J. Am. Vet. Med. Assoc.*, *155*:1621 (1969).
37. L. M. Grubb and J. K. Baxter, U.S. Patent 3,318,769 (1967).
38. J. E. Miller, N. F. Baker, and E. L. Colburn, *Am. J. Vet. Res.*, *38*:923 (1977).
39. A. Miller and J. G. Morales, U.S. Patent 3,944,662 (1976).

11

Pesticide Delivery

Robert C. Koestler   *Atochem North America, Bryan, Texas*

George Janes   *Marine Test Stations, Akron, Ohio*

J. Allen Miller   *Knipling-Bushland U.S. Livestock Insects Research Laboratory, U.S. Department of Agriculture, Kerrville, Texas*

INTRODUCTION

History of Controlled Release

The controlled delivery of pesticides was not an obvious technology in the beginning. Technology grows slowly, driven by demand, advances in other areas, and a dissatisfaction with current ways of doing things. Pesticide delivery in the 1930s was sometimes no more than a farmer shaking a bag of calcium arsenate by hand in the middle of a field, trying to get the powder to blow in the wind in order to widen its distribution. Such a technique today would bring forth gasps of horror. In the 1930s, however, it seemed a miracle to be able to prevent insect damage to crops simply by dusting them with calcium arsenate. Residues were not a problem. The rain would wash off the poison, and analysis by most methods would indicate that no pesticide remained. And so, through the use of pesticides, yields of crops began to increase, causing food prices to remain low. Additional advances in technology continued the increase in yields of food needed to feed the growing population. The introduction of DDT gave the world an all-purpose insecticide, seemingly nontoxic to mammals while ridding man of his insect pests. DDT has been credited with saving 20 million lives during the three decades of its use. The price of DDT, owing to very large-scale production, was so low that its use became ubiquitous. For a long time DDT was cheap, effective, and seemed to have no deleterious side effects. It was also long-lasting and virtually indestructible in the environment. The tremendous tonnage used was in fact the downfall of DDT. Its effectiveness began to drop off as insects exposed to less than lethal doses began to develop resistance. Higher doses had to be used to achieve control. Thus, it became more expensive to use. Finally, it turned out that there *were* undesirable side effects, and in areas that had not been imagined before.

When DDT was banned, farmers were committed to using chemical pest control. It was too expensive not to. But the switch to nonpersistent, highly toxic organophosphates was a real shock. A walk in a newly sprayed field could mean death due to dermal adsorption of an organophosphate. And the new materials were truly nonpersistent. So nonpersistent that you had to keep spraying, maybe as often as every other day, to raise a good crop. Also, residues could now be detected in parts per billion on the crops, in the air, and in the groundwater. Clearly, the workings of the organophosphate pesticides left much to be desired.

In the 1950s and 1960s, the problems were attacked by trying to synthesize pesticidal molecules that would work in a more preferable way. As the cost of such syntheses rose, and upwards of 10,000 compounds had to be made to obtain one commercial hit, reformulation of existing materials began to look like a more promising way to develop a product.

The ideal insecticide would be nontoxic to mammals, noncontaminating to the environment, highly selective against the target species, always potent enough to give a fatal dose, and by that prevent resistance and residual for the required lifetime. Formulation is the means by which the pure form (technical) of the pesticide is converted into a usable state so that it may be applied from existing commercial equipment, on, under, or around the plant, to give it protection. Formulation, which prevents the immediate release of a material to the environment, is called controlled delivery or controlled release. Many different techniques have been devised to attain controlled delivery, but in this chapter we shall examine mostly microencapsulation. We will give a physical basis for the reasons that controlled delivery gives superior results compared to standard formulations.

THEORY

Loss of Active Ingredient

A material, when released into the environment into a concentrated localized area, is caused by natural laws to be dissipated. It moves from areas of high concentration to areas of low concentration. Applied at a high dose, it soon is too dissipated to function. Pesticides can be lost due to evaporation, diffusion, photolysis, adsorption, hydrolysis, bacterial decomposition, or just washed away.

In order to reduce the premature loss of a pesticide, it is necessary to determine the critical-loss pathways that are operating in a given situation. Once this piece of information is in hand, the course of work to be done and the means of measuring success become apparent. Generally, similar materials used in similar ways have similar critical-loss pathways. The organophosphates, which replaced DDT, were found to be very labile materials, not effective for longer than a few days or weeks at the most. They are sensitive via all of the loss pathways mentioned above. However, for a majority of organophosphorus pesticides for foliar use, the mechanism of loss that is most important is *evaporation*, because it occurs rapidly and is a perhaps unexpected route.

Problem: If the evaporation rate is proportional to surface area, calculate the loss per day due to evaporation for 1.0 kg of material spread over an acre if the loss per square inch of surface area is 0.1 mg/day.

Solution:

$$1 \text{ acre} = 6.26 \times 10^6 \text{ in.}^2 \tag{1}$$

$$\text{Loss} = \frac{\text{square inches}}{\text{acre}} \times k$$

where k is the rate constant in milligrains per day. Therefore,

Evaporation loss $= (0.1 \text{ mg/in.}^2) \times 6.26 \times 10^6 = 626$ g/day

The same materials in the soil present a much more complicated situation. Evaporation is practically the only route that is not important. The loss pathways in the soil vary in importance relative to time. Adsorption and diffusion are virtually immediate. Hydrolysis follows first-order kinetics, but losses due to bacterial action are exponential in nature following a period of inactivity called the lag time. The history of pesticide application to the soil of a field can have a profound effect on subsequent application of the same pesticide. The first application results in an explosion of the population of bacteria that can metabolize the material. A long lag time ensues while the bacteria build up sufficient population to substantially reduce the concentration of the pesticide. However, the application of the same pesticide the next season may find a large, hungry population of the same organism just waiting to be fed. The result is a much faster disappearance of the chemical than after the first application.

Kinetics of Loss

The use of a pesticide in the environment by normal application methods is very inefficient and potentially damaging to the environment; it also has the additional capacity to cause development of resistance. Typically, the first-order decrease in concentration of an applied pesticide with an effective lifetime in the field of 3 days gives a plot like Fig. 1. The area above the 3-day concentration line represents wasted pesticide, because it is more than that necessary for control. The material present after 3 days is not enough for control.

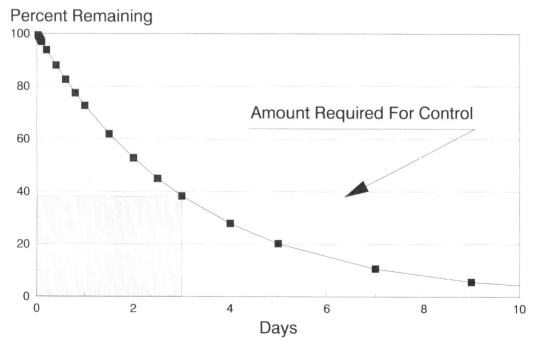

Figure 1   First- order loss versus time. A hypothetical curve demonstrating the amount of pesticide remaining in the environment versus time, with the control period ending at 3 days after treatment.

The material used efficiently is represented by the shaded rectangle below the control line and from 0 to 3 days, which obviously is a rather small percentage of the total application. The very long tail, after 3 days, unfortunately accounts for an appreciable percentage of the applied pesticide, and during this period selective pressure is put on the insect population. Individuals sensitive to the insecticide may be killed off, leaving less sensitive ones to breed. Thus, in a relatively short number of years of pesticide use, resistance develops.

In the 1960s, it was apparent that something had to be done to improve the efficiency of pesticide applications, but it did not become a necessity until DDT and other persistent organohalide pesticides were banned. DDT was relatively nontoxic to humans, cheap, and long-lasting. The switch to the labile, short-lived organophosphates was not without problems. Not only were they acutely toxic by ingestion, but they were also toxic by dermal adsorption. Initially, there were fatalities due to unsuspecting people walking through fields newly sprayed with such materials as methyl or ethyl parathion. These materials were dangerous to handle, dangerous in the field, and so short-lived that crops such as cotton sometimes had to be sprayed every other day. Clearly, a method of formulation was needed that would change the kinetics of disappearance in the field.

A formulation that would solve the problem would dispense the pesticide at a zero-order release rate. Theoretically, the simplest way to achieve such a rate is through use of a reservoir system. The rate of release for a reservoir system is controlled by the rate of diffusion of the reservoir contents through the walls of the container. The concentration of the material in the reservoir is unity until it is practically exhausted. Microencapsulation provided a successful mean of making a reservoir system to hold the labile pesticides, with the decided advantage that the product can be used in mechanical systems already in place and used by farmers and professional pesticide applicators for treating crops.

Unlike microencapsulation, most methods have suffered because their formulations cannot be used in existing equipment with existing application techniques. It is unlikely that applicators would buy new equipment and learn new techniques of application just to use a new product, no matter that the new product promises a more efficient use of material or even better pest control leading to a better-grade product for the farmer. In fact, at the present time, the major selling point is that use of the controlled-delivery product will save the farmer money. One factor impeding the development of the techniques that incorporate really desirable advantages is that they do not save the user any money. This point will be covered in a subsequent paragraph. Some controlled-delivery systems, which have been hindered by not conforming to existing application techniques, are laminates and hollow-fiber technology.

Microencapsulation can be accomplished by a number of techniques. For the preparation of carbonless carbon paper, both interfacial polycondensation [1] and coacervation [2] have been used. For use with pesticides, coacervation may be too expensive; interfacial methods are therefore favored. Microencapsulation by interfacial polycondensation is accomplished by dispersing a pesticide containing a water-insoluble monomer in an aqueous solution of a dispersing agent and then causing polymerization to occur at the interface by adding a complementary monomer to the aqueous phase. The thin spherical shell of polymer that forms around the pesticide droplet (Fig. 2) is the rate-determining membrane required for zero-order release. The internal phase of liquid pesticide is the reservoir. Patent number 3,577,515 granted to Pennwalt Corporation, discloses the invention of this technique, and particularly, the art of how to control the diffusion rate across the wall (membrane). Interfacial polymerization has never been a good way to make high-quality polymer; the molecular weights of the polymer are very low, and the films formed have

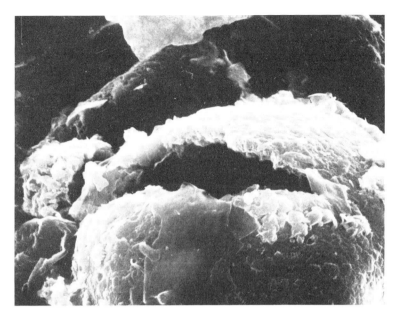

Figure 2   A scanning electron micrograph of a capsule prepared by interfacial polymerization. The fracture shows the thin spherical-shell nature of the wall. (Photo courtesy of Atochem North America.)

little strength. The Pennwalt patent overcame this difficulty through use of a cross-linking agent, which caused the formation of a stronger, much-higher-molecular-weight polymer wall.

ENCAPSULATION RECIPES

The following examples of microencapsulation by interfacial polycondensation and co-acervation can be conducted in the lab to get a feeling for the technique.

Recipe for encapsulation by interfacial polycondensation [3]: In this example, a polyfunctional isocyanate is one of the complementary intermediates; it also is the sole cross-linking agent needed to produce a polyurea capsule wall. The encapsulation is carried out in a 2000-ml indented three-neck flask. A Premier Dispersator is used to agitate and disperse the organic liquid. (Any high-shear agitator can be used; standard laboratory stirrers will give capsule sizes above 100 μm.)

In the flask:

600 ml 0.5% Elvanol 51-05 (polyvinyl alcohol) in water

First addition:

240 g Diazinon
50 g polymethylene polyphenylisocyanate (Papi)

Second addition:

19.5 g sodium carbonate
22.5 g ethylenediamine

22.5 g diethylenetriamine
200 g distilled water

The first addition is made rapidly, while stirring at top speed. After 10–30 s, the second addition is made. Stirring is reduced, and the reaction is allowed to run for 2 h. Hydrochloric acid is added to neutralize the excess base. If the capsules are to be sprayed, oversize material should be removed by passage through a 50-mesh sieve.

In the interest of safety, xylene may be substituted for diazinon to practice making this recipe.

Recipe for encapsulation by coacervation: The following recipe is interesting because it prepares pearlescent capsules [4].

1. The internal phase is prepared from 588 g of white mineral oil and 12 g of mica particles coated with titanium dioxide. The mica particles are first made wet with the mineral oil and then added to the rest of the mineral oil with stirring. Stirring is continued for 1 h at ambient temperature.

2. Into a 4-liter beaker are added 1760 g of water, 40 g of gum arabic, and 40 g of gelatin. The resulting mixture is heated at 50°C while agitating at low speed.

3. The mineral oil-mica particle dispersion, 600 g, is added to the gelatin-gum arabic phase and stirred at a rate that gives a 1400- to 2500-μm dispersion. The pH of the batch is adjusted to 4.3, and the temperature is slowly lowered to 27°C, whereupon the oil droplets become coated with the coacervate. The temperature is further lowered to 10°C. Glutaraldehyde, 10 ml of 50% aqueous solution, is added, and the mixture is stirred overnight to harden the walls.

4. Further hardening of the capsule walls may be accomplished with urea-formaldehyde.

Although the microencapsulation reservoir system was designed to give zero-order release, release of the active ingredient was not found to be zero-order.

Examination of release-rate data of microcapsules prepared by interfacial polycondensation showed the release rate to be first-order (Figs. 3 and 4). The reason for this apparent paradox is based on the multiplicity of capsules. One lone capsule will indeed release in accordance with zero-order kinetics, and similar capsules will display similar kinetics. The method of production of these capsules, however, gives rise to a population of widely differing capsules. Some have imperfect walls and release rapidly, while some have lesser imperfections and release much more slowly. The sum of the widely varying release profiles approximates first-order.

If the kinetics of release of the microcapsule are the same order as that in classical emulsifiable concentrate applications, how do we account for the benefits that we observe through the application of these formulations?

WHY DO MICROCAPSULES WORK?

Wall Thickness

Microcapsules are thin-walled spheres generally filled with an organic liquid pesticide. The pesticide may be diluted in or dissolved by an organic solvent in order to facilitate encapsulation. The average diameter of microcapsules to be sprayed by conventional equipment generally should be less than 100 μm and greater than 10 μm. The upper limit is set in part by the requirement that the capsules must pass through the screens and

Percent Remaining

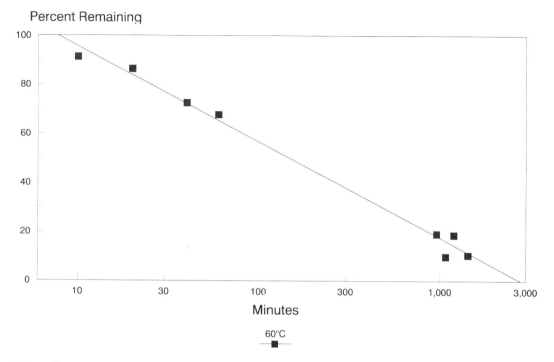

Figure 3 Release of methyl parathion from Penncap-M capsules, in a laboratory experiment a 60°C, plotted against log of time. The data show a linear relationship, which agrees with a first-order release mechanism.

Percent Remaining

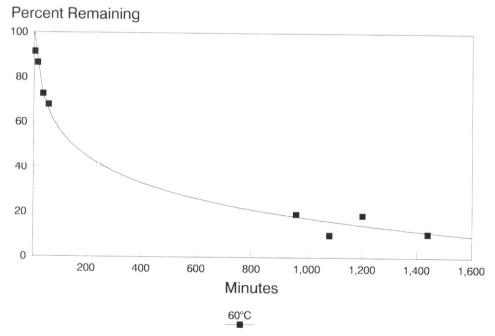

Figure 4 Release of methyl parathion from Penncap-M capsules, in a laboratory experiment at 60°C, plotted against time. The typical first-order decay curve is evident.

nozzles of spraying equipment. The lower limit is set in part to prevent excessive drift of the small capsules and to avoid possible inhalation hazards.

Capsule wall thickness and surface area as well as the absolute and relative payload are important parameters, which must be studied before their effects can be understood. If we assume that the density of the encapsulate and the wall polymer are the same and equal to 1.0, the following relationships between some important parameters can be derived:

$$W = r_1 - r_2 \tag{2}$$

where W = wall thickness, r_1 = capsule radius, and r_2 = radius of the liquid of a single capsule.

$$V_1 = \frac{4}{3} \pi r_1{}^3 \tag{3}$$

$$V_2 = \frac{4}{3} \pi r_2{}^3 \tag{4}$$

Where V_1 = volume of capsule and V_2 = volume of the encapsulated liquid. The volume of the polymer V_p can be calculated from its weight and

$$V_p = V_1 - V_2 \tag{5}$$

$$V_2 = V_1 - \frac{(100 - \% \text{ wall})}{100} \tag{6}$$

Then

$$W = \sqrt[3]{\frac{3V_1}{4\pi}} - \sqrt[3]{\frac{100 - \% \text{ wall}}{100}\left(\frac{3V_1}{4\pi}\right)} \tag{7}$$

Problem: Calculate the wall thickness of a 30-μm capsule having a 10% wall.
Answer: Using Eq. (3):

$$V_1 = \frac{4}{3} \pi (15)^3 = 14,137 \ \mu\text{m}^3$$

Then, using Eq. (7),

$$W = \sqrt[3]{\frac{3}{4\pi}(14,137)} - \sqrt[3]{\left(100 - \frac{10}{100}\right)\frac{3}{4\pi}(14,137)}$$
$$= 15.000 - 14.482 = 0.518 \ \mu\text{m} \tag{8}$$

A plot of Eq. (7) is given in Fig. 5, which shows the change in wall thickness versus capsule diameter for capsules having payloads varying from 90% to 60%. An average capsule of 30 μm diameter having a wall of 20% by weight has a wall thickness of about 1 μm. One gram of encapsulate would be distributed among approximately 100,000,000 capsules.

The large number of capsules with the smaller capsule diameters provides a very large surface area. In fact, formulations of capsules having very porous walls can actually dissipate some active ingredients faster by evaporation than standard formulations owing to the large capsule surface area. It is important to note that, at a constant ratio of wall material to organic phase, as the particle size of the capsules is decreased, wall thickness also decreases (Fig. 6).

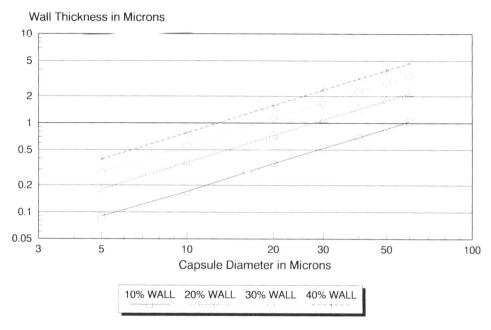

Figure 5 Graph of wall thickness versus capsule diameter for capsules consisting of 10% to 40% wall loading and payloads of 90% to 60%. Equation (7) was used to calculate these data.

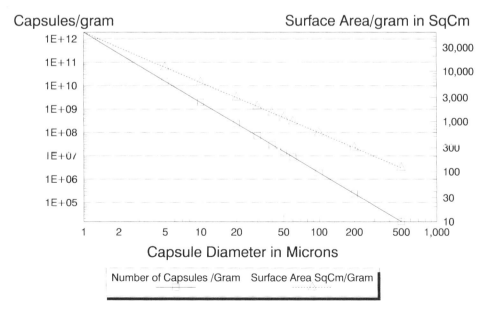

Figure 6 Capsule diameter plotted against the number of capsules per gram of capsules and against the surface area per gram of capsules. Equation (3) was used.

Problem: Calculate the number of capsules per gram of capsules of 30-μm diameter, assuming a density of 1.

Solution: Use the inverse of Eq. (3); the number of capsules per gram, C_g, is

$$C_g = 3 \frac{k}{4\pi} \left(\frac{2}{D}\right)^3 \tag{9}$$

where D is the capsule diameter in micrometers, and k is the conversion constant of 10^{12} cubic micrometers/cubic centimeter.

$$C_g = \frac{3}{4*\pi} \left(\frac{2}{30}\right) * 10^{12} = 70.7 \times 10^6 \frac{capsules}{gram} \tag{10}$$

Geometry

Study of the geometry of the system will help in understanding the relation between efficiency and persistence of encapsulated insecticides [5].

The Concentration Effect

In normal application of an emulsifiable concentrate pesticide appropriately diluted with water, one attempts to attain 100% coverage of the sprayed area. The area covered by the application of microcapsules is a function of the capsule diameter and the application rate:

$$A = \pi r^2 (10^{12}) \tag{11}$$

where A is the area covered by one capsule in square centimeters, and r is the capsule radius in micrometer. To calculate the percent coverage,

$$\% \text{ coverage} = \frac{(A)(C)(D)}{T} (100) \tag{12}$$

where C is the number of capsules per gram, D is the dose in grams per unit area, and T is the total area.

Figure 7 shows the percent area covered versus capsule diameter.

Problem: What is the percent of area covered by capsules having a diameter of 30 μm treated at a dose rate of 1 lb/acre?

Solution: Read from the graph of Fig. 7: percent area = 0.6%.

Very small capsules (below 1 or 2 μm) have percent-area coverage figures that are large, so very small capsules do not have the advantage of highly concentrating the application.

The concentration of material within the spherical geometry of the capsule helps protect the pesticide in a number of ways and thereby increases its effectiveness. High local concentrations are able to satisfy substrate adsorption, while the capsule wall protects the sensitive payload from photochemical degradation. Figure 7 also shows that the larger the capsules are, the greater the concentration effect they exhibit. Then why are microcapsules of 20 to 50 μm used instead of capsules in the 1000- to 2000-μm range, where the concentration effect is much greater? Obviously, if all the active ingredient were put into one large capsule and placed in the middle of a field, it would be a very high concentration of material and be very persistent, but it would not kill very many insects. Another factor, which functions in opposition to the concentration effect, must also be considered.

Percent Coverage

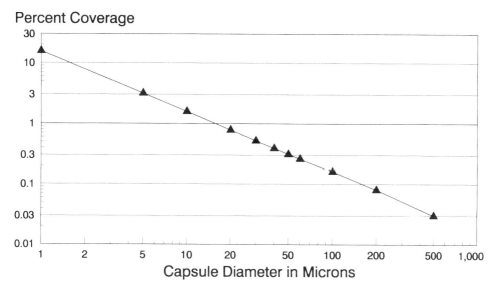

Figure 7 Capsule diameter versus theoretical percent area coverage at an application rate of 1 lb/acre.

The Capsule Density Effect

The greater the concentration effect, the lower the distributed density of capsules. If an insect seldom comes upon an insecticide-laden capsule, insecticidal efficiency of the formulation will be low (Fig. 8).

Problem: Calculate the space between the capsules put down on a substrate by a

Capsule Spacing, microns

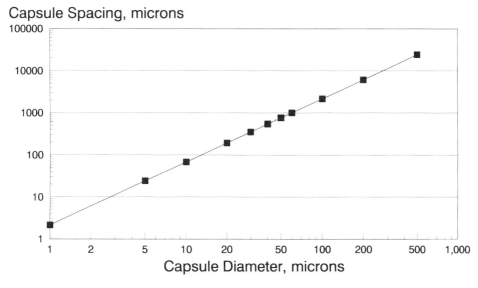

Figure 8 Theoretical capsule spacing in micrometers versus capsule diameter for an application rate of 1 lb/acre.

theoretically perfect application, if the application rate is 1 lb/acre and the average capsule diameter is 30 μm.

$$\frac{\text{Microns}}{\text{Capsule}} = \frac{\sqrt{\text{ft}^2/\text{acre}} * 30.48 \text{ cm/ft} * 10^4 \text{ μm/cm}}{\sqrt{\text{capsules/g}} * 453 \text{ g/lb}} \tag{13}$$

Answer:

$$\frac{\text{μm}}{\text{Capsule}} = \frac{\sqrt{43555} (30.48)(10^4)}{\sqrt{70.7 \times 10^6}(453)} \tag{14}$$
$$= 355 \text{ μm/capsule}$$

There are 355 μm or 0.355 millimeter between capsules.

The objective in creating the microencapsulated diffusion-controlled reservoir system was to attain zero-order release and thereby solve the problems associated with first-order release. It seems, however, that insecticides microencapsulated by interfacial polycondensation do not release by zero-order kinetics but rather by pseudo-first-order kinetics. If we could measure the release from one capsule, it would no doubt be zero-order. The interfacial method, however, gives capsules that vary tremendously in size and in quality. Inferior capsules break or leak out their contents very quickly. Other capsules that lack defects retain their contents for a long time. Overall, the net release from all the capsules turns out to be first-order.

As explained earlier, the properties of a model microencapsulated insecticide have accomplished just what creation of a zero-order system would be expected to do. To further explain the observed results of using microcapsules in real-world applications, we must look closer at the geometry involved and closer still at the kinetics of first-order release.

First-order kinetics define the decrease in concentration with respect to time as $dc/dt = kC$. Integration gives $C = e^{-kt}$. Plotting concentration versus time gives the characteristic first-order decay curve (Fig. 9). In order to maintain a concentration required for pest control for an appreciable time, substantially more than that concentration must be applied. In Fig. 9, line A-B represents the concentration required for control of a certain pest. If it is desired to maintain that concentration for 100 time units, the area of the shaded rectangle is the useful area under the decay curve. All area above the line AB represents overkill, while area to the right of the 100 time line denotes insufficient pest control, which can lead to the development of resistance by the pest.

To determine the maximum efficiency of the first-order treatment, we can plot the area of the shaded rectangle versus position of the intersection with the first-order decay curve: $A = te^{-kt}$. Differentiating and setting $da/dt = 0$ gives the maximum area at $t = 1/k$. The concentration component of the maximum area $= e^{-kk}$ or $e^{-1} = 0.38$.

Thus maximum efficiency is attained when the amount required for control is 38% of the application rate. Also, since the solution of the maximum area gives a constant that is independent of k, the 38% maximum efficiency holds for any first-order decay process involving pesticide application.

Although the efficiency of the first-order decay process is not changed by microencapsulation, the useful lifetime of the formulation certainly can be changed. As k becomes smaller, the first-order decay curve becomes less steep and the concentration remains high for a longer time. It then becomes necessary to distinguish between what has come out of the capsule and what remains. If pesticide inside the capsule is unavailable to the pest,

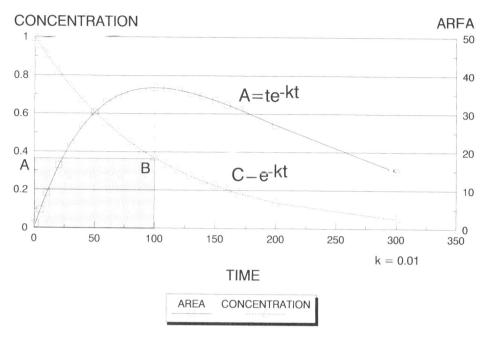

Figure 9 A first-order release curve. The shaded box represents the most efficient pesticide use. To realize an effective concentration of 0.38, for time = 100, pesticide must be applied at level 1.0. For a first-order system the maximum efficiency will always be at the level of 38% of the original application. Thus, a 62% excess is required to obtain maximum efficiency.

then what is outside the capsule must be relied upon. Unfortunately, the amount of material outside the capsule is not the amount indicated by the first-order decay curve, because the material that has come out is much more volatile (which is why it was encapsulated in the first place) and therefore evaporates.

$$\text{Encapsulated pesticide} \Rightarrow \frac{\text{release}}{\text{slow}} \Rightarrow \frac{\text{evaporation}}{\text{fast}} \qquad (15)$$

As a result, the popular conception of slow release from the capsules resulting in a lethal concentration of insecticide on the outside of the capsule holds only for rapidly releasing formulations. If the release is slow compared to the loss of material after it is out of the capsule, then there is not enough concentration of active ingredient on the outside of the capsule to give control. Therefore, very-slow-releasing capsules must be insecticidally active by a different mechanism.

Very-slow-releasing capsules can be effective only if the target insect itself causes the toxicant to be released. Three mechanisms may function. First, capsule breakage may be insect-initiated, as proposed by Tsuji et al. [6], wherein it was shown that cockroaches running over capsules containing the insecticide Sumithion broke many of the capsules, thereby releasing the poison. Second, insects crawling over the deposit of capsules are contaminated by the capsules sticking to them. Later, grooming activities cause the capsules to break, thereby delivering a massive dose of the insecticide. This is most likely the mechanism by which insects that are resistant to the pesticide are killed. Third, insecticidal action will be obtained by changing the environment of the capsule to one that will cause a release of the internal phase.

The first mechanism is difficult to attain, since an increase in the strength of the wall necessary to retain the toxicant may also make the capsule too tough for an insect to break. However, most commercial microencapsulated products, especially those that are used indoors, probably employ this mechanism to some extent. For instance, a nonencapsulated treatment (diazinon) at 1% will largely dissipate in 3 to 5 days. One treatment of microencapsulated diazinon at the 1% spraying concentration can last 2 months. Therefore, some type of insect-initiated capsule-breaking mechanism must be functioning.

ENVIRONMENTAL FATE OF CONTROLLED-RELEASE PESTICIDE FORMULATIONS

Indoors

The environmental fate of pesticides used indoors is a newly emerging area of concern [7]. It is widely known that most organophosphate pesticides, when sprayed indoors, will give insect control for only a few days. It is less commonly known what causes the loss of control. Organophosphate insecticides, when sprayed on indoor surfaces such as baseboards in a house, will lose effectiveness through two modes. Mode one is adsorption onto exposed surfaces. Mode two is by evaporation into the air. Mode two ultimately leads to adsorption of the organophosphate from the air onto most indoor surfaces that it contacts. Thus, organophosphates sprayed in high concentration within a limited space become redistributed and adsorbed onto most of the indoor surfaces. The adsorption and redistribution take place in just a few days at ambient room temperatures. Toxic deposits are then no longer accessible to insects, and no longer given control; instead, they slowly desorb and contaminate the indoor air. To summarize, standard pesticide application indoors results in a short insecticidal period followed by a long period of contamination.

Microencapsulation of organophosphates decreases both surface adsorption and redistribution of the active ingredient due to evaporation. The magnitude of decrease of the two factors is a function of the tightness of the capsule wall.

Outdoors

The environmental fate of controlled-release pesticide formulations used out of doors has two aspects.

Foliar Applications

Pesticide may be lost by evaporation or by substrate adsorption. It may also be washed off crops by rain. Photochemical degradation is another possibility, as is bacterial action.

Most of these losses may be diminished by controlled-delivery formulations. The carrier system is also of concern. The formulation must be registered pursuant to EPA labeling, and an explanation of the environmental fate of the inactive ingredients in the formulation generally will be required. Specific analyses may also be required for the environmental detection of these materials as well as measurements of their inherent toxicity.

Soil Applications

The environmental fate of soil applications of controlled-release formulations predominantly is loss of the active ingredient by bacterial degradation. Hydrolysis in the soil, a common loss pathway, is effective in breaking down many pesticides. A major difference between hydrolysis and bacterial action is that hydrolysis is much more predictable, causing loss by first-order kinetics. Bacterial action depends on the particular bacteria present. The loss curve associated with bacteria is the mirror image of the first-order curve characterized by a slow "lag period" and then a period of exponential acceleration [8]. Currently there is much concern over metabolites of pesticides in the environment, and a number of materials have been regulated or had their registrations canceled because of the perceived danger of their metabolites.

RATIONALE FOR PESTICIDE ENCAPSULATION

It is a mistake to believe that all pesticides can benefit from encapsulation. The number one consideration before undertaking the task of changing the environmental availability of a material is to determine the problem to be solved. With a sufficient knowledge of the technology, a decision must then be made as to how a change in release profile can help the situation. Two classes of problems need to be considered:

1. Problems dealing with the physical chemistry of encapsulation itself and the compatibility of the release method with the toxicant
2. The inadequacy of the pesticide to be remedied to solve a particular application problem

Problems to Be Solved

Physical Problems Concerning the Encapsulate

Before a decision is made to encapsulate a pesticide, careful assessment of the properties of the toxicant should be made to determine the suitability of the method of formulation. There are many methods of formulation and many types of controlled-delivery techniques, with each method having its own strong points and weaknesses. For instance, the centrifugal encapsulation technique can deal with water solutions or water-soluble materials, but it cannot be used efficiently to fabricate very small capsules (under 100 μm). The interfacial technique can be used to make large amounts of microcapsules quickly and at reasonable cost, but it cannot be used to encapsulate most amines, most drugs, or water-soluble materials. See Table 1.

When a material can be formulated by a number of different methods, final goals

Table 1 Comparison of Various Controlled Delivery Techniques

Encapsulate	Interfacial	Starch	Coacervation	Hollow fiber	Centrifugal
Water soluble	No	No	No	No	Yes
Reactive	No	No	Yes	Yes	Yes
High melting point	If dissolved	If dissolved	As solid	No	Yes
Organic soluble	Yes	Yes	Yes	Organic liquid	Yes
Water soluble	No	No	No	No	Yes
Particle size	1 – 1000 μm	>1000 μm	1 – 1000 μm	Fibers	>100 μm
Solids	Yes	No	Yes	No	No

must be kept in mind before committing to one method or another. Sometimes, what seems at first to be a disadvantage can be turned into an advantage. For example, the process output from an interfacial process is in water containing perhaps 30% solids. It is extremely difficult (expensive) to dry something from 30% solids to 90–95% solids to get a free-flowing powder. Fortunately, many agricultural formulations are flowables sold in the 30% solids range. Thus, no drying or capsule isolation step is needed to go from the reaction mixture to the final product. Because the interfacial method works best on pesticide-type materials (water-insoluble, organic, liquid), coupled with the fact that the process output is directly usable in existing commercial spraying equipment, this process is both economical and competitive.

The encapsulation of water or of water solutions is impossible for the interfacial process, since it is commonly run in water. The centrifugal process of the Southwest Research Institute has been used to prepare water capsules, which can hold the water for years [9]. These capsules are mixed with gypsum to form a pressure-activated grout used in the ceilings of mines. While not a pesticide application, it does illustrate the importance of finding the right controlled-delivery system for a specific application.

Application Problems to Be Solved by Controlled Delivery

How strong is a capsule? Individual microcapsules are easily broken by insects and therefore must be quite fragile. In bulk, however, they are amazingly strong, because an applied force is shared by a large number of capsules. In an experiment to determine what would happen if spilled methyl parathion capsules were walked upon, the product was poured onto a wooden floor and a person wearing rubber boots trampled around on the material for a while. Surprisingly, no significant release of the toxicant was observed [10]. To our knowledge, the actual strength of one microcapsule has not been measured, so the following problem is hypothetical.

Problem: If one capsule of 30 μm can be broken by a force of 0.001 g, how much force will it take to break all the capsules in 1 g simultaneously?

Solution:

$$F_t = C_g \times F_c \tag{16}$$

Where F_t = total force, C_g = number of capsules per gram, and F_c = force required to break one capsule.

$$F_t = (7.07 \times 10^7 \text{ capsules/g}) \times (0.001g) \tag{17}$$
$$= 7.07 \times 10^5 \text{ g or 156 lb}$$

High volatility may be improved by many controlled-delivery techniques. But one must bear in mind the problem being caused by the high volatility. The problem of completely stopping volatilization is quite different from that of maintaining a desired vapor concentration in the air.

Cost of formulation depends on many factors. The particular technique used sets certain limits. Some methods require rather expensive and sophisticated equipment even to do bench-scale runs. The centrifugal method requires precisely machined head, jets, pumps, and a vacant room. The interfacial method, on the other hand, can be carried out in a 1-oz bottle using hand shaking to make the dispersion. Labor-intensive and equipment-intensive operations such as air-suspension coating raise the cost. Water or solvent removal is energy- and time-demanding and therefore expensive. Formulation loading is important. Higher active-ingredient loading gives a more efficient use of resources, including savings on packaging material, storage fees, and shipping costs. Higher loading also requires less inactive ingredients, thereby lowering raw material costs.

Economics are very important because competition in the pesticide field is intense. With EPA labeling requirements for pesticide formulations, the user has assurance that a product labeled for a particular use will perform well when used as per label specifications. In fact, it is not legal to do otherwise. The critical figure is the cost to the user of the formulation per acre, not an easy figure to estimate. Data inputs include the number of applications required, treatment level, pest types involved, type of application, government regulations, cultural practices of different areas, crops to be treated, and even public opinion. The best product for an application with a lower price than that of your competitors will do very well in the marketplace.

If a formulation fails, the *mode of failure* must be determined and a rapid test must then be devised to quantitate it. Only then can one predict the results of efforts to make improvements. For instance, the literature indicates that trifluralin is very prone to photodecomposition upon exposure to sunlight, thereby mandating use of ultraviolet absorbers. But weatherometer tests on some formulations showed no correlation of trifluralin loss with ultraviolet exposure. It soon became apparent that vaporization was the major loss pathway for trifluralin, and that water accelerated the loss. Accordingly, volatilization of the active ingredient in the presence of moisture become the factor to monitor.

Established tolerance defines the amount of a pesticide that may remain on a particular crop at a certain time after application. For this reason use of a very long-lasting, first-order, slow-release formulation may not be a good idea in a case where an established tolerance exists for a classical emulsifiable concentrate formulation.

Phytotoxicity: It is sometimes possible to gain entrance to a market forbidden to a standard formulation. Methyl parathion is very phytotoxic when used on apples and causes blemishes on the fruit. Standard formulations of methyl parathion are not used on that crop for this reason. When microencapsulated formulations are applied, the capsules come in contact with less than 1% of the crop and show no phytotoxicity. The cost of using microencapsulated methyl parathion, specifically, Penncap-M insecticide on apples, is very economical. The product has had substantial success in this market (Fig. 10).

Air contamination reduction is a proved benefit of controlled-release formulation [11]. Airborne methyl parathion was found to be seven-fold less adjacent to a field sprayed with encapsulated methyl parathion than next to a field sprayed with a non-controlled-release formulation. Microencapsulation has been shown to reduce the amount of diazinon in the air when compared to its standard formulation (Fig. 11). As stated earlier, this is important in order to prevent the active ingredient from redistributing and becoming slow release and too spread out to be insecticidal.

Water contamination is decreased by controlled-delivery formulations simply because the active ingredient is retained in the desired area. Pesticide runoff should be reduced if the formulation does not release when diluted in water.

Soil contamination is also reduced, because the capsules do not easily flow through soil but tend to remain where originally placed.

Bacterial decomposition of the active ingredient, in soil use, will be retarded if the active ingredient is not released. In cases where it is released slowly, effectiveness may be less than that of nonencapsulated formulations [12]. Bacteria cannot attack the encapsulate until the capsule wall is broken. If, however, the material is slowly released from the formulation, bacteria in the soil will be able to metabolize the pesticide as fast as it emerges. This would be a good reason *not* to use slow-release products in soil.

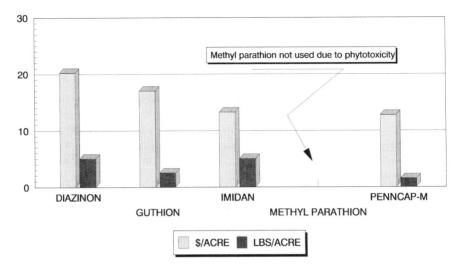

Figure 10 Codling moth control on apples. A comparison of the application rate and cost of various pesticides used.

Photochemical decomposition may be decreased by a controlled-delivery formulation. Ultraviolet absorbers have been incorporated into the walls of capsules for this purpose, and many patents have been issued on these materials [13].

Hydrolysis upon storage and general storage stability are important and must be considered. Aqueous preparations of microcapsules have been found to be exceedingly stable even though the same active ingredients were not stable in the presence of water.

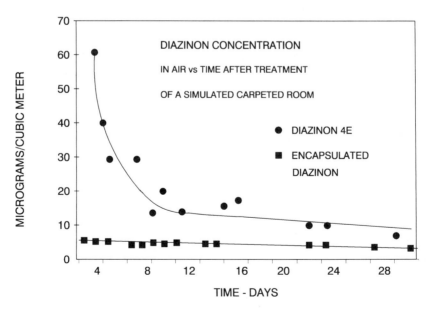

Figure 11 Microencapsulated diazinon gives a much lower level of air contamination than does the standard emulsifiable concentrate.

Knox Out 2FM, a product of Atochem North America, is the only commercial preparation of diazinon sold in water, yet free diazinon in the presence of much water hydrolyzes to innocuous pyrimidines and phosphates. In the presence of traces of water, diazinon hydrolyzes to the very toxic S,S-TEPP and O,S-TEPP [14,15]. Obviously, diazinon sold as an emulsifiable concentrate must be kept free of water. When traces of water are present, either in the original product or by absorption from the air through normal opening and closing of the container, toxic impurities will form. In this manner, the toxicity of standard formulations of diazinon will increase drastically with time, creating a very dangerous situation. Analytic studies have shown that Knox Out 2FM does not contain the aforementioned impurities. Thus, even though Knox Out 2FM is stored in water, its shelf life is projected to be about 10 years at reasonable storage temperatures, without the formation of toxic impurities.

Reduced pesticide load on the environment is accomplished because with the increased efficiency of a controlled-delivery preparation, less pesticide is used in each application and fewer applications may be required.

Substrate absorption is reduced because very little of the substrate comes in contact with the formulation. As calculated earlier, for a 30-μm capsule sprayed at the rate of 1 lb per acre, only 0.6% of the substrate is contacted by the formulation.

Drift reduction through use of controlled-delivery formulations could come about several ways. Evaporation of the toxicant does not occur as it falls to earth, an extremely important consideration when aerial spraying is employed. The controlled particle size of the formulation will also have an effect. A spray droplet cannot be smaller than the smallest capsule. Actually, each spray droplet probably contains many capsules, and as the droplet falls through the air, the water may evaporate. All the capsules in a droplet will then agglomerate together and fall as a group. As a group, the tiny capsules will probably not tend to drift away.

VETERINARY PRODUCTS*

A wide variety of drugs/chemicals is used in the veterinary field to maintain the health and productivity of livestock and pets. Worldwide, these chemicals have an estimated annual value in excess of $6 billion. In the United States, approximately $1.5 billion are spent annually and account for approximately 18% of the total cost of animal production. Chemicals are used to combat and prevent disease, to enhance growth, to improve nutrition, to synchronize breeding, to speed healing of wounds, and to control endo- and ectoparasites. The arsenal of chemicals includes antibiotics, growth promotants, trace elements, hormones, antibacterial agents, anthelmintics, and pesticides. Although research continues to discover new drugs/chemicals, the pace of introductions is likely to decrease because of the escalating cost of development and registration. For this reason, there is growing interest in development of controlled-release technology to enable optimal use of existing compounds.

Although having many similarities, the design, development, and marketing of controlled-release delivery systems in the veterinary field differs significantly from that in

---

* This part of the chapter was prepared by J. Allen Miller, a U.S. government employee, as part of his official duties, and cannot legally be copyrighted. Mention of a pesticide does not constitute a recommendation for use by the USDA, nor does it imply registration under FIFRA as amended. Also, mention of a commercial or a proprietary product in this section does not constitute an endorsement by the USDA.

human medicine. Aside from the more obvious differences in anatomy and physiology of the target species, the differences are primarily in the priority given to such factors as cost, convenience (ease and frequency of application), safety (side effects and residues), and elegance of design. Cost considerations are a major concern in the application of controlled-release technology in the livestock area. Producers are often operating on a small profit margin and must therefore know that the cost-benefit ratio of any management practice is in their favor. Convenience of treatment and frequency of dosing are also very important. Unlike in human medicine, where there is an intelligent, responsible and accessible patient, livestock must be gathered for treatment. Often, cattle are held in very large pastures and the cost of gathering, in terms of labor and stress to animals, can exceed the cost of the treatment. Controlled-release delivery can provide an important advantage over conventional treatments, which might require multiple gatherings.

Controlled-release formulations have been used to deliver a variety of veterinary drugs/chemicals successfully; however, the present discussion will focus only on delivery of pesticides (excluding anthelmintics) used for livestock and pets. Details concerning principles of controlled release are left mainly to other chapters.

Livestock Applications

The health and production efficiency of livestock are adversely affected by a variety of ectoparasites (insects, ticks, and mites). Damage due to these pests is in the form of irritation, loss of blood, annoyance, behavior modification, loss of grazing time, secondary infections, and disease transmission. Losses to the livestock industry include reduced feed-conversion efficiency; decreased weight gains; reduced production of wool, mohair, and milk; reduced value of meat and hides; reduced rate of maturation; and, in extreme cases, even death of the animal. Although difficult to quantitate in dollars, these losses to the U.S. livestock industry are estimated at $3 billion to $5 billion annually [16]. The major pests of cattle include horn flies, stable flies, horse flies, face flies, cattle grubs, Gulf Coast ticks, and lone star ticks.

Conventional methods of control of ectoparasites involve repeated applications of pesticides as sprays, dips, and pour-ons [17]. Repeated treatment of livestock is expensive in terms of labor, debilitation of animals, and costs of insecticide. To compensate for the rapid degradation of the insecticide on animals and to maintain protection for an extended period, the producer must apply larger quantities of insecticide than necessary to control the immediate pest population. Such a practice is wasteful of insecticide, results in greater environmental contamination, and increases the probability of toxicity to the animal and of residues in animal product. In order to reduce the cost of gathering cattle, self-treatment devices such as automatic sprayers, dustbags, and backrubbers were developed. With the automatic sprayers, cattle walking through a narrow passageway trigger a sprayer, which applies a preset quantity of insecticide to the animal's hair cost. Dustbags consist of a porous bag containing an insecticidal dust. Cattle walking beneath the bag brush against the lower edge of the bag, causing dust to be deposited onto the hair coat. The backrubber is made of wicking material that packed into a fabric sleeve and saturated with an insecticide in a light oil formulation. As the cattle rub on the device, insecticide is applied to the hair coat. These self-treating devices are most effective if the animals are forced to use them daily. They are least effective in large-pasture and open-range conditions. Automatic sprayers, dustbags, and backrubbers are still available in the marketplace, and they offer the advantage of being low in cost and in labor of maintenance. They are, in fact, the forerunners of controlled/sustained-release delivery systems in livestock pest control.

Controlled-release delivery systems for livestock applications can be divided into three broad categories on the basis of site of delivery: (1) intraruminal devices, (2) injections and implants, and (3) external attachments. Each of these will be discussed and examples given.

Intraruminal Devices (IRDs)

Controlled-release oral dosage forms have been used primarily in ruminants (cattle, sheep, and goats) because of their unique digestive system. The ruminant has a stomach made up of four compartments, the rumen, the reticulum, the omasum, and the abomasum. In these compartments the food is subjected to digestion by microorganisms before continuing through the digestive tract. The largest of these compartments is the rumen; the volume of bovine rumen can be 50 to 60 liters. Within this large mixing chamber the breakdown of the cellulosic food stuff begins. The microbial degradation is most efficient when the particle size is small. The animal works at reducing the particle size by chewing, swallowing, regurgitating, and chewing again.

The rumino-reticular compartments provide a unique repository for a controlled-release device. The primary problem is that of retention of the device. Not only must the device not be passed on through the remainder of the digestive system, but it must survive the rumination processes. Retention can be accomplished in one of two ways, either by density of the device or shape of the device. Research has shown that a specific gravity of at least 1.6 is required for retention in the rumen and 2.0 for retention in the reticulum [18]. When shape is used for retention, the minimum diameter of the device in the rumino-reticular compartment must be significantly greater than the diameter of the reticular-omasal orifice or the esophagus.

One of the simplest approaches is to use a high-density composition that slowly erodes in the rumen or reticulum. This approach may have originated with the development of the so-called cobalt bullet [19]. To provide a trace element supplement to livestock in cobalt-deficient areas, a cobalt oxide pellet is placed in the reticulum. The cobalt bullet has had a significant economic impact on the grazing industry in Australia. Similar devices are available for delivery of other trace elements such as selenium, copper, zinc, and magnesium.

Two erodible boluses are being marketed for control of dung-breeding flies of cattle (Fig. 12). The Vigilante by American Cyanamid and the Inhibitor by Zoecon deliver

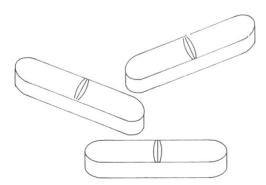

Figure 12   Erodible boluses are used to deliver insect growth regulators to cattle for control of dung-breeding flies.

insect growth regulators (IGRs). The boluses are made of a blend of monostearin, carnauba wax, and the IGR, with barium sulfate as the weighting agent to increase the specific gravity to greater than 2.0 [20]. These boluses lodge in the reticulum, where they slowly erode, releasing the active agent into the digestive tract and ultimately into the manure. The presence of the IGR in the manure dropping of cattle prevents the development of horn flies and face flies in the manure. The Vigilante bolus contains diflubenzuron, a chitin inhibitor, and the Inhibitor contains methoprene, a juvenile hormone mimic. These boluses provide 10 to 16 weeks of control of horn fly and face fly development in the manure of treated cattle [21,22] and, in addition, the Vigilante bolus also controls face fly development. Formulations can be developed that are active for 24 to 30 weeks. As might be expected, release from these systems is not zero-order or uniform. As the bolus erodes, surface area decreases, resulting in a decline in the rate of delivery over time.

Zero-order delivery can be achieved with a bolus through the use of devices such as the Paratect bolus, the Paratect Flex bolus, and the osmotic pump described in Chapter 10. Although the Paratect bolus and the push-melt osmotic pump offer the advantage of zero-order delivery, they do have the disadvantage of being more expensive, and the spent devices remain permanently in the rumen. Such remnants can pose a serious problem to equipment in conventional slaughter operations.

The use of shape for retention in the rumen is an interesting concept being pursued both in the United States and Australia [23,24] and would certainly have application for delivery of larvicides and IGRs. The concept is perhaps not yet as far advanced as the use of high density for retention, but it appears to have potential.

Trilaminate systems, developed several years ago for zero-order delivery of drugs including pesticides and pheromones, are finding application in the veterinary area. The trilaminate has a polymer core containing the drug sandwiched between two rate-controlling films. In one configuration, the rate-controlling films are permeable, allowing the drug to move from the core to the surface. In a sense, this configuration is very much like a reservoir system. A second configuration uses impermeable barriers on either side of the drug-loaded core. Holes through all three layers provide the release path. Rate can be controlled by the number and diameter of holes in the system and the solubility of the active agent [25]. While this technology is currently being used for delivery of an anthelmintic (see Paratect Flex bolus, Chapter 10), it has not yet been applied to the delivery of IGRs, larvicides, or systemically active pesticides. The trilaminate is tightly rolled and held in a cylindrical shape by water-soluble tape. Upon arrival in the rumen, it opens to a slab configuration and is retained by virtue of its shape.

Injectables and Implants

When the objective is systemic activity, the use of a controlled-release injectable or implant can be more efficient than delivery to the rumen. Because of practical limitations in the volume of injectable or implant, these systems have generally been most successful in delivery of drugs that are active at a very low dose.

Several controlled-release injectable or implantable systems are used in the veterinary field for delivery of antibiotics and growth promotants; however, because of the larger required doses, few systems are available for delivery of pesticides. One of the simplest techniques is the injection of a formulation that is insoluble at the injection site and therefore is slowly absorbed from the site. This technique has been used for long-acting penicillin products such as procaine and benzathine. Several other possibilities for injectables including the use of prodrugs, microcapsules, microspheres, and liposomes [26]. Because of the current level of research in these areas, it is likely that we will see greater use of these injectable formulations for controlled delivery of animal drugs.

Implantable systems can also be used for the controlled delivery of chemicals in the veterinary area. These can take on a variety of forms, such as rods, tubes or reservoirs, and laminates. Controlled-release implantable systems have been used for the delivery of growth promotants and hormones for estrus synchronization in cattle. Interest in the use of implantable systems for delivery of pesticides has increased largely because of the recent availability of drugs that are efficacious at extremely low doses.

Implantable rods and microspheres formulated of either poly(d,l-lactic acid) or copolymers of lactide/glycolide containing insect growth regulators were used to deliver systemically active amounts against cattle grubs [27,28]. Microporous polycaprolactone tubing has been used as an implantable reservoir for the controlled delivery of insect steroid analogs against ticks [29]. A sustained-release implant has been developed to deliver invermectin, a potent antiparasitic agent, to cattle. The implant, containing 20% invermectin, was formulated by dissolving the technical drug in a high-molecular-weight polyethylene glycol [30]. When the implants were installed subcutaneously in Hereford steers at a dosage of 200 µg/kg body weight, more than 70% control of adult *Amblyomma cajennense* (F.) and more than 85% control of *A. americanum* feeding on the steers was observed for 7 weeks. In contrast, when steers were treated with a single subcutaneous injection of the commercially formulated drug at 200 µg/kg, the effect on ticks was evident for only 1 week post-treatment. A similar formulation containing 30% ivermectin and administered at a rate of 400 µg/kg provided up to 11 weeks' control.

The increasing availability of drugs which are effective at very low doses (in the microgram per kilogram body-weight range) will undoubtedly result in increased research emphasis in the use of implantable systems for pesticide delivery.

External Attachments

Most people are familiar with the tick and flea collars developed several years ago for use on pets. A similar concept has been applied to the control of ectoparasites on cattle and, to a limited extent, on horses. These external attachments can take the form of ear tags, neckbands, or tail tags. Studies using a visible dye have been used to identify those body surfaces contacted by such attachments [31]. Because the distribution of insecticide from these various devices is not uniform over the body surface of the animal, the device must be matched to the behavior and habits of the target pest so that insecticide will be applied to preferred feeding or resting sites on the host (Fig. 13).

The insecticidal ear tag (Fig. 14) was introduced in the mid-1970s. Although originally developed for control of the Gulf Coast ear tick, the tag has found its greatest use in the control of horn flies, a much more serious pest of cattle. Albeit there are approximately a half-dozen major manufacturers, until recently, the tags were basically alike in that a pyrethroid was incorporated into a plasticized polyvinyl chloride matrix. The insecticide moves to the surface of the tag by a diffusion process. The insecticide is then rubbed from the surface of the tag to the hair coat of the animal, where it is contacted by adult horn flies. Because the horn flies move over various areas of the animal in response to animal movement, sun, shade, and temperature, it is not necessary that the insecticide

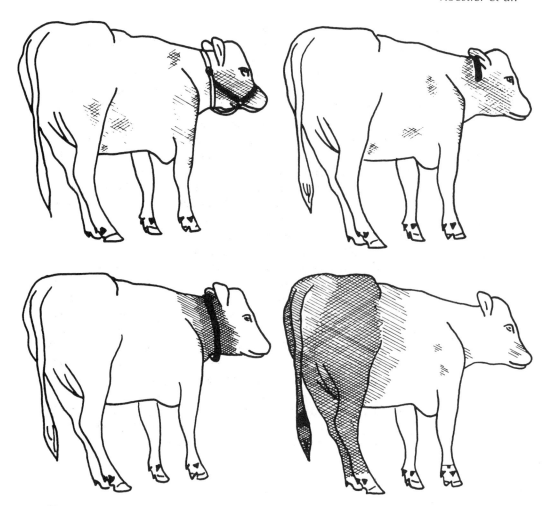

Figure 13 Dye transferred to hair coat of a cow from (clockwise from upper left) a halter, ear tag, tail tag, and neckband.

be uniformly distributed over the body of the cow. The tags are capable of controlling horn flies on cattle for 16 to 24 wks.

The release-rate profiles of these monolithic tags containing pyrethroids have been described [32]. Figure 15 shows the cumulative fraction of pesticide released and the release rate over time for an 8% fenvalerate tag and a 10% permethrin tag under normal use on cattle. The minimum effective delivery rate for control of horn flies on cattle was determined to be 1.0 mg/day for the fenvalerate tag and 1.9 mg/day for the permethrin tag. The tags tend to lose their effectiveness when 50% of the insecticide has been released. The release of the pesticide was found to conform to the Baker and Lonsdale model [33] based on Fick's law of diffusion applied to a slab configuration. As discussed in Chapter 3, the release of the first 60% of the solute is a function of the square root of time and is approximated by

Figure 14 Cows are fitted with insecticidal ear tags for control of horn flies, face flies, and ear ticks. (Courtesy Fermenta Animal Health.)

$$\frac{M_t}{M_\infty} = 4 \left(\frac{D\,t}{\pi\,l^2} \right)^{1/2} \qquad \text{for } 0 \le \frac{M_t}{M_\infty} \le 0.6 \tag{18}$$

whereas the release of the last 60% is approximated by

$$\frac{M_t}{M_\infty} = 1 - \frac{8}{\pi^2} \exp^- \left(\frac{\pi^2 Dt}{l^2} \right) \qquad \text{for } 0.4 \le \frac{M_t}{M_\infty} \le 1.0 \tag{19}$$

where

M_t = cumulative mass of solute released at time t

M_∞ = total possible release at infinite time

D = diffusion coefficient (cm^2/time) of the permeator

l = one-half the thickness of the slab

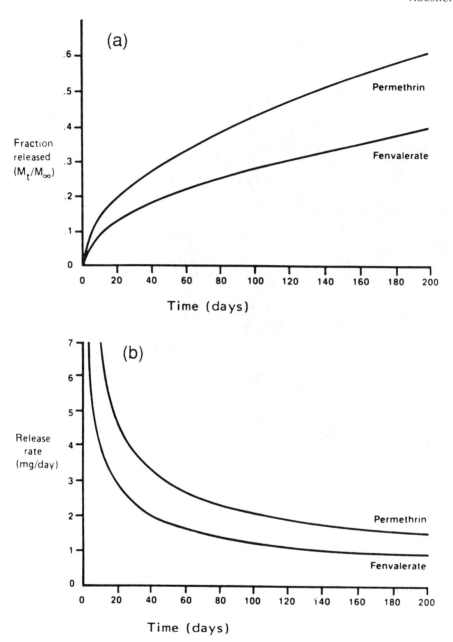

Figure 15 Fraction released (a) and release rate (b) from fenvalerate and permethrin insecticidal ear tags.

The dynamics of the diffusion process occurring within the tags is illustrated in Fig. 16, which shows the concentration profile from surface to center within the fenvalerate and permethrin tags prior to use and after 24 weeks on cattle.

Because of a problem encountered with horn fly resistance to the pyrethroids, new tags are being marketed with combinations of pyrethroids and organophosphates or organophosphates only. Some also include a synergist for the pyrethroids.

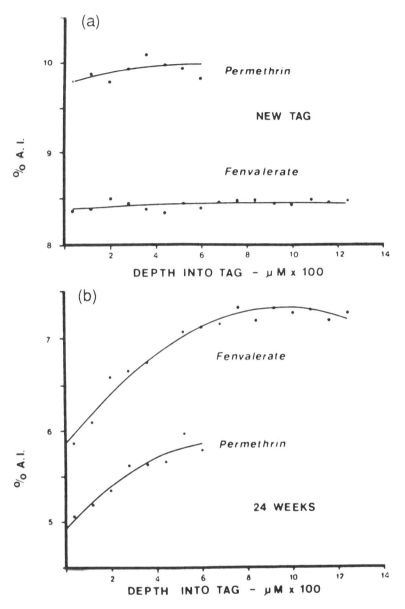

Figure 16 Profile of pesticide concentration within new tags (a) and after 24 weeks (b) on cattle.

Several other controlled-release systems are being used or explored for use as insecticidal ear tags. A membrane system developed by Bend Research and Conseps Membranes has been applied to ear tags [34]. Essentially, an insecticide reservoir, one side of which is a rate-controlling membrane, is formed on the surface of an ear tag. The release rate is controlled by the permeability, area, and thickness of the membrane. Unlike the monolithic tags, this configuration gives nearly constant release.

The trilaminate system has been used at least experimentally for ear tags. In this configuration, the insecticide-impregnated core is sandwiched between two rate-controlling membrane films. Such a configuration has been shown in field trials on cattle to provide near zero-order release until ca. 70% of the load is dispensed. Systems dispensing pyrethroids have been successfully field-tested against horn flies on cattle.

Both monolithic and reservoir neckband systems have been used for controlled delivery of insecticides to cattle. One of the important advantages of a neckband is the ability to carry greater loads of active agent. The ear tag is limited to an approximate total weight of 17 g.; greater than 17 g results in the tag enlarging the hole in the ear and eventually being lost. Tags are usually 8–15% AI. In contrast, a neckband can be as heavy as 100–200 g without difficulty. Because of the current problem with horn fly resistance to pyrethroids, it may be necessary to change to a class of insecticides with a different mode of action. The organophosphates are a logical choice, but because they are generally less toxic and more rapidly degraded, it will be necessary to deliver larger quantities of these insecticides. Using thin-wall PVC tubing as a reservoir in a neckband for cattle, organophosphates can be delivered at rates of 20–25 mg/day [35]. The variety of materials available for fabrication of reservoirs means that release can be tailored to practically any need.

Problem: An insecticide is dissolved in a plasticized polyvinyl chloride and injection-molded into an ear tag for cattle. The diffusion coefficient on the insecticide in this matrix is found to be 1.795×10^{-6} cm^2/day. The 10-g tag contains 8.7% insecticide and is a uniform 2194 ± 7 μm thick. The tag is found to lose its efficacy against horn flies on cattle when the release rate reaches 1 mg/day. Determine the expected duration of efficacy on cattle and the fraction of the original quantity of insecticide remaining in the tag when it is no longer efficacious.

Solution: Take the first derivative of the Baker and Lonsdale equation (see text) for early-time approximation of diffusion of the active ingredient from a polymeric slab to obtain the release-rate equation:

$$\frac{\partial M_t}{\partial t} = 2M_\infty \left(\frac{D}{\pi l^2} \right)^{1/2} \tag{20}$$

Solving for t when the derivative equals 1 mg/day results in t = 144 days, that is, the expected duration of efficacy.

Going back to the Baker and Lonsdale equation, the fraction release by day 144 can be found to be 0.33; therefore, 67% of the original quantity of insecticide remains in the tag at the time the tag loses its efficacy against horn flies.

(Note: these values are those determined for the fenvalerate ear tag discussed in the text.)

Another interesting concept that has not received much attention but that appears to have potential is the use of insecticidal tail tags. Tail tags have been used effectively for both control of horn flies and lone star ticks on cattle. Cattle switch their tails as part of their defense against horn flies, and therefore the tail is a logical insecticide applicator. One of the main difficulties is the development of a system for retention of the tag in the tail switch. The combination of a neckband and a tail tag have been used successfully for the control of lone star ticks on cattle. With this particular species, 80% of the ticks attach in the area of the neck, dewlap, and brisket area, and in the escutcheon area from the tail head to the udder of cattle. Therefore, controlled-release devices that impact those areas can provide good tick control.

Pet Applications

The close bond between humans and their pet animals has resulted in a multibillion dollar pet-care industry in the United States. The control of fleas and ticks, the primary ecto-parasites on cats and dogs, is an important component of that industry. Fleas and ticks feed on the blood of these animals and cause irritation and annoyance, scratching, and in some cases allergic-type reactions. Because of the proximity of people to their pets, both of these parasites frequently bite and irritate humans. In addition, fleas and ticks can transmit diseases to other animals as well as to humans.

Conventional means of control include the use of pesticides formulated as dips, dusts, or powders, and pressurized sprays applied directly to the pet animals. Premises treatments using powders or sprays are also used to treat the bedding and loafing areas. In severe cases the whole ranging area, either the interior of the home or the entire lawn, may be treated. Such treatments are usually of short duration (< 2 weeks) and must be repeated for effective control.

Controlled-release technology has been used effectively for delivery of pesticides for control of parasites on cats and dogs. Flea and tick collars have proven to be a popular means of control with pet owners. In this particular market, convenience is of higher priority than cost of control, and despite the close contact between people and pets, convenience of treatment is a major consideration. Chapter 10 contains a discussion of the state of the art in applications of controlled-release technology to the delivery of pesticides to companion animals.

The Resistance Problem

A discussion of controlled-release delivery systems for veterinary applications would be incomplete without a brief reference to the resistance problem mentioned earlier. The development of resistance to the pyrethroid ear tags by the horn fly serves as a good case study [36]. The development of the insecticidal ear tag provided the producer with an innovative tool for control of the horn fly. Because the ear tag could provide season-long control without repeated handling of cattle, the technology was rapidly accepted by the producer. As a result, the pyrethroid tags were used extensively; for example, it is estimated that in 1984, >50% of all cattle in Texas were tagged. Horn fly populations, particularly in the southern United States, were subjected to greater selection pressure by a single class of insecticides than ever before. The continuous exposure of multiple generations (as many as 13 to 14 generations in a single season) to the heavy selection pressure exerted by the tags resulted in horn flies developing significant resistance to pyrethroids in as little as 2 to 3 years.

Researchers are actively seeking solutions to the problem through resistance man-agement strategies. Such strategies range from a periodic change of the class of insecticides used in delivery systems, the complete elimination of the use controlled-release delivery and reversion to the periodic applications of earlier days, and alteration of chemical and nonchemical strategies. As one might well imagine, the application of controlled-release technology is near the center of the controversy. The needs of the producer for long-lasting, inexpensive ectoparasite control are being balanced against the reality of resistance development.

Certainly controlled-release delivery systems will continue to provide important ben-efits to the veterinary area. However, the future development of such systems should consider the potential for development of resistance against the target pest.

The Future

The development and application of controlled-release delivery systems for veterinary applications is a fascinating field with tremendous opportunities. In some cases, techniques developed in human medicine can be transferred directly to the veterinary area; but in other cases, work in the veterinary area can pioneer techniques that will ultimately benefit human medicine. The introduction of new products is expected to increase rapidly. Even if no new controlled-release technology were developed, the application of existing technology to the veterinary area offers both great challenge and great potential. Many current veterinary applications are yet in their early stages of development. However, the major challenge will remain that of providing useful systems with a favorable cost-benefit ratio. The use of intraruminal devices for delivery of pesticides and anthelmintics is expected to increase. Pulse release in the rumen appears to be a practical and achievable goal, worthy of development. As new, more potent parasiticides are developed, many of the delivery systems presently being used for antibiotics, growth promotants, and hormones will find application for pest control. In the short term, progress is expected in the use of implants, microspheres, and microcapsules as delivery systems.

In addition to the challenge of developing practical controlled-release delivery systems, the practitioner will face the challenge of how these systems can best be used in an effective, strategic pest management approach. There is growing concern as to the potential for resistance development and the role of controlled-release technology in its creation or avoidance.

PHEROMONES

Pheromones have been investigated over the past 20 years as insect behavior-modifying chemicals. They are moderately complex organic molecules, which are generally very labile in the environment. For example, the sex attractant pheromone of the gypsy moth, *Lymantria dispar* (L.), which is (2S-cis)-2-decyl-3-(5-methylhexyl)oxirane and known as (+)-disparlure [37], is not long-lasting when free in the environment. Pheromones containing conjugated diene systems are sensitive to ultraviolet light and thus decompose photochemically. The main component of the sex pheromone of the Egyptian cotton leafworm, *Spodoptera littoralis*, is (Z,E)-9,11-tetradecadien-1-ol acetate, which is quickly isomerized by exposure to ultraviolet light [38]. Other pheromones often contain groups such as aldehydes, ketones, and esters, which may be altered by environmental exposure.

Means of applying protecting pheromones were required to increase the efficiency of their use and thus bring the price of their use to an acceptable level. Among the many systems tried to control the release of pheromones are silicone rubber, regular rubber, polymers such as polyvinyl acetate and polyvinyl chloride, foams, capsules, multilayer laminates, and hollow fibers.

Use of pheromones has always been one of the great promises of integrated pest management techniques. Perhaps these materials could replace or at least minimize the use of the poisons we are now forced to use. Potentially, they appeared to be nonresistance-promoting, noncarcinogenic, nonharmful to the environment, and just generally innocuous.

Applications of Controlled-Release Pheromones

Controlled-release pheromones are used in three different ways.

1. *Monitoring insect populations*. Pheromone traps are set up in areas to determine the presence and number of a certain species of insect in a given area. The gypsy moth

detection program has been going on for 50 years in the United States [39]. Before the synthetic pheromone was developed in the early 1970s, female moths were used to bait the traps [40].

2. *Mating disruption*. High concentrations of pheromones applied by air or from ground-applied dispensers confuse the male insect in his attempt to find a female. Recently, the tomato pin worm pheromone has been employed in hollow fibers glued to tomato stakes. Damage to the tomatoes from the pin worm was reduced [41].

3. *Mass trapping*. Many pheromone traps are distributed into an area in the attempt to trap a large proportion of a population. This can result in reduced damage in a particular area. The spruce bark beetle, *Ips typographus* (L.), was mass trapped in a large program in Norway in 1979–1980 [42].

Application Problems

Regardless of the means used to apply pheromones, there are some profound problems to be overcome in their use, and not just chemical problems. For instance, sex attractants, being pheromones, modify behavior, usually of the male of the species. They are at present used very successfully in monitoring traps and serve a very important purpose. To detect the Mediterranean fruit fly, *Ceratitis capitata* (Wiedemann), in a particular area, one would use a trap containing trimedlure. Although trimedlure is a synthetic lure and not a true pheromone, it suffers from high volatility, leading to lack of environmental persistance. The performance of the trap is such that as the population of the target insect decreases, the greater is the percentage of the population that will be caught. The reason is that if the trap attractant is the only attractant around, a male stands a good chance of locating it. However, if many females are emitting pheromone at the same time, many males will find females and not the trap. In effect, the females would then be using the confusing technique on your monitoring trap. The situation can be turned around, however. If an entire area is "saturated" with the attractant, then the males will not be able to home in on the females (theoretically). As stated above, efficiency is highest at low populations. Attractant or pheromone saturation of an area can be extremely effective as the insect population approaches zero. In areas of high insect population, the technique does not work at all. Such is the case with the gypsy moth. In areas containing many moths, where the males don't have to fly more than a few feet to see a female, the confusion technique will not work at all. It works quite well if there are a few hundred moths per square mile. This is done only for the purpose of eradication of incipient infestations in areas far removed from existing gypsy moth infestations.

Another problem with pheromone application can be illustrated by the case of the pink boll worm, *Pectinophora gossypiella* (Saunders), one of the most serious pests on cotton [43]. A certain release rate is required of a controlled-release formulation in order to have the desired effect on the insect. The pink boll worm moths are night-flying insects, and so the pheromone level must be maintained at night, when the temperature is low. During the daytime, however, field temperatures in California can reach 125°F, causing premature release of most of the pheromone when it is not needed. First-order-releasing formulations are not good in this situation.

A partial solution to this problem has been found in the application of hollow-fiber dispensers for controlled release of the pink boll worm pheromone, gossyplure [44]. The mass released versus time curve initially has a steep slope followed by an extended region which approximates zero-order kinetics. The mass released is proportional to $\sqrt{t}$. Hollow fibers approximating zero-order release, therefore, waste less pheromone during the hot day, and at night the proper amount of attractant is dispensed.

Successful use of the pheromone gossyplure on cotton in California also presents another situation relating to the use of pheromones. In California, the pink boll worm is the only serious pest of cotton. With pheromone treatment successful, insecticide sprays are not used or needed because there are no other damaging insects. However, this is a rare situation. In most areas of the country, the insect complex is wide and varied. With many species of insect pests present on a crop, treatment specific for only one pest is not normally economical.

Pheromone Controlled-Release Devices

More techniques have been employed for controlling the release of pheromones than for any other pest-control agent. See Table 2. This is most likely due to the high cost of the pheromones, which makes the cost of the delivery fabrication technique insignificant by comparison.

AQUATICS

Antifouling Systems

Introduction

Marine fouling refers to the attachment of sessile organisms to artificial objects in the marine environment. These same organisms, when attached to natural objects, are a part of the ecosystem. It is the interference with human use of the oceans and waterways that dictates the use of the term "fouling" in connection with marine growth. Marine plants and animals attach to ship hulls, buoys, pilings, and any other objects under the sea. They destroy the streamlines that are essential to efficient movement through water. They accumulate relentlessly, adding tons of weight which can swamp small craft and sink channel-marker buoys. Prior to the development of modern antifouling paints, it was not uncommon for a single ship to be dry-docked with 300 tons of organisms clinging to the hull [45].

Toredoes and limnoria destroy wood. Barnacles and other calcareous growth cut through protective coatings as they grow, and their secretions chemically etch the surfaces to which they are attached. This, in turn, exposes the structural material to a host of physical, chemical, and biological activities that can destroy the integrity of the structure and make it unfit for service.

Table 2 Materials Used for Controlled-Release Pheromone Dispensers

Laminates	Rubber stoppers
Cigarette filters	Corncobs
PVC disks	Rope
Hollow fibers	laminate flakes
Rubber septa	Nylon microcapsules
Polythen capsules	Glass capillaries
Polyethylene vials	Gelatin microcapsules
Silicone rubber	PVC tubes

Corrosion of metals is perhaps the most destructive consequence of marine fouling. Fouling can cut through an anticorrosive coating and expose the underlying metal to the salt water. Often corrosion is concentrated in small areas; under the worst conditions, it can generate a hole in a ship's hull plate in a few months.

The accumulation of marine growth will block sea-water conduits of all types and often causes damage to engines, pumps, and other related equipment. Ship valves are often severely affected, and any submerged moving part can become obstructed and inoperative.

Mussels are particularly bothersome to utility plants that get their cooling water from the oceans. The intakes are large, and the flow of water provides a continuous supply of food, an ideal situation for the mussels. Left alone, they will coat the walls of the intakes, settling on top of each other as well as on other fouling species, to a thickness of 3 ft or more in a single season. Two hundred and sixty-six tons of fouling organisms were removed from a New England utility tunnel after 1 year of accumulation [46]. The flow of cooling water is seriously restricted by the marine growth. In addition, individual shells fall off and are carried to small condenser tubes, where they may completely block the flow of cooling water; tube ruptures and costly repairs follow.

Cleaning marine fouling from surfaces is expensive, difficult, and unpleasant. Also, it is increasingly difficult to dispose of large amounts of dead and rotting marine organisms following removal. Estimates of the worldwide total annual cost of fouling to the maritime community is billions of dollars. The U.S. Navy, looking only at increased fuel consumption resulting from hull fouling, figures the wasted fuel cost to be $150 million per year [47]. Saving a small percentage of the cost of this single item would justify a development program for longer-lasting antifouling coatings. If one extended this one factor to include freighters, oil tankers, and pleasure craft, just in the United States the cost would be truly significant.

Background

Marine fouling has been a serious economic, military, and safety problem since people first began using the oceans. The ancient Phoenicians and Carthaginians coated their ship hulls with pitch and possibly copper sheathing to provide a measure of protection. Wooden ships were beached and cleaned regularly. Fresh-water ports were used to kill salt-water growth and to provide a safe haven from further infestation. Naval tactics included consideration of length of time at sea, the effect of fouling accumulation on speed and maneuverability, and the threat of sinking due to damage by wood-boring toredoes.

However, it was the advent of metal ships and fossil-fuel propulsion that emphasized the problem of fouling. As ships grew tremendously in size and cost, it became increasingly difficult to clean fouling off the hulls. The ships had the size, speed, range, and reliability to sail the world with ease, but they were held back by tiny marine plants and animals that began to accumulate on the hull as soon as it touched the water. As late as 1952, the British Admiralty figured the loss of efficiency to be $1/2$% per day for a ship after it was cleaned, painted, and put back in service [48].

Many items used in the ocean suffer more than a loss of efficiency as a result of fouling accumulation. Consider an underwater port or window that would be useless with a small amount of fouling. Sonar has been shown to lose up to 50% of its incident energy due to fouling accumulation after only 165 days' exposure [49]. At this level of performance, sonar has little or no value. Communication cables that are cut, broken, or punctured due to fouling fail and must be raised and repaired. Wood structures and ships are completely destroyed by ship worms if not protected.

There is extensive literature on historical efforts to prevent fouling on ships. While interesting, a review of this history serves no purpose for this chapter other than to emphasize the effort expended on the fouling problem over the years. However, a few examples are noted below.

Copper sheathing has been a continuing, if somewhat intermittently used, material. Arsenic and sulfur mixed with oil was used in 412 B.C. [50]. Lead sheeting was used as early as the third century B.C. by the Greeks. It was attached to the ship's hull with copper or gilt nails, usually over an insulating layer of paper or cloth, suggesting recognition of the corrosive effect of lead on iron [51].

The Romans used lead sheathing; and several of their ships, with the sheathing intact, have been recovered [52]. Lead was forgotten for several centuries but was used in fifteenth-century England. Indications are that it was copied from Spanish ships [53]. In 1500, Leonardo da Vinci designed a rolling mill for making sheet lead [54]. In 1682, the use of lead was discontinued by the Admiralty, and in 1761 copper sheeting was successfully introduced [55].

In 1883, a patent was granted for a reservoir placed in a convenient part of the ship containing a poison combined with an oily substance that was to ooze constantly through perforated pipes along the bottom and sides of the ship [56]. Yet another patent was for bags to be filled with compounds of phosphorous, which were to be dragged with ropes along the hull [57]. In 1915, a patent was granted for an antifouling paint containing radioactive materials [58]. Through the history of antifouling development, an unending variety of toxic and common materials were tried in paint formulations. By the 1930s, paint systems were the only practical answer to the fouling problem. They were adequate to the extent that people had adjusted to them.

It can be argued with some conviction that controlled release has been used to control marine fouling for centuries, perhaps starting with seafarers as early as the twelfth century B.C. There is no question that many of the efforts achieved elements of controlled release in one or more of its many forms. It also seems likely that over such a long period of time, the concept of ''release'' must have been envisioned, but it does not seem to have made an impact on the thinking of the time.

Looking at history from another point of view, we see an early recognition of the use of a barrier film to prevent toredo infestation and damage. Lead and copper were used for this purpose because they were available and also demonstrated antifouling qualities. Perhaps it was believed that the antifouling was necessary to control the toredoes, but we now know that a simple plastic film will work as well.

A marine paint or coating was also needed to seal and protect the ships from water penetration. The emphasis was on the development of a durable coating film loaded with antifouling, or later, anticorrosive ingredients. It was, in effect, an attempt to develop a coating that would act as a barrier film. It was a paint mentality that developed and not controlled release.

During the period leading up to World War I, the U.S. Navy initiated research to obtain more fundamental knowledge of how to prevent fouling. Attention ultimately focused on the question of what property of the paint is responsible for the antifouling action. In 1945, it was determined that the antifouling action of currently successful shipbottom paints depends on the rate of solution of the toxic material [59].

The first controlled-release formulation was developed in 1964 in response to a fouling problem with sonar domes. The state-of-the-art domes were constructed of reinforced rubber, which protects the transducer from the open sea and provides a window that is invisible to the sonar signal. Antifouling paints applied to these domes have a very short life due to the high power transmission of the acoustical signal. Calcareous fouling on the dome creates cavitation as the ship moves through the water, leading to interference and distortion of the signal. The concept of adding a toxicant to the rubber-molded dome was found to be feasible. An organotin, bis(tri-*n*-butyltin)oxide, was incorporated into the compound as an antifouling agent. It was soluble in the rubber matrix and both invisible and impervious to the sonar signal. The toxicant moves by the *diffusion-dissolution mechanism* to maintain a foul-free surface for up to 10 years.

Antifouling Environment

The most important and obvious design criteria for a controlled-release antifouling coating is that it must work in an aqueous environment. There are almost endless variations of exposure. Fouling organisms differ with geographic location, water temperature, salinity, water depth, tidal range, and other factors. These variations must be considered when developing an antifouling coating.

The U.S. Navy currently tests in the tropical Atlantic, tropical Pacific, and a cold water (60°F) site. This selection of test sites provides a good cross section of fouling species as well as a check on how temperature affects the performance of the coating. Low temperature will slow the release rate of CR systems. This can be very beneficial in extending the life of the coating by reducing the release of toxicant when the ship is in cold water where no fouling occurs, or when the ship is in deeper and thus colder water.

Coatings designed to provide long-lived protection in tropical ports, where water temperature can be above 90°F, may have a release rate that will fall below the fouling threshold in cold-water areas such as the west coast of United States. When this happens, the coating will fail even though the active agent is available.

In service, antifouling coatings may be static on buoys and oil-drilling platforms or intermittently moving on ships. There are also dynamic areas such as rudders and around propellers, or hydrofoils on high-performance craft. On submarines, coatings are subjected to alternate periods of static and moving service plus the pressure of deep ocean submergence. The top side of a submarine must be antifouling despite its being out of the water for long periods of time, exposed to direct sunlight, and then exposed to the pressure of deep ocean submergence.

The waterline area on a ship (known as the ''boot top'') moves up and down with the weight that the ship is carrying. On large container ships and tankers, the variation between the high and low waterlines can be more than 20 ft. The alternating wet and dry exposure, coupled with the exposure to physical damage from floating objects and chemical damage from chemical pollutants, make this a difficult area to protect.

There are several types of panel tests to aid in the evaluation of coatings. Test protocols, designed to address the various types of exposures outlined above and to indicate how the coating might perform in service, fall into two categories:

1. Dynamic testing, where the panel is subjected to the wear and tear of flowing water
2. Static testing, where the panel is exposed to marine fouling accretion

Dynamic tests take many forms, depending on the severity of exposure desired. The most common is a rotating-drum device that exposes the paint sample to a specified peripheral speed in the water. A typical test protocol of repetitive exposures consists of

30 days on the rotating drum alternated with 30 days' static exposure to marine fouling, this over a 12-month period (ASTM D 4939).

An alternative is to erode the test panel with high-velocity water. In this case, the panel is static and the water is pumped and concentrated to give the desired forces. Cavitation erosion is conducted in a closed water chamber where a disk is rotated at high speed to create the implosive forces.

Static testing of antifouling coatings consists of suspending the test samples in the fouling environment and recording the results. Variations arise due to vertical zonation of fouling species and the special problems that the water-air interface poses for coatings. The basic approach is to hang panels from a specially designed dock so that they are at least 1 ft below the surface at all times.

The dock may be of a floating or stationary design. If the dock is floating, then the panels remain at a constant depth. If the dock is stationary, then the panels vary in depth between 1 ft at low tide to the greater depth determined by the high tide. In tropical areas, the tide range is approximately $2\frac{1}{2}$ ft twice each day.

Tidal fluctuation is beneficial, as it subjects the panels to a broader fouling zone. On the negative side, some seasonal adjustment of lines is necessary due to variations in the mean tide level. Also, at more northern sites, the tide range is so large as to make stationary docks impractical. Static immersion testing is covered by ASTM D 3623.

Variations of static exposure testing include partial submergence with the water line on the test specimen at a specific point on the panel at all times, a test that is helpful in evaluating the effects of chemical floating on the surface. In tidal zone exposure, the panel is immersed in the water at high tide and exposed to the atmosphere at low tide. For splash zone exposure, the panel is not immersed but is in constant contact with the wave splash. For atmospheric exposure, the panel is positioned well above the water but exposed to the corrosive sea atmosphere and salt-water spray.

Finally, antifouling coatings must be designed to resist chemical attack. If we were a more perfect society, this would not be an important factor. But as things stand today, harbors, boatyards, marinas, and ports are sources of pollution; in addition, they are located in or near urban areas, which are major sources of chemical pollution, as the many objects floating in harbors attest. Even static tests are subject to a tidal flow of 3 to 5 knots and wave action. Floating objects along with silt and sand in suspension will damage and erode antifouling coatings.

Controlled-Release Systems

A paint has three main ingredients: a polymer to develop the film; a solvent or carrier; color and additives to enhance physical properties. An antifouling paint requires a fourth item, the biological agent.

To be effective, the biological agent must work against all fouling forms. It must combine in a dried polymeric film in a manner that will enable it to release over an extended period of time at a rate above the fouling threshold. The combined coating must have and retain physical qualities that will enable it to endure the service exposure. Finally, the biological agent must combine in a film that will adhere to the surface to be protected. The release rate can be considered as a weakness in the paint structure. It is not uncommon for this to affect the adhesion.

To be practical, a biological agent must be equal to or more effective than agents currently in use. The cost of the final product must be competitive, and the active ingredient must be environmentally acceptable.

It is possible to clear the water of fouling organisms in small, select areas. Chlorine is fed into utility-cooling tunnels for this purpose, and it has also been used in floating bags around the hulls of small pleasure boats when they are in dock. Biocides have been released around the hulls of ships while they are in heavily fouling ports. An electrical charge maintained continually through the metal hull will also prevent fouling.

However, for the most part, the oceans are too great, with too little active agent available and too many fouling organisms, to attempt to create a foul-free environment. The only practical approach is an antifouling coating.

As ships become larger and more expensive, it is increasingly important to extend their service time, keep their speed up, and hold their fuel consumption down. A small boat can be pulled from the water at almost any place or time. However, there are only a few dry docks in the world that can accommodate the large oil tankers. It is therefore very expensive to haul these ships out of the water for repair and painting. Thus, it is important that the antifouling protection last as long as possible to keep the ships in service and operating at high efficiency.

Twenty years ago an antifouling paint was considered good if it kept the hull clean for a year. Today, as the result of controlled-release technology, little interest can be generated for a coating that does not last 5 years with promise of perhaps a 10-year effective life.

A long-lived antifouling coating is not forgiving. The low release rate does not provide a radius of biological activity. Areas not properly covered in the application process will fail, and even slight problems with mixing and dispersion of ingredients are likely to show up as failures. Often overlooked in an antifouling development program, coating adhesion and film integrity must also be equal to the long life expectancy of the controlled-release biocide.

Most antifouling paints in service today follow the lead of the U.S. Navy, with cuprous oxide being the toxicant of choice. The mechanism of release is leaching, wherein the toxicant molecule releases copper ions upon contact with sea water. The leaching mechanism depends on an expanding pore structure so that water can reach deep into the film as the surface molecules are depleted. The system requires a high copper oxide concentration to support the system dynamics. Most copper paints have a cuprous oxide loading of 85–92%; in those below 70%, leaching will shut down and fouling will occur. The system will also shut down when the leaching pathways through the film become so complex as to restrict transfer of the copper ions to the film surface.

In a second type of system, the matrix as well as the toxicant may be lost from the paint. Three possible mechanisms by which this takes place are

1. Dissolution of the matrix
2. Bacterial action on the matrix
3. Mechanical erosion of the matrix

With this system there are more variables to work with, and the goal is to adjust the loss of matrix to the leaching rate of the cuprous oxide toxicant. Unlike the leach-only mechanism, the effective life of this system can be extended by increasing film thickness.

At the extreme limit in matrix loss of antifouling systems are the exfoliating coatings that contain little or no active agent. These systems rely on shedding of film to keep the surface in operating condition.

The *diffusion-dissolution mechanism* was introduced to antifouling research with the invention of antifouling rubber [60]. Various organotins were incorporated in elastomers, and the leach-rate mechanism was utterly different from that associated with antifouling paint films. It had been universally observed by those in the field that the foul-free lifetime of paint films was neither predictable nor consistent. Table 3 is a typical rubber formulation.

With the diffusion-dissolution system, no measurable gradient was observed in the partially depleted coating. The dissolution rate decreases with falling concentration, but a gradient effect is not present, save possibly right at the water interface. There also is no toxic loss due to velocity. Toxic loss cannot exceed the surface replenishment rate, so speed and water turbulence make little difference [17].

It was possible to predict lifetime based on the diffusion-dissolution mechanism. Assuming true diffusion, the following equation was developed:

$$T = \frac{a}{k} \ln \left(\frac{C_0}{C_t} \right) \tag{21}$$

In this formula, T, the foul-free lifetime, is determined by the logarithmic ratio of C_0, the initial toxic concentration, to C_t, the attachment threshold; a is the sheet rubber thickness and K is a constant that depends on the *compounded* rubber.

Scientists at the U.S. Naval Research and Development Center synthesized organometallic polymers which were applied directly and in combination with other base materials to marine test panels [18]. It was reported that "an optimal antifouling performance material exhibiting a minimal amount of leaching can be produced by varying the ratio of organometallic monomer to inert co-monomer along the copolymer back-bone." Effective antifouling properties are observed with a magnitude (or better) reduction in tin loss. The original concept was to produce a "nontoxic" toxicant with effects apparent only on surfaces treated with it. This research appears to have made great strides toward that goal. Many controlled-release techniques have been evaluated for antifouling coatings with reasonable success. Microcapsules have proven helpful in attaining high loadings of liquid toxicants. Reservoir systems have maintained toxicant levels in working coatings and extended useful life. Paint systems that vary the permeability and toxic loadings of various coats have helped with problems of adhesion and delivery.

Antifouling Conclusions

It is only during the last 30 years that fouling has been subject to systematic scientific inquiry. Its importance, however, has been recognized from very ancient times. The

Table 3 Antifouling Rubber Formulation

Ingredient	Composition (by parts)
Neoprene WRT	100
Zinc oxide	5
Magnesium oxide	4
Petroleum wax	0.5
FEF black	12–20
PBNA	2
MBT	1
Lauric acid	0–4
bis(Tri-*n*-butyltin) oxide	0.02–20

written records of ship bottom treatment date back to the 5th century B.C., but the search began even earlier.

Historically, antifouling development progressed in three stages:

1. The repeated introduction and use of metallic sheathing culminating in the discovery of copper sheathing as an effective antifouling surface
2. The failure of copper on iron ships, following their development, because of galvanic action
3. The successful development of antifouling paints that could be applied to metal hulls through use of an anticorrosive coating

The quest for more fundamental knowledge on means of preventing fouling led to the discovery that the antifouling action of copper paints depends on the rate of solution of the toxic material. Speculation on how to protect Sonar domes from fouling prompted an investigation of toxicants and use of the *diffusion-dissolution mechanism* to obtain controlled release that is effective for 5 to 10 years.

We have come a long way, but to the people in the Great Lakes region facing the zebra mussel (*Dreissena polymorpha*) and serious fouling for the first time, the situation appears hopeless. There appear to be no answers, only opportunities to come up with a new effective antifouling agent to reverse the situation.

Chronicity Phenomenon

The control of aquatic weeds is a major problem in the United States and throughout the world. The ever-increasing mobility of people has expanded interest in exotic plant and animal species. In almost all instances, aquatic weeds are a problem because they have been introduced into areas that are favorable for their development and where they have no natural enemies to stress the plants and check their growth. With no natural restraints, they quickly multiply and become an esthetic and economic disaster in almost all untreated areas. They clog irrigation systems, block navigable waterways, interfere with drainage and flood control, eliminate the recreational uses of lakes and streams, and provide a breeding site for pests and disease vectors.

Control of aquatic weeds may be accomplished by biological, mechanical, or chemical methods. However, technology and economics of mechanical control have not reached the point where they are practical in most large-scale programs. Biological control is difficult because of the research time required to find a control that, when introduced against the target species, will not alternately, or even preferentially, destroy desirable plants. Even then, if it is successful, might not another weed species immediately fill the void left by the biological control?

Consequently, the major effort toward controlling aquatic weed pests has been, and will continue to be, chemical control. Governmental agencies and environmental groups express concern over this approach, and subject large-scale applications of chemicals to many restraints. Controlled release is viewed as a means of enhancing chemical control and reducing the environmental impact.

Herbicides are conventionally applied in acute concentrations to aquatic plants that have reached the problem stage. Near 100% mortality must be achieved in a few days before natural detoxification occurs; otherwise, a retreatment would be required. The logistics and economics are therefore in favor of applying too much rather than too little chemical.

Though the threat of a large dose of herbicide is real, it is a short-lived threat. A heavy infestation of water weeds is extremely detrimental to an aquatic ecosystem. Application of the herbicide is followed by an equally destructive period during which tons of vegetation decay. The cycle is repeated at least annually, so the aquatic system is essentially never in balance.

Controlled Release

Laboratory studies by the Creative Biology Laboratory and others have shown that biologically active chemicals can be released from specially compounded elastomers at relatively constant rates over long periods of time. The chemicals or "pesticides" thus released demonstrate the same biological activity as if they were conventionally applied. Controlled release, however, offers advantages or variations in approach that are not possible or at best are impractical with conventional application. Phytozone treatment, contact area control, and sublethal control show promise. However, chronic intoxication of pest aquatic weeds with a constant low-level herbicide dosage is perhaps the most important concept in this area, and it is the "chronicity phenomenon" that makes this a valid approach. Obviously, the high cost of conventional spreading or spraying herbicides precludes daily applications of ultralow concentrations and dictates that a single acute dose be used with this methodology.

Chronic intoxication of aquatic weeds, by itself, does not favor use of controlled release. Ultimately, there must be an economic or environmental advantage. Thus, it is not sufficient simply to extend the toxicant disbursal time. There must be an accompanying enhancement of efficacy, an extension of the time between applications, a reduction in toxicant usage, or some other factor that can be converted into a monetary savings or environmental benefit.

Chronicity Phenomenon

It was noted in laboratory studies of controlled-release herbicides used against aquatic weeds that mortality, in many instances, was not proportional to dosage; i.e., if a dose of 0.1 ppm/day killed in 1 week and the concentration * time (Ct) equation were accurate, then a dose of 0.01 ppm/day should give the same results in 10 weeks. In both laboratory and pond tests, this was not so. The 0.01-ppm/day dose killed in about 1.7 weeks. Thus, it was hypothesized that whereas a Ct relationship probably held for *acute* dosages in conventional applications when the time period was confined to several days, the ultralow agent concentrations dispensed via slow-release methodology led to a terminal *chronic* intoxication whose mechanism of action was different.

It is also suspected that the *slow-release mechanism* of the CR material results in a true solution of active ingredient in the water envelope, whereas conventional use of granules, emulsions, etc., lead to molecular aggregates. The statistics of contact, as well as the absorptivity of the agent by the plant, may be significantly different. The chronic effect is particularly meaningful for the control of aquatic weeds because of the dilution factor in the water environment. It is theoretically possible to maintain almost any concentration of active agent through controlled release. However, the inert or material controlling adds expense and bulk to the product.

Pest aquatic plants grow in remote as well as in developed areas, and in water of varying depth and flow rates. It can be difficult to apply herbicides. Spraying of liquids and the broadcast distribution of pellets is most common. These operations can be conducted from boats, planes, trucks, or by individuals with backpacks. This same methodology must be used with controlled-release materials. When the dilution factor is high and the time frame for control is long, the cost of the inert (release-controlling) ingredients and the transport of the increased bulk to the site could be prohibitive. On the other hand, if the Ct phenomenon does not apply, and control can be achieved with only a small time penalty at ultralow dose levels, then the cost and bulk of the controlled-release formulation is not a problem. Truly long-term formulations with an effective life of a year or longer may then be practical.

The "chronicity" phenomenon was investigated by comparing two treatments. In one, the herbicide was added daily, and the toxicant was allowed to accumulate. In the other, the herbicide was also added daily, but the test water was changed daily too, so that the agent concentration was held fairly constant.

Watermilfoil, *Elodea*, *Cabomba*, *Vallisneria*, and Southern naiad were evaluated in 1-gal jars, each with 3 liters of water. Plants were potted, three to the jar, in 200-ml cups. Gro-lux lighting was used in indoor tests, with intensity adjusted for optimum growth. Other plants were tested in 5-gal plastic-lined containers. Plants were conditioned for 4 to 8 weeks prior to start of treatment.

Test containers were observed daily and plant mortality rated subjectively on a scale of 100 (healthy) to 0 (mortality) scale, with the degree of thinning and browning serving as the rating criteria.

Evaluations performed during the course of this effort tended to confirm the hypothesis that the Ct relationship does not hold when ultralow herbicide concentrations are maintained for extended periods of time. This is illustrated in Tables 4 and 5.

Table 4 Effect of Five Toxicants on Watermilfoil, Dose Accumulative

		Days to given % mortality		
Toxicant	Dose (ppmw)	50%	90%	100%
Diquat 1.0	9	13	14	
	0.1	9	12	13
	0.01	10	16	19
	0.001	8	24	38
Fenac	1.0	14	19	21
	0.1	22	42	43
	0.01	24	43	48
	0.001	Never	Never	Never
Silvex 1.0	8	13	18	
	0.1	18	23	27
	0.01	18	20	21
	0.001	Never	Never	Recovery
2,4-D acid	1.0	8	12	14
	0.1	8	14	18
	0.01	13	20	4
	0.001	21	Never	Recovery
2,4-D BEE	1.0	10	13	15
	0.1	10	15	17
	0.01	10	14	18
	0.001	10	18	22

Water controls average 6% mortality; solvent controls average 16% mortality.

Table 5 Effect of Five Toxicants on Watermilfoil, Dose Constant

Toxicant	Dose (ppmw)	Days to given % mortality		
		50%	90%	100%
Diquat 1.0	8	10	11	
	0.1	9	14	19
	0.01	9	13	16
	0.001	23	27	32
Fenac	1.0	19	Never	Recovery
	0.1	Never	Never	Recovery
	0.01	23	Never	Recovery
	0.001	35	Never	Recovery
Silvex 1.0	8	11	15	
	0.1	20	Never	Never
	0.01	21	Never	Never
	0.001	Never	Never	Never
2,4-D acid	1.0	12	18	20
	0.1	20	Never	Never
	0.01	Never	Never	Never
	0.001	Never	Never	Never
2,4-D BEE	1.0	7	10	13
	0.1	6	13	19
	0.01	13	22	24
	0.001	20	38	Never

Water controls average 31% mortality; solvent controls average 35% mortality.

The preceding data represent an average reading of four test aquaria, or a total of 12 plants, for each toxicant at each concentration. Control data are the average of 12 jars or 36 plants each.

A controlled-release formulation of copper sulfate monohydrate with a measured release rate of only a fraction of 1% of total available toxicant per day was evaluated against *Vallisneria*, *Cabomba*, duckweed, watermilfoil, and algae to see if the "chronicity" phenomenon was operative [61]. See Table 6.

Planting, preparation, and rating were the same as in the previous experiments. Toxicant pellets were added at 10 ppm, 50 ppm, and 100 ppm by rubber (carrier) weight with respective copper ion contents of 1.75 ppm, 8.75 ppm, and 17.5 ppm. Test aquaria were also treated with a 0.03-ppm copper-ion solution to approximate the actual release from a 100-ppm pellet. Table 6 shows the data for the *Cabomba* exposure. The "chronicity" phenomenon is pronounced here in that a difference of only 5 days is noted for a 100% kill between the lowest and highest rates [62].

Table 7 shows the results of a 1-ppm controlled-release copper sulfate monohydrate pellet test against *Vallisneria amerocama* and *Myriophyllum spicatum* (watermilfoil). The total copper ion available for the 1-ppm pellet is 0.175 ppm. The total amount released during the 60-day test period would be considerably less, as the formulation has an effective release of over 6 months. The test data indicates sublethal control.

Historically, the only approach to weed control has been the destruction of the weed. Anything less was economically unsound. Pest weeds are so persistent and competitive that the term "eradication" has long been dropped from weed control jargon. But weed control or "management" still means to destroy as many as possible.

Table 6 Controlled-Release Copper Sulfate[a] Versus *Cabomba caroliniana*, Percent Mortality at a Given Time (day)[b]

Day	Control	Cu^{2+}, 0.03 ppm/day	10 ppm	50 ppm	100 ppm
5	3%	10%	23%	37%	26%
10	10	30	33	60	40
15	24	35	50	82	60
20	30	40	66	85	78
25	30	45	80	85	78
30	30	65	90	92	90
35	40	65	93	100	100
40	30	90	100	—	—
45	25	100	—	—	—
50	25	—	—	—	—

[a] E-51 1-ppm controlled-release copper sulfate monohydrate pellets.
[b] Average of replicates.

Table 7 shows the results of a 1-ppm controlled-release copper sulfate monohydrate pellet test against *Vallisneria amerocama* and *Myriophyllum spicatum* (watermilfoil).

It is obvious that the low-level copper ion has a profound sublethal effect on the two plants. A similar effect has been demonstrated by many herbicides against various pest weeds, although the sublethal level varies from one species of plant to another, as well as from one chemical to another.

Though the plants are not killed outright, a significant measure of control is achieved. With controlled release, it is possible to maintain a level of chemical in the water course that is marginally unacceptable to pest plants.

Chronicity Conclusions

Studies indicate that very low doses of herbicides will produce chronic intoxication of aquatic plants and will ultimately destroy pest weeds given a long enough period of time.

Table 7 E-51[a] Versus Vallisneria and Milfoil, Percent Mortality at a Given Time (day)[b]

Day	Control	Vallisneria	Milfoil
5	5%	0%	7%
10	4	13	18
15	8	33	29
20	9	44	45
25	10	58	51
30	10	62	33
35	10	65	36
45	10	68	37
50	9	69	40
55	9	72	54
60	8	74	74

[a] Controlled-release copper sulfate monohydrate pellets (1 ppm).
[b] Average of replicates.

In many instances, there is a beneficial chronic effect which minimizes the time penalty. This "chronicity phenomenon" indicates that, compared to current practice, control with slow-release materials may significantly reduce environmental contamination as well as lower costs and reduce the field hazards of handling toxicants.

HERBICIDES

Controlled-delivery products for herbicides have been slow to develop. The mechanisms that work to make slow-release insecticides useful do not function well in the soil. For instance, the geometry that works well for insecticides on surfaces (two dimensions) is spread over three dimensions in the soil. The mobility of insects that brings them in contact with the capsules and contaminates them is not nearly as useful in the soil. Insect-initiated release of very-slow-releasing formulations is also not easily translated into the realm of soil and plants. Even the decrease in toxicity observed with insecticides is of lesser importance with herbicides, because herbicides are seldom highly toxic.

Stauffer (now part of ICI) has marketed microencapsulated Sutan, a herbicide for rice, prepared by interfacial polymerization. Monsanto has microencapsulated Lasso and Bullet. The microencapsulation of trifluralin by interfacial polymerization techniques produces a sprayable formulation that will not lose its active ingredients rapidly by vaporization and therefore allows the speed and economy of aerial application. Sufficient time is also available to the farmer for incorporating the pesticide into the soil without a substantial loss in pesticidal activity [63].

Biobarrier Technology

The intrusion of roots into sewers, septic tanks and drain fields, sidewalks, streets, tennis courts, roadways, swimming pools, waste sites, and landfills has been a problem for decades. Controlled-delivery herbicides are presently being developed to deal with the problem. Formulations have been advanced that will continuously release the herbicide trifluralin for periods of up to 100 years. These products, basically polyethylene/trifluralin combinations, are being developed by the U.S. Department of Energy's Pacific Northwest Laboratory. Products thus far developed are Root-Shield, a root-repelling sewer gasket; Biobarrier, a spun-bonded polypropylene fabric with equally spaced pellets, for burial in the ground to prevent root growth along the edges of sidewalks, airport runways, and other landscaped areas; Root-Guard, an impregnated plastic drip-irrigation pipe; and Grow Guard, a herbicide-impregnated polymeric cord, for filling cracks or joints to inhibit plant and seed growth.

Controlled-release formulations of the herbicide 2,4-D reacted with sawdust have been reported to give excellent results in preventing weed competition with seedling trees in reforestation projects [64].

Herbicides encapsulated using starch have given excellent results in many field tests [65]. Herbicides investigated include DBCP (1,1,2-dibromo-3-chloropropane), EPTC (S-ethyldipropylthiocarbamate), butylate (S-ethyldiisobutylthiocarbamate), and trifluralin. There are currently many modifications of the original starch encapsulation process, which proceeded through the xanthate.

Equation (22) shows the preparation of starch xanthate and subsequent oxidation to the disulfide:

$$\text{starch—OH} + CS_2 + NaOH \rightarrow \text{starch—OC(}{=}\text{S)SNa}$$

$$\xrightarrow{H_2O_2} \text{starch—OC(}{=}\text{S)S—SC(}{=}\text{S)O—starch} \qquad (22)$$

The effectiveness of starch encapsulation formulations depends on the rate of release as well as other parameters discussed relative to microencapsulation, such as particle size, percent coverage, and capsule spacing. Even so, there are a number of fundamental functional differences between microencapsulation and starch encapsulation. The starch controlled-release technique leads to a granular-type product having a particle size greater than 1000 μm. The material is produced by a grinding step and sieving to secure the desired size range. Particles too small make the product dusty and must be recycled. This requirement incurs a significant cost, which will hurt the product's competitiveness in today's market. Most competing granular formulations utilize very inexpensive carriers, such as ground corncobs or ground clay.

Starch-encapsulated products are dry granular formulations. They cannot be sprayed, but they can be applied using the same equipment as is used for conventional granular formulations. Unlike microencapsulated formulations, starch granular formulations release when wet, but do not do so when dry. For this reason, starch granular formulations are believed to function in the soil, whereas microcapsules under the same conditions do not. For example, when soil is dry and weeds are not growing rapidly, the release of herbicide would be futile. When the soil is wet and weeds are growing rapidly, herbicides are released from starch formulations at a rate proportional to their water solubility. Absolute release rates can be controlled by the methodology employed in cross-linking the xanthate [66].

CONSUMER PRODUCTS

Microencapsulated methyl parathion has been sold by Pennwalt Corporation since 1974 under the name Penncap-M Microencapsulated Insecticide [67].* It was the first microencapsulated product registered for commercial marketing by the EPA. Initial registration was on cotton, alfalfa, and sweet corn. The number of crops cleared for use with Penncap-M rapidly increased. In 1977, the label included use on 16 major crops. In 1989, the product was registered for use in the United States, France, Spain, Italy, Greece, the Netherlands, Morocco, Hungary, Cyprus, Algeria, and other countries. The wall material of the capsules is a cross-linked polyamide-polyurea polymer. As previously stated, cross-linking increases the molecular weight of the polymer. The wall material is not soluble in any solvent and does not have a melting point, characteristics typical of cross-linked materials. The average size of the capsules produced is about 30 μm. The capsule walls are about 1 μm thick. (Figure 2 shows a broken capsule and the shell nature of the wall.) Such capsules have a volume of approximately 10^{-8} ml, which indicates that 1 g of the insecticide is distributed among 100 million capsules. The product is supplied as a flowable material with a viscosity of about 800 cp. A gallon of concentrate contains 2 lb of active. Like other agricultural concentrates, the material is diluted with water by the amounts specified on the label and then sprayed from conventional aerial or ground-based equipment.

---

* Penncap-M Microencapsulated Insecticide—a product of Atochem North America.

Penncap-M's advantages gave farmers incentive to use this product.

1. *Increased residual*. The farmer saves money where multiple sprayings a season are required, the longer residual of Penncap-M allowing a reduction in the number of applications. In other applications, less pesticide may be used, which may or may not save money, but it will definitely be a more desirable situation from the standpoint of environmental pollution (Fig. 17).

2. *Decreased mammalian toxicity*. One of the most dramatic results of microencapsulating methyl parathion is the great decrease in toxicity. The water barrier and polymer barrier around the toxicant limit its contact with skin and increase its dermal LD_{50} from 100 mg/kg to 1200 mg/kg on rabbits (Fig. 18). Similarly, the oral LD_{50} (mice) is raised from LD_{50} 10–20 mg/kg to greater than 100 mg/kg. In spite of the dramatic reductions in toxicity exhibited by many microencapsulated formulations, it is not easy to put a price tag on this feature, because farmers seldom wish to pay more for the benefits of increased safety. However, governmental regulations regarding the use of highly toxic materials, and regulations regarding their shipping and storage, may make use of less hazardous formulations less costly.

3. *Decreased phytotoxicity*. Emulsifiable concentrates of methyl parathion are not used on apples because of phytotoxicity, which blemishes the fruit. Microencapsulated methyl parathion does not have this defect, and is labeled for and used very successfully on apples. The probable reason for the decreased phytotoxicity is the concentration effect. As explained earlier, less than 1/100th of the apple surface is contacted when 1 lb of microencapsulated formulation is applied per acre.

Two problems of note regarding the use of Penncap-M were encountered. The first was that it *was* the first such material to be labeled, and owing to its great departure from the usual agricultural formulations, the EPA treated it as a new pesticide. Not only were tests required of the formulation as a whole, but it was necessary to check the toxicity

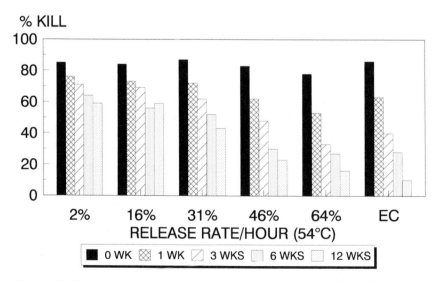

Figure 17 The bioassay of a group of experimental microencapsulated diazinon formulations of different release rates. The release rates equate to the percent diazinon lost from the formulation in 1 h at 54°C.

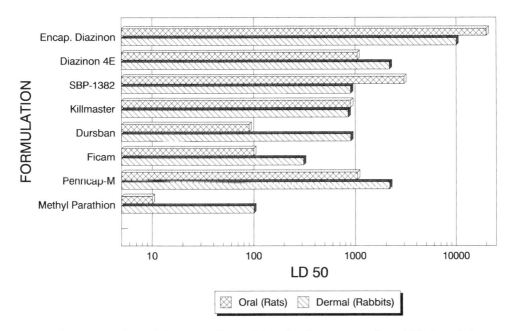

Figure 18 A comparison of the mammalian toxicity of various common insecticide formulations. Both oral and dermal toxicities are important. Oral toxicities were conducted on rats and dermal toxicities on rabbits. LD_{50} is the dose lethal to 50% of the test animals.

of the wall material itself. An exemption from tolerance for use on food crops was finally granted for the wall polymer. The pesticidal properties, however, were finally assumed to be those of methyl parathion. The residue tolerances for Penncap-M were to be the same as for the parent pesticide, meaning that the amount of residue pesticide on the harvested crop could be no more than that of any other formulation of methyl parathion. And therein lies a paradox. Since the release of methyl parathion was required to be relatively rapid to meet the established tolerances, why is the material so much more effective in the field than the normal first-order-releasing formulations?

The second problem involved a nontarget species, bees. Although methyl parathion itself is very toxic to bees, the microencapsulated formulation is not. The microcapsules are in the same particle-size range as pollen. It was soon found that if blooming crops were sprayed, the bees would inadvertently pick up the microcapsules which were contaminating pollen and bring them back to the hive. The pollen was used as food for the newly hatched bees, and the loaded capsules would persist in the pollen for many months. (Little or no pesticide is found in the honey.) Many hives were wiped out before the applicators were educated not to spray blooming crops. In recent years, complaints of beekeepers have been very few, indicating that microencapsulated pesticides and honeybees can coexist. Incidentally, these materials could be a potent weapon against the *killer bees* when the need arises.

Microencapsulated diazinon, Knox Out 2FM Flowable Microencapsulated Insecticide* was developed by Pennwalt Corporation for commercial marketing and awarded

---

* Knox Out 2FM Flowable Microencapsulated Insecticide, a product of Atochem North America.

an EPA label in 1978. It differs from Penncap-M in that it is basically an indoor-use product; if used out-of-doors, it is not used on food crops and therefore is not subject to residue tolerances. The product is effective for up to 2 months indoors against ants, roaches, fleas, flies, etc. The label provides for use in structures (houses, apartments, barns, etc.) and food establishments. Its benefits include advantages similar to those possessed by microencapsulated methyl parathion. Encapsulated technical diazinon is one of the least toxic consumer insecticide products. See Table 8, comparisons of toxicity.

Other microencapsulated pesticides made by interfacial polymerization are the following:

1. Sectrol encapsulated natural pyrethrum made by 3M Company
2. Dursban ME microencapsulated chlorpyrifos, a product of Dow Chemical Company, containing 11.7% active ingredient
3. Empire microencapsulated chlorpyrifos, a product of Dow Chemical Company, containing 20% active ingredient
4. Flea Halt microencapsulated chlorpyifos, a product of Farnam Companies, containing 2.5% active ingredient
5. Capsolane microencapsulated EPTC, a product of Stauffer Chemical Company, containing 36% active ingredient
6. Fenitrothion 400 ME microencapsulated fenitrothion, a product of Pennwalt France, containing 36% active ingredient
7. Microtech microencapsulated Lasso, a product of Monsanto Corporation
8. Fonofos Seed Treatment microencapsulated dyfonate, a product of Stauffer Chemical Company, containing 43% active ingredient

One of the first and most successful controlled-release insecticide products was the Shell No Pest Strip. The strip was a polymer matrix containing the insecticide DDVP,

Table 8 Comparisons of Toxicity, Based on Formulation, LD_{50} in mg/kg

Formulation		Technical	Encapsulated
Knox Out 2FM[a]			
(diazinon)			
	Oral (rat)	110–270	>21,000
	Dermal (rabbit)	300	>10,000
Penncap-M Insecticide			
(methyl parathion)			
	Oral (rat)	10–20	1,048
	Dermal (rabbit)	100	2,160
Fenitrothion 400 M.E.[a]			
(fenitrothion)			
	Oral (rat)	800	5,400
	Dermal (rabbit)	1,300	2,160
Fonofos Seed Treatment[b]			
(dyfonate)			
	Oral (rat)	8–17	2,370
	Dermal (rabbit)	25	1,500

[a] A product of Atochem North America.

[b] A product of Stauffer Chemical Company.

which vaporizes readily. Thus, closed-in areas of surprisingly large proportions could with one of these strips be rendered insect-free for up to 3 months. Unfortunately, health experts came to the conclusion that humans and animals who breathe such concentrations of DDVP may be harmed, and the product was discontinued.

During the period when the Shell No Pest Strip was in use, another controlled-release device was employed in virtually every restaurant in the United States. The lindane vaporizer, mounted high on a wall, had a small integral heater that slowly evaporated a charge of lindane.* Ultimately, the level of toxicant vapors was adjudged harmful, and the lindane vaporizers, along with lindane itself, were removed from use.

PROBLEMS AND STUDY QUESTIONS

1. What are the advantages (or disadvantages) of using controlled-delivery pesticides over classical formulations in the following areas: cost, economy, application frequency, application rate, environmental contamination, safety, public acceptance, governmental acceptance, effectiveness, residual, development of resistance, phytotoxicity, EPA registration, product lifetime, temperature dependence, sales appeal, product renovation from a marketing viewpoint, first-order versus zero-order release?

2. At a constant percentage of wall, what happens to the wall thickness as the microcapsules are made smaller?

3. It is desired to encapsulate a water-soluble material and obtain a dry powder as the final product. What encapsulation techniques would be suitable? If a particle size of less than 100 μm is desired, what techniques could be used?

4. How many capsules would be required to hold enough active ingredient to treat 1 acre if the average particle size was 100 μm and the dose rate was $\frac{1}{2}$ lb/acre?

5. List at least three ways that a highly persistent material such as DDT could benefit from controlled-release technology.

6. Why, in your opinion, have so many different substrates been used in the attempt to make controlled-delivery formulations of pheromones?

7. What are some of the problems that could be encountered by complete replacement of insecticides by pheromones? What evolutionary changes in insect behavior could lead to the development of resistance to pheromones?

8. Most fungicides are solids, insoluble in water and sparingly soluble in petroleum solvents. Would controlled-delivery techniques be of use in formulating them? How would the active ingredient get out of the slow-release matrix? Would liquid fungicides work?

9. What is the single most important factor limiting the effective life a cuprous oxide-leaching paint system?

10. Why is matrix dissolution more beneficial for leaching paint than it is for a diffusion-dissolution system?

11. Why does a leaching copper paint not attain the same extension or service life from an increase in film thickness as a leaching copper paint with water-soluble matrix?

---

* Lindane is the γ-isomer of benzene hexachloride.

12. What is the concept of a "nontoxic" toxicant?
13. Sublethal control envisions a mildly toxic environment that stresses aquatic plants to the extent that they will not thrive and be a problem. Why is this not feasible with conventional herbicides?
14. Why is the "chronicity" phenomenon important to the control of aquatic weeds when using controlled release and not important in antifouling?
15. What is the significance of the "chronicity" phenomenon?

SELECTED ANSWERS

9. As depletion occurs and the degree of pore structure increases, the pathways lengthen and it becomes more difficult for the leaching process to continue. It will eventually shut down.
10. In a copper-leaching paint the water must dissolve the copper and lift the ions to the surface. If the paint is not very heavily loaded, the copper molecules will be isolated by the film and not be available. Ideally, the matrix will dissolve at the same rate as the toxicant constantly exposing a fresh surface. In the diffusion-dissolution system, the toxicant is constant throughout, and loss of matrix will waste toxicant.
11. Matrix solubility is used to expose toxicant that would otherwise be unavailable. If the film is thicker, then the solubility process can continue for a longer period. A leaching system fails when the water cannot reach the toxicant. A thicker film would not help.
12. The goal is a system in which there is no release of toxicant to the environment, and activity is initiated only when contact is made with a fouling organism. Realistically, the release is so low that it will have no adverse affects.
13. The low dose level must be maintained throughout the water during the control period. This can be accomplished by controlled-release formulation disbursed throughout the site. Conventional treatment would require oft-repeated applications.
14. Antifouling coatings protect the surface of the marine objects. There is no attempt to bring the water to a specified dose level. The opposite is true for aquatic weed control; all effort is directed toward keeping the toxicant in place.
15. The economics of treating a large body of water for an entire growing season or longer are sensitive to the dose level. The lower levels compensate for the longer exposure and added cost of the CR formulation.

SUGGESTED READING

Cardinal, J. R., *J. Controlled Release*, 2:393–403 (1985).

Kwan, L. C., in *Drug Delivery Devices* (P. Tyle, Ed.), Marcel Dekker, New York, pp. 519–547 (1988).

Pope, D. G., and J. D. Baggot, The basis for selection of the dosage form, in *Formulation of Veterinary Dosage Forms* (J. Blodinger, Ed.), Marcel Dekker, New York, pp. 1–70 (1983).

Pope, D. G., Specialized dose dispensing equipment, in *Formulation of Veterinary Dosage Forms* (J. Blodinger, Ed.), Marcel Dekker, New York, pp. 1–70 (1983).

REFERENCES

1. E. E. Ivy, Penncap-M: An improved methyl parathion formulation, *J. Econ. Entomol.*, 65:473 (1972).

2. G. Markin and S. Hill, Microencapsulated oil bait for control of the imported fire ant, *J. Econ. Entomol.*, *61*:193 (1971).

3. J. E. Vandegaer, U.S. Patent 3,577,515 (1971).

4. N. Maarinelli, U.S. Patent 4,115,315 (1978).

5. R. C. Koestler, A theory of a mechanism of action of encapsulated herbicides and insecticides, in *Proc. Int. Control. Rel. Soc.*, N. F. Cardarelli, Ed.), University of Akron, Ohio, p. 8.1 (1976).

6. K. Tsuji, S. Tsuda, T. Ohtsubo, H. Kawada, Y. Manabe, N. Kishibuchi, and G. Shinjo, *Proc. Int. Symp. Control. Rel. Biol. Mater.*, *13*:44 (1986).

7. *The Inside Story—A Guide to Indoor Air Quality*, EPA/400/1-88/004, U.S. Environmental Protection Agency, Washington, D.C. (September 1988).

8. G. S. Hartley, Herbicide behavior in the soil, in *The Physiology and Biochemistry of Herbicides* (L. J. Audus, Ed.), Academic Press, New York, chap. 4 (1964).

9. R. E. Simpson, U.S. Patent 4,096,944 (1978).

10. Unpublished experiment, Pennwalt Corporation (1975).

11. M. D. Jackson, and R. G. Lewis, Effect of physical form on volatilization of methyl parathion from treated fields, Paper 6, Pesticide Division, in Abstracts of Papers, 172nd American Chemical Society National Meeting, San Francisco, 8/29–9/3 (1976).

12. R. C. Koestler, The particle size effect of microencapsulated pesticides in the soil, *Proc. Int. Control. Rel. Soc.*, Ft. Lauderdale, Fla. (1981).

13. L. L. Barber, Jr., A. J. Lucas, and R. Y. Wen, U.S. Patent 4,056,610 (1977).

14. J. W. Ralls, D. R. Gilmore, and A. Cortes, The fate of radioactive OO-diethyl O-(2-isopropyl-4-methylpyrimidine-6-yl) phosphorothioate on field-grown experimental crops, *J. Agr. Food Chem.*, *14*:387–392 (1966).

15. D. O. Eberle and D. Novak, Fate of diazinon in field-sprayed agricultural crops, soil and olive oil, *J. Assoc. Offic. Anal. Chem.*, *52*:1067–1074 (1969).

16. S. E. Kunz, K. D. Murrell, G. Lambert, L. James, and C. Terrill, in *CRC Handbook of Pest Management in Agriculture* (D. Pimentel, Ed.), CRC Press, Boca Raton, Fla. (in press, 1990).

17. R. O. Drummond, *Veterinary Parasitology*, *18*:111–119 (1985).

18. J. L. Riner, R. L. Byford, L. G. Stratton, and J. A. Hair. *Am. J. Vet. Res.*, *43*:2023–2030 (1982).

19. R. H. Marston, U.S. Patent 3,056,724 (1962).

20. J. A. Miller, M. L. Beadles, and R. O. Drummond, U.S. Patent 4,166,107 (1979).

21. J. A. Miller, F. W. Knapp, R. W. Miller, and C. W. Pitt, *Southw. Entomol.*, *4*:195–200 (1979).

22. J. A. Miller, F. W. Knapp, R. W. Miller, C. W. Pitts, and J. Winetraub, *J. Agric. Entomol.*, *3*:48–55 (1986).

23. B. E. Simpson, U.S. Patent 4,416,659 (1983).

24. R. H. Laby, U.S. Patent 3,844,285 (1974).

25. W. A. Boettner, A. J. Aguiar, J. R. Cardinal, A. C. Curtiss, G. R. Ranade, J. A. Richards, and W. F. Sokol, *J. Control. Rel.*, *8*:23–30 (1988).

26. D. H. Carter, M. Luttinger, and D. L. Gardner, *J. Control. Rel.*, *8*:15–22 (1988).

27. H. Jaffe, P. A. Giang, and J. A. Miller, Implants of methoprene in poly(lactic acid) against cattle grubs, *Proc. 5th Int. Symp. Control. Rel. Bioact. Mater.*, Gaithersburg, Md., pp. 5.5–5.11 (1978).

28. H. Jaffe, J. A. Miller, P. A. Giang, and D. K. Hayes, Implantable systems for delivery of insect growth regulators in livestock, *Proc. 7th Int. Symp. Cont. Rel. Bioact. Mater.*, Ft. Lauderdale, Fla., pp. 97–100 (1980).

29. H. Jaffe and D. K. Hayes, Controlled release reservoir systems for the delivery of insect steroid analogues against ticks, *Proc. 10th Int. Symp. Control. Rel. Bioact. Mater.*, San Francisco, pp. 67–70 (1983).

30. J. A. Miller, R. O. Drummond, and D. D. Oehler, in *Controlled Release Delivery Systems* (T. J. Roseman and S. Z. Mansdorf, Eds.), Marcel Dekker, New York, pp. 223–236 (1983).

31. M. L. Beadles, A. R. Gingrich, and J. A. Miller, *J., Econ., Entomol., 70*:72–75 (1977).

32. J. A. Miller, D. D. Oehler, and S. E. Kunz, *J. Econ. Entomol., 76*:1335–1340 (1983).

33. R. W. Baker and H. K. Lonsdale, in *Controlled Release of Biologically Active Agents* (A. C. Tanquary and R. E. Lacy, Eds., Plenum Press, New York, pp. 15–71 (1974).

34. S. M. Herbig and K. L. Smith, *J. Control. Rel., 8*:63–72 (1988).

35. J. A. Miller and D. D. Oehler, *J. Control. Rel., 8*:73–78 (1988).

36. J. A. Miller, Controlled release technology and the horn fly resistance problem, *Proc. 12th Int. Symp. Control. Rel. Bioact. Mater.*, Geneva, Switzerland, pp. 187–188 (1985).

37. J. R. Plimmer, C. P. Schwalbe, E. C. Paszek, B. A. Bierl, R. E. Webb, S. Marumo, and S. Iwaki, *Environ. Entomol., 6*:518–522 (1977).

38. R. Ideses, A. Shani, and J. T. Klug, *J. Chem. Ecol., 8*:978 (1982).

39. C. P. Schwalbe, The Gypsy Moth: Research Toward Integrated Pest Management, USDA Technical Bulletin 1584, C. C. Doane and M. L. McManus, Eds. (1981).

40. M. Beroza, E. C. Paszek, E. R. Mitchell, B. A. Bierl, and J. R. McLaughlin, *Environ. Entomol., 3*:361–365 (1974).

41. J. W. Jenkins, C. C. Doane, D. J. Schuster, J. R. McLaughlin, and M. Jimenez, in *Behavior-Modifying Chemicals for Insect Management: Applications of Pheromones and Other Attractants* (R. L. Ridgeway, R. M. Silverstein, and M. N. Inscoe, Eds.), Marcel Dekker, New York (in press).

42. A. Bakke, Insect pheromone technology in chemistry and applications, *ACS Symp. Ser.* 190 (B. A. Leonhardt and M. Beroza, Eds.), American Chemical Society, Washington, D.C., pp. 219–230 (1982).

43. C. L. Metcalf, W. P. Flint, and R. L. Metcalf, *Destructive and Useful Insects, Their Habits and Control*, McGraw-Hill, New York, p. 587 (1962).

44. T. W. Brooks, and R. L. Kitterman, Gossyplure H. F.—Pink bollworm suppression with male sex attractant pheromone released from hollow fibers—1976 experiments, *Proc. 1977 Beltwide Cotton Production-Mechanization Conf.*, National Cotton Growers of America, Memphis, p. 79 (1977).

45. Porter, G., The end of the free ride, Natl. Bur. Stds., *Dimensions* (March 1980).

46. J. G. Dobson, The control of fouling organisms in fresh and salt-water circuits, *Trans. Am. Soc. Mech. Eng.*, 247–265 (April 1946).

47. V. J. Castelli and W. L. Yeager, Organometalic polymers: Development of controlled release antifoulants, American Chemical Society 171st Meeting, New York (April 7–9, 1976).

48. Various, Marine fouling and its prevention, *Woods Hole Oceanographic Inst.*, p. 21 (1952).

49. J. W. Fitzgerald, M. S. Davis, and G. B. Hurdle. Corrosion and fouling of sonar equipment, *Pt. I. Report No. S2477*, NRL (March 1947).

50. H. B. Culver and G. Grant, *The Book of Old Ships*, London (1928).

51. L. B. Alberti, *De Re Aedificatoria* (1470) German trans. by V. Hoffman, Frankenberg (1883).

52. Sir George C. V. Holmes, *Ancient and Modern Ships*, London (1906).

53. E. K. Chatterton, *Sailing Ships and Their Story*, London (1914).

54. J. W. Higgins, The forge of Vulcan, Nature overlooked, *Worcester National Historical Soc., 1*:23–29 (1943).

55. Maurer, M. Copper bottoms for the United States Navy 1794–1803, *U.S. Naval Inst. Proc., 71*:693–699 (1945).

56. A. Ellisen, Br. Patent 1503 (May 1857).

57. E. H. Monkton, British Patent 1503 (May 21, 1867).

58. D. F. Comstock, U.S. Patent 113687 (January 5, 1915).

59. B. H. Ketchum, J. D. Ferry, A. C. Redfield, A. E. Burns, Evaluation of antifouling by leaching rate determinations, *Eng. and Eng. and Eng. Chem. Ind. Ed.*, (1945).

60. N. F. Cardarelli and S. J. Caprette, Antifouling covering, U.S. Patent 3,426,473 (1969).

61. G. A. Janes and S. M. Bille, Chronicity phenomenon and controlled release copper, Weed Sci. Soc. Am., 16th Meeting (February 2–5, 1976).

62. G. A. Janes, Control of aquatic weeds by chronic intoxication, Am. Chem. Soc., 171st National Meeting, New York, April 7–9, 1976).

63. R. C. Koestler, Microencapsulated trifluralin, U.S. Patent 4,360,376 (1982).

64. G. G. Allan, J. W. Beer, M. J. Cousin, and R. A. Mikels, The biodegradative controlled release of pesticides from polymeric substrates, in *Controlled Release Technologies: Methods, Theory and Applications*, (A. F. Kydonieus, Ed.), CRC Press, Boca Raton, Fla. (1980).

65. B. S. Shasha, W. M. Doane, and C. R. Russell, Starch encapsulated pesticides for slow release, *J. Polym. Sci.*, *14*:417 91976).

66. B. S. Shasha, Starch and other polyols as encapsulating matrices for pesticides, in *Controlled Release Technologies: Methods, Theory and Applications* (A. F. Kydonieus, Ed.), CRC Press, Boca Raton, Fla. (1980).

67. C. B. Desavigny, U. S. Patent 3,959,464 (1976).

Index